Lecture Notes in Computer Science 3340

Commenced Publication in 1973
Founding and Former Series Editors:
Gerhard Goos, Juris Hartmanis, and Jan van Leeuwen

Cristian S. Calude   Elena Calude
Michael J. Dinneen (Eds.)

Developments in Language Theory

8th International Conference, DLT 2004
Auckland, New Zealand, December 13-17, 2004
Proceedings

 Springer

Volume Editors

Cristian S. Calude
Michael J. Dinneen
University of Auckland
Department of Computer Science
Auckland, New Zealand
E-mail:{cristian,mjd}@cs.auckland.ac.nz

Elena Calude
Massey University Albany
Computer Science
Institute of Information and Mathematical Sciences
Auckland, New Zealand
E-mail: e.calude@massey.ac.nz

Library of Congress Control Number: 2004116341

CR Subject Classification (1998): F.4.3, F.4.2, F.4, F.3, F.1, G.2

ISSN 0302-9743
ISBN 3-540-24014-4 Springer Berlin Heidelberg New York

Springer is a part of Springer Science+Business Media

springeronline.com

© Springer-Verlag Berlin Heidelberg 2004
Printed in Germany

Typesetting: Camera-ready by author, data conversion by Scientific Publishing Services, Chennai, India
Printed on acid-free paper   SPIN: 11358855   06/3142   5 4 3 2 1 0

Preface

The main subjects of the Developments in Language Theory (DLT) conference series are formal languages, automata, conventional and unconventional computation theory, and applications of automata and language theory. Typical, but not exclusive, topics of interest include: grammars and acceptors for strings, graphs, and arrays; efficient text algorithms; combinatorial and algebraic properties of languages; decision problems; relations to complexity theory and logic; picture description and analysis; cryptography; concurrency; and DNA and quantum computing.

The members of the steering committee of DLT are: J. Berstel (Paris), M. Ito (Kyoto), W. Kuich (Vienna), G. Păun (Bucharest and Seville), A. Restivo (Palermo), G. Rozenberg (chair, Leiden), A. Salomaa (Turku) and W. Thomas (Aachen).

The first DLT conference was organized by G. Rozenberg and A. Salomaa in Turku in 1993. After this, the DLT conferences were held in every odd year: Magdeburg (1995), Thessaloniki (1997), Aachen (1999) and Vienna (2001). Since 2001, a DLT conference has been organized in every odd year in Europe and in every even year outside Europe. The last two DLT conferences were organized in Kyoto, Japan in 2002 and Szeged, Hungary in 2003. The titles of the volumes of the past DLT conferences are the following:

1. *Developments in Language Theory. At the Crossroads of Mathematics, Computer Science and Biology* (edited by G. Rozenberg and A. Salomaa) (1994) (World Scientific)
2. *Developments in Language Theory II. At the Crossroads of Mathematics, Computer Science and Biology* (edited by J. Dassow, G. Rozenberg and A. Salomaa) (1996) (World Scientific)
3. *Proceedings of the Third International Conference on Developments in Language Theory* (edited by S. Bozapalidis) (1997) (Aristotle University of Thessaloniki)
4. *Developments in Language Theory. Foundations, Applications and Perspectives* (edited by G. Rozenberg and W. Thomas) (2000) (World Scientific)
5. *Developments in Language Theory*, Lecture Notes in Computer Science 2295 (edited by W. Kuich, G. Rozenberg and A. Salomaa) (2002) (Springer)
6. *Developments in Language Theory*, Lecture Notes in Computer Science 2450 (edited by M. Ito and M. Toyama) (2003) (Springer)
7. *Proceedings of the Developments in Language Theory Conference*, Lecture Notes in Computer Science 2710 (edited by Z. Ésik and Z. Fülöp) (2003) (Springer)

The latest conference, DLT 2004, which ran under the auspices of the European Association for Theoretical Computer Science (EATCS), was supported by the New Zealand Royal Society. It was jointly organized by Massey University at Albany and the Centre for Discrete Mathematics and Theoretical Computer Science of the University of Auckland and was held at the Auckland Campus of Massey University at Albany in Auckland, New Zealand in the period 13–17 December 2004. The conference was accompanied by two thematic workshops, the *International Workshop on Automata, Structures and Logic* (organized by B. Khoussainov) and the *International Workshop on Tilings and Cellular Automata* (organized by M. Margenstern).

The five invited speakers of the conference were: Bruno Courcelle (Bordeaux, France), Rodney Downey (Wellington, New Zealand), Nataša Jonoska (Tampa, USA), Anca Muscholl (Paris, France) and Grzegorz Rozenberg (Leiden, Netherlands).

The Programme Committee thanks the paper reviewers of the conference for their much appreciated work. These experts were:

Azat Arslanov	Juraj Hromkovič	George Păun
Jean Berstel	Oscar Ibarra	M.J. Perez-Jimenez
Cristian S. Calude	Lucian Ilie	Ion Petre
Elena Calude	Masami Ito	Bala Ravikumar
Cezar Câmpeanu	Heath James	Bernd Reichel
Julien Cassaigne	Nataša Jonoska	Antonio Restivo
Bruno Courcelle	Lila Kari	Branislav Rovan
E. Csuhaj-Varju	Bakh Khoussainov	Sasha Rubin
Mark Daley	R. Klempien-Hinrichs	Yasubumi Sakakibara
Sylvain Degeilh	Peter Knirsch	Kai Salomaa
Volker Diekert	Satoshi Kobayashi	F. Sancho-Caparrini
Michael J. Dinneen	Hans-Joerg Kreowski	Nicolae Santean
Michael Domaratzki	Werner Kuich	Pavel Semukhin
Pal Domosi	Chang Li	Ulrich Speidel
Rod Downey	Markus Lohrey	Ludwig Staiger
Frank Drewes	Maurice Margenstern	Karl Svozil
Allen Emerson	Giancarlo Mauri	Rick Thomas
Zoltan Ésik	Melanija Mitrovič	Nicholas Tran
Henning Fernau	André Nies	Gyorgy Vaszil
Claudio Ferretti	Enno Ohlebusch	Todd Wareham
Vesa Halava	Alexander Okhotin	Takashi Yokomori
Tero Harju	Friedrich Otto	Sheng Yu
Ken Hawick	Holger Petersen	Shyr-Shen Yu

The Programme Committee, consisting of J. Berstel (Paris, France), C.S. Calude (chair; Auckland, New Zealand), J. Cassaigne (Marseille, France), V. Diekert (Stuttgart, Germany), M.J. Dinneen (secretary; Auckland, New Zealand), Z. Ésik (Szeged, Hungary), T. Harju (Turku, Finland), J. Hromkovič (ETHZ, Switzerland), O. Ibarra (Santa Barbara, USA), M. Ito (Kyoto, Japan), B. Khoussainov (Auckland, New Zealand), H.-J. Kreowski (Bremen,

Germany), W. Kuich (Vienna, Austria), M. Margenstern (Metz, France), G. Păun (Bucharest, Romania), I. Petre (Turku, Finland), A. Restivo (Palermo, Italy), Y. Sakakibara (Tokyo, Japan), K. Salomaa (Kingston, Canada), L. Staiger (Halle, Germany) and S. Yu (London, Canada), selected 30 papers (out of 47) to be presented as regular contributions and 5 other special CDMTCS papers which appeared in the *CDMTCS Research Report* 252 at http://www.cs.auckland.ac.nz/CDMTCS/.

Finally, we want to acknowledge the tremendous work and dedication of the DLT 2004 Conference Committee, which consisted of P. Barry, M. Bowers, E. Calude (chair), S. Ford, V. Harris, J. Hunter, H. James, P. Kay, A. Lai, N. Luke, R. McKibbin, L. O'Brien, D. Parsons, H. Sarrafzadeh, C. Scogings, U. Scogings, I. Sofat, U. Speidel, D. Viehland and Y.-T. Yeh.

The editors also thank Alfred Hofmann, Anna Kramer and Ingrid Beyer from Springer, Heidelberg, for producing this volume in the Lecture Notes in Computer Science series.

October 2004

C.S. Calude
E. Calude
M.J. Dinneen

Table of Contents

Invited Papers

Contributed Papers

Recognizable Sets of Graphs, Hypergraphs and Relational Structures: A Survey

Bruno Courcelle*

LABRI, Université Bordeaux 1 and CNRS, Talence, France
`bruno.courcelle@labri.fr`

Abstract. New results on the recognizability of sets of finite graphs, hypergraphs and relational structures are presented. The general framework of this research which associates tightly algebraic notions (equational and recognizable sets) and Monadic Second-Order logic (for defining sets and transformations of graphs, hypergraphs and relational structures) is reviewed. The lecture [3] is based on two submitted but nevertheless available articles [1,4] ; the present text is an informal overview. The numerous definitions and results can be found in the two articles.

1   Introduction

The description of sets of finite words (called *languages*) and of their transformations (called *transductions*) was the original goal of the Theory of Formal Languages. This theory now extends its scope to infinite words, to finite and infinite trees (modelling finite and infinite algebraic terms), and more recently, to finite and infinite graphs, hypergraphs and related structures like partial orders and traces. Unless otherwise specified, we will use "graph" as a generic term covering directed and undirected, labelled and unlabelled graphs and hypergraphs. All these objects are conveniently handled as *relational structures, i.e.*, as logical structures with no function symbol except possibly nullary ones.

In addition to classical tools like grammars, automata and transducers, First-Order and Monadic Second-Order logic have proved to be useful to describe sets of words and trees. For dealing with graphs, logic is also essential, not only for defining sets of graphs but also for defining graph transformations. However, the variety of types of graphs, and consequently of operations on them (that generalize the concatenation of words), makes it necessary to use also some unifying concepts provided by Universal Algebra.

The basic notion of a context-free language can be characterized in terms of least solutions of equation systems (equivalent to context-free grammars). That of a regular language can be characterized in terms of congruences with finitely many classes (equivalent to finite deterministic automata). These algebraic definitions are interesting in that they apply to every algebra. They are

---

* http://www.labri.fr/ courcell

C.S. Calude, E. Calude and M.J. Dinneen (Eds.): DLT 2004, LNCS 3340, pp. 1–11, 2004.

thus appropriate for dealing with various types of sets of graphs, but they are useful for other reasons. Context-free sets of graphs can be defined as *equational sets*, i.e. as components of least solutions of systems of recursive set equations, in a much easier way than in terms of graph rewriting sequences. Since there is no good notion of graph automaton except in very particular cases, the notion of a finite congruence is the only way to obtain a workable generalization of regularity to sets of graphs: a *recognizable set* is a set which is saturated for a congruence with finitely many classes of each sort (we use many-sorted algebras).

An equational set of graphs can be specified in a readable way by an equation system. But specifying a recognizable set of graphs by a congruence is no more convenient than specifying in this way a regular language, because even in simple cases, congruences tend to have many classes. *Monadic Second-Order logic* is here especially useful as a specification language. Since a graph is nothing but a relational structure, every closed formula, either first-order or second-order specifies a set of graphs, namely the set of its finite models. (We only consider finite objects in this survey). Furthermore most graph properties can be expressed easily by logical formulas. And every set of graphs characterized by a Monadic Second-Order formula, i.e. a formula where quantified variables denote individual elements (typically vertices, but also edges) or sets thereof is recognizable. In particular, basic classes of graphs like trees, connected graphs, planar graphs are monadic second-order definable.

Hence Monadic Second-Order logic (MS logic in short) is an appropriate language for specifying sets of graphs. It can be seen as an alternative to the non-existing notion of graph automaton. Furthermore, it can be used to specify graph transformations, called *MS transductions* that are as useful for studying sets of graphs as are rational transductions for languages. Applications to the construction of linear algorithms for hard (NP complete) problems restricted to certain equational sets of graphs substanciate the claim of usefulness of Monadic Second-Order logic.

We hope to convince the reader that the algebraic notions of a recognizable and of an equational set of graphs on the one hand, and the logical notions based on MS logic on the other build a coherent and robust framework for extending to graphs the notions and results of Formal Language Theory. The following table summarizes the main results. (MST means MS transduction).

Algebraic notions	Algebraic characterizations	Logical characterizations	Closure properties
EQ	Equation systems, val(REC(Terms))	MST(Trees)	Union, ∩Rec, Homomorphisms, MST
REC	Finite Congruences	MS-definable (⊂ REC)	Boolean operations, Inverse homomorphisms, Inverse MST

2 Notions from Universal Algebra

For dealing with graphs one needs to use *many-sorted algebras*, with countably many sorts. A sort is here a finite set of labels (which is a subset of a fixed countable set of labels). These labels are used to specify in a nonambigous way "graph concatenation" operations. Since the combinatorial structure of graphs is much more complicated than that of words, one cannot limit oneself to concatenation operations (or graph construction operations) based on a uniformly bounded number of labels. (One can actually generate all finite graphs from 6 operations with 2 sorts in a somewhat artificial way but the interesting algorithmic results discussed below do not work for these operations. See [4].)

In a many-sorted algebra, an *equational set* is a set of elements of the same sort that is a component of the least solution of a (finite) system of recursive set equations. An example of such a system is:

$$\{X = f(X,Y) \cup \{a\}; Y = g(X,Y,Y) \cup f(Y,Y) \cup \{b\}\},$$

where X and Y denote sets and f and g denote the set extensions of functions belonging to the signature. A *recognizable set* is a set of elements of the same sort, that is a union of classes for a finite congruence, i.e., a congruence such that any two equivalent elements are of the same sort and which has finitely many classes of each sort. (See Courcelle [2]). These two notions depend on the signature. We refer to F-equational and to F-recognizable sets, forming the classes $EQ(F)$ and $REC(F)$ respectively, when we need to specify the signature F.

In some cases, the notion of a recognizable set somehow degenerates. If the signature is "poor" (for example if it consists only of constants and a finite set of unary operations closed under composition), then every set is recognizable. If on the opposite it is "too rich" then the only recognizable sets are the empty set and the set of all elements of a same sort. This is the case of the set of positive integers equipped with the successor and the predecessor function.

In every algebra, even having an infinite signature, the following properties hold:

Property 1: The class of recognizable sets is closed under union, intersection, difference and inverse homomorphisms. In particular, for every finite subsignature F of the considered signature, the set of (finite) F-terms, the value of which belongs to a recognizable set is recognizable, hence is definable by a finite deterministic tree-automaton.

Property 2: The class of equational sets is closed under union, homomorphisms, and intersection with recognizable sets ; it is closed under the operations of the signature.

Property 3: One can decide the emptiness of an equational set given by a system of equations.

Property 4: A set is equational iff it is the set of values of a recognizable set of terms over a finite subset of the signature.

For two signatures F and K on a same set M, such that F is a subsignature of K, we get immediately from the definitions.

Property 5: Every F-equational set is K-equational. Every K-recognizable set is F-recognizable.

We will say that the signatures F and K are *equivalent* if the corresponding classes of equational and recognizable sets are the same. If K is F enriched with operations that are defined by finite F-terms, then F and K are equivalent. But there are examples of equivalent signatures not of this type. Consider the set of words over a finite alphabet, where F consists of concatenation, empty word and letters, and K is the same together with the mirror image operation. Then the signatures F and K are equivalent, but the mirror image is not expressible as a composition of operations of F.

We will be interested by the comparison of various signatures of graph operations. We will see that no more than three signatures have to be considered, each of them having many equivalent variants. This indicates the robustness of this algebraic approach.

Algorithmic applications of recognizability can be described from the above algebraic properties, using the fact that an MS definable set of graphs is recognizable. Let us consider an equational set L given by an equation system. Let K be a recognizable set which is "effectively given", for instance by means of a homomorphism into an algebra each domain of which is finite, or by an MS formula φ. By Property 2, the set $L \cap K$ is an equational set for which one can construct an equation system. It follows from Property 3 that one can decide the emptiness of $L \cap K$. By applying this to the case of an equational set of graphs L and a set K defined as the set of finite models of a closed MS formula φ, one obtains that one can decide whether there exists in L a graph satisfying φ. This decision problem, called the *Monadic Second-Order satisfiability problem for L* is non trivial: it is undecidable L is the (non-equational) set of all finite graphs, and even for first-order formulas φ.

By using the second assertion of Property 1 one can decide in linear time if an F-term (where F is finite subsignature of the considered global signature) has for value an object belonging to the recognizable set K defined by an MS formula φ. It follows that if a graph G in L is given by a term over the finitely many operations that occur in the defining system for L (say a *derivation tree* of G relative to the context-free graph grammar represented by the equation system) then one can decide in time proportional to the size of this term whether G satisfies the MS formula φ. This applies to *NP complete problems* expressible by MS formulas, like 3-vertex colorability.

3   Graph Operations

The use of algebraic notions is based the definition of the operations on graphs that form the signature. There are actually two main (non-equivalent) signatures of interest, that we call HR and VR because the corresponding equational sets have been defined independently and previousy by "Hyperedge Replacement" Context-free Graph Grammars and, respectively, by context-free graph grammars based on "Vertex Replacement". (See the book edited by G. Rozenberg on graph grammars [7]). They are robust in the sense that many variants of the definitions yield the same equational sets. Furthermore, the corresponding classes of equational sets are closed under MS transductions. This is analogue to the closure of the class of context-free languages under rational transductions.

The HR operations deal with graphs and hypergraphs having distinguished vertices called *sources* designated by *labels*. (There is only one source for each label.) The HR operations are the *parallel composition* of two graphs or hypergraphs (one takes the disjoint union of the two and one fuses the sources with same labels), operations that *change the labels* of sources, and operations that *"forget"* sources (forgetting the a-source means that the vertex designated by a is no longer distinguished, but is made "ordinary"). Basic graphs or hypergraphs are those with a single edge or hyperedge (possibly with loops), and isolated vertices.

The VR operations deal with graphs (not with hypergraphs) with labelled vertices. Each vertex has one and only one label but several vertices may have the same label. The VR operations are the *disjoint union* of two labelled graphs, some operations that *modify labels* in a uniform way (every vertex labelled by p is relabelled by q) and operations that *add edges* between every vertex labelled by p and every vertex labelled by q, for fixed vertex labels p and q. (These last operations produce graphs having complete bipartite subgraphs $K_{n,m}$ (with $n + m$ vertices) where n is the number of p-labelled vertices and m is the number of q-labelled vertices. Since n and m are not bounded, such graphs have unbounded tree-width and the VR operations cannot be replaced by compositions of HR-operations.) The basic graphs are those with a single vertex.

In both cases, every graph (or hypergraph in the case of HR) can be generated by these operations by using one label for each vertex. By bounding the allowed number of labels, one obtains particular classes of graphs and hypergraphs forming two infinite hierarchies. We obtain also the families of *HR-equational, VR-equational, HR-recognizable* and *VR-recognizable sets* of graphs and hypergraphs. They correspond to the families of context-free and regular languages, but we have two notions in the case of graphs.

There are other significant differences with the case of words. The set of all words on a fixed finite alphabet is context-free, whereas the set of all finite graphs (say, of simple undirected unlabelled graphs to get a precise statement) is neither HR-equational nor VR-equational. This is due to the necessity of using infinitely many operations to generate all graphs (whereas the unique operation of concatenation suffices to generate all words). Since an equation system is by definition finite, none can define all graphs. From this observation, it follows that a set of graphs may be non-HR-equational for two reasons: either because it has unbounded tree-width (this the case of the set of finite planar graphs) or because it has an "irregular" internal structure (this is the case of the set of strings, the length of which is a prime number). Note however that classical non-context-free languages like $a^n b^n c^n$ are equational as sets of vertex labelled graphs.

Classes of recognizable sets of graphs (here we discuss labelled, directed or undirected graphs, not hypergraphs) are associated with the signatures HR and VR. Again, due to the infiniteness of the signatures, some properties of recognizable sets of words do not extend to graphs. In particular there are uncountably many recognizable sets (every set of square grids is HR-recognizable as well as VR-recognizable ; see [4]).

There exist two complexity measures on graphs, called *tree-width* and *clique-width* defined respectively as the minimum number of labels necessary to construct the considered graphs or hypergraphs with HR and VR operations. A set of graphs has bounded tree-width (resp. bounded clique-width) iff it is a subset of an HR-equational (resp. of a VR-equational) set. Tree-width has been introduced independently of graph grammars by Robertson and Seymour, and this parameter is essential in the *theory of parametrized complexity* developed by Downey and Fellows [5]. Clique-width is also useful for constructing polynomial algorithms for hard problems for particular classes of graphs.

Up to now we have only presented two signatures HR and VR. The first one concerns graphs *stricto sensu* as well as hypergraphs, whereas the second concerns only graphs or slightly more generally, *binary relational structures* (i.e., structures with relations of arity 1 and 2). We will now consider general relational structure, which correspond exactly to *directed, ranked, hyperedge-labelled hypergraphs*, simply called *hypergraphs* in the sequel. The *arity* of a hypergraph is the maximal arity of the (relation) symbols labelling its hyperedges.

For dealing with them, we introduce a many-sorted signature STR. We fix a set of relation symbols with countably many relations of each arity. Every finite subset Σ of this set is a sort and the corresponding domain is the set $STR(\Sigma)$ of finite Σ-structures. The operations are the disjoint union and all the unary operations that transform a structure into another one by means of quantifier-free conditions. The transformations performed by these operations can delete elements (for example, in a graph one may want to remove all vertices incident to no edge or loop), and/or redefine relations (for example, for defining the edge-complement of a graph). The unary VR operations are of this latter form. The HR operations also, if we denote "sources" by nullary function symbols. Technical details are omitted here.

Although there exist infinitely many quantifier-free formulas written with finitely many variables and relation symbols, there are only finitely many quantifier-free formula up to logical equivalence, and this equivalence is decidable. It follows that there are only finitely many quantifier-free operations $STR(\Sigma) \longrightarrow STR(\Gamma)$ where Σ and Γ are finite sets of relations. We obtain the inclusions of signatures:

$$\text{HRg} \subseteq \text{VR} \subseteq \text{STR}$$

where HRg denotes the restriction of HR to graphs obtained by taking graphs and not hypergraphs as basic objects. Then follow the inclusions:

$$\text{EQ(HRg)} \subseteq \text{EQ(VR)} \subseteq \text{EQ(STR)}$$

and the reverse inclusions for the corresponding classes of recognizable sets of graphs by Property 5.

The inclusion EQ(HRg) $\subseteq$ EQ(VR) is proper: the set of all cliques is VR-equational (easy to see ; cliques have clique-width 2) but not HR-equational because cliques have unbounded tree-width. If a set of graphs is VR-equational but is "without large complete bipartite subgraphs" which means that for some large enough n, no graph in L has a subgraph isomorphic to $K_{n,n}$ then it is HR-equational. The intuition about this result is the following: the operation in the VR signature which is not expressible in terms of HR operations is that which adds to a graph edges forming a complete bipartite subgraph. If L has no graph containing large $K_{n,n}$'s, this means that this operation is not used in a crucial way hence that each of its occurrence can be replaced by a composition of HR-operations.

A result proved in [4] establishes a similar result for recognizable sets of graphs. First the inclusion REC(VR) $\subseteq$ REC(HR) is proper because every set of cliques is HR-recognizable, but the set of cliques of size n such that n belongs to a set of positive integers which is not recognizable (like the set of prime numbers), is not VR-recognizable. Second, we have:

Theorem 1 [4]: *If set of graphs without large complete bipartite subgraphs is HR-recognizable, then it is VR-recognizable.*

Hence the same combinatorial condition collapses simultaneously the two proper inclusion of EQ(HR) in EQ(VR) and of REC(VR) in REC(HR). However, the proofs of the two results are different.

We now discuss the inclusion of VR in STR. We have the following:

Theorem 2 [4]: *A set of graphs is VR-equational iff it is STR-equational, and it is VR-recognizable iff it is STR-recognizable.*

The first result is a direct consequence of characterizations of VR-equational and STR-equational sets in terms MS transductions in Theorem 7 below. The second one is proved in [4]. We only explain here the meanings of these results.

To generate graphs by means of VR-operations, one uses only structures over the following relation symbols: a single binary relation *edg* representing edges linking vertices, and an unbounded number of unary auxiliary relations for representing vertex labels. These auxiliary relations need not occur in the generated graphs. They are only useful at intermediate stages of the generation process to establish edges. Among the quantifier-free unary operations used in STR are operations that may delete relations. This means that if one uses the operations of STR to generate graphs, some intermediate generated objects may be hypergraphs (represented by non-binary relational structures). One might think that because of this richer signature, the family EQ(STR) would contain sets of graphs not in EQ(VR), but this not the case. The signature VR is "strong enough" to yield the same equational sets as the apparently more powerful signature STR, which uses domains (sets of hypergraphs) not in the VR algebra of graphs. Hence in order to generate graphs represented by relational structures with one relation of arity 2, the auxiliary relations may be limited to arity 1. A similar statement can be made for recognizable sets: to establish that a set of graphs is STR-recognizable, it is enough to produce a congruence for the VR-operations, without having to extend it to a congruence on all the domains $STR(\Sigma)$ for all Σ's.

The above discussion showing that for generating graphs, one need not use auxiliary relations of arity more than 1 can be repeated for each maximal arity or even size: for generating hypergraphs with hyperedges with at most n distinct vertices, one need not use auxiliary hyperedges of arity more than $n-1$. This is formally defined and proved in [1]. Letting STR_n denote the restriction of STR to relations (and the corresponding domains and operations) of arity at most n we get the following statement:

Theorem 3 [1]: *A set of hypergraphs of arity at most n is STR_n-equational iff it is STR-equational. It is STR_n-recognizable iff it is STR-recognizable.*

Hence, with respect to sets of hypergraphs of maximal arity n, the full signature STR is equivalent to STR_n, its restriction to structures of arity at most n. Stronger formulations and their proofs can be found in [1].

4   Monadic Second-Order Logic and Graph Properties

Sets of graphs can be specified either recursively, in terms of base graphs and application of operations: this is what yields an equation system. But they are also frequently specified by *characteristic properties*. We will use logical formulas to write formally these properties, which is possible since a graph can be defined as or represented by a logical structure with relations representing adjacency or incidence.

First-Order logic can only express local properties like bounds on the degrees of vertices, hence is here of limited interest. But *Monadic Second-Order logic*, (*MS logic* in short) *i.e.*, the extension of First-Order logic with variables denoting subsets of the domains of the considered structures is quite powerful. It can express both coloring properties (for instance that a graph is 3-vertex colorable, an NP complete property), and path properties (like connectivity, existence of cycles, planarity via Kuratowski's theorem). But a property based on the existence of bijections, like the existence in a graph of a nontrivial automorphism, is provably not MS expressible.

A fundamental theorem says that:

Theorem 4: Every MS definable set of graphs is VR-recognizable. More generally, every MS-definable set of relational structures is STR-recognizable.

The algorithmic applications of recognizability reviewed in the first section are based on this result. One can actually improve Theorem 4 and its algorithmic consequences as follows. If instead of representing a graph by a relational structure, the domain of which is the set of vertices and which has a binary relation representing the edges, we use its incidence graph, *i.e.*, the relational structure the domain of which consists of vertices and edges, and equipped with an incidence relation, then, MS logic becomes more expressive because one can denote sets of edges by set variables. The same can be done for hypergraphs, and we denote by MS_2 the use of MS logic with this representation of graphs or hypergraphs. That a graph has a Hamiltonian circuit is MS_2 expressible but provably not MS expressible. Then we get the following result:

Theorem 5: Every MS_2 definable set of graphs or hypergraphs is HR-recognizable.

If we compare Theorems 4 and 5 in the perspective of their algorithmic applications, we can see that Theorem 5 concerns more properties, namely the MS_2 definable properties instead of the more restricted MS definable ones, but less families of graphs, namely those of bounded tree-width (generated by finite subsignatures of HR) instead of bounded clique-width (generated by finite subsignatures of VR).

5   Monadic Second-Order Transductions

Transformations of words and trees are based on *finite automata, equipped with output functions (sequential machines, tree-transducers)*, or on *rational expressions* and homomorphisms (*rational transductions* between languages). Since for graphs we have neither automata nor rational expressions, we must base *graph transformations* on another model. Monadic Second-Order logic offers the appropriate alternative.

An *MS transduction*: $STR(\Sigma) \longrightarrow STR(\Gamma)$ is a partial multivalued function specified by a finite sequence of MS formulas, using (possibly) set variables called *parameters*, and forming its *definition scheme*. A structure S in $STR(\Sigma)$ is transformed into a structure T in $STR(\Gamma)$ as follows: one select values for the parameters that satisfy a formula, the first one in the definition scheme. Then one builds a structure S' consisting k disjoint "marked" copies of S (k is fixed in the definition scheme). The output structure T is defined inside S' by restricting the domain and by defining its Γ−relations from the Σ−relations in S' and the "marks". These restrictions and definitions are done by MS formulas depending on the parameters. The transformation is multivalued because in general several choices of parameters can be made. An MS transduction defined by a parameterless definition scheme is a partial function. An important difference with rational transductions is the fact that the inverse of an MS transduction is not always an MS transduction whereas that of a rational transduction is a rational transduction. With respect to the equational and recognizable sets, MS transductions behave like homomorphisms as we will see.

The fundamental property of MS transductions is, with the above notation:

Theorem 6: The monadic second-order properties of the output structure T can be expressed by monadic second-order formulas in the input structure S in terms of the parameters used to define T from S.

It follows that the composition of two MS transductions is an MS transduction (a result much more important than the closure on inverse). It follows also that if a set of structures has a decidable monadic second-order satisfiability problem, then so has its image under an MS transduction.

The mapping from a term in $T(HR)$ (the set of finite HR-terms), $T(VR)$ or $T(STR)$ to the corresponding graph or hypergraph is an MS transduction. It follows that an equational set of graphs is the image under an MS transduction of a recognizable set of terms, equivalently (we omit details) of the set of finite binary trees. This result have a very important converse:

Theorem 7: If a set of graphs (resp. a set of hypergraphs) (resp. the set of incidence graphs of a set L of hypergraphs) is the image of the set of finite binary trees under an MS transduction, then this set is VR-equational (resp. is STR-equational) (resp. is HR-equational).

This theorem is somewhat similar to the one saying that the context-free languages are the images under rational transductions of the Dyck language, which is actually a coding of trees by words. An important consequence is the following:

Theorem 8: The image of a VR-equational set (resp. of an STR-equational set) (resp. of an HR-equational set) under an MS transduction of appropriate type is VR-equational (resp. STR-equational) (resp. HR-equational).

"Appropriate" means that it produces graphs from graphs in the first case, and that it transforms hypergraphs through their incidence graphs in the third case. Hence, with respect to equational sets, MS transductions behave like homomorphisms. They do the same with respect to recognizable sets since we have:

Theorem 9 [1]: *The inverse image of an STR-recognizable set under an MS transduction is STR-recognizable.*

This result is not very surprizing, because we know already from Theorem 6 that the inverse image of an MS definable set is MS definable, which yields Theorem 9 for those particular recognizable sets that are MS definable. However it shows how algebraic and logical notions are tightly linked.

6   Open Questions and Research Directions

We only mention a few questions related to the notions discussed in this overview.

1. Which quantifier-free operations on relational structures preserve recognizability?
2. Can one define a complexity measure on relational structures generalizing clique-width and that is linked to the signature STR or rather, to a subsignature equivalent to it like STR_n?
3. How can one enrich the signature STR into a larger signature equivalent to it?

Concerning Question 1: Not all quantifier-free operations preserve recognizability: the operation : $STR(\Sigma) \longrightarrow STR(\Sigma - \{R\})$ that deletes a relation R of arity at least 2 does not. Concerning question 2, clique-width is a complexity measure linked to VR, which is a subsignature of STR equivalent to it for graphs. Answers to question 3 can be found in [1] but they do not close the question.

References

More references can be found in [1,4].

[1] A. Blumensath, B. Courcelle: Recognizability and hypergraph operations using local informations, July 2004, http://www.labri.fr/Perso/ courcell/Textes/ BC-Blumensath-submitted(2004).pdf

[2] B. Courcelle: Basic notions of Universal Algebra for Language Theory and Graph Grammars, Theoretical Computer Science 163 (1996) 1-54.

[3] B. Courcelle: Recognizable sets of graphs, hypergraphs and relational structures: a survey, http://www.labri.fr/Perso/ courcell/Conferences/ExpoDLT04.pdf, (slides).

[4] B. Courcelle, P. Weil: The recognizability of sets of graphs is a robust property, December 2003, http://www.labri.fr/Perso/ courcell/Textes/BC-Weil-submitted (2004).pdf

[5] R. Downey, M. Fellows: Parametrized complexity, Springer, 1999.

[6] J. Mezei, J. B. Wright, Algebraic automata and context-free sets, Information and Control 11 (1967) 3-29.

[7] G. Rozenberg ed., Handbook of graph grammars and computing by graph transformations, Volume 1, World Scientific, 1997.

Some New Directions and Questions in Parameterized Complexity

Rodney G. Downey[1] and Catherine McCartin[2]

[1] Victoria University, Wellington, New Zealand
Rod.Downey@mcs.vuw.ac.nz
[2] Massey University, Palmerston North, New Zealand
C.M.McCartin@massey.ac.nz

Abstract. Recently there have been some new initiatives in the field of parameterized complexity. In this paper, we will report on some of these, concentrating on some open questions, and also looking at some current investigations aimed towards applying ideas of parameterized complexity in the field of online model theory.

1 Introduction

Paremeterized complexity was developed as a tool to address practical issues in algorithmic complexity. The basic idea is that the combinatorial explosion that occurs in exact algorithms for many intractable problems can be systematically addressed by seeking parameters that can be exploited to contain this explosion. The main idea is that for *natural, practical* problems often there are one, or more, parameters that are naturally "small", and if it is these that "make" the problem intractable then, for a limited range of these parameters, the problem can be efficiently solved.

For example, suppose we are analyzing data arising as, for instance, the conflict graph of some problem in computational biology. Because of the nature of the data we know that it is likely the conflicts are at most about 50 or so, but the data set is large, maybe 10^8 points. We wish to eliminate the conflicts, by identifying those 50 or fewer points. Let's examine the problem depending on whether the identification turns out to be a dominating set problem or a vertex cover problem. These are the classic two, opposing, examples usually quoted here.

DOMINATING SET. Essentially the only known algorithm for this problem is to try all possibilities. Since we are looking at subsets of size 50 or less then we will need to examine all $(10^8)^{50}$ many possibilities. Of course, this is completely impossible.

VERTEX COVER. There is now an algorithm running in time $O(1.286^k + k|G|)$ ([12]) for determining if a graph G has a vertex cover of size k. This has been implemented and is practical for $|G|$ of unlimited size and k up to around 400 [41].

C.S. Calude, E. Calude and M.J. Dinneen (Eds.): DLT 2004, LNCS 3340, pp. 12–26, 2004.

The issue is *the manner by which the running time for a fixed k depends on that k*. Critically, is k in the exponent of the size of the problem, or independent of the size of the problem? Consider the situation of a running time of $\Omega(n^k)$ vs $2^k n$. If k is small, then a running time of $2^k n$ may be feasible, even for very large n, while a running time of $\Omega(n^k)$ will be impractical.

There are myriads of natural implicit, and explicit, parameters which can be exploited to obtain problems admitting this kind of *exponential in the parameter only* running time behaviour. These include familiar graph-width metrics such as pathwidth, treewidth, and cliquewidth, as well as logical restrications.

To investigate the complexity of such *parameterized* problems, we use the framework of *parameterized complexity theory*, introduced by Downey and Fellows [18]. We review the main concepts and definitions of the theory here.

We remind the reader that a parameterized language L is a subset of $\Sigma^* \times \Sigma^*$. If L is a parameterized language and $\langle \sigma, k \rangle \in L$ then we refer to σ as the *main part* and k as the *parameter*. The basic notion of tractability is *fixed parameter tractability* (FPT). Intuitively, we say that a parameterized problem is fixed-parameter tractable (FPT) if we can somehow confine the any "bad" complexity behaviour to some limited aspect of the problem, the parameter. Formally, we say that a parameterized language, L, is fixed-parameter tractable if there is a computable function f, an algorithm A, and a constant c such that for all k, $\langle x, k \rangle \in L$ iff $A(x, k) = 1$, and $A(x, k)$ runs in time $f(k)|x|^c$ (c is independent of k). For instance, k-VERTEX COVER, introduced above, is solvable in time $\mathcal{O}(|x|)$. On the other hand, for k-TURING MACHINE ACCEPTANCE, the problem of deciding if a nondeterministic Turing machine with arbitrarily large fanout has a k-step accepting path, we are in the same situation as for k-DOMINATING SET. The only known algorithm is to try all possibilities, and this takes time $\Omega(|x|^k)$. This situation, akin to NP-completeness, is described by hardness classes, and reductions. A parameterized reduction, L to L', is a transformation which takes $\langle x, k \rangle$ to $\langle x', k' \rangle$, running in time $g(k)|x|^c$, with $k \mapsto k'$ a function purely of k.

Downey and Fellows [18] observed that these reductions gave rise to a hierarchy called the W-hierarchy.

$$FPT \subset W[1] \subseteq W[2] \subseteq \ldots \subseteq W[t] \subseteq \ldots .$$

The core problem for $W[1]$ is k-TURING MACHINE ACCEPTANCE, which is equivalent to the problem WEIGHTED 3SAT. The input for WEIGHTED 3SAT is a 3CNF formula, φ and the problem is to determine whether or not φ has a satisfying assignment of Hamming weight k. $W[2]$ has the same core problem except that φ is in CNF form, with no bound on the clause size. In general, $W[t]$ has as its core problem the weighted satisfiability problem for φ of the form "products of sums of products of ..." of depth t. It is conjectured that the W-hierarchy is proper, and from $W[1]$ onwards, all parametrically intractable.

There have recently been exciting new methods developed for establishing parameterized tractability. In this paper we will not address these. Instead, we point towards Rolf Niedermeier's survey [37], as well as Fellows' survey [22], and Downey [16]. These is an upcoming book [38] which will be devoted to methods of parameterized *tractability*.

There have also been some new developments in the application of parameterized *intractability* to gaining understanding of when PTAS's are likely to be feasible. These, and similar applications, highlight the use of parameterized complexity for exploring the boundary of feasibility *within polynomial time*. There have been very interesting developments exploring the running times of exact exponential algorithms, including use of the "Mini-" classes of Downey et. al. [17]. These have been used, in particular, by Fellows and his co-authors. This area is very promising and is the basis for an upcoming Dahstuhl meeting in 2005, as well as the recently held *International Workshop in Parameterized and Exact Computation*. We will examine this material in Section 2.

In Section 3, we examine new classes generated by the notion of EPT, introduced by Flum, Grohe and Weyer [25]. In this paper, our first major goal will be to examine these new notions and to mention a number of natural open questions that they suggest. A solution to any of these problems would be significant and the material gives us an interesting case study in a more-or-less neglected arena: structural parameterized complexity. We will examine these ideas in Section 3.

The last goal of this article is to articulate a new program of the authors [19, 20], devoted to applying the ideas of parameterized complexity, and topological graph theory, to online algorithms and online model theory. Our underlying idea is to provide a proper theoretical foundation for the study of online algorithms on online structures. As a case study we will look at online colourings of online graphs. Again, we will highlight a number of open questions in this area. This section is especially relevant to the present conference in view of the interest in *automatic* structures (such as Rubin's Thesis [39]). Here we have structures, such as graphs, whose domains and operations are *presented* by automata, and these are natural examples of online structures since the domains and other aspects of their diagrams are given one point at a time. We expand on this in Section 4.

2 $M[1]$, ETH and PTAS's

2.1   PTAS's

Let's re-examine the notion of P: classical *polynomial time*. Classical polynomial time allows for polynomials which can in no way be regarded as "tractable". For instance, a running time of n^{9000} is certainly not feasible.

This fact has certainly been recognized since the dawn of complexity theory. The main argument used is that P is a robust class (in its closure properties) and that "practical" problems in P have feasible running times. Certainly, in the early 70's this point of view was correct, but recently developed general tools for establishing times in P have given rise to bad running times.

When you are given some problem, a classical approach is to either find a polynomial time algorithm to solve it, or to demonstrate that it is NP-hard. The latter would then suggest that no exact polynomial time algorithm exists.

The question now is, suppose that you have a problem that *is* in P, but with a running time that is hideous. What can you do? Parameterized complexity has been shown to be useful here. This is particularly true for bad running times in PTAS's (polynomial time approximation schemes). As per Garey and Johnson [28], polynomial time approximation schemes (PTAS's) are one of the main traditional methods used for battling intractability. Many ingenious polynomial time approximation schemes have been invented for this reason. Often, the wonderful PCP theorem of Arora *et al.* [6] shows that no such approximation exists (assuming $P \neq NP$), but sometimes they do. Let's look at some recent examples, taken from some recent major conferences such as STOC, FOCS and SODA. (See Downey [16], and Fellows [22] for more examples.)

- Arora [4] gave a $O(n^{\frac{3000}{\epsilon}})$ PTAS for EUCLIDEAN TSP
- Chekuri and Khanna gave a $O(n^{12(\log(1/\epsilon)/\epsilon^8)})$ PTAS for MULTIPLE KNAPSACK
- Shamir and Tsur [40] gave a $O(n^{2^{2^{\frac{1}{\epsilon}}}-1})$ PTAS for MAXIMUM SUBFOREST
- Chen and Miranda gave a $O(n^{(3mm!)^{\frac{m}{\epsilon}+1}})$ PTAS for GENERAL MULTIPROCESSOR JOB SCHEDULING
- Erlebach *et al.* gave a $O(n^{\frac{4}{\pi}(\frac{1}{\epsilon^2}+1)^2(\frac{1}{\epsilon^2}+2)^2})$ PTAS for MAXIMUM INDEPENDENT SET for geometric graphs.

Table 1 below calculates some running times for these PTAS's with a 20% error.

Table 1. The Running Times for Some Recent PTAS's with 20% Error

Reference	Running Time for a 20% Error
Arora [4]	$O(n^{15000})$
Chekuri and Khanna [11]	$O(n^{9,375,000})$
Shamir and Tsur [40]	$O(n^{958,267,391})$
Chen and Miranda [14]	$> O(n^{10^{60}})$ (4 processors)
Erlebach *et al.* [21]	$O(n^{523,804})$

After the first author presented the table above at a recent conference (Downey [16]), one worker from the audience remarked to him "so that's why my code did not work," having tried to implement the Chen-Miranda algorithm!

Now sometimes the algorithms can be improved. For instance, Arora [5] also came up with another PTAS for EUCLIDEAN TSP, but this time it was nearly linear and practical. The crucial question is: having found the algorithms above and being unable to find better algorithms, how do we show that there are no practical PTAS's? Remember that we are *in P*, so lower bounds are hard to come by.

If the reader studies the examples above, they will realize that a source of the appalling running times is the $\frac{1}{\epsilon}$ in the exponent. We can define an optimization problem Π that has an *efficient P-time approximation scheme* (EPTAS) if it can be approximated to a goodness of $(1 + \epsilon)$ of optimal in time $f(k)n^c$ where c is a constant. Now, if we set $k = 1/\epsilon$ as the parameter, and then produce a reduction to the PTAS from some parametrically hard problem we can, in essence, demonstrate that no EPTAS exists.

Here is one early result in this program:

Theorem 1 (Bazgan [7], also Cai and Chen [8]). *Suppose that Π_{opt} is an optimization problem, and that Π_{param} is the corresponding parameterized problem, where the parameter is the value of an optimal solution. Then Π_{param} is fixed-parameter tractable if Π_{opt} has an EPTAS.*

Here is one recent application of Bazgan's Theorem taken from Fellows, Cai, Juedes and Rosamond [9]. In a well-known paper, Khanna and Motwani introduced three planar logic problems towards an explanation of PTAS-approximability. Their suggestion is that "hidden planar structure" in the logic of an optimization problem is what allows PTASs to be developed (Khanna and Motwani [32].) One of their core problems was the following.

PLANAR TMIN
Input: A collection of Boolean formulas in sum-of-products form, with all literals positive, where the associated bipartite graph is planar (this graph has a vertex for each formula and a vertex for each variable, and an edge between two such vertices if the variable occurs in the formula).
Output: A truth assignment of minimum weight (i.e., a minimum number of variables set to *true*) that satisfies all the formulas.

Theorem 2 (Fellows, Cai, Juedes and Rosamond [9]). PLANAR TMIN *is hard for $W[1]$ and therefore does not have an EPTAS unless $FPT = W[1]$.*

Fellows *et al.* [9] also show that the other two core problems of Khanna and Motwani [32] are $W[1]$ hard and hence have no EPTAS's.

For most of the problems in Table 1 it is open which have EPTAS's. It is an open *project* to understand when problems such as those in [2], can have real EPTAS's rather than just PTAS's which have unrealistic running times. Recently Chen, Huang, Kanj, and Xia [13] have made significant progress by provding an exact parameterized classification for problems with *fully polynomial time approximation schemes*, for a wide class of problems (scalable problems):

Theorem 3 (Chen et. al. [13]). *Suppose that Q is a scalable NP optimization problem. Then Q has a FPTAS iff Q is in the class of PFTP of parameterized problems which can be solved by an algorithm whose running time is polynomial in both $|x|$ and k.*

It seems reasonable that something of a similar ilk will be true for PTAS's. There are a number of similar applications using parameterized complexity for *practical* lower bounds in polynomial time, for example, Alekhnovich and Razborov [3].

2.2   ETH

The previous section demonstrates that, using parameterized complexity, we can address the classical issue of polynomial time approximation and whether there is an *efficient* polynomial time approximation. Recently, the increased computational power available has shown that there are a lot of problems for which exponential time algorithms can be useful in practice. This is particularly the case if the problem has an exponential time algorithm which is significantly better than $\text{DTIME}(2^{o(n)})$. However, there seems to be a "hard core" of problems for which, not only do we think that there is no polynomial time algorithm admitted, but, in fact, that there is no subexponential time one either. This is the *exponential time hypothesis* first articulated in Impagliazzo and Paturi [29].

- (ETH) n-variable 3-SATISFIABILITY is not solvable in $\text{DTIME}(2^{o(n)})$.

Again, it has been shown that parameterized techniques are extremely useful in showing that problems are likely not in $\text{DTIME}(2^{o(n)})$. Actually, the idea that subexponential time is intimately related to parameterized complexity is relatively old, going back to Abrahamson, Downey and Fellows [1]. However, there has been a lot of interest in this area recently, especially after the realization of the central importance of the *Sparsification Lemma* of Impagliazzo, Paturi and Zane [30] by Cai and Juedes [10]. As we will see, showing that a problem is $W[1]$-hard would likely be enough (depending on the reduction). To address these issues, Downey, Estivill-Castro, Fellows, Prieto-Rodriguez and Rosamond [17] introduced a new set of classes, the MINI-classes. Here the problem itself is parameterized. The notion is most easily explained by the following example.

MINI-3SAT
Input: A 3-CNF formula φ.
Parameter: k.
Question: If φ has size $\leq k \log n$ is it satisfiable?

The point here is that the solution of the problem is akin to classical NP-completeness in that we are asking for an unrestricted solution to a restricted problem. We can similarly miniaturize any combinatorial problem. For instance, it is easy to see that MINI-CLIQUE is FPT. To remain in the mini-classes you need to make sure that the reductions are small. The core machine problem is a circuit problem: MINI-CIRCUIT SAT. This allows us to generate the class $M[1]$ of miniaturized problems. It is not hard to show that MINI-INDEPENDENT SET and MINI-VERTEX COVER are $M[1]$-complete (Cai and Juedes [10], Downey et al. [17].) It is unknown if MINI-TURING MACHINE ACCEPTANCE is $M[1]$ complete, as the usual reductions are *not* linear, and we know of no such reduction.

It is not hard to show that $FPT \subseteq M[1] \subseteq W[1]$. The relevance of this class to subexponential time is the following.

Theorem 4 (Cai and Juedes [10], Chor, Fellows and Juedes [23], Downey et. al. [17]). *The $M[1]$-complete problems such as* MINI-3SAT *are in FPT iff the exponential time hypothesis fails.*

For a recent survey summarizing this material with new proofs, we refer the reader to Flum and Grohe [24].

3   EPT and FPT

Now, let's re-examine the notion of *fixed parameter tractability*. Recall that $L \subset \Sigma^* \times \Sigma^*$ is FPT iff there is an algorithm deciding $\langle x, k \rangle \in L$ running in time $f(k)|x|^c$, with f an arbitrary (computable) function, and c fixed, independent of k.

The criticisms of polynomial time can also be leveled, perhaps with even more force, at the notion of FPT, since the function f can be arbitrary. Remember, one of the claims of the theory is that it tries to address "practical" complexity. How does that claim stack up? One of the key methods of demonstrating *abstract* parameterized tractability is the use of logical methods such as Courcelle's Theorem. Here, one demonstrates that a given problem is for graphs of bounded treewidth and is definable in monadic second order logic, perhaps with counting. Then Courcelle's Theorem says the problem is linear time fixed-parameter tractable.

However, FPT is only really a general first approximation to feasability. The kinds of constants we get from applying Courcelle's Theorem are towers of 2's, roughly of the order of the number of alternations of the monadic quantifiers. This was proven by Frick and Grohe [26]. These types of constants stand in contrast to constants obtained using elementary methods such as *bounded search trees*, and *crown reduction rules* and *kernelization*. Here, the parameter constants are managable, more like 2^k.

Flum, Grohe, and Weyer [25] recently introduced a new class to perhaps better address when a problem is likely to have a *practical* FPT algorithm.

Definition 1 (Flum, Grohe and Weyer [25]). *A parameterized problem is in EPT iff it is solvable in time $2^{0(k)}|x|^c$.*

For example, k-VERTEX COVER is in EPT. Now the surprise. The reductions that Flum, Frick and Grohe use are *not* parameterized ones. The appropriate parameterized reductions to keep within the class would be linear in k, and FPT in $|x|$. Flum, Frick and Grohe introduced their notion of an EPT reduction as being one that is *not parametric*, rather

$$L \leq_{EPT} L' \text{ iff } \langle x, k \rangle \in L \text{ iff } \langle x', k' \rangle \in L',$$

where $x \mapsto x'$ in time $2^{0(k)}|x|^c$, *but*

$$\langle k, x \rangle \mapsto k' \text{ has } k' \leq k \log |x|.$$

This means that, as the size of the problem grows, the slice of L' used for the reduction can slowly grow. Clearly, it is easy to prove that the two notions of reduction are distinct. Technically, the addition of the $\log |x|$ in the reduction allows for counters to be used in the reduction.

Frick, Flum and Grohe then go on to define a new hierarchy based upon this reduction notion called the E-hierarchy,

$$EPT \subseteq E[1] \subseteq E[2] \subseteq \dots,$$

which is robust for $t \geq 2$.

What is the point of all of this? Frick, Flum and Grohe demonstrate that a number of FPT problem lie at higher levels of this hierarchy and are therefore likely *not* EPT. For instance, classes of model-checking problems which Flum and Grohe showed did not have FPT algorithms with elementary parameter dependence are complete for various levels of the class. Another example of the phenomenom is the following.

Theorem 5 (Frick, Grohe and Weyer [25]). k-VAPNIK-CHERVONENKIS DIMENSION *is complete for* $E[3]$ *under EPT reductions.*

We remark that k-VAPNIK-CHERVONENKIS DIMENSION was known to be $W[1]$ complete under FPT reductions.

There are a number of very important FPT problems which have no known FPT algorithms which are also single exponential in the parameter. Showing that any of the following is likely *not* EPT would be very significant: k-TREEWIDTH, k-BRANCHWIDTH, and k-CUTWIDTH. The same is true for many problems which only have FPT algorithms via Courcelle's Theorem or have been proven FPT (or PTIME) by treewidth methods.

An interesting technical question is whether k-INDEPENDENT SET is in $M[1]$ under EPT reductions. And what about randomization in this context?

We believe that the methods that are surely lacking here are suitable analogs of the PCP techniques.

4   Online Algorithms

In this section, we discuss ideas from a new project of the authors which has several goals. They include (i) providing a theoretical foundation for *online model theory*, (ii) trying to apply methods from parameterized complexity to online problems, and (iii) seeking to understand the use of "promises" in this area.

The last 20 years has seen a revolution in the development of graph algorithms. This revolution has been driven by the systematic use of ideas from topological graph theory, with the use of graph width metrics emerging as a fundamental paradigm in such investigations. The role of graph width metrics, such as treewidth, pathwidth, and cliquewidth, is now seen as central in both algorithm design and the delineation of what is algorithmically possible. In turn, these advances cause us to focus upon the "shape" of much real life data. Indeed, for many real life situations, worst case, or even average case, analysis no longer seems appropriate, since the data is known to have a highly regular form, especially when considered from the parameterized point of view.

The authors have begun a project that attempts to systematically apply ideas from classical topological graph theory to online problems. In turn, this has given rise to new parameters which would seem to have relevance in both the online and offline arenas.

Online problems come in a number of varieties. There seems to be no mathematical foundation for this area along the lines of finite model theory and it is our intention to develop such a theory. The basic idea is that the input for an online problem is given as a sequence of "small" units, one unit per time step, and our process must build the desired object according to this local knowledge. The following gives a first definition for two such online structures.

Definition 2. *Let $\mathcal{A} = \langle A, R_1, \ldots, R_n \rangle$ be a structure. (We represent functions as relations for simplicity.)*

(i) *We will say that a collection $\mathcal{A}_s \subset \mathcal{A}_{s+1} \ldots$ of finite substructures of $\mathcal{A}$ is a monotone online presentation where $\mathcal{A}_s$ denotes the restriction of $\mathcal{A}$ to the domain A_s.*

(ii) *More generally we can have $\mathcal{A} = \lim_s \mathcal{A}_s$, as above, but non-monotonically.*

For instance, an online graph could be one where we are given the graph one vertex at a time, and for each new vertex, we need to be told which of the vertices seen so far are adjacent to the new vertex. In the non-montonic version, vertices may be added and then subtracted. Monotonic online structures model the situation where more and more information is given and we need to cope; for instance, bin packing; and the non-monotonic situation is more akin to a database that is changing over time.

An *online procedure* on an online presentation $\{\mathcal{A}_s : s \leq n\}$ is a computable function f, and a collection of structures $\{\mathcal{B}_s :\leq n\}$, such that

(i) $f : \mathcal{A}_s \mapsto \mathcal{B}_s$, and
(ii) $\mathcal{B}_s$ is an expansion of $\mathcal{A}_s$, and
(iii) f_{s+1} extends f_s.

The idea is that $\mathcal{B}$ would have some extra relation(s) which would need to be constructed in some online fashion. A nice example to consider is colouring an online graph. Here we need to assign a colour to each vertex as it is presented. The colour used must differ from the colour of each of the neighbouring vertices, but we get to use only the information given by the finite graph coloured so far. The point is that the online situation is very different from the offline situation. Every planar graph is 4-colourable, but there are trees on n vertices having online presentations which require any online procedure to use at least $\log n$ many colours to colour them. The *performance ratio* compares the offline and the online performances.

There is a constellation of questions here. For instance, suppose that $\mathcal{B}$ is the expansion of $\mathcal{A}$ by a single relation R, and that this relation is first order definable. First order properties are essentially local. Hence, one would expect that there is a general theorem which will show that for any such first order R, there is an online procedure with reasonable performance ratio.

What might be achieved for structures of bounded width? We could ask that the Gaifman graph of the structure have bounded treewidth (or branchwidth, or pathwidth.) There is almost nothing known here.

There is a wonderful analogy with computable structure theory, since there we are dealing with strategies in games against a, perhaps hostile, universe. In some general sense, in an online situation it is very hard to use the finiteness of, say, a graph within the algorithm, since we don't know *how big* the graph might be. Information is only local. The recent work (at this conference!) on *automatic structures*, where the algorithms in question deal with structures where the relations are given in an online way *by automata*, are examples of this area. (See e.g. Rubin [39].)

Online colouring of monotonic online graphs is one area that has historically received some attention. Of course colouring is related to scheduling, where one can expect inputs that are naturally "long" and "narrow". For instance, consider the online scheduling of some large collection of tasks onto a small number of processors. One might reasonably expect a pattern of precedence constraints that gives rise to only small clusters of interdependent tasks, with each cluster presented more or less contiguously. Alternatively, one might expect a pattern of precedence constraints giving rise to just a few long chains of tasks, with only a small number of dependencies between chains, where each chain is presented more or less in order. Hence, *pathwidth* would seem to be a natural parameter to exploit in relation to online graph colouring and the authors have considered the situation where the pathwidth of the input graph is bounded.

Kierstead and Trotter [35] considered online colouring for *interval graphs*. A graph $G = (V, E)$ is an *interval graph* if there is a function ψ which maps each vertex of V to an interval of the real line, such that for each $u, v \in V$ with $u \neq v$, $\psi(u) \cap \psi(v) \neq \emptyset \Leftrightarrow (u, v) \in E$. It is not hard to show that G has pathwidth k iff G is a subgraph of an interval graph with clique size bounded by $k + 1$.

If an interval graph is presented via an ordering of the vertices that reflects a *path decomposition of width* k, then it can be coloured online using only $k + 1$ colours, using the simple strategy of *First-Fit*. First-Fit is a simple, but important, example of an online algorithm which colours the vertices of the input graph $G = (V, E)$ with an initial sequence of the colours $\{1, 2, \ldots\}$ by assigning to each vertex v_i, $i = 1, \ldots, |V|$, the least colour that has not already been assigned to any vertex previously presented that is adjacent to v_i.

Note, however, that the topological fact that the graph has pathwidth k will guarantee good performance no matter how the graph is presented. Kierstead and Trotter [35] have given an online algorithm that colors *any* online presentation of an interval graph having maximum clique size at most $k + 1$, using at most $3k + 1$ colors. Thus, any online presentation of a graph of pathwidth

k can be colored using at most $3k + 1$ colors. If we insist upon sticking with the simple strategy of First-Fit we can at least achieve a constant performance ratio on graphs of pathwidth k. Kierstead [33] has shown that for every online interval graph, with maximum clique size at most $k + 1$, First-Fit will use at most $40(k + 1)$ colours. In [34] Kierstead and Qin have improved the constant here to 25.72.

Chrobak and Slusarek [15] have shown that there exists an online interval graph G, and a constant c, where c is the maximum clique size of G, such that First-Fit will require at least $4.4\,c$ colours to colour G. Thus, the performance ratio of First-Fit on graphs of pathwidth k must be at least 4.4. It is open as to what the correct lower bound is here.

In [19], the authors have shown that for trees of pathwidth k, First-Fit will do much better, needing only $3k + 1$ many colours. It is unclear how the parameters of treewidth and pathwidth interact on number of colours needed. Related is Sany Irani's [31] notion of a *d-inductive* graph. A graph is d-inductive if there is an ordering $v_1, \ldots, v_n$ of its vertices such that for all i, v_i is adjacent to at most d vertices amongst $\{v_{i+1}, \ldots, v_n\}$. For instance, a planar graph is 5-inductive. To see this, any planar graph must have a vertex of degree 5 or less. Call this v_1. Remove it, and edges adjacent to it. Repeat. Similarly, all graphs of treewidth k are k-inductive. This notion generalizes the notions of bounded treewidth, bounded degree, and planarity. Irani gives an upper bound on the number of colours needed for First-Fit acting on d-inductive graphs.

Theorem 6 (Irani [31]). *If $G = (V, E)$ is d-inductive, then First-Fit will colour any monotonic online presentation of G with at most $O(d \log |V|)$ many colours.*

This bound is tight for the class, as this second theorem from Irani [31] shows.

Theorem 7 (Irani [31]). *For every online graph colouring algorithm A, and for every $d > 0$, there is a family of d-inductive graphs $\mathcal{G}$ such that for every $n > d^3$, there is a $G \in \mathcal{G}$ where G has n vertices and $A(G) = \Omega(d \log n)$.*

We give here a weaker lower bound for First-Fit acting on bounded treewidth graphs.

Theorem 8. *For each $k > 0$, there is a family of graphs $\mathcal{G}$ of treewidth k, such that for every $n > k$, there is a $G \in \mathcal{G}$ where G has n vertices and an online presentation of G on which First-Fit will use $\Omega(\frac{k}{\log(k+1)} \log n)$ many colours.*

Proof. We build $\mathcal{G}$ by describing a single online presentation. Each G_n will be defined by a prefix of this presentation.

We first present $k + 1$ vertices, $v_1, \ldots, v_{k+1}$, forming a clique at each step. First-Fit will colour these vertices, in order, with colours $c_1, \ldots, c_{k+1}$. We then present an isolated vertex v^1 which First-Fit will colour using c_1. To force First-Fit to colour the next vertex, v_{k+2}, using colour c_{k+2}, we will ensure that v_{k+2} is adjacent to $v_2, \ldots, v_{k+1}$ and also to v^1.

Note that the graph presented so far, which we denote as G_{k+2}, has treewidth k as required, with vertices $v_2, \ldots, v_{k+1}, v_{k+2}$, coloured using $c_2, \ldots, c_{k+1}, c_{k+2}$, forming a $k + 1$ clique in G_{k+2}.

We now proceed inductively. Suppose that we have presented a graph G_m, of treewidth k, on which First-Fit has been forced to use colours c_1 through c_m, and that G_m contains a $k+1$ clique consisting of vertices $v_{m-k}, \ldots, v_m$, coloured using $c_{m-k}, \ldots, c_m$. We form G_{m+1}, having treewidth k, on which First-Fit is forced to use colour c_{m+1} as follows:

We first present a copy of $G_{m-k+1} \backslash v_{m-k+1}$, the graph up to the point where v_{m-k+1} was presented (and coloured with c_{m-k+1}), but with v_{m-k+1} left out. We then present a new vertex v_{m+1} and make it adjacent to $v_{m-k+1}, \ldots, v_m$ and also adjacent to all those vertices in the copy of $G_{m-k+1} \backslash v_{m-k+1}$ which are (copies of) neighbours of v_{m-k+1}. Since First-Fit was forced to use colour c_{m-k+1} on v_{m-k+1} in G_{m-k+1} it must have been the case that the neighbours of v_{m-k+1} in G_{m-k+1} used all the colours c_1 through c_{m-k}. Thus, the neighbours of v_{m+1} use all the colours c_1 through c_m, and so First-Fit will be forced to colour v_{m+1} using c_{m+1}.

We now show that G_{m+1} has treewidth k by constructing a tree decomposition of width k for G_{m+1}. Since $v_{m-k}, \ldots, v_m$ form a clique in G_m, it must be the case that in any tree decomposition of G_m these vertices will appear together in some bag, which we denote by B_m. To create a tree decomposition T_{m+1} of G_{m+1} having width k, we start with a tree decomposition T_m of width k for G_m. We create a new bag containing $v_{m-k+1}, \ldots, v_m, v_{m+1}$, denoted by B_{m+1}. We add an edge from B_m in T_m to B_{m+1}. We then replicate a tree decomposition T_{m-k+1} of width k for G_{m-k+1} and replace every occurrence of v_{m-k+1} in the bags of T_{m-k+1} by v_{m+1}. Finally, we add an edge from some bag in T_{m-k+1} that contains v_{m-k+1} to B_{m+1}. Note that, apart from the vertex v_{m+1}, there are no vertices in common between T_m and our modified T_{m-k+1}, so this completes the construction.

Now, if G_m consists of n_m vertices then G_{m+1} consists of $n_m + n_{m-k+1}$ vertices, G_{m+2} consists of $n_m + n_{m-k+1} + n_{m-k+2}$ vertices, and so on. G_{m+k} consists of $n_m + n_{m-k+1} + n_{m-k+2} + \cdots + n_m$ vertices. Thus, $n_{m+k} \le (k+1)n_m$, giving $n_{1+k.d} \le (k+1)^d$, for any $k > 0$, $d \ge 0$.

Rearranging, we get $1 + k.d > \frac{k}{\log(k+1)} \log n_{1+k.d}$, which gives us the required lower bound.   □

There are still many unknowns here. For instance, what can be said on average? To make a *bad* online presentation of a graph G having low pathwidth one seems to need to *begin at the outer bags of some path decomposition for G and work in*. This would seem to be a rare event. John Fouhy [27] has run some simulations and found that, in general, for random pathwith k graphs, we only ever seem to need $3k + 1$ colours using First-Fit. This is not understood. Along with 0-1 behaviour, it is also suggestive of a more general theorem.

Finally, this material invokes ideas relating back to classical topological graph theory. The intuition behind a graph having low pathwidth is that it is "pathlike." In practice, if we are thinking of some process which is pathlike in its graphical representation then we would not think of it as a "fuzzy ball."

To make this idea precise, the authors introduced a new notion. We say that a path decomposition of width k, in which every vertex of the underlying graph belongs to at most l nodes of the path, has pathwidth k and *persistence* l, and say that a graph that admits such a decomposition has *bounded persistence pathwidth*. We believe that this natural notion truly captures the intuition behind the notion of pathwidth.

A graph that can be presented in the form of a path decomposition with both low width and low persistence is properly pathlike, whereas graphs that have *high* persistence are, in some sense, "unnatural" or pathological. Consider the graph G presented in Figure 1. G is not really path-like, but still has a path decomposition of width only two. The reason for this is reflected in the presence of vertex a in *every* node of the path decomposition. Our underlying idea is that a pathwidth 2 graph should look more like a "long 2-path" than a "fuzzy ball".

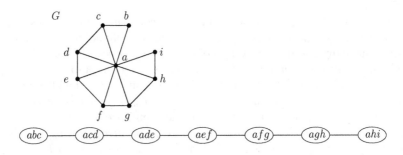

Fig. 1. A graph G having low pathwidth but high persistence

What is interesting is that this notion is hard to recognize. Bounded persistence pathwidth is a $W[t]$-hard property to recognize in general, and even recognizing domino pathwidth, where we restrict every vertex of the underlying graph to belong to at most 2 nodes of the path decomposition is $W[2]$-hard (Downey and McCartin [20, 36]). Nevertheless, it seems that *if* we *know* that a graph has low persistence for a given pathwidth, we ought to be able to get better performance for algorithms. This is a very interesting situation where, because we know the "shape" of the input, we can explore algorithms which might be fast for the kinds of input we might expect, yet could be slow in general. This idea remains to be explored.

References

1. K. Abrahamson, R. Downey and M. Fellows: *Fixed Parameter Tractability and Completeness IV: On Completeness for $W[P]$ and PSPACE Analogs*, Annals of Pure and Applied Logic, 73, pp.235–276, 1995.

2. G. Ausiello, P. Crescenzi, G. Gambosi, V. Kann, A. Marchetti-Spaccamela and M. Protasi: *Complexity and Approximation*, Springer-Verlag, 1999.

3. M. Alekhnovich and A. Razborov: *Resolution is Not Automatizable Unless W[P] is Tractable*, Proc. of the 42nd IEEE FOCS, pp.210-219, 2001.

4. S. Arora: *Polynomial Time Approximation Schemes for Euclidean TSP and Other Geometric Problems*, In Proceedings of the 37th IEEE Symposium on Foundations of Computer Science, pp. 2–12, 1996.

5. S. Arora: *Nearly Linear Time Approximation Schemes for Euclidean TSP and Other Geometric Problems*, Proc. 38th Annual IEEE Symposium on the Foundations of Computing (FOCS' 97), IEEE Press, pp.554-563, 1997.

6. S. Arora, C. Lund, R. Motwani, M. Sudan and M. Szegedy: *Proof Verification and Intractability of Approximation Algorithms*, Proceedings of the IEEE Symposium on the Foundations of Computer Science, 1992.

7. C. Bazgan: *Schémas d'approximation et complexité paramétrée*, Rapport de stage de DEA d'Informatique à Orsay, 1995.

8. Liming Cai and J. Chen: *On Fixed-Parameter Tractability and Approximability of NP-Hard Optimization Problems*, J. Computer and Systems Sciences 54, pp.465–474, 1997.

9. Liming Cai, M. Fellows, D. Juedes and F. Rosamond: *Efficient Polynomial-Time Approximation Schemes for Problems on Planar Stru ctures: Upper and Lower Bounds*, manuscript, 2001.

10. L. Cai and D. Juedes: *On the existence of subexponential time parameterized algorithms*, J. Comput. Sys. Sci., Vol. 67, pp.789-807, 2003.

11. C. Chekuri and S. Khanna: *A PTAS for the Multiple Knapsack Problem*, Proceedings of the ACM-SIAM Symposium on Discrete Algorithms (SODA 2000), pp. 213-222, 2000.

12. J. Chen, I.A. Kanj and W. Jia: *Vertex Cover: Further Observations and Further Improvements*, Journal of Algorithms, 41, pp.280–301, 2001.

13. J. Chen, X. Huang, I. Kanj, and G. Xia: *Polynomial time approximation schemes and parameterized complexity*, in Mathematical Foundations of Computer Science, 29th Annual Meeting, 2004, (J. Faila, V. Koubek, and J. Kratochvil, eds.) Springer-Verlag Lecture Notes in Computer Science 3153, pp.500-512, 2004.

14. J. Chen and A. Miranda: *A Polynomial-Time Approximation Scheme for General Multiprocessor Scheduling*, Proc. ACM Symposium on Theory of Computing (STOC '99), ACM Press, pp. 418–427, 1999.

15. M. Chrobak and M. Slusarek: *On some packing problems related to dynamic storage allogation*, RARIO Inform. Theor. Appl., Vol. 22, pp.487-499, 1988.

16. R. Downey: *Parameterized complexity for the skeptic*, in Proceeding of the 18th Annual IEEE Conference on Comutational Complexity, 2003.

17. R. Downey, V. Estivill-Castro, M. Fellows, E. Prieto-Rodriguez and F. Rosamond: *Cutting Up Is Hard To Do: the Parameterized Complexity of k-Cut and Related Problems*, in Proc. Australian Theory Symposium, CATS, 2003.

18. R. G. Downey, M. R. Fellows: *Parameterized Complexity*, Springer-Verlag, 1999.

19. R. G. Downey, C.M.McCartin: *Online Problems, Pathwidth, and Persistence*, to appear in Proceedings of IWPEC 2004.

20. R. G. Downey, C.M.McCartin: *Bounded Persistence Pathwidth*, to appear.

21. T. Erlebach, K. Jansen and E. Seidel: *Polynomial Time Approximation Schemes for Geometric Graphs*, Proc. ACM Symposium on Discrete Algorithms (SODA'01), pp. 671–67, 2001.

22. M. Fellows: *Parameterized complexity: the main ideas and connections to practical computing.* In: R. Fleischer et al. (Eds.) Experimental Algorithmics, LNCS 2547, pp.51–77, 2002.

23. M. Fellows: Personal communication, March 2003.

24. J. Flum and M. Grohe: *Parameterized complexity and subexponential time,* to appear Bulletin EATCS.

25. J. Flum, M. Grohe, and M. Weyer: *Bounded fixed-parameter tractability and* $\log^2$ *nondeterministic bits,* to appear Journal of Comput. Sys. Sci.

26. M. Frick and M. Grohe: *The Complexity of First-Order and Monadic Second-Order Logic Revisited,* in LICS, pp.215-224, 2002.

27. J. Fouhy: *Computational Experiments on Graph Width Metrics,* MSc Thesis, Victoria University, Wellington, 2003.

28. M. Garey and D. Johnson: *Computers and Intractability: A Guide to the Theory of* NP-*completeness,* W.H. Freeman, San Francisco, 1979.

29. R. Impagliazzo and R. Paturi: *On the complexity of K-SAT,* JCSS, 62(2), pp.367-375, March 2001.

30. R. Impagliazzo, R. Paturi and F. Zane: *Which problems have strongly exponential complexity?,* JCSS 63(4), pp.512–530, 2001.

31. S. Irani: *Coloring Inductive Graphs On-Line,* Algorithmica, vol.11, pp.53-72, 1994.

32. S. Khanna and R. Motwani: *Towards a Syntactic Characterization of PTAS,* in Proc. STOC 1996, ACM Press, pp.329–337, 1996.

33. H. A. Kierstead:
The Linearity of First Fit Coloring of Interval Graphs,
SIAM J. on Discrete MAth, Vol 1, No. 4, pp 526-530, 1988.

34. H. Kierstead and J. Qin: *Colouring interval graphs with First-Fit,* Discrete Math. Vol. 144, pp.47-57, 1995.

35. H. Kierstead and W. Trotter: *The linearity of first-fit colouring of interval graphs,* SIAM J. on Discrete Math., Vol. 1, pp.528-530, 1988.

36. C. McCartin: *Contributions to Parameterized Complexity,* PhD Thesis, Victoria University, Wellington, 2003.

37. R. Niedermeier: *Ubiquitous parameterization: invitation to fixed-parameter algorithms,* in Mathematical Foundations of Computer Science, 29th Annual Meeting, 2004, (J. Faila, V. Koubek, and J. Kratochvil, eds.) Springer-Verlag Lecture Notes in Computer Science 3153, pp.84-103, 2004.

38. R. Niedermeier: *Fixed Parameter Algorithms,* Oxford University Press, in preparation.

39. S. Rubin: *Automatic Structures,* PhD Diss, Auckland University, 2004.

40. R. Shamir and D. Tzur: *The Maximum Subforest Problem: Approximation and Exact Algorithms,* Proc. ACM Symposium on Discrete Algorithms (SODA'98), ACM Press, pp.394–399, 1998.

41. U. Stege, *Resolving Conflicts in Problems in Computational Biochemistry,* Ph.D. dissertation, ETH, 2000.

Basic Notions of Reaction Systems

A. Ehrenfeucht[1] and G. Rozenberg[1,2,⋆]

[1] Department of Computer Science, University of Colorado at Boulder,
Boulder, CO, U.S.A.
[2] Leiden Institute of Advanced Computer Science (LIACS),
Leiden University, Leiden, The Netherlands
`rozenber@liacs.nl`

Natural Computing is a general term referring to computing taking place in nature and computing inspired by nature. There is a huge surge of research on natural computing, and one of the reasons for it is that two powerful and growing research trends happen at the same time (and actually strengthen and influence each other). These two trends are:

(1) trying to understand the functioning of a living cell from the cell-as-a-whole perspective,

(2) trying to free the theory of computation from classical paradigms (the ongoing transition to the so called "non-classical computation") in order to explore a much broader notion of computation. This broader notion should take into account not only the original/classical point of view of "computation as calculation" but should also account for (be inspired by) processes, e.g., life processes, taking place in nature.

Considering chemical reactions as metaphors for computing from both more applied (programming) and more theoretical point of view has already an established tradition in computer science, see, e.g., [1], [2] and [3].

In our talk (paper) we present a computational model based on chemical reactions but geared more towards the chemistry of living cells. The underlying idea of our model is that cell life consists of thousands of individual chemical reactions (generally referred to as metabolism) and all chemical reactions in a cell are regulated, where the regulation is achieved by two opposing effects:

(1) facilitation/acceleration, and
(2) inhibition/retardation.

Therefore the basic notion of our theory is the notion of reaction defined as follows.

Definition 1. *A reaction is a 3-tuple $a = (X, Y, Z)$ of finite sets. If S is a set such that $X, Y, Z \subseteq S$, then we say that a is a reaction in S.*

The set X, also denoted by X_a, is the *reactant (set)* of a, the set Y, also denoted by Y_a, is the *inhibitor (set)* of a, and the set Z, also denoted by Z_a, is the *result (set)* of a.

⋆ All correspondence to the author's address in Leiden.

C.S. Calude, E. Calude and M.J. Dinneen (Eds.): DLT 2004, LNCS 3340, pp. 27–29, 2004.
© Springer-Verlag Berlin Heidelberg 2004

Definition 2. *For a reaction a and a set K, the result of a on K, denoted $res_a(K)$, is defined by: $res_a(K) = Z_a$ if $X_a \subseteq K$ and $Y_a \cap K = \emptyset$, and $res_a(K) = \emptyset$ otherwise.*

If $X_a \subseteq K$ and $Y_a \cap K = \emptyset$, then we say that a is *enabled* in K; otherwise a is *not enabled* in K.

For a set of reactions A and a set K, the *result of A on K*, denoted $res_A(K)$, is defined by: $res_A(K) = \bigcup_{a \in A} res_A(K)$.

The way that we define the result of a set of reactions on a set of elements formalizes the following two assumptions that we make about the chemistry of a cell.

(i) We assume that we have the "threshold" supply of elements (molecules) - either an element is present and then we have "enough" of it, or an element is not present. Therefore we deal with a qualitative rather than quantitative (e.g., multisets) calculus.

(ii) We do not have the "permanence" feature in our model: if nothing happes to an element then it remains/survives (status quo approach). On the contrary, in our model, an element remains/survives only if there is a reaction sustaining it.

We are ready now to define the basic object of our theory. It formalizes the fact that every reaction in a cell interacts directly with other reactions. Thus, e.g., an inhibitor of one particular reaction is almost always an element of some other reaction, or a result of one reaction is a reactant of another reaction. That is, the thousands of reactions in a cell form a network.

Definition 3. *A reaction system is an ordered pair $\mathcal{A} = (S, A)$ such that A is a set of reactions in S.*

The set S is called the *background* of $\mathcal{A}$.

Definition 4. *For a reaction system $\mathcal{A} = (S, A)$ and a set $K \subseteq S$, the result of $\mathcal{A}$ on K, denoted $res_\mathcal{A}(K)$, is the result of A on K.*

Thus, $res_\mathcal{A}(K) = res_A(K)$.

In our talk we discuss the basic notions and properties of reaction systems, and the place of the theory of reaction systems within the current research on natural computing.

Acknowledgements

We are indebted to D.M. Prescott for enlightening discussions about cell biology. The second author acknowledges the support by NSF grant 0121422 and the European Research Training Network Segravis.

References

[1] J.-P. Banâtre, P. Fradet, D. Le Metayer, Gamma and the chemical reaction model: Fifteen years later, *in* C. Calude, Gh. Păun, G. Rozenberg, and A. Salomaa (eds.), Lecture Notes in Computer Science, 2235, pp. 17–44, Springer–Verlag, 2001.

[2] G. Berry and G. Boudol, The chemical abstract machine, Theoretical Computer Science, 96, pp. 217–248, 1992.

[3] A. Ehrenfeucht and G. Rozenberg, Forbidding–enforcing systems, Theoretical Computer Science, 292, pp. 611–638, 2003.

A Kleene Theorem for a Class of Communicating Automata with Effective Algorithms

Blaise Genest[1], Anca Muscholl[1], and Dietrich Kuske[2]

[1] LIAFA, Université Paris 7, France
`muscholl@liafa.jussieu.fr`
[2] Institut für Algebra, Technische Universität Dresden, Germany

Abstract. Existential bounded communication of a communicating finite-state machine means that runs can be scheduled in such a way that message channels are always bounded in size by a value that depends only on the machine. This notion leads to regular sets of representative executions, which allows to get effective algorithms. We show in this paper the equivalence of several formalisms over existentially bounded models: monadic second order logic, communicating automata and globally-cooperative compositional MSC-graphs.

1 Introduction

Communicating finite-state machines (CFM for short), or equivalently, FIFO channel systems/nets, are a fundamental model for concurrent systems. Despite the undecidability of basic questions even for two processes [7], such as e.g. reachability, several papers aim at identifying subclasses or heuristics allowing to solve model-checking problems. For example, for lossy FIFO systems, reachability is shown to be decidable (albeit of non-primitive recursive complexity [26]) using well-quasi-orderings on channel contents [1, 10].

One of the techniques used by several papers [3–5] is based on the computation of the set of reachable configurations, hence of all channel contents, by some regular device. Often this approach means that one has to relax the operations on channels, which yields an overapproximation of the result.

The approach taken by our paper goes beyond regular languages. We use partial-order methods for describing the behavior of a CFM. The formal model are *Message sequence charts* (MSC for short), a diagram notation described by the ITU norm Z.120. The advantage of reasoning about CFMs using MSCs is both succintness and comprehension, since a single diagram subsumes a set of words representing channel contents. Moreover, MSCs are a partial-order formalism, and we specify CFM and MSC properties using partial-order logics such as monadic-second order logic.

The MSC model has become popular in software development through its visual representation, depicting the involved processes as vertical lines, and each message as an arrow between the source and the target processes, according to

C.S. Calude, E. Calude and M.J. Dinneen (Eds.): DLT 2004, LNCS 3340, pp. 30–48, 2004.

their occurrence order. The Z.120 standard has also extended the notation to *MSC-Graphs*, which consist of finite transition systems, where each state embeds a single MSC.

An early line of research considered a decidable subclass of MSC-based formalisms, corresponding to CFMs using universally-bounded channels. This means that every CFM run can be executed with channels of fixed size, no matter how actions are scheduled. Such a constraint allows to represent reachable configurations by regular sets, but is very restrictive. Whereas deciding whether a CFM is universally-bounded is undecidable, recently heuristics have been considered for this problem [19]. For MSC-graphs, universal-boundedness is guaranteed by syntactic restrictions [2, 23] that imply regularity at the same time. Over universally-bounded models, several characterizations have been obtained [15] (see also [20], [17]), showing that CFMs, MSO and regular MSC languages have the same expressive power.

As already mentioned, universally-bounded channels lack expressive power. Many basic protocols of producer-consumer type (such as e.g. the asynchronous protocol [27]) are not universally-bounded. However, we can relax the restriction on channels in order to capture interesting behaviors such as the asynchronous protocol. The idea is to require an *existential bound* on channels. This means roughly that every run of a protocol must have *some* scheduling that respects a given channel size (other schedulings might exceed the bound). In other words, runs *can* be executed with bounded channels, provided that we schedule the send/receive actions conveniently. Note that this requirement is perfectly legitimate in practice, since real life protocols must be executable within bounded channels. Existential channel bounds appear implicitly in [12] (realizable CHM-SCs). They allow to use representatives for e.g. solving the model-checking problem of CMSC-graphs against MSO [21]. A set of representatives is actually an abstract representation of the set of MSCs representing the exact behavior (of the CFM, MSC-graph, MSO-formula, etc). For properties expressed by globally-cooperative MSC-graphs, regular sets of representatives can be used for deciding model-checking within the same complexity bounds as for universally-bounded channels [13].

In a nutshell, we have two objectives here: First, we would like to have a robust class of MSC-models, i.e., to provide equivalent characterizations in terms of logics, regular expressions and automata. Second, we aim at providing decidability for the model-checking problem.

The main result of the paper is that CFM, MSO and globally-cooperative CMSC-graphs are equivalent over existentially-bounded MSCs. The MSO logic here uses the partial order and the message relation as in [21], and is a priori more powerful than the logic used in [6]. A consequence of the main result is that several interesting model-checking instances are decidable in this setting. For instance, we can check whether an existentially-bounded CFM satisfies an MSO formula or if a safe CMSC-graph is included in (intersects, respectively) a CFM. Finally, we answer in particular to the question left open by

[13], where we ask whether globally-cooperative MSC-graphs are always implementable.

Overview. In Section 2 we define the formalisms used in the paper – message sequence charts, communicating automata, MSO, and Mazurkiewicz traces. Section 4 establishes the main tool used for existential bounded communication, that is, the connection to Mazurkiewicz traces. Then we describe the way model-checking works using representative executions. In Section 5 we state the main characterizations and we sketch the proof. The difficult part is the construction of a CFM from a regular set of representative runs, since this means that we have to distribute the control. Finally, Section 7 contains some additional results.

Related work. Our paper generalizes several results about expressivity and model-checking for MSCs with universally-bounded channels [2, 23, 15, 17, 20]. Representative executions in model-checking MSO properties have been used in [21], whereas model-checking MSC properties has been considered in [13]. Recently, the equivalence between CFM and existential MSO has been shown in [6] without channel restriction. However, the logic used by [6] is weak, for it allows only the message relation and the immediate successor on each process. In contrast, MSO as used in our paper and by [21, 15] refers to the partial order.

2   Definitions

2.1   Message Sequence Charts

The communication framework used in our paper is based on sequential processes that exchange asynchronously messages over point-to-point, error-free FIFO channels. Let $\mathcal{P}$ be a finite set of process names that we fix throughout this paper. Processes act by either sending a message, that is denoted by $p!q$ meaning that process p sends to process q, or by receiving a message, that is denoted by $p?q$, meaning that process p receives from process q. Thus we do not use different message contents in our notation. In the same line, we do not consider local events, that is events which are neither send nor receive. This is done for convenience and the reader might convince himself/herself that proofs work (with small alterations) in the more general setting as well.

For $p \in \mathcal{P}$, we define a local alphabet $\Sigma_p = \{p!q, p?q \mid q \in \mathcal{P} \setminus \{p\}\}$ and set $\Sigma = \bigcup_{p \in \mathcal{P}} \Sigma_p$. For the rest of the paper, whenever a pair of processes $p, q \in \mathcal{P}$ communicates, we will implicitly assume that $p \neq q$.

We introduce now the notation of *(compositional) message sequence charts*, that is usually employed for describing scenarios of communication. The *message sequence chart* notation (MSC for short) corresponds to the Z.120 standard of the ITU. Theoretical work has revealed several deficiencies of the standard notation of MSCs and MSC-graphs, which motivated the extended notation of *compositional message sequence charts, CMSC* for short. We will be mainly interested in MSCs as a complete formalism, but we will use CMSCs as a kind of technical tool.

Definition 2.1. *[12] A compositional message sequence chart (CMSC) is a tuple $M = (E, \lambda, \mathrm{msg}, (<_p)_{p \in \mathcal{P}})$ where*

- *E is a finite set of events*
- *$\lambda : E \to \Sigma$ maps each event to a type, and we set*
  - *$E_p = \{e \in E \mid \lambda(e) \in \Sigma_p\}$ the set of events of process p,*
  - *$S = \{e \in E \mid \exists p, q \in \mathcal{P} : \lambda(s) = p!q\}$ the set of send events, and*
  - *$R = E \setminus S$ the set of receive events*
- *$<_p$ is a total order on E_p for any $p \in \mathcal{P}$*
- *$\mathrm{msg} : S \to R$ is a partially defined, injective mapping satisfying*
  - *if $\mathrm{msg}(s) = r$, then there are $p, q \in \mathcal{P}$ distinct such that $\lambda(s) = p!q$ and $\lambda(r) = q?p$,*
  - *if $s_1 <_p s_2$ and $\lambda(s_1) = \lambda(s_2) = p!q$, then $\mathrm{msg}(s_1) <_q \mathrm{msg}(s_2)$ if defined (FIFO),*

such that the relation $< := \bigcup_{p \in \mathcal{P}} <_p \cup \{(s, \mathrm{msg}(s)) \mid s \in S\}$ is acyclic.

A message sequence chart (MSC for short [16]) is a CMSC $(E, \lambda, \mathrm{msg}, (<_p)_{p \in \mathcal{P}})$ such that the message mapping $\mathrm{msg} : S \to R$ is one-to-one, i.e., in particular defined everywhere.

For a CMSC M, we will write $P(e) = p$ if $e \in E_p$, i.e., $\lambda(e) \in \Sigma_p$. Moreover, we write $e <_p f$ if e is the immediate predecessor of f on process p, i.e., $e <_p f$ and $e <_p g \leq_p f$ implies $g = f$.

Figure 1 depicts an MSC. In that picture, there are two processes named 1 and 2. Further, $a = 1!2$ and $b = 2?1$ are two actions. Hence, the MSC from Figure 1 denotes the sending and receiving of three messages from process 1 to process 2. It does not specify the content of these messages.

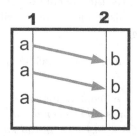

Fig. 1. A Message Sequence Chart (MSC)

Let $M = (E, \lambda, \mathrm{msg}, (<_p)_{p \in \mathcal{P}})$ be a CMSC. Then the relation $<$ being acyclic implies that its transitive and reflexive closure $\leq$ is a partial order on E, called the *visual order*.[1] Thus, we can look at a CMSC as a poset $(E, \leq, \lambda)$. Any linear extension of $\leq$ is called a *linearization* of M. We represent it as a word $u = u_1 \cdots u_n$ over the alphabet Σ. Thus, the set $\mathrm{Lin}(M)$ of linearizations of the

---

[1] Note that $< \neq \{(e, f) \in E^2 \mid e \leq f, e \neq f\}$.

CMSC M is a subset of Σ^*. For a set (or language) of CMSCs $\mathcal{M}$, we write $\mathrm{Lin}(\mathcal{M}) = \bigcup_{M \in \mathcal{M}} \mathrm{Lin}(M)$.

Any linearization of an MSC is simply called a *linearization*. A word (linearization) $w \in \Sigma^*$ is *B-bounded* for some positive integer B if for any prefix u of w and any $p, q \in \mathcal{P}$, the number of occurrences of $p!q$ in u exceeds that of occurrences of $q?p$ in u by at most B. An MSC M is *existentially B-bounded* ($\exists$-*B-bounded* for short) if it has some B-bounded linearization $w \in \mathrm{Lin}(M)$. Let $\mathrm{Lin}^B(M) \subseteq \mathrm{Lin}(M)$ denote the set of B-bounded linearizations of M – by definition, this set is non-empty iff M is $\exists$-B-bounded. An MSC M is *universally-B-bounded* if $\mathrm{Lin}(M) = \mathrm{Lin}^B(M)$ (where $B \in \mathbb{N}$).

The set of all CMSCs, resp. MSCs and $\exists$-B-bounded MSCs, will be denoted $\mathbb{CMSC}$, resp. $\mathbb{MSC}$ and $\mathbb{MSC}^B$. For checking whether an MSC $M = (E, \lambda, \mathrm{msg}, (<_p)_{p \in \mathcal{P}})$ is $\exists$-B-bounded one can use the following relation $\prec_B \subseteq E \times E$ [18]:

Let $\prec_B = \mathrm{msg} \cup \mathrm{rev} \cup (<_p)_{p \in \mathcal{P}}$, where

$$\mathrm{rev}(r) = s' \text{ iff } \quad \mathrm{msg}(s) = r, \lambda(s) = \lambda(s'), \text{ and}$$
$$|\{x \in E \mid s <_p x \leq_p s', \lambda(s) = \lambda(x)\}| = B$$

That is, rev maps a receive r with $r = \mathrm{msg}(s)$ to the send s' that is the B-th event with $\lambda(s') = \lambda(s)$ and $s < s'$ (if such an event exists).

Lemma 2.2. *[18] An MSC M is $\exists$-B-bounded iff the relation $\prec_B = \mathrm{msg} \cup \mathrm{rev} \cup (<_p)_{p \in c\mathcal{P}}$ is acyclic.*

2.2   Communicating Finite-State Machines

The most natural formalism to describe scenarios of communication protocols, i.e., sets of MSCs, are communicating finite-state machines (CFM for short) that we define in this section. CFMs are a realistic model for distributed algorithms based on asynchronous message passing.

Definition 2.3. *[7] A communicating finite state machine (CFM) is a tuple $\mathcal{A} = (C, (\mathcal{A}_p)_{p \in \mathcal{P}}, F)$ such that*

- *C is a finite set of* message contents *or* control message.
- *$\mathcal{A}_p = (S_p, \rightarrow_p, \iota_p)$ is a finite transition system over the alphabet $\Sigma_p \times C$ for any $p \in \mathcal{P}$ (i.e., $\rightarrow_p \subseteq S_p \times (\Sigma_p \times C) \times S_p$) with initial state $\iota_p \in S_p$.*
- *$F \subseteq \prod_{p \in \mathcal{P}} S_p$ is a set of global final states.*

The usual way to define the behavior of a CFM considers these machines as sequential devices that accept linearizations of MCSs. The details are as follows: Let $\mathcal{A} = (C, (\mathcal{A}_p)_{p \in \mathcal{P}}, F)$ be a CFM. A configuration of $\mathcal{A}$ consists of a tuple of local states and of buffer contents, i.e., it is an element $((s_p)_{p \in \mathcal{P}}, (w_{p,q})_{p,q \in \mathcal{P}})$ of $\prod_{p \in \mathcal{P}} S_p \times \prod_{p,q \in \mathcal{P}} C^*$. For two configurations, an action $a \in \Sigma_p$, and a control message $c \in C$, we have

$$((s_p^1)_{p \in \mathcal{P}}, (w_{p,q}^1)_{p,q \in \mathcal{P}}) \xrightarrow{a,c} ((s_p^2)_{p \in \mathcal{P}}, (w_{p,q}^2)_{p,q \in \mathcal{P}})$$

if

- $s_p^1 \xrightarrow{a,c} s_p^2$ is a transition of the local machine $\mathcal{A}_p$ and $s_q^1 = s_q^2$ for $q \neq p$.
- if $a = p!q$, then $w_{p,q}^2 = w_{p,q}^1 c$ (i.e., the control message c is appended to the buffer from p to q) and $w_{p',q'}^2 = w_{p',q'}^1$ for $(p',q') \in \mathcal{P}^2 \setminus \{(p,q)\}$ (i.e., all other buffers are unchanged)
- if $a = p?q$, then $w_{q,p}^1 = cw_{q,p}^2$ (i.e., the control message c is deleted from the front of the buffer from q to p) and $w_{q',p'}^1 = w_{q',p'}^2$ for $(q',p') \in \mathcal{P}^2 \setminus \{(q,p)\}$ (i.e., all other buffers are unchanged).

A *sequential run* of $\mathcal{A}$ is a path $d_1, a_1, d_2, a_2, \ldots, a_n, d_{n+1}$ with $d_i \xrightarrow{a_i, c_i} d_{i+1}$ for some control messages $c_i \in C$. It is accepting if $d_1 = ((\iota_p)_{p \in \mathcal{P}}, (\varepsilon)_{p,q \in \mathcal{P}})$ and $d_n = (f, (\varepsilon)_{p,q \in \mathcal{P}})$ for some $f \in F$. Finally, $L(\mathcal{A}) \subseteq \Sigma^*$ is the set of words $a_1 a_2 \cdots a_n$ such that there exists an accepting sequential run $d_1, a_1, d_2, a_2, \ldots, a_n, d_{n+1}$.

A natural alternative definition of the semantics of a CFM $\mathcal{A}$ uses MSCs for representing succesful runs. To this purpose, let $M = (E, \lambda, \mathrm{msg}, (<_p)_{p \in \mathcal{P}})$ be an MSC and $\rho : E \to \bigcup_{p \in \mathcal{P}} S_p$ be a mapping. For this mapping, we define a second mapping $\rho^- : E \to \bigcup_{p \in \mathcal{P}} S_p$ as follows. Let $e \in E_p$. If there is $e' \in E_p$ such that $e' \lessdot_p e$ then $\rho^-(e) = \rho(e')$. Otherwise (i.e., if e is minimal in $(E_p, <_p)$), we set $\rho^-(e) = \iota_p$. The idea behind these definitions is that $\rho(e)$ denotes the state of the local machine $\mathcal{A}_p$ *after* executing e and that $\rho^-(e)$ describes the state the local machine was in *before* executing e. Then the mapping ρ is a *run* if for any $s \in E$ with $\lambda(s) = p!q$ and $\mathrm{msg}(s) = r$, there is some control message $c \in C$ such that $\rho^-(s) \xrightarrow{\lambda(s),c}_p \rho(s)$ and $\rho^-(r) \xrightarrow{\lambda(r),c}_q \rho(r)$.

Now let ρ be a run on the MSC M. If $E_p \neq \emptyset$, let $s_p = \rho(e_p)$ where e_p is the maximal event in $(E_p, <_p)$. Otherwise, define $s_p = \iota_p$. The run ρ is *succesful* if the tuple $(s_p)_{p \in \mathcal{P}}$ belongs to the set of global final states F. An MSC is accepted by the CFM $\mathcal{A}$ if it admits a succesful run. We will denote by $\mathcal{L}(\mathcal{A})$ the set of MSCs accepted by $\mathcal{A}$. Notice that a MSC can admit several (accepting) runs of a CFM.

Now it is straightforward, but technically cumbersome to prove the relation between the two languages of a CFM:

Proposition 2.4. *Let $\mathcal{A}$ be a CFM. Then $L(\mathcal{A}) = \mathrm{Lin}(\mathcal{L}(\mathcal{A}))$.*

2.3   CMSC-Graphs

An MSC stands for a single scenario, so for describing sets of scenarios one needs some formalism for composing single scenarios. A simple way to do this was proposed in the Z.120 standard through high-level MSCs (denoted here as MSC-graphs). We define now CMSC-graphs [12], that can be viewed as a kind of regular expressions over communication events.

We need first to define the composition of two CMSCs. Intuitively, to compose CMSCs M_1 and M_2, we glue the corresponding process lines together and draw the second CMSC below the first one. In order to make this formal, we need the restriction of a CMSC $M = (E, \lambda, \mathrm{msg}, (<_p)_{p \in \mathcal{P}})$ to a subset $F \subseteq E$: It is the CMSC $M|_F = (F, \lambda^F, \mathrm{msg}^F, (<_p^F)_{p \in \mathcal{P}})$ with $\lambda^F = \lambda|_F$, $\mathrm{msg}^F(s) = r$ if $\mathrm{msg}(s) = r$ and $s, r \in F$, and $<_p^F = <_p \cap (F \times F)$.

Definition 2.5. *Let $M_i = (E^i, \lambda^i, \mathrm{msg}^i, (<_p^i)_{p \in \mathcal{P}})$ for $i = 1, 2$ be CMSCs. The composition $M_1 \cdot M_2$ is the set of CMSCs $M = (E_1 \uplus E_2, \lambda, \mathrm{msg}, (<_p)_{p \in \mathcal{P}})$ such that*

- *$M|_{E_i} = M_i$ for $i = 1, 2$, and*
- *$e \in E_2$ and $e <_p e'$ imply $e' \in E_2$ for any $e, e' \in E_1 \uplus E_2$.*

This composition can naturally be extended to a binary operation on languages of CMSCs by $\mathcal{M}_1 \cdot \mathcal{M}_2 = \bigcup_{M_1 \in \mathcal{M}_1, M_2 \in \mathcal{M}_2} M_1 \cdot M_2$. This operation turns out to be associative. Notice that, in general, the product $M_1 \cdots M_n$ contains more than one CMSC. However, it may contain at most one MSC, since the message mapping must be FIFO.

Definition 2.6. *A CMSC-graph is a labeled graph $G = (V, \to, \lambda, V^0, V^f)$ with*

- *V is the finite set of vertices.*
- *$V^0, V^f \subseteq V$ are finite sets of initial/final vertices respectively.*
- *$\to \subseteq V \times V$ is the set of edges.*
- *$\lambda : V \to \mathbb{CMSC}$ labels a node v with the CMSC $\lambda(v)$.*

A path in the CMSC-graph G is a sequence $v_1, \ldots, v_n$ of nodes in V such that $v_i \to v_{i+1}$ for all i. It is accepting if $v_1 \in V^0$, $v_n \in V^f$ and the product $\lambda(v_1) \cdots \lambda(v_n)$ contains an MSC M. This MSC is said to be accepted by G. The set of all MSCs accepted by G is denoted $\mathcal{L}(G)$.

Figure 2 depicts a CMSC-graph. Here, we have two processes named "host" and "function". In the leftmost node of the CMSC-graph, one finds a CMSC. This CMSC describes that host sends messages "send" and "m" that are received by the process function. Immediately after receiving a message, process function sends an acknowledgment. While the sending of these acknowledgments is part of the current CMSC node, their receiving by the host is located in the next node. Thus, after executing this CMSC, the buffer from function to host contains two acknowledgments while the other buffer is empty. Altogether, the CMSC-graph describes all MSCs where process function immediately acknowledges messages it gets. The first message from host to function initializes the transfer from host to function. Host sends the message m for the first time. Since it does not get the acknowledgement in time, it must send again m, before getting the first acknowledgement. Then it iterates between sending a message and receiving the acknowledgement from the message before, and at the end, clears the buffer.

The size $|M|$ of a CMSC M is the number of its events. The size $|G|$ of a CMSG G is $\sum_{v \in V} |\lambda(v)|$. The size of $\mathcal{P}$ is the number $\wp$ of processes.

2.4 Monadic Second Order Logic

Yet another way of specifying sets of MSCs (after CFMs and CMSC-graphs) are logical formulas. We consider here monadic second order logic, that is the

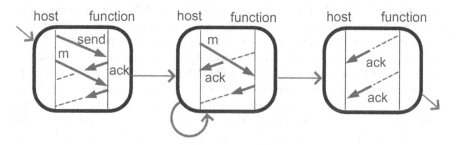

Fig. 2. A CMSC-Graph specifying transactions of usb 1.1

classical formalism when characterizing regularity of sets of words, trees, traces, etc. The syntax is defined as follows. Let $\mathcal{R}$ be a set of binary relations.

Definition 2.7. MSO($\mathcal{R}$)-formulas over the alphabet Γ *are defined by the syntax*

$$\varphi ::= v_a(x) \mid R(x,y) \mid \neg\varphi \mid \varphi \vee \varphi \mid \exists X\varphi \mid \exists x\varphi \mid x \in X$$

where $R \in \mathcal{R}$, $a \in \Gamma$, x, y *are first order variables, and* X *is a second order variable.*

The relations in $\mathcal{R}$ used in the paper are the message relation msg, the visual order $\leq$, the process order $<_p$ and the immediate successor $\lessdot_p$ on $p \in \mathcal{P}$. An MSO($\mathcal{R}$)-formula over the alphabet Σ is interpreted on an MSC $M = (E, \lambda, \text{msg}, (<_p)_{p \in \mathcal{P}})$ as expected. We have $M \models v_a(x)$ if $\lambda(x) = a$, $M \models x \leq y$ if $x \leq y$ in the visual order on M, and $M \models \text{msg}(x, y)$ if $x \in S$ and $\text{msg}(x) = y$. Moreover, $M \models (x <_p y)$ if $x <_p y$ and $M \models (x \lessdot_p y)$ if y is the immediate successor of x w.r.t. $<_p$.

For an MSO-formula φ over Σ without free variables, let $\mathcal{L}(\varphi)$ denote the set of MSCs that satisfy φ.

An MSO($\leq$)-formula over an alphabet Γ can be interpreted on Γ-labeled partial orders $M = (E, \leq, \lambda)$ with $\lambda : E \to \Gamma$ as usual, by letting $M \models v_a(x)$ if $\lambda(x) = a$ and $M \models x \leq y$ if $x \leq y$. Note that words over Γ can be considered in a natural way as Γ-labeled linear orders. Using this interpretation, we write $w \models \varphi$ to denote that the word w (more precisely: the associated linear order) satisfies φ. Let $L(\varphi)$ denote the set of words over Γ that satisfy φ.

We will also consider existential monadic second-order logic (EMSO). An EMSO formula is of the form $\exists X_1 \ldots X_n \varphi$ with φ a first order formula (that is, without second order quantifications).

2.5   Traces

So far, we described four formalisms for the specification of sets of MSCs: finite automata that accept linearizations, communicating finite-state machines that generate MSCs, CMSC-graphs, and MSO over MSCs or pomsets. Later, we will relate the expressive power of these formalisms. A crucial tool in these investigations are Mazurkiewicz traces [22] that are introduced next.

A *trace alphabet* is a pair (Ω, I) consisting of an alphabet Ω and a symmetric and irreflexive relation $I \subseteq \Omega^2$. The relation I will be refered to as the *independence relation*; its complement $D = \Omega^2 \setminus I$ is the *dependence relation*.

Let $\sim_I \subseteq \Omega^* \times \Omega^*$ be the congruence on the free monoid Ω^* generated by the equations $ab \sim_I ba$ for all $(a, b) \in I$. A *trace* is an equivalence class $[w]_I$ of this equivalence relation. Further, the *I-closure* of a set $L \subseteq \Omega^*$ is the set $[L]_I = \bigcup_{w \in L}[w]_I$ of all words that are $\sim_I$-equivalent to same element of L. If $L = [L]_I$, we say that L is closed under I-commutation (or I-closed for short).

An alternative way to define traces is via labeled partially ordered sets. Any word $u = a_1 a_2 \cdots a_n$ with $a_i \in \Omega$ defines a labeled poset $t_u = (E, \leq_I, \lambda)$ with

- $E = \{1, \ldots, n\}$.
- $\lambda(i) = u_i$, and
- $\leq_I$ is the least partial order on E such that $i \leq_I j$ whenever $i < j$ and $\lambda(i) D \lambda(j)$.

It belongs to the very basics of trace theory that two words u and v are equivalent w.r.t. $\sim_I$ iff the two labeled partial orders t_u and t_v are isomorphic. This implies in particular that $[u]_I$ is the set of linearizations of the labeled poset t_u. At places, it will be useful to consider a trace not as an equivalence class of words, but as (an isomorphism class of) labeled partial orders t_u. This allows in particular to interpret MSO($\leq_I$)-formulas in a trace and thereby to define notions like $[u]_I \models \varphi$ for a trace $[u]_I$.

An Ω-labeled partially ordered set $(E, \leq, \lambda)$ is isomorphic to some t_u if we have for any $r, s \in E$

- $r \lessdot s$ implies $(\lambda(r), \lambda(s)) \in D$, and
- if r and s are incomparable, then $(\lambda(r), \lambda(s)) \in I$.

We end this section by recalling some fundamental results from Mazurkiewicz trace theory (cf. [8]). Below, we call an automaton $\mathcal{A}$ *I-loop-connected* if for every loop of $\mathcal{A}$, the set of letters labeling the loop induces a connected subgraph of (Ω, D). Asynchronous automata are defined in Section 6.

Theorem 2.8. *Let (Ω, I) be a trace alphabet and let $L \subseteq \Omega^*$ be I-closed.*

1. *(Ochmański's theorem [24]) L is regular iff there exists some I-loop-connected automaton $\mathcal{A}$ with $L = [L(\mathcal{A})]_I$.*
2. *(Zielonka's theorem [29]) L is regular iff it is accepted by a deterministic asynchronous automaton.*
3. *L is regular iff $L = \{u \in \Omega^* \mid t_u \models \varphi\}$ for some MSO formula φ [28, 9].*

3   Regular MSC Languages

Regularity is a fundamental concept in many settings - strings, trees, graphs etc, and plays a crucial role in verificaton, ensuring the effectiveness of several operations. The most basic operations are intersection $\mathcal{M} \cap \mathcal{M}' = \emptyset$ and inclusion $\mathcal{M} \subseteq \mathcal{M}'$ of two models $\mathcal{M}, \mathcal{M}'$. We can view these operations as model-checking a model $\mathcal{M}$ against a negative property $\mathcal{M}'$ (intersection) or a positive one (inclusion). For models $\mathcal{M}, \mathcal{M}'$ expressed by unrestricted MSC-graphs these two questions are undecidable:

Theorem 3.1. *[2, 23] Let $\mathcal{M}, \mathcal{M}'$ be sets of MSCs. Then both questions $\mathcal{M} \cap \mathcal{M}' = \emptyset$ and $\mathcal{M} \subseteq \mathcal{M}'$ are undecidable, provided that the model $\mathcal{M}$ is generated by an MSC-graph and the property $\mathcal{M}'$ is generated either by an MSC-graph or an LTL-formula defining a set of linearizations.*

For languages expressed by MSC-graphs a syntactic condition recalling Ochmański's Theorem on Mazurkiewicz traces was proposed in order to preserve regularity:

Definition 3.2. *The communication graph $CG^M = \langle P, \rightarrow \rangle$ of an MSC M contains the processes $P \subseteq \mathcal{P}$ that occur in M, and with edges $p \rightarrow q \in E$ if there is a message from p to q in M.*

Definition 3.3. *[2, 23] An MSC-graph G is regular, if for each loop σ in the graph of H, the communication graph CG^M of the MSC M labeling σ is strongly connected.*

The definition of regular MSC-graphs is syntactic, and can be checked in co-NP [23, 2]. Model checking becomes decidable for regular MSC-graphs, since their languages are regular [23, 2]. More precisely:

Theorem 3.4. *[14] A set of MSCs $\mathcal{M}$ is the language of a regular MSC-graph if and only if it has a regular set of linearizations $\mathrm{Lin}(\mathcal{M})$ and is generated by a finite set of MSCs.*

The price that has to be paid for preserving the regularity of $\mathrm{Lin}(\mathcal{M})$ is that communication channels must be universally-bounded. This is an important restriction, that prevents the description of many known protocols (see next section).

Proposition 3.5. *For a regular MSC-graph G, its MSC language $\mathcal{L}(G)$ is universally $(s+1)\wp n$-bounded where s is the number of nodes of G and n the maximal number of messages in a node.*

The next proposition shows an exponential bound on the size of an automaton for $\mathrm{Lin}(G)$, with G being regular. A similar proof appears in [2]. Note that this bound on the automaton size would imply an exponential channel bound, whereas Proposition 3.5 provides a polynomial bound.

Proposition 3.6. *For a regular MSC-graph G, the set of linearization $\mathrm{Lin}(G)$ is accepted by an automaton of size at most $2^{O(\wp^2(s+1))} n^{\wp^2}$, where s is the number of nodes of G and n the maximal number of messages in a node.*

Regular MSC languages have been shown to be a robust class of specifications:

Theorem 3.7. *[15] Let $\mathcal{M}$ be an $\forall$-B-bounded set of MSCs. Then the following assertions are equivalent:*

1. $\mathcal{M} = \mathcal{L}(\mathcal{A})$ *for some CFM $\mathcal{A}$.*
2. $\mathcal{M} = \mathcal{L}(\varphi)$ *for some MSO$(\leq, \mathrm{msg})$ formula φ.*
3. $\mathrm{Lin}(\mathcal{M})$ *is a regular set.*

4   Existential Boundedness

4.1   Why?

This section presents the main technical restrictions that we use in order to obtain a robust characterization of a class of CFMs. In a nutshell, we have two objectives: first we would like to have equivalent characterizations in terms of logics, regular expressions and automata. Second we aim at providing decidability for the model-checking problem. One of the main restrictions that we impose is an existential bound on channels. Note that this requirement is perfectly legitimate in practice, since real life protocols must be executed with bounded communication buffers. Of course, this only means that any protocol run must have *some* bounded linearization (other linearizations might exceed the bound), for some bound that depends on the protocol, only.

A closely related notion is that of *representative linearizations*. For a set of MSCs $\mathcal{M}$, we call $X \subseteq \mathrm{Lin}(\mathcal{M})$ a *set of representatives* if for every MSC $M \in \mathcal{M}$ we have $X \cap \mathrm{Lin}(M) \neq \emptyset$. For a *regular* set X of representatives of $\mathcal{M}$, it is easy to see that X is composed only of B-bounded linearizations for some B [21], so $\mathcal{M}$ is $\exists$-B-bounded. Conversely, for any $\exists$-B-bounded set $\mathcal{M}$, the set $\mathrm{Lin}^B(\mathcal{M})$ of B-bounded linearizations is a set of representatives.

Representatives based on B-bounded linearizations can be used e.g. to do model-checking beyond regular MSC languages, as shown in [21, 13]. The proposition below summarizes the basic ideas:

Proposition 4.1. *Let $\mathcal{M}$ be a set of $\exists$-B-bounded MSCs, and let $\mathcal{M}'$ be an arbitrary set of MSCs.*

1. *We have $\mathcal{M} \cap \mathcal{M}' = \emptyset$ iff $\mathrm{Lin}^B(\mathcal{M}) \cap \mathrm{Lin}^B(\mathcal{M}') = \emptyset$, and $\mathcal{M} \subseteq \mathcal{M}'$ iff $\mathrm{Lin}^B(\mathcal{M}) \subseteq \mathrm{Lin}^B(\mathcal{M}')$.*
2. *Assume that $\mathrm{Lin}^B(\mathcal{M}')$ is regular (but not necessarily a representative set) and that $\mathcal{M}$ has a regular set of B-bounded representatives. Then negative model-checking $\mathcal{M} \cap \mathcal{M}' = \emptyset$ and positive model-checking $\mathcal{M} \subseteq \mathcal{M}'$ are decidable.*

For CMSC-graphs we can obtain existentially-bounded representatives via a simple syntactic restriction:

Definition 4.2. *A CMSC-graph G is safe if every sequence of CMSCs labeling a path from some initial to some final node of G admits one composition that is an MSC (i.e., is accepting).*

For example, the CMSC-graph in Figure 2 is safe. For any CMSC-graph $G = (V, \rightarrow, \lambda, V^0, V^f)$ we can define a regular set K_G of representatives by choosing a linearization $\ell(v)$ for every CMSC $\lambda(v)$ and letting

$$K_G = \{\ell(v_1) \cdots \ell(v_n) \mid v_1, \ldots, v_n \text{ is an accepting path of } G\}$$

The important observation is that a *safe* CMSC-graph G is $\exists$-$|G|$-bounded[2]. Moreover, K_G is a (regular) set of $|G|$-bounded representatives for G.

We also have that $\text{Lin}^B(\mathcal{M}')$ is regular for all B if $\mathcal{M}' = \mathcal{L}(\mathcal{A})$ for some CFM $\mathcal{A}$, or if $\mathcal{M}'$ is defined by an MSO formula. The first claim is trivial, whereas the second claim is shown using for example the translation of MSO over traces to MSO over words [28, 9]. Together with Proposition 4.1 this shows the decidability of the different instances of the model-checking problem below (where we identify a device with its language, e.g., a CFM is $\exists$-B-bounded if its language is):

Proposition 4.3. *All questions below are decidable:*

1. *Checking that an $\exists$-B-bounded CFM or a safe CMSC-graph satisfies a formula of MSO($\leq$, msg).*
2. *Checking that a CFM has empty intersection with a safe CMSC-graph.*
3. *Checking that a safe CMSC-graph is included in a CFM.*

As shown by [21, 25], having a set of representatives K_G for a CMSC-graph G suffices for doing model-checking against a property expressed by a partial order logics. More precisely, $\text{Lin}^B(\mathcal{M})$ (with $\mathcal{M}$ the set of MSCs satisfying such a formula φ) is an effective regular set. The existential bound B is given by the graph G.

However, if we want to do model-checking of two safe CMSC-graphs, we need that $\text{Lin}^B(G)$ is regular for at least one of them. We can use again the notion of communication graph. Note however that we only require *weak* connectedness:

Definition 4.4. *The* communication graph *of a set $A \subseteq \Sigma$ is a graph whose vertices are the processes involved in A, and there is an (undirected) edge between vertices p, q iff A contains both a send $p!q$ from p to q and a receive $q?p$ on q from p. The* communication graph *of a CMSC $M = (V, \lambda, \text{msg}, (<_p)_{p \in \mathcal{P}})$ is the communication graph of $A = \lambda(E)$. A loop in a CMSC-graph is* connected *if the union of the communication graphs of the CMSCs labeling it, is connected. A CMSC-graph G is* loop-connected *if every loop of G is connected.*

A CMSC-graph is globally-cooperative *(gc-CMSG for short) if it is safe and loop-connected [13].*

For example, the CMSC-graph in Figure 2 is globally-cooperative. We will show that $\text{Lin}^C(G)$ is regular for all $C \geq |G|$ for any globally-cooperative CMSC-graph G in Proposition 5.4.

---

[2] The reason is that any loop on an accepting path has equally many sends $p!q$ and receives $q?p$, for every p, q.

4.2   Existential Bounds and Traces

Let B be a positive integer that we fix for this section. We define as in [17] a trace alphabet (Ω, I) with $\Omega_p = \Sigma_p \times \{0, \dots, B-1\}$ for $p \in \mathcal{P}$ and $\Omega = \bigcup_{p \in \mathcal{P}} \Omega_p$. The dependence relation $D \subseteq \Omega \times \Omega$ is given by $(x, i) D(y, j)$ if either $P(x) = P(y)$ or $\{(x, i), (y, j)\} = \{(p!q, n), (q?p, n)\}$ for some p, q, n. Then $I = \Omega^2 \backslash D$ is symmetric and irreflexive, hence (Ω, I) is a trace alphabet.

We define a mapping $\tilde{\ } : \Sigma^\star \to \Omega^\star$ by numbering the events of the same type modulo B. Let $\widetilde{x_1 \cdots x_m} = (x_1, n_1) \dots (x_m, n_m)$, with $n_i = |\{j \leq i \mid x_j = x_i\}|$ mod B, i.e., modulo B, there are n_i occurrences of the letter x_i in the prefix $x_1 x_2 \dots x_i$. We also consider the projection $\pi : \Omega^* \to \Sigma^*$ given by $\pi(x, n) = x$ for $(x, n) \in \Omega$. A word $u \in \Omega^*$ is B-bounded if $\pi(u) \in \Sigma^*$ is B-bounded. We denote $\tilde{L} = \{\tilde{u} \mid u \in L\}$.

Let $M = (E, \lambda, \mathrm{msg}, (<_p)_{p \in \mathcal{P}})$ be an MSC. For $e \in E$, let $\lambda_I(e) = (\lambda(e), n)$ with $n = |\{f \in E \mid f \leq_p e, \lambda(f) = \lambda(e)\}|$ mod B (i.e., n is the number of events before e labeled by the same element of Σ). We associate with M the structure $\mathrm{tr}(M) = (E, \prec_B^*, \lambda_I)^3$. Figure 3 depicts the result when applying this operation to the MSC from Figure 1 with $B = 2$. As before, a stands for 1!2 and b for 2?1. Note that there is one additional edge from the first occurence of $(b, 1)$ to the first second occurence of $(a, 1)$. If M is $\exists$-B-bounded, then the relation $\prec_B$ is acyclic by Lemma 2.2, i.e., in this case $\mathrm{tr}(M) := (E, \prec_B^*, \lambda_I)$ is a Ω-labeled partial order.

Lemma 4.5. *Let* $M = (E, \lambda, \mathrm{msg}, (<_p)_{p \in \mathcal{P}})$ *be an* $\exists$-B-*bounded MSC.*

1. *The labeled poset* $\mathrm{tr}(M)$ *is a trace over* (Ω, I).
2. *If* u *is a* B-*bounded linearization of* M, *then the set of linearizations of* $\mathrm{tr}(M)$ *equals* $[\tilde{u}]_I$. *Finally,* $\mathrm{Lin}^B(M) = \pi([\tilde{u}]_I)$.

5   The Main Result

The main result is stated in the following theorem, which generalizes the results of [17, 15, 20] from universally-bounded to existentially-bounded sets of MSCs. We use a unified proof technique, interpreting MSCs as traces and applying known constructions for traces.

Theorem 5.1. *Let* $\mathcal{M}$ *be an existentially-bounded set of MSCs. Then the following assertions are equivalent:*

1. $\mathcal{M} = \mathcal{L}(\mathcal{A})$ *for some CFM* $\mathcal{A}$.
2. $\mathcal{M} = \mathcal{L}(\varphi)$ *for some EMSO($<_p$, msg) formula* φ.
3. $\mathcal{M} = \mathcal{L}(\varphi)$ *for some MSO($\leq$, msg) formula* φ.
4. $\mathcal{M} = \mathcal{L}(G)$ *for some gc-CMSC-graph* G.
5. $\mathrm{Lin}^B(\mathcal{M})$ *is a regular set of representatives for* $\mathcal{M}$, *for some* $B \in \mathbb{N}$.

---

³ Recall that $\prec_B = \mathrm{msg} \cup \mathrm{rev} \cup (<_p)_{p \in \mathcal{P}}$.

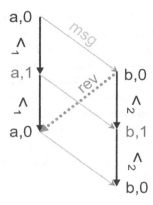

Fig. 3. Trace pomset associated with the MSC of figure 1

Remark 5.2. Since the set of B-bounded linearizations of MSCs (over a fixed set of processes) is obviously regular, the last item of Theorem 5.1 implies in particular that there exist a CFM (an EMSO($<_p$, msg) formula, a gc-CMSC-graph, resp.) that generates precisely the set $\mathbb{MSC}^B$ of $\exists$-B-bounded MSCs. Constructing this CFM is the most difficult part of the proof.

It is easy to show that (1) implies (2) by the usual techniques (actually, [6] shows that (1) and (2) are equivalent over unrestricted sets of MSCs). The implication (2) to (3) is immediate. We will show that (3) implies (5), that (4) and (5) are equivalent, and finally that (5) implies (1). The proofs use the trace alphabet (Ω, I) and in particular Theorem 2.8 at crucial points: For showing that (3) implies (5), we use the equivalence between MSO and regular sets of traces [28, 9], i.e., Theorem 2.8(3). To prove that (4) and (5) are equivalent, we will use a slight variation of Ochmański's Theorem 2.8(1) [24]. Finally, to prove (5) implies (1), we will use Zielonka's Theorem 2.8(2) [29] and simulate asynchronous automata by CFMs. More precisely, we first build a CFM $\mathcal{A}$ such that $\mathcal{L}(\mathcal{A}) \cap \mathbb{MSC}^B = \mathcal{M}$, where $\mathcal{M}$ is a set of $\exists$-B-bounded MSCs with regular set of representatives $\mathrm{Lin}^B(\mathcal{M})$. As mentioned above, the difficult part is to add a control CFM that ensures that the generated MSC is $\exists$-B-bounded (and deadlocks if this is not the case).

We first demonstrate the implication (5)$\Rightarrow$(4).

Proposition 5.3. *Let $\mathcal{M}$ be a set of $\exists$-B-bounded MSCs such that $\mathrm{Lin}^B(\mathcal{M}) \subseteq \Sigma^*$ is regular. Then there exists a globally-cooperative CMSC-graph G with $\mathcal{L}(G) = \mathcal{M}$.*

Proof. Since the mapping ˜ is sequential, the set $L = \widetilde{\mathrm{Lin}^B}(\mathcal{M}) \subseteq \Omega^*$ is regular as well. By Lemma 4.5, the set L is I-closed. Thus, we can apply Ochmański's Theorem 2.8 (1) for obtaining an I-loop-connected automaton $\mathcal{B}$ with $[L(\mathcal{B})]_I = L$. Replacing each label $(a, n) \in \Omega$ by a yields an automaton $\mathcal{A}$ over the alphabet Σ. Since all words in L are of the form $\tilde{u}$ for some $u \in \Sigma^*$, we get $\widetilde{L(\mathcal{A})} = L(\mathcal{B}) \subseteq$

L and therefore $L(\mathcal{A}) \subseteq \mathrm{Lin}^B(\mathcal{M})$. Thus, all successful paths in $\mathcal{A}$ are labeled by B-bounded linearizations of MSCs from $\mathcal{M}$. Conversely, if $M \in \mathcal{M}$, then there is $u \in \mathrm{Lin}^B(M)$ with $\tilde{u} \in L(\mathcal{B})$ and therefore $u \in L(\mathcal{A})$. Thus, $L(\mathcal{A}) = \mathrm{Lin}^B(\mathcal{M})$.

Now let ρ be a loop in the automaton $\mathcal{A}$ and let $A \subseteq \Sigma$ be the set of labels appearing in this loop. Then ρ is also a loop in the automaton $\mathcal{B}$ with label set $A' \subseteq \Omega$. Since $\mathcal{B}$ is I-loop-connected, A' is I-connected. Since $\mathcal{B}$ accepts only words of the form $\tilde{u}$ for some linearization u of an MSC, we have that $(p!q, n) \in A'$ iff $(p?q, n) \in A'$. Hence the I-connectedness of A' implies the connectedness of the communication graph of $\{a \mid \exists n : (a, n) \in A'\}$.

To obtain a CMSC-graph, we transform $\mathcal{A}$ in such a way that labels move from transitions to nodes. The resulting CMSC-graph G is safe since $\mathcal{A}$ accepts only linearizations of MSCs, and globally-cooperative since loops in $\mathcal{A}$ are labeled by sets $A \subseteq \Omega$ whose communication graph is connected. □

The following proposition does not only claim the implication (4)⇒(5), but it also gives an upper bound on the size of the automaton accepting $\mathrm{Lin}^B(\mathcal{M})$ in terms of the gc-CMSC-graph generating $\mathcal{M}$.

Proposition 5.4. *Let G be a gc-CMSC-graph. Then $\mathrm{Lin}^B(G)$ is regular and one can construct an automaton of size at most $|G|^{O(\wp^4 B^2 |G|)}$ recognizing it.*

6   From Regular Representatives to CFM

In order to obtain a CFM from a regular set of representatives, we will use Zielonka's theorem, which characterizes regular trace languages by a distributed automaton model, namely by asynchronous automata[4].

Definition 6.1. *An asynchronous automaton over the trace alphabet (Ω, I) is a tuple $\mathcal{B} = ((K_e, \delta_e, k_e^0)_{e\in\Omega}, \mathrm{Acc})$ such that for any $e \in \Omega$:*

- *K_e is a finite set of local states,*
- *$\delta_e : \prod_{(e,f)\in D} K_f \to K_e$ is a local transition function,*
- *$k_e^0 \in K_e$ is a local initial state,*

and $\mathrm{Acc} \subseteq \prod_{e\in\Omega} K_e$ is a set of global accepting states.

The idea is that an asynchronous automaton consists of local components, one for each letter $e \in \Omega$. When the e-component executes the action e, its new state results from the current states corresponding to the letters depending on e.[5] Only at the very end of a run, there is a global synchronization through final states.

---

[4] The definition we give below actually corresponds to deterministic asynchronous cellular automata.

[5] In terms of parallel algorithms, this corresponds to a concurrent-read-owner-write mechanism.

Next we define runs of asynchronous automata. Intuitively, a run can be seen as a labeling of the pomset by local states, that is consistent with the transition relations. Let $(E, \leq, \lambda_I)$ be a trace over (Ω, I), $\theta : E \to \bigcup_{e \in \Omega} K_e$ a mapping, and $t \in E$ with $\lambda_I(t) = e$. For $f \in \Omega$ with $(e, f) \in D$, we define $\theta_f^-(t) = \theta(t_f)$ if t_f is the maximal f-labeled event of E properly below t. If no such event exists, $\theta_f^-(t) = k_f^0$ is the f-component of the initial state of $\mathcal{B}$. The mapping θ is a *run* if for any $t \in E$ with $\lambda(t) = e$, we have $\theta(t) = \delta_e((\theta_f^-(t))_{(e,f) \in D})$. Next, for $e \in \Omega$ let $k_e = \theta(t_e)$ where t_e is the maximal e-labeled event of E (if such an event exists), and $k_e = k_e^0$ otherwise. The run θ is *successful* provided that $(k_e)_{e \in \Omega} \in \mathrm{Acc}$ is a (global) accepting state. The set of traces $\mathcal{L}(\mathcal{B})$ accepted by $\mathcal{B}$ is the set of traces that admit a successful run. By Thm. 2.8, an I-closed set of words L is regular iff there exists an asynchronous automaton $\mathcal{B}$ with $L = \mathrm{Lin}(\mathcal{L}(\mathcal{B}))$.

6.1   A CFM Almost Computing a Regular Set of Representatives

Given a set of bounded MSCs $\mathcal{M}$ with $\mathrm{Lin}^B(\mathcal{M})$ regular, in this section we construct a CFM that checks whether a given $\exists$-B-bounded MSC M belongs to $\mathcal{M}$ (the behavior on MSCs that are not $\exists$-B-bounded is of no concern in this section). To this aim, we use the set of traces $\mathrm{tr}(\mathcal{M} \cap \mathrm{MSC}^B)$ that can be accepted by an asynchronous automaton $\mathcal{B}$, this asynchronous automaton will be simulated by the CFM $\mathcal{A}$. The simulation of $\mathcal{B}$ at receive-events $p?q$ is easy: the CFM has access to its previous state at process p and to the state at the matching send event $q!p$. A problem arises at send-events $p!q$: because of the rev-edges in $\mathrm{tr}(M)$, the asynchronous automaton has access to the state at the send-event event s that is connected to the current event by an rev-edge. But the CFM cannot read the state at this node. To overcome this problem, the CFM will guess the state at s. The notion of a *good labeling* captures the idea that these guesses shall be correct. In order to check the correctness by the CFM, guesses will be sent to process q. Process q will keep track of previous states and check that the guess was correct.

Let K be a finite set, $M = (E, \lambda, \mathrm{msg}, (<_p)_{p \in \mathcal{P}})$ an MSC. Furthermore, let $\gamma : E \to K$ be a mapping and $k_p^0 \in K$ for $p \in \mathcal{P}$. Then, for $t \in E_p$, $\gamma^-(t) = \gamma(s)$ if s is the predecessor of t on process p. If no such predecessor exists, $\gamma^-(t) = k_p^0$. Furthermore, we define $\gamma^m(t) = \gamma(s)$ if $\mathrm{msg}(s) = t$. If there is $r \in E$ with $\mathrm{rev}(r) = t$, then $\gamma^m(t) = \gamma(r)$. If $\lambda(t) = p!q$ is a send-event, but there is no $s \in E_q$ with $\mathrm{rev}(s) = t$, then $\gamma^m(t) = k_q^0$.

We say that $\gamma : E \to K$ is a *good labeling* of M with respect to the mapping $update : \Sigma \times K \times K \to 2^K$ and the values $(k_p^0)_{p \in \mathcal{P}}$ if $\gamma(t) \in update(\lambda(t), \gamma^-(t), \gamma^m(t))$ for any $t \in E$.

Proposition 6.2. *Let* $update : \Sigma \times K^2 \to K$ *be a mapping and* $k_p^0 \in K$ *for* $p \in \mathcal{P}$. *Then there exists a CFM* $\mathcal{A}$ *with local state set* $\Omega_p \times K^{\Omega_p} \times \{0, \ldots, B-1\}_p^{\Sigma}$ *for* $p \in \mathcal{P}$ *and all global states accepting with the following properties for any MSC* M:

(1) if ρ is a run of $\mathcal{A}$ on M, then the labeling γ defined by $\gamma(t) = \text{mem}(e)$ if $\rho(t) = (e, \text{mem}, \text{cnt})$ is good with respect to update and $(k_p^0)_{p\in\mathcal{P}}$.

(2) if γ' is a good labeling with respect to update and $(k_p^0)_{p\in\mathcal{P}}$, then there exists a run ρ with $\gamma = \gamma'$ (where γ is defined as in (1)).

Proposition 6.3. *Let $B \in \mathbb{N}$ and $\mathcal{M}$ a set of $\exists$-B-bounded MSCs with $\text{Lin}^B(\mathcal{M})$ regular. Then there exists a CFM $\mathcal{A}'$ with $\mathcal{L}(\mathcal{A}) \cap \mathbb{MSC}^B = \mathcal{M}$.*

Thus, we succeed in building a CFM that checks membership in $\mathcal{M}$ provided that the input is $\exists$-B-bounded. The construction of a CFM $\mathcal{A}'$ which accepts precisely $\mathbb{MSC}^B$ is technically involved and can be found in the full version of this paper. Taking the direct product of these two machines shows that $\mathcal{M}$ can be accepted by a CFM whenever $\text{Lin}^B(\mathcal{M})$ is a regular set of representatives.

Theorem 6.4. *Let $\mathcal{M}$ be a set of MSCs with $\text{Lin}^B(\mathcal{M})$ a regular set of representatives of $\mathcal{M}$. Then there exists a CFM $\mathcal{A}$ with $\mathcal{L}(\mathcal{A}) = \mathcal{M}$.*

7   Further Results

In section 4.1, we explained how to do model-checking of two models when their sets of B-bounded linearizations are regular sets of representatives. Our Main Theorem 5.1 describes the models for which the set of B-bounded linearizations is a regular set of representatives, in terms of (E)MSO logics, communicating automata and CMSC-graphs.

We can even test negative model-checking - $\mathcal{L}(\mathcal{A}) \cap \mathcal{L}(\mathcal{B}) = \emptyset$ - when there exists a regular set of representatives $X \subseteq \text{Lin}^B(\mathcal{A})$ and $\text{Lin}^B(\mathcal{B})$ is regular (but might not be a set of representatives). The proposition below summarizes several decidable model-checking instances:

Proposition 7.1. *Let $\mathcal{A}$ and $\mathcal{B}$ be two MSC formalisms. Then both $\mathcal{L}(\mathcal{A}) \cap \mathcal{L}(\mathcal{B}) = \emptyset$ and $\mathcal{L}(\mathcal{A}) \subseteq \mathcal{L}(\mathcal{B})$ are decidable when*

  − $\mathcal{A}$ *is a safe CMSC-graph or an $\exists$-B-bounded CFM, for some $B > 0$.*
  − $\mathcal{B}$ *is either a CFM, or an $MSO(\leq, \text{msg})$ formula, or a globally-cooperative CMSC-graph.*

Among the model-checking instances covered by the proposition above, the only case that was already known is model-checking of a safe CMSC-graph against an $MSO(\leq, \text{msg})$ formula [21]. This problem has non-elementary complexity, and let us briefly discuss the complexity of the remaining instances. First, given a bound B and a CFM $\mathcal{A}$, we obtain for $\text{Lin}^B(\mathcal{A})$ an automaton that is exponential only in the number $\wp$ of processes. For a gc-CMSC-graph G we get an automaton that is exponential in $|G|$ (Proposition 5.4). Thus, for $\mathcal{B}$ either a CFM or a gc-CMSC-graph, negative model-checking $\mathcal{L}(\mathcal{A}) \cap \mathcal{L}(\mathcal{B}) = \emptyset$ is in PSPACE and positive model-checking $\mathcal{L}(\mathcal{A}) \subseteq \mathcal{L}(\mathcal{B})$ is in EXPSPACE.

We also showed that gc-CMSC-graphs are quite expressive, since they are equivalent to existentially bounded CFMs. An equivalent gc-CMSC-graph of size

$\wp!(B|\mathcal{A}|)^\wp$ can be constructed from an $\exists$-B-bounded CFM $\mathcal{A}$, that is exponential only in the number of processes $\wp$.

Moreover, Theorem 5.1 shows that the logics EMSO($\leq_p$, msg) and MSO(($\leq$, msg) are expressively equivalent over $\mathbb{MSC}^B$. By the proposition below, one obtains even the equivalence of the logics EMSO($\leq_p$, msg) and EMSO($\leq$, msg).

Proposition 7.2. *Let φ be an (E)MSO($\leq_p$, msg) formula with $\mathcal{L}(\varphi) \subseteq \mathbb{MSC}^B$. Then there exists an (E)MSO($\leq$) formula φ' with $\mathcal{L}(\varphi) = \mathcal{L}(\varphi')$.*

8   Conclusion

We proved an extension of Kleene's theorem for $\exists$-bounded CFM, with a unified technique using Mazurkiewicz traces.

One side result is that, since we can easily complement an MSO formula, gc-CMSC-graphs and existentially B-bounded CFM are both closed under complementation.

To sum up, the hardest part of the proof is to go from regular sets of representatives to CFMs, which uses at a crucial stage Zielonka's theorem. While this result allows theoretically to implement MSO formulae and gc-CMSC-graphs into CFMs, the result is of poor practical use, since deadlocks are unavoidable in the implementation. Anyway, this is not a special issue with our implementation, since deadlock-free CFMs are too weak to model even some simple regular languages. A further study of the expressivity of deadlock-free CFMs in term of logics and CMSC-graphs would be interesting.

References

1. P. Abdulla and B. Jonsson. Verifying programs with unreliable channels. *Information and Computation*, 127(2):91–101, 1996.
2. R. Alur and M. Yannakakis. Model checking of message sequence charts. In *CONCUR'99*, LNCS 1664, pp. 114-129, 1999.
3. A. Bouajjani and P. Habermehl. Symbolic Reachability Analysis of FIFO-Channel Systems with Nonregular Sets of Configurations. *Theoretical Computer Science*, 221(1-2):211–250, 1999.
4. B. Boigelot, P. Godefroid, B. Willems and P. Wolper. The Power of QDDs. *SAS'97*, pp. 172-186, 1997.
5. B. Boigelot and P. Godefroid. Symbolic Verification of Communication Protocols with Infinite State Spaces using QDDs. *Formal Methods in System Design*, 14(3):237-255 (1999).
6. B. Bollig and M. Leucker. Message-Passing Automata are expressively equivalent to EMSO Logic. In *CONCUR'04*, LNCS 3170, pp. 146-160, 2004.
7. D. Brand and P. Zafiropulo. On communicating finite-state machines. *Journal of the ACM*, 30(2):pp.323-342, 1983.
8. V. Diekert and G. Rozenberg, editors. *The Book of Traces*. World Scientific, Singapore, 1995.

9. W. Ebinger and A. Muscholl. Logical definability on infinite traces. In *Theoretical Computer Science* 154:67–84 (1996).

10. A. Finkel and Ph. Schnoebelen. Well-structured transition systems everywhere! *Theoretical Computer Science*, 256 (1,2):63–92, 2001.

11. B. Genest, M. Minea, A. Muscholl, and D. Peled. Specifying and verifying partial order properties using template MSCs. In *FoSSaCS'04*, LNCS 2987, pp. 195-210, 2004.

12. E. Gunter, A. Muscholl, and D. Peled. Compositional Message Sequence Charts. In *TACAS'01*, LNCS 2031, pp. 496–511, 2001. Journal version *International Journal on Software Tools for Technology Transfer (STTT)* 5(1): 78-89 (2003).

13. B. Genest, A. Muscholl, H. Seidl, and M. Zeitoun. Infinite-state High-level MSCs: Model-checking and realizability. In *ICALP'02*, LNCS 2380, pp.657-668, 2002. Journal version to appear in *JCSS*.

14. J. G. Henriksen, M. Mukund, K. Narayan Kumar, and P. Thiagarajan. On Message Sequence Graphs and finitely generated regular MSC languages. In *ICALP'00*, LNCS 1853, pp. 675–686, 2000.

15. J. G. Henriksen, M. Mukund, K. Narayan Kumar, M. Sohoni and P. Thiagarajan. A Theory of Regular MSC Languages. To appear in *Information and Computation*, available at http://www.comp.nus.edu.sg/thiagu/icregmsc.pdf.

16. ITU-TS recommendation Z.120, Message Sequence Charts, Geneva, 1999.

17. D. Kuske. Regular sets of infinite message sequence charts. *Information and Computation*, (187):80–109, 2003.

18. M. Lohrey and A. Muscholl. Bounded MSC communication. *Information and Computation*, (189):135–263, 2004.

19. S. Leue, R. Mayr, and W. Wei. A scalable incomplete test for the boundedness of UML-RT models. In *TACAS'04*, LNCS 2988, pp. 327–341, 2004.

20. R. Morin. Recognizable Sets of Message Sequence Charts. In *STACS'02*, LNCS 2285, pp. 523-534, 2002.

21. P. Madhusudan and B. Meenakshi. Beyond Message Sequence Graphs In *FSTTCS'01*, LNCS 2245, pp. 256-267, 2001.

22. A. Mazurkiewicz. Concurrent program schemes and their interpretation. Technical report, DAIMI Report PB-78, Aarhus University, 1977.

23. A. Muscholl and D. Peled. Message Sequence Graphs and decision problems on Mazurkiewicz traces. In *MFCS'99*, LNCS 1672, pp. 81-91, 1999.

24. E. Ochmański. Regular behaviour of concurrent systems In *Bulletin of the EATCS* 27, pp.56-67, 1985.

25. D. Peled. Specification and verification of Message Sequence Charts. In *FORTE/PSTV'00*, pp. 139-154, 2000.

26. Ph. Schnoebelen. Verifying lossy channel systems has nonprimitive recursive complexity. *Information Processing Letter*, 83(5):251–261, 2002.

27. USB 1.1 specification, http://www.usb.org/developers/docs/usbspec.zip

28. W. Thomas. On logical definability of trace languages. In V. Diekert, editor, *Proceedings of a workshop of the ESPRIT BRA No 3166: Algebraic and Syntactic Methods in Computer Science (ASMICS) 1989*, Report TUM-I9002, Technical University of Munich, pages 172–182, 1990.

29. W. Zielonka. Note on finite asynchronous automata, R.A.I.R.O. In *Informatique Thorique et Applications*, 21:pp.99-135, 1987.

Algebraic and Topological Models for DNA Recombinant Processes

Nataša Jonoska* and Masahico Saito**

University of South Florida, Department of Mathematics,
Tampa, FL, USA
{jonoska, saito}@math.usf.edu

Abstract. We propose algebraic and topological techniques that could be employed for studying recombinant DNA processes. We show that sequence of splices performed by the operation *hi*-excision/reinsertion on one DNA molecule can be modeled by the elements of a semidirect product of n-copies of the cyclic group of order 2 and the symmetric group S_n. We associate a surface in space-time (subspace of $\mathbb{R}^3 \times [0,1]$) with a sequence of splicings on circular molecules and present examples of applications of this model.

1 Introduction

There have been extensive formal language theory models for various DNA recombinant processes. Most of these processes are considered as new models for computation and are investigated for their computational power. In this aspect much of the attention was put on the input/output results without studying the whole process that occurs between the input and the output. In this paper we suggest several ways of investigating these processes from algebraic and topological point of view, but not necessarily looking into the computational power of these processes.

In living organisms such as ciliates, DNA molecules undergo certain recombinant processes to reorganize themselves. A detailed description of the recombination from the micronuclear DNA into the macronuclear DNA in ciliates can be found in the recent book [3]. These processes were proposed as a model for DNA computing [14], and later it was shown that they have computational power equivalent to a universal Turing machine [15]. One operation that has been the subject of various investigations is DNA splicing (see for example [8,20]). An accessible account of the DNA structure and splicing by the endonucleases is presented in the introductory chapter of [20]. A particular splicing is depicted in Fig. 1, and symbolically, it can be written as follows:

$$(a_0va_1, b_0vb_1) \xrightarrow{p} (a_0vb_1, b_0va_1).$$

---

* Supported in part by NSF Grants CCF #0432009 and EIA#0086015.
** Supported in part by NSF Grant DMS #0301089.

C.S. Calude, E. Calude and M.J. Dinneen (Eds.): DLT 2004, LNCS 3340, pp. 49–62, 2004.

In this expression, the symbols that make up the sequences of words $a_0, a_1, b_0,$ b_1, v are A-T, C-G pairs. An endonuclease enzyme recognizes the subsequence v in the pair of larger sequences $(a_0 v a_1, b_0 v b_1)$, cuts at the site v, and splices them after switching their positions as shown in Fig. 1. Another enzyme, a ligase, joins the phosphodiester bonds of the nicked molecules. In general, different enzymes may have the same overhangs, such that this recombination is possible with different enzymes. In the above description we may assume that v represents the overhang produced by the enzyme. The above specific pair of sequences may be segments of two DNA molecules involved in the process, or may be part of a single molecule.

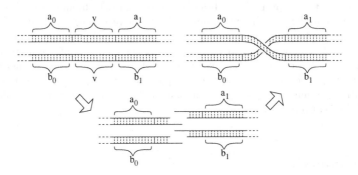

Fig. 1. Splicing DNA molecules by an enzyme

With this paper we propose algebraic and topological models for investigating similar recombinant processes among DNA molecules. We concentrate on the sequence of individual steps of such processes, rather than just inputs and outputs involved and propose mathematical models that can be used for studying such processes. If we denote $h = (a_0 v a_1, b_0 v b_1)$ and $h' = (a_0 v b_1, b_0 v a_1)$, the above splicing operation can be written as $h \xrightarrow{p} h'$. Then a process of gene assemblies in a living organism or in DNA computing that is based on the splicing operation can be written as a sequence:

$$p : h_0 \xrightarrow{p_1} h_1 \xrightarrow{p_2} \dots \xrightarrow{p_n} h_n$$

where n is the number of individual steps leading from the input h_0 to the output h_n (splicings in this particular case). The structure and length of sequences $p_1, p_2, \dots, p_n$ that for a given input produce the same output is the subject of this discussion.

Our motivation for considering processes and their equivalences is two fold:

– In living organisms, not only the inputs and outputs, but also the whole process of producing outputs from given inputs are of importance. When theoretically there are more than one way of producing the same output (for example the functional genes in the macronuclear DNA in ciliates) from a given input (for example the micronuclear DNA in ciliates), it is of interest to

understand which one of the possible processes happened. In fact, many experiments are designed to study the intermediate processes of recombination since both the input and the output are already known.
- In biomolecular computing, complexities of computing processes [7] have been considered. Consider sequences of individual steps that produce the same outputs from a given input. For evaluating or minimizing computational complexities, it is important to list these steps algebraically and/or topologically, and by examining their properties to study the relationships among different processes.

In mathematics, these problems can be related to the problem of finding the shortest word representation in groups, similarly as in combinatorial group theory (see for example [16]), or to characterize mathematical objects by means of category theory (see background in [17]). The purpose of this paper is to show how recombinant DNA processes can be modeled from a point of view of category theory or group theory, and to examine certain type of splicing processes from this perspective.

Our second purpose is to propose studying DNA processes topologically. Several applications of knot theory to DNA are already known [2, 4, 24]. When a sequence of splicings appear among circular DNA molecules, it is natural to investigate topological aspects of such processes, and study rewriting processes topologically. In fact, the number of crossings that are necessary to obtain one knot from another is discussed from a point of view of recombinant DNA processes in [2]. We point out here that a natural topological setting to represent DNA recombinant processes already exists within the theory of surfaces in the space-time. We describe how this theory applies to the processes of DNA molecules.

The paper contains two parts. The first part introduces algebraic ideas and shows how concepts from group theory and category theory can be used to study recombinant DNA processes. We mainly concentrate on the operation of excision/reinsertion (hi) that was originally introduced as one of the operations that are performed during the gene assembly process in ciliates (for details see [3]). We show that there is a one-to-one correspondence between the elements of the semi-direct product $H_n = \mathbb{Z}_2^n \rtimes S_n$ and the number of non-equivalent sequences of hi-operations on one molecule. Then we show how category theory can be used in studying similar processes between different molecules.

The second part of the paper describes how a surface in space-time can be associated with a sequence of splicings of circular molecules. This surface in general may be knotted and we show how results from knot theory can be directly applied to studying differences in such processes. We show how the genus of the surface can be used as a complexity measure, how to estimate the number of individual steps in a given process and how on a first glance "same" processes can be topologically different.

2　Algebraic Models

In this section, we show how group theory, in particular braid groups and, more generally category theory, can be used to model some DNA splicing processes. An alphabet Δ is a finite non-empty set of symbols. We concentrate on the special case when the alphabet is $\{A, G, C, T\}$ representing the DNA nucleotides. A word u over Δ is a finite sequence of symbols in Δ. We denote by Δ^* the set of all words over Δ, including the empty word 1.

The mapping $\nu : \Delta \to \Delta$ defined by $\nu(A) = T$, $\nu(T) = A$, $\nu(C) = G$, $\nu(G) = C$ is an involution on Δ and can be extended to a morphic involution of Δ^*. Since the Watson-Crick complementarity appears in a reverse orientation, we consider another involution $\rho : \Delta^* \to \Delta^*$ defined inductively, $\rho(s) = s$ for $s \in \Delta$ and $\rho(us) = \rho(s)\rho(u) = s\rho(u)$ for all $s \in \Delta$ and $u \in \Delta^*$. This involution is antimorphism such that $\rho(uv) = \rho(v)\rho(u)$. The Watson-Crick complement $\overleftarrow{u}$ of a DNA strand u, then, is obtained by the antimorphic involution $\rho\nu(u) = \nu\rho(u) = \overleftarrow{u}$. The involution ρ reverses the order of the letters in a word and as such is used in the rest of the paper.

2.1　Braid and Symmetric Groups

This first subsection concentrates on a single molecule recombination process, in particular the splicing operation called hi in [3] (see also references there). This operation (called hi-excision/reinsertion) is considered to be one of the three operations involved in the assembly of the macro-nucleus from the micronucleus in ciliates. First we concentrate on processes that involve only this operation. Let $v \in \Delta^*$ represent the recognition site or a pointer for the excision. The operation hi relative v is defined in the following way:

$$XvA\,\overleftarrow{v}\,Y \xrightarrow{hi} Xv\,\overleftarrow{A}\overleftarrow{v}\,y$$

Consider a single linear DNA molecule $M \in \Delta^*$ with a sequence represented by $M = XvAvB\,\overleftarrow{v}\,Y$. The splicing by hi operation depicted in Fig. 2, performed at the first and the last site v changes this sequence to $M' = Xv\,\overleftarrow{B}\overleftarrow{v}\,\overleftarrow{A}\overleftarrow{v}\,Y$. Hence the order of A and B is switched and the words are reversed. Since most enzymes have palindromic recognition sites, we will assume that $v = \overleftarrow{v}$, and will

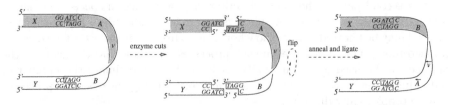

Fig. 2. The operation hi relative v on molecule $M = XvAvB\,\overleftarrow{v}\,Y$. The enzyme recognition site is $v = {}^{5'}GGATCC^{3'}$ with a cut that leaves ATC as a 3' overhang

write $M' = Xv\ \overleftarrow{B}\ v\ \overleftarrow{A}\ vY$. We denote this transformation by σ and write $M \xrightarrow{\sigma} M'$.

If the splicing occurs at the first and the second recognition site then the result is $M'' = Xv\ \overleftarrow{A}\ vBvY$. In this case the sequence A is substituted by its Watson-Crick complement. We denote this transformation by τ and write $M \xrightarrow{\tau} M''$.

Consider a single molecule $M = XvA_1vA_2v\ldots vA_{i-1}vA_ivA_{i+1}v\ldots vA_nvY$. Denote by σ_i the transformation of type σ that swaps A_i with A_{i+1} and replaces them by their Watson-Crick complement, i.e.

$$Xv\ldots vA_ivA_{i+1}v\ldots vY \xrightarrow{\sigma_i} Xv\ldots v\ \overleftarrow{A}_{i+1}\ v\ \overleftarrow{A}_i\ v\ldots vY$$

This σ_i can be considered as the ith generator of the braid group B_n (see the left of Fig. 3).

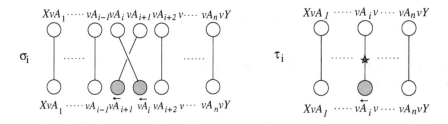

Fig. 3. Graphical representation of the action of the two generators, σ_i in the left and τ_i in the right

The braid group B_n is defined by a presentation

$$B_n = \langle \sigma_1,\ldots,\sigma_{n-1} \mid \sigma_i\sigma_{i+1}\sigma_i = \sigma_{i+1}\sigma_i\sigma_{i+1},\ i = 1,\ldots,n-2,$$
$$\sigma_i\sigma_j = \sigma_j\sigma_i,\ |i-j| > 1 \rangle$$

where σ_i denote the generators. Graphical interpretations of generators and relations of B_n are depicted in Fig. 3. Since $\sigma_i\sigma_i$ (i.e. performing the same operation twice) brings back the original sequence, it is convenient in studying sequences to substitute the braid group with the symmetric group S_n whose presentation is obtained from the presentation of B_n by adding relations $\sigma_i^2 = 1$ for all $i = 1,\ldots,n-1$.

Denote with τ_i the transformation that replaces A_i with its Watson-Crick complement. It is clear that by performing τ_i twice we obtain the original sequence back. Also, the order in which τ-transformations are performed is irrelevant. Hence we have that these transformations are generators of a direct product of n cyclic groups of order 2: $\mathbb{Z}_2^n$. Schematically τ_i is represented by a $\star$ in the line that corresponds to A_i and we denote the two Watson-Crick complements with a clear and shaded circle (see figure 3 to the right).

Besides the individual relations within the symmetric group and the product of $\mathbb{Z}_2$ there are relations between σ's and τ's. It is not difficult to see that

$$\tau_i \sigma_i = \sigma_i \tau_{i+1} \quad \text{and} \quad \tau_{i+1}\sigma_i = \sigma_i \tau_i.$$

This relator is schematically shown in Fig. 4. Furthermore, σ's and τ's commute if they are far apart.

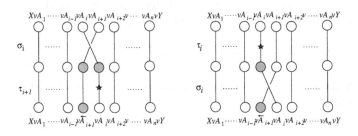

Fig. 4. The relator connecting generators in S_n and in $\mathbb{Z}_2^n$

Denote with H_n the group with generators σ_i $(i = 1, \ldots n-1)$, τ_j $(j = 1, \ldots n)$ and the above described relations:

$$\sigma_i\sigma_{i+1}\sigma_i = \sigma_{i+1}\sigma_i\sigma_{i+1} \ (i = 1, \ldots, n-2), \quad \sigma_i\sigma_j = \sigma_j\sigma_i \ (|i-j| > 1),$$
$$\sigma_i^2 = \tau_i^2 = 1, \quad \tau_i\tau_j = \tau_j\tau_i \ (i, j = 1, \ldots, n), \quad \tau_i\sigma_j = \sigma_j\tau_i \ (|i-j| > 1),$$
$$\tau_i\sigma_i = \sigma_i\tau_{i+1}, \quad (i = 1, \ldots, n-1).$$

In fact, it is not difficult to see that $\mathbb{Z}_2^n$ is normal in H_n and we have the short exact sequence:

$$0 \longrightarrow \mathbb{Z}_2^n \rightarrowtail H_n \twoheadrightarrow S_n \longrightarrow 1$$

which shows that H_n is a semi-direct product of $H_n = \mathbb{Z}_2^n \rtimes S_n$.

Let $M = XvA_1v \ldots vA_nvY$ be a molecule and v be a palindromic recognition site for an enzyme. Assume further that A_i are distinct ($A_i \neq A_j$ for $i \neq j$) and they are not palindromic subsequences ($A_i \neq \overleftarrow{A_i}$). A sequence of hi-operations $\alpha = h_1h_2 \cdots h_k$ changes M into another molecule M' denoted with $M \stackrel{\alpha}{\to} M'$. We say that two sequences α and α' are equivalent if $M \stackrel{\alpha}{\to} M'$ and $M \stackrel{\alpha'}{\to} M''$ implies that $M' = M''$. Let $S(M, v)$ be the set of all equivalence classes of such sequences. By the above discussion we have the following proposition.

Proposition 1. *For a DNA molecule $M = XvA_1v \ldots vA_nvY$ such that $A_i \neq A_j$ and $A_i \neq \overleftarrow{A_i}$ there is a one-to-one correspondence between the elements in $S(M, v)$ and $H_n = \mathbb{Z}_2^n \rtimes S_n$.*

Moreover, consider the set $\mathcal{H}(M, v, hi) = \{M' \mid M \stackrel{\alpha}{\to} M', \alpha = h_1 \cdots h_s\}$ where each h_j is an hi-operation. Then we have the following corollary.

Corollary 1. $\#\mathcal{H}(M, v, hi) = \#H_n$ *where* $\#$ *stands for cardinality.*

We note a couple of possible applications. The group H_n is finite and hence the word problem is solvable. So it is possible to determine whether two given processes are equivalent in this model. The minimal length of a given element when written as a sequence of generators can be determined. In that sense, processes with minimal number of steps can be identified.

2.2   Category Theory

The above observations model only one type of operation obtained as a type of DNA recombinant process. In [3] there are two additional operations, and some of these processes involve interactions of different DNA segments, even segments with graph structures. Furthermore, some processes can be performed on certain molecules and not on all molecules. Thus we propose to use aspects from category theory as a more suitable model. Roughly speaking, a category consists of a class of objects $\mathcal{O}$, and arrows between objects, called morphisms. More specifically, for every ordered pair (X, Y) of objects, there is a set of morphisms $\text{Hom}(X, Y)$ with composition of morphisms defined. The composition is defined for every triple (X, Y, Z) of objects as a map

$$\text{Hom}(X, Y) \times \text{Hom}(Y, Z) \to \text{Hom}(X, Z),$$

The image of (f, g) under this map is denoted by $g \circ f$. A morphism $f \in \text{Hom}(X, Y)$ is also denoted by $f : X \to Y$ as usual. There are some conditions to be satisfied by these, such as associativity.

For DNA recombinant processes in consideration, a category can be constructed as follows.

- (*Objects.*) Let $\mathcal{M} = \{M_i \,|\, i \in I\}$ be a set of DNA molecules. We assume that it is a collection of DNA molecules (which include all possible inputs, outputs, and all the molecules that appear in the middle steps of processes in consideration). Then $\mathcal{M}$ is the set of objects of the category.
- (*Morphisms.*) Let $\mathcal{P}$ be the set of all possible operations (one-step recombinant processes) between DNA molecules. For example, if a splicing operation is represented with $h \xrightarrow{p} h'$ then this is an element of $\mathcal{P}$. For such an operation, the input h and output h' are uniquely specified. Denote them by $\iota(p)$ and $\tau(p)$ respectively. Then the set of operations p with $\iota(p) = A$ and $\tau(p) = B$ is the set of morphisms between A and B, where $A, B \in \mathcal{M}$. The trivial operation, of doing nothing to a given molecule, is the identity morphism.

The goal of using categories is to generalize the H_n group approach to disjoint union of linear DNA molecules and graph structures. In the previous section, the action of hi-operation on a single linear molecule is associated to an H_n group. A natural question is to identify the groups that are used for disjoint unions of linear segments, and maybe even graph structures. Groups that are analogous to H_n can be used, and such associated groups are of interest. It is expected that these groups could be found from the corresponding category.

3   Topological Models

To simplify our argument, in this section, we assume that all DNA molecules are circular. hus, in this section. a symbol representing a DNA molecule denotes a knotted and linked circles in 3-space. Our first goal is to describe our topological model.

Claim. Surfaces embedded in the space-time can give a topological model for DNA recombinant processes.

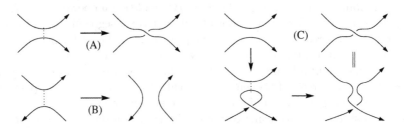

Fig. 5. A different view of a splicing

We explain our claim step by step in a series of figures.

First we explain how a splicing process corresponds to changes on a surface. Figure 5 (A) depicts a splicing process, where arcs represent part of dsDNA molecules. They are not necessarily oriented, but to help visualization the orientations are locally given and represented by arrows. The recognition site is at the dotted line and the molecules are lined such that both sites have overhangs at the same direction. When applied to single stranded DNA or RNA, then the arrows in the figure define the orientation of the molecules. In Fig. 5 (B), the same process is depicted from a different perspective. In this case the overhangs of the cleavages are oppositely oriented. The figure (C) explains why process (B) follows from (A), and vice versa. The process (B) is called a *surgery* between arcs.

Next we observe that the surgery corresponds to a saddle point of a surface as depicted in the left of Fig. 6. When the graph of such a surface with a saddle is sliced by "level surfaces," the top slice and bottom slices are as depicted in the right of Fig. 6. Thus the process of splicing of DNA molecules can be topologically identified with the continuous cross sections of a surface of a saddle.

Similarly, discarding or adding a small loop of DNA molecules can be regarded as minimal and maximal points of a surface (a bowl and an up-side down bowl). These three types of points, maximal, minimal, and saddle points are called (*non-degenerate*) *critical points* of a surface with respect to a height function (the z-coordinate direction).

A process of circular DNA splicing consists of a sequence of splices, discarding/adding a small loops, which can be considered as the critical points on the surface. Thus as a whole, a process corresponds to a surface in $\mathbb{R}^3 \times [0,1]$, the space-time, which is a 4-dimensional space. The unit interval $[0,1]$ represents

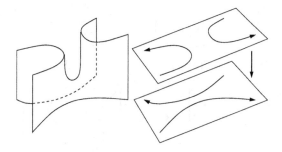

Fig. 6. A splicing is a saddle point

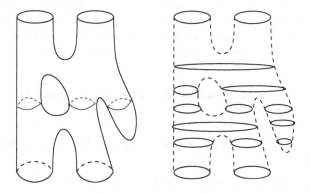

Fig. 7. A splicing process represents a surface

the time direction, and can be regarded as if the process occurred between the time $t = 0$ and $t = 1$. Figure 7 depicts an example, where the corresponding surface is depicted in the left. The circular DNA molecules that appear during the splicing process are depicted in the right.

A caution is needed in Figs. 6 and 7. Although the surfaces are depicted as objects in 3-space in these figures, in general they are lying in the space-time, or 4-dimensional space, $\mathbb{R}^3 \times [0,1]$. In fact, the cross sectional curves depicted should be regarded as curves in 3-space, instead of a plane. In general they could be knotted or linked. Since it is impossible to draw figures in dimension 4, we used the convention that the cross sections are drawn on the plane, and the surfaces are depicted in $\mathbb{R}^2 \times [0,1]$. An alternate way to see the figures is to regard the surfaces as projections of surfaces in $\mathbb{R}^3 \times [0,1]$ into $\mathbb{R}^2 \times [0,1]$. See [1] for more details.

We note that surfaces are classified by its genus, orientability and the number of boundary components. More precisely (see for example [18]):

Two compact surfaces F and F' are homeomorphic if and only if both are orientable or both are non-orientable, they have the same genus $g(F) = g(F')$, and the same number of boundary components $b(F) = b(F')$.

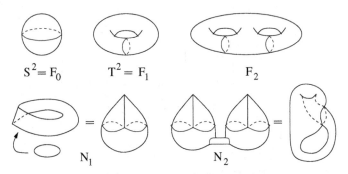

Fig. 8. Classification of surfaces. At the top: a sphere, a torus and a double torus which are orientable surfaces with genus 0,1, 2, respectively. At the bottom: a projective plane (obtained by cupping off a Möbius strip with a disk), and the Klein bottle (obtained by tubing two projective planes). These are non-orientable surfaces of genus 1 and 2, respectively. All of these surfaces have no boundary components

We present a few applications of this topological model in the form of examples.

3.1   The Bounding Genus as Topological Complexity in Splicing DNA

Suppose that a process of splicing DNA $p : h_1 \xrightarrow{p_1} h_2 \xrightarrow{p_2} \cdots \xrightarrow{p_n} h_{n+1}$ consists of circular DNAs at every step so that it can be modeled by a surface in $\mathbb{R}^3 \times [0, 1]$ as proposed above. Denote the corresponding surface F and consider its boundary $\partial F \cap \mathbb{R}^3 \times \{i\} = L_i$, for $i = 0, 1$. Thus $\partial F = L_0 \cup L_1$, that are the input ($L_0$) and the output ($L_1$) of the process.

We propose to use the genus of F, $g(F)$, as a topological complexity measure of such processes. We call $g(F)$ the *genus* of the given process p. For example, the process depicted in Fig. 7 has genus 1. The genus can be also regarded as a "distance" between two DNA molecular configurations, say input/output, and it can define a "metric" among molecules. This problem has already been considered when the splicing is performed by "crossing" changes instead of surgery. The distances and classifications were given in [2].

We formulate a problem that is of interest from both points of view, topologically and for DNA computing:

Problem: For given input and output sets of circular molecules, determine the minimal genus for splicing processes leading the input to the output. In other words, find the minimal genus of the surfaces cobounded by the links corresponding to the DNA molecules of input and output.

The corresponding problem for "crossing change" was considered in [2] but little seems to be known when both surgery and crossing, or just surgery changes are allowed.

For example, if the input is the unknot and the output is a trefoil knot, then it is known that the invariant called the *signature* of knots determines that the minimal genus of the process is 1. As in this example, if the input is the unknot, then the problem is to determine the minimal genus of surfaces spanned in 4-dimensional ball for a given knot or a link L on the boundary, S^3, of the 4-ball. Such a genus is called the 4-*ball genus*, or the *slice genus* of L. There is a standard argument to reduce the above formulated problem for an input link L_0 and an output link L_1, to the slice genus problem for the link $L_0 \# (rL_1)$ obtained by the connected sum (a link obtained by two links by connecting them by a band). There are extensive studies on slice genus, and significant progress has been made recently (see [13, 19, 21], for example).

In summary, we pointed out here that the slice genus can be used as a computational complexity in DNA computing, and recent results in this area of knot theory can be directly applied.

3.2   Estimating Numbers of Steps Necessary in DNA Splicing

In this example we consider situations where an input is a circular DNA molecule K_0 and so is an output, K_1. By our topological model, a process that changes K_0 to K_1, consisting of steps that are either splicing, adding or deleting small loops, corresponds to a surface in $\mathbb{R}^3 \times [0, 1]$ bounded by $K_0 \subset \mathbb{R} \times \{0\}$ and $K_1 \subset \mathbb{R} \times \{1\}$.

It was proved in [9] that the number $c(F)$ of critical points of a surface F in the upper 4-space bounded by a given knot K in the 3-space (which is the boundary of the upper 4-space) is bounded from below by $m(K) + 1$,

$$c(F) \geq m(K) + 1,$$

where the surface F is assumed to be orientable, and $m(K)$ denotes the minimal size of square matrices, called Alexander matrices, associated to a given knot K. See [12] for the case of links (with multi-components).

If K_1 is unknotted, then this result can be applied directly, as follows. Cap off K_1 with a standard disk D in $\mathbb{R}^3 \times \{1\}$ to obtain a surface $F_0 = F \cup D$ in $\mathbb{R}^3 \times [0, \infty)$ bounded by K_0, to which we apply the above result. Then we have $c(F_0) \geq m(K_0) + 1$, and regarding D as a maximum, we have $c(F_0) = c(F) + 1$. Thus we obtain $c(F) \geq m(K_0)$. As an example, take a square knot K_s, the connected sum of a trefoil knot (the simplest knot with three crossing points) and its mirror image. Let $K_s^{(n)}$ be the n-fold connected sum of K_s. It is known that K_s bounds a disk in the upper 4-space, so does K, so that the genus as complexity does not apply to this example (the slice genus of $K_s^{(n)}$ is zero). It is known that $m(K_s^{(n)}) = 2n$ (the fact that all summands are either trefoil and its mirror image is critical here). Hence for $K_0 = K_s^{(n)}$ and $K_1 =$ unknot, we have $c(F) \geq 2n$. In fact, it is easy to construct a surface F with $c(F) = 2n$, so that the minimal number of steps necessary is precisely determined to be $2n$.

More generally, denote by $c(K)$ the minimal number of $c(F)$ for a given K over all possible surfaces F bounded by K. Let F be a surface bounded by K_0 and

K_1, then we have $c(K_0) \leq c(F) + c(F_1)$, so that we obtain $c(F) \geq c(K_0) - c(F_1)$, where F_1 is a surface in $\mathbb{R}^3 \times [1, \infty)$ bounded by K_1. Thus, finding small value of $c(F_1)$ would give better estimates of $c(F)$, the minimal number of steps necessary for a given process.

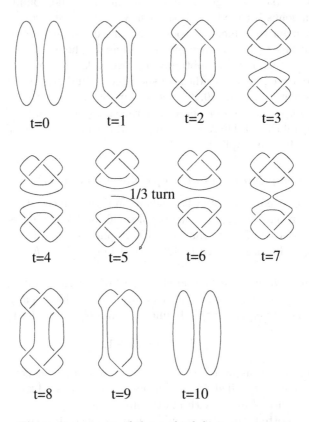

Fig. 9. Turning a trefoil one-third during a process

3.3   Topological Inequivalence Between Processes

In this section we present examples of two processes with the same number of steps, and with the same knot and links at every instance of time, but topologically distinct.

Consider the example depicted in Fig. 9, which is found in [23]. At time $t = 0$, the process starts with two component unknotted, unlinked circles, and at $t = 10$, the process ends with the same unlink, after going through two steps of splicings. These splices occur between time $t = 3$ and $t = 4$, and $t = 6$ and $t = 7$. Between time $t = 4$ and $t = 5$, one of the trefoil knot undergoes a one-third twist before it is reattached to the other trefoil by splicing. Consider this process P, and another process P_0 where no one-third twist has occurred. From the processes P and P_0, we obtain two embedded tori T and T_0, respectively, by capping off the top and the bottom unlinks by standard disks. It is known that T_0 is indeed knotted, while T is topologically equivalent to unknotted, standard torus. In particular, T and T_0 are not topologically equivalent, and so the processes P and P_0 are not equivalent either.

We conclude this section with two remarks that arise from this example: (1) Even if two DNA processes have the same knots and links at every given time, as a whole, the processes may not be topologically equivalent. (2) Topological methods in theory of knotted surfaces can be directly applied to studies of topological aspects of DNA processes.

References

1. Carter, J.S., Saito, M., *Knotted surfaces and their diagrams,* the American Mathematical Society, 1998.
2. Darcy, I. K., *Biological distances on DNA knots and links: applications to XER recombination.* Knots in Hellas '98, Volume 2 (Delphi). J. Knot Theory Ramifications **10** No. 2 (2001) 269–294.
3. Ehrenfeucht, A., Harju, T., Petre, I., Prescot, D.M., Rozenberg, G., *Computation in living cells-gene assembly in ciliates,* Springer-Verlag 2004.
4. Flapan, E., *When topology meets chemistry. A topological look at molecular chirality. Outlooks.* Cambridge University Press, Cambridge; Mathematical Association of America, Washington, DC, 2000.
5. Freyd, P.J., Yetter, D.N., *Braided compact closed categories with applications to low-dimensional topology.* Adv. Math. **77** No. 2 (1989) 156–182.
6. Garside, F., *The braid group and other groups,* Quart. J. Math. Oxford, (2) **20** (1969) 235-254.
7. Garzon, M., Jonoska, N., Karl S.A., *Bounded Complexity of DNA Computing BioSystems,* **52** (1999) 63-72.
8. Head, T., Paun, Gh., Pixton, D., *Language theory and molecular genetics* in *Handbook of formal languages, Vol.II* (G. Rozenberg, A. Salomaa editors) Springer Verlag (1997) 295-358.
9. Hosokawa, F.; Kawauchi, A.; Nakanishi, Y.; Sakuma, M., *Note on critical points of surfaces in 4-space.* Kobe J. Math. **1** No. 2 (1984) 151–152.
10. Joyal, A., Street, R., *Braided tensor categories.* Adv. Math. **102** No. 1 (1993) 20–78.
11. Kawauchi, A., *A survey of knot theory.* Translated and revised from the 1990 Japanese original by the author. Birkhuser Verlag, Basel, 1996.
12. Kawauchi, A., *On the integral homology of infinite cyclic coverings of links.* Kobe J. Math. **4** No. 1 (1987) 31–41.
13. Kronheimer, P.B., Mrowka, T.S., *Gauge theory for embedded surfaces, I,* Topology **32** (1993) 773–826.
14. Landweber, L.F., Kari, L., *The evolution of cellular computing: nature's solution to a computational problem.* BioSystems **52** (1999) 3-13.
15. Landweber, L.F., Kari, L., *Universal molecular computation in ciliates.* In: Evolution as Computation, Landweber, L., Winfree, E. (eds), Springer Verlag, Berlin, Heidelberg, 1999.
16. Magnus, W., Karrass, A., Solitar, D., *Combinatorial group theory. Presentations of groups in terms of generators and relations.* Second revised edition. Dover Publications, Inc., New York, 1976.

17. Mac Lane, S., *Categories for the working mathematician*. Second edition. Graduate Texts in Mathematics, 5. Springer-Verlag, New York, 1998
18. Munkres, J.R., *Topology,* second edition, Prentice-Hall, 2000 (first edition 1975).
19. Ozsvath, P., Szabo, Z., Knot Floer homology and the four-ball genus, Geom. Topol. **7** (2003) 615–639.
20. Paun, Gh, Rozenberg, G., Salomaa, A., *DNA Computing, new computing paradigms*, Springer Verlag 1998.
21. Rasmussen, J.A., *Khovanov homology and the slice genus*, Preprint, available at: http://xxx.lanl.gov/abs/math.GT0402131.
22. Rolfsen, D., *Knots and links.* Mathematics Lecture Series, No. 7. Publish or Perish, Inc., Berkeley, Calif., 1976.
23. Roseman, D., *Twisting and turning in four dimensions,* a video made at Geometry Center, 1993.
24. Sumners, D.W., *Untangling DNA*. Math. Intelligencer **12** No. 3 (1990) 71–80.

Regular Expressions for Two-Dimensional Languages Over One-Letter Alphabet*

Marcella Anselmo[1], Dora Giammarresi[2], and Maria Madonia[3]

[1] Dipartimento di Informatica ed Applicazioni, Università di Salerno, Italy
anselmo@dia.unisa.it
[2] Dipartimento di Matematica. Università di Roma "Tor Vergata", Roma, Italy
giammarr@mat.uniroma2.it
[3] Dip. Matematica e Informatica, Università di Catania, Catania, Italy
madonia@dmi.unict.it

Abstract. The aim of this paper is to give regular expressions for two-dimensional picture languages. The paper focuses on a one-letter alphabet case, that corresponds to the study of "shapes" of families of pictures. A new diagonal concatenation operation is defined. Languages denoted by regular expressions with union, diagonal concatenation and its closure are characterized both in terms of rational relations and in terms of two-dimensional automata moving only right and down. The class of languages denoted by regular expressions with union, column, row and diagonal concatenation, and their closures are included in REC and strictly contains languages defined by three-way automata, but they are not comparable with ones defined by four-way automata. In order to encompass a wider class of languages, we propose some new operations that define languages that still lie in REC.

1 Introduction

A picture or two-dimensional string is a rectangular array of symbols taken from a finite alphabet. Two-dimensional languages are nowadays a rich field of investigation. Many approaches have been presented in the literature in order to generalize formal languages theory to two dimensions. In [6] an unifying point of view is presented: the family REC of picture languages is proposed as the candidate to be "the" generalization of the class of regular one-dimensional languages. Indeed REC is well characterized from very different points of view and thus inherits several properties from the class of regular string (one-dimensional) languages. It is characterized in terms of projections of local languages (tiling systems), of some finite-state automata, of logic formulas and of regular expressions with alphabetic mapping. The approach by regular expressions is indeed not completely satisfactory: the concatenation operations there involved are partial functions

---

* Work partially supported by MIUR Cofin: *Linguaggi Formali e Automi: Metodi, Modelli e Applicazioni*.

C.S. Calude, E. Calude and M.J. Dinneen (Eds.): DLT 2004, LNCS 3340, pp. 63–75, 2004.
© Springer-Verlag Berlin Heidelberg 2004

and moreover an external operation of alphabetic mapping is needed. Then, in [6], the problem of a Kleene-like theorem for the theory of recognizable picture languages remains open.

Several papers were recently devoted to find a better formulation for regular expressions for two-dimensional languages. In [14], O. Matz affords the problem of finding some more powerful expressions to represent recognizable picture languages and suggests some regular expressions where the iteration is over combinations of operators, rather than over languages. The author shows that the power of these expressions does not exceed the family REC, but it remains open whether or not it exhausts it. In [16] some tiling operation is introduced as extension of the Kleene star to pictures and a characterization of REC is given that involves some morphism and the intersection. The paper [17] compares star-free picture expressions with first-order logic.

The aim of this paper is to look for a homogenous notion of regular expression that could extend more naturally the concept of regular expression of one-dimensional languages. In this framework, we propose some new operations on pictures and picture languages and study the families of languages that can be generated using old and new operations.

The paper focuses on one-letter alphabets. This is a particular case of the more general case of several letters alphabets. However this is not only a simpler case to handle, but it is a necessary and meaningful case to start. Indeed studying two-dimensional languages on one-letter alphabets means to study the "shapes" of pictures: if a picture language is in REC then also the language of its shapes is in REC. Such approach allows us to separate the twofold nature of a picture: its shape and its content.

Classical concatenation operations on pictures and picture languages are the row and column concatenations and their closures. Regular expressions that use only Boolean operations and this kind of concatenations and closure however cannot define a large number of two-dimensional languages in REC. As an example, take the simple language of "squares" (that is pictures with number of rows equal to the number of columns). The major problem with this kind of regular expressions is that they cannot describe any relationship existing between the two dimensions of the pictures. Such operations are useful to express some regularity either on the number of rows or on the number of columns but not between them. This is the reason we introduce, in the one-letter case, a new concatenation operation between pictures: the *diagonal concatenation*. The diagonal concatenation introduces the possibility of constructing new pictures forcing some dependence between their dimensions. Moreover an important aspect of the diagonal concatenation is that it is a total function between pictures. This allows to find a quite clean double characterization of *D-regular languages*, the picture languages denoted by regular expressions containing union, diagonal concatenation and its closure: they are exactly those picture languages in which the dimensions are related by a rational relation and also exactly those picture languages recognizable by particular automata moving only right and down. Relationships between rational relation and REC were also studied in [3].

Unfortunately, an analogous situation does not hold anymore, when we also introduce row and column concatenations in regular expressions, essentially because they are partial functions. The class of *CRD-regular languages*, the languages denoted by regular expressions with union, column, row and diagonal concatenations and their closures, strictly lies between the class of languages recognized by three-way deterministic automata and REC. Further it is not comparable with another remarkable class lying in that set, that is the class of languages recognized by four-way automata. These results can be proved using a necessary condition for CRD-regular languages. This necessary condition regards the possible *extensions* of a picture inside the language it belongs, that are the pictures of the language containing it as a sub-picture. In a CRD-regular language an infinite sequence of extensions of a picture in that language has some kind of regularity: it necessarily contains a subsequence obtained iterating the concatenation of a same sub-picture as many times as we want.

Examining some examples of languages not captured by CRD formalism, we find that the extensions of a picture cannot be obtained by iterating the concatenation of a same picture, and this independently from the picture to what we concatenate. On the contrary the extensions grow in a non-uniform way, indeed depending from the picture just obtained. Such considerations show the necessity of a more complex definition for regular expressions in order to denote a wider class of two-dimensional languages in REC. We propose some other new definitions of operations on pictures. They allow to capture a wider class of languages, that still remain inside the class REC.

All definitions are given in such a way to synchronize the extensions of a picture with the picture just constructed. We also provide a collection of examples classically considered in the literature, specifying for each of them its belonging or not to the classes of picture languages considered throughout the paper.

The paper is organized as follows. In Section 2 we recall some preliminary definitions and results later used in the paper. Section 3 contains the main results. It presents our proposals for possible classes of regular expressions in three different subsections. Moreover, we show a table summarizing a wide collection of examples. Section 4 draws some conclusions.

2   Preliminaries

In this section we recall terminology for two-dimensional languages. We briefly describe some machine models for two-dimensional languages; all models reduce to conventional ones when restricted to operate on one-row pictures. Finally we summarize all major results concerning the class of *Recognizable Two-Dimensional Languages*, that is the one that seems to generalize better the family of regular string languages to two dimensions. The notations used, as far as more details, can be mainly found in [6].

Let Σ be a finite alphabet. A *two-dimensional string* (or a *picture*) over Σ is a two-dimensional rectangular array of elements of Σ. The set of all two-dimensional strings over Σ is denoted by Σ^{**}. A *two-dimensional language* over

Σ is a subset of Σ^{**}. Given a picture $p \in \Sigma^{**}$, let $\ell_1(p)$ denote the number of rows of p and $\ell_2(p)$ denote the number of columns of p. The pair $(\ell_1(p), \ell_2(p))$ is called the *size* of the picture p. Differently from the one-dimensional case, we can define an infinite number of empty pictures namely all the pictures of size $(m, 0)$ and of size $(0, n)$, for all $m, n \geq 0$, that we call *empty columns* and *empty rows*, and denote by $\lambda_{0,n}$ and $\lambda_{m,0}$ respectively. The *empty picture* is the only picture of size $(0, 0)$ and it will be denoted by $\lambda_{0,0}$. We indicate by Λ_{col} and Λ_{row} the language of all empty columns and of all empty rows, respectively.

The *column concatenation* of p and q (denoted by $p \oplus q$) and the *row concatenation* of p and q (denoted by $p \ominus q$) are partial operations, defined only if $\ell_1(p) = \ell_1(q)$ and if $\ell_2(p) = \ell_2(q)$, respectively, and are given by: $p \oplus q = \boxed{p \mid q}$

and $p \ominus q = \boxed{\dfrac{p}{q}}$. Moreover we set that $p \oplus \lambda_{m,0} = p$ and $p \ominus \lambda_{0,n} = p$ that is the empty columns and the empty rows are the neutral elements for the column and the row concatenation operations, respectively. As in the string language theory, these definitions of pictures concatenation can be extended to set of pictures. Let $L_1, L_2, L \subseteq \Sigma^{**}$. The *column concatenation* and the *row concatenation* of L_1 and L_2 are defined respectively by $L_1 \oplus L_2 = \{p \oplus q \mid p \in L_1, q \in L_2\}$ and $L_1 \ominus L_2 = \{p \ominus q \mid p \in L_1, q \in L_2\}$. By iterating the concatenation operations, we can define their *transitive closures*, which are somehow "two-dimensional Kleene stars". The *column closure (star)* and the *row closure (star)* of L are defined as $L^{*\oplus} = \bigcup_{i \geq 0} L^{i\oplus}$ and $L^{*\ominus} = \bigcup_{i \geq 0} L^{i\ominus}$, where $L^{0\oplus} = \Lambda_{col}$, $L^{1\oplus} = L$, $L^{i\oplus} = L \oplus L^{(i-1)\oplus}$ for $i > 1$ and $L^{0\ominus} = \Lambda_{row}$, $L^{1\ominus} = L$, $L^{i\ominus} = L \ominus L^{(i-1)\ominus}$ for $i > 1$.

One of the first attempts at formalizing the concept of "recognizable picture language" was the introduction of a finite automaton that reads a two-dimensional tape (cf. [2]). A deterministic (non-deterministic) four-way automaton, denoted by 4DFA (4NFA), is defined as extension of the two-way automaton for strings (cf. [7]) by allowing it to move in four directions: *Left, Right, Up, Down*. The families of picture languages recognized by some 4DFA and 4NFA are denoted by $\mathcal{L}(\text{4DFA})$ and $\mathcal{L}(\text{4NFA})$ respectively. Unlike in the one-dimensional case, $\mathcal{L}(\text{4DFA})$ is strictly included in $\mathcal{L}(\text{4NFA})$ (cf. [2]). Both families $\mathcal{L}(\text{4DFA})$ and $\mathcal{L}(\text{4NFA})$ are closed under Boolean union and intersection operations. The family $\mathcal{L}(\text{4DFA})$ is also closed under complement, while for $\mathcal{L}(\text{4NFA})$ this is not known. On the other hand, $\mathcal{L}(\text{4DFA})$ and $\mathcal{L}(\text{4NFA})$ are not closed under row and column concatenation and closure operations [11].

In [13], it is considered also a weaker model called three-way automaton (3NFA) that is allowed to move right, left and down only. The family $\mathcal{L}(\text{3NFA})$ is strictly included in $\mathcal{L}(\text{4NFA})$. Another interesting model is the *two-dimensional on-line tessellation acceptor*, denoted by 2-OTA (see [8]).

A different way to define (recognize) picture languages was introduced in [5]. It generalizes the characterization of regular languages by means of local strings languages and alphabetic mapping to two dimensions. A local picture language L over an alphabet Γ is defined by means of a finite set Θ of pictures of size $(2, 2)$ (called *tiles*) that represents all allowed sub-pictures for the pictures in L. A *tiling system* for a language L over Σ is a pair of a local language over an alphabet Γ and an alphabetic mapping $\pi : \Gamma \rightarrow \Sigma$. The mapping π can be extended in the obvious way to pictures and picture languages over Γ. Then, we say that a language $L \subseteq \Sigma^{**}$ is *recognizable by tiling systems* if there exist a local language L' over Γ and a mapping $\pi : \Gamma \rightarrow \Sigma$ such that $L = \pi(L')$. The family of two-dimensional languages recognizable by tiling systems is denoted by *REC*. The family REC is closed under Boolean union and intersection but not under complement. It is also closed under all row and column concatenations and stars. Moreover, by definition, it is closed under alphabetic mappings. This notion of recognizability by tiling systems turns out to be very *robust*: we have that $REC = \mathcal{L}(2\text{-OTA})$ ([10]) and that the family REC and the family of languages defined by existential monadic second order formulas coincide (as a generalization of Büchi's theorem for strings to two-dimensional languages; cf. [6]).

3   Two-Dimensional Regular Expressions on One-Letter

The characterizations of the family REC show that it captures in some sense the idea of unification of the concept of recognizability from the two different points of view of descriptive and computational models, that is one of the main properties of the class of recognizable string languages. It seems thus natural to ask whether one can prove also a sort of two-dimensional Kleene's Theorem, but the definitions of regular operations that consider row and column concatenations and stars seem not to be satisfactory for this aim. One reason of this fact could be attributed to the fact that the row and column concatenation operations are partial functions. Using concatenation and closure operations, it is possible to express two-dimensional languages by means of simpler languages. Nevertheless it can be easily observed that row and column concatenations and stars cannot be used to define any relation between the two dimensions of a picture: it can be shown (cf. [6]) that to describe the whole class REC we need also the alphabet mapping between the regular operations. For example, the simple language of "squares" (see Section 2) cannot be described by a regular expression that uses this kind of operations. This is also a clear sign that, going from one to two dimensions, we find a very rich family of languages that need a non-straightforward generalization of the one-dimensional definitions and techniques.

In this section we propose some different types of regular expressions comparing the resulting classes of denoted languages with known families of picture languages. Through all the section, we assume to be in the case of languages over a one-letter alphabet $\Sigma = \{a\}$. Observe that this corresponds to consider the "shapes" of pictures. Indeed if $L \subseteq \Sigma^{**}$, with $|\Sigma| \geq 2$, is in REC then the language obtained by mapping Σ into a one-letter alphabet $\{a\}$, is still in REC, since REC is closed under alphabetic mappings. All the families of languages over a one-letter alphabet will be denoted by adding a superscript "(1)". For example all languages in REC over $\Sigma = \{a\}$ will be denoted by $REC^{(1)}$.

Remark 1. When a one-letter alphabet Σ is considered, any picture $p \in \Sigma^{**}$ is characterized only by its size. Therefore it can be equivalently represented either by a pair of words in Σ^*, where the first one is equal to the first column of p and the second one to the first row of p, i.e. $(a^{\ell_1(p)}, a^{\ell_2(p)})$, or simpler by its size, i.e. $(\ell_1(p), \ell_2(p))$.

3.1   Diagonal Concatenation and D-Regular Expressions

We introduce a new simple definition of concatenation of two pictures in the particular case of a one-letter alphabet. The definition is motivated by the necessity of an operation between pictures that could express some relationship existing between the dimensions of the pictures. We use this new concatenation to construct some regular expressions and to define a class of languages. This class is characterized both in terms of the relations between the dimensions of the pictures and in terms of four-way automata. Let $\Sigma = \{a\}$.

Definition 1. *The* diagonal concatenation *of $p = (m, n)$ and $q = (m', n')$ (denoted by $p \oslash q$) is the picture of size $(m + m', n + n')$. It is represented by*

$$p \oslash q = \boxed{\begin{array}{|c|c|} \hline p & \\ \hline & q \\ \hline \end{array}}$$

Observe that, differently from the classical row and column concatenation, the diagonal concatenation is a total operation. As usual, it can be extended to diagonal concatenation between languages. Moreover the Kleene closure of $\oslash$ can be defined as follows. Let L be a picture language over a one-letter alphabet.

Definition 2. *The* diagonal closure *or* star *of $L \subseteq \{a\}^{**}$ (denoted by $L^{*\oslash}$) is defined as $L^{*\oslash} = \bigcup_{i \geq 0} L^{i\oslash}$ where $L^{0\oslash} = \lambda_{0,0}$, $L^{1\oslash} = L$, $L^{i\oslash} = L \oslash L^{(i-1)\oslash}$ for $i > 1$.*

Example 1. Let $L_{n,n} = \{p \mid \ell_1(p) = \ell_2(p) \geq 1\}$, the language of squares. It can be easily shown that $L_{n,n} = \{(1,1)\}^{*\oslash} = \{\lambda_{0,1} \oslash \lambda_{1,0}\}^{*\oslash}$.

Proposition 1. *The family $REC^{(1)}$ is closed under diagonal concatenation and diagonal star.*

Proof. The proof uses similar techniques to the one for the closure of REC under row (or column) concatenation and star ([5]). A tiling system for $L_1 \oslash L_2$ can be defined from the ones for L_1 and L_2, adding some tiles to "glue" bottom-right corners of pictures in L_1 to top-left corners of pictures in L_2.

The diagonal concatenation can be used to generate families of picture languages, starting from atomic languages. Formally, let us denote $D = \{\cup, \oslash, *\oslash\}$; the elements of D are called *diagonal-regular operations* or *D-regular operations*.

Definition 3. *A diagonal-regular expression (D-RE) is defined recursively as:*

1. $\emptyset$, $\{\lambda_{0,1}\}$, $\{\lambda_{1,0}\}$ *are D-RE*
2. *if α, β are D-RE then $(\alpha) \cup (\beta)$, $(\alpha) \oslash (\beta)$, $(\alpha)^{*\oslash}$ are D-RE.*

Every D-RE denotes a language as usual. Languages denoted by D-RE are called *diagonal-regular languages*, briefly *D-regular languages*. The class of D-regular languages is denoted $\mathcal{L}(D)$. Observe that languages containing a single picture (n, m) can be denoted by the D-RE $E_{n,m} = (\lambda_{1,0}^{n\bigcirc}) \oslash (\lambda_{0,1}^{m\bigcirc})$.

We will now characterize D-regular languages both in terms of rational relations and in terms of some 4NFA. Let us recall that (see [1]) a *rational relation* over alphabets Σ and Δ is a rational subset of the monoid $(\Sigma^* \times \Delta^*, ., (\lambda, \lambda))$, where the operation . is the componentwise product defined by $(u_1, v_1).(u_2, v_2) = (u_1 u_2, v_1 v_2)$ for any $(u_1, v_1), (u_2, v_2) \in \Sigma^* \times \Delta^*$. When the alphabet $\Sigma = \Delta = \{a\}$ is considered, there is natural correspondence between pictures over Σ and relations over $\Sigma \times \Sigma$. For any relation $T \subseteq \Sigma^* \times \Sigma^*$ we can define the picture language $L(T) = \{p \in \Sigma^{**} \mid \ell_1(p) = |r_1| \text{ and } \ell_2(p) = |r_2| \text{ for some } (r_1, r_2) \in T\}$. Vice versa, for any picture language $L \subseteq \Sigma^{**}$ we can define the relation $R(L) = \{(r_1, r_2) \in \Sigma^* \times \Sigma^* \mid |r_1| = \ell_1(p) \text{ and } |r_2| = \ell_2(p) \text{ for some } p \in L\}$.

Remark 2. We recall that a 4NFA M over a one-letter alphabet is equivalent to a two-way two-tape automaton M_1 (cf. [9]). In fact, let H_1 and H_2 be the heads of M_1, then M_1 simulates M as follows. If the input head H of M moves down (up) one square, M_1 moves H_1 right (left) one square without moving H_2, and if H moves right (left) one square, M_1 moves H_2 right (left) without moving H_1.

Proposition 2. *Let $L \subseteq \{a\}^{**}$. Then $L \in \mathcal{L}(D)$ if and only if $L = L(T)$ for some rational relation $T \subseteq \Sigma^* \times \Sigma^*$ if and only if $L = \mathcal{L}(A)$ for some 4NFA A that moves only right and down.*

Proof. Let $M = \Sigma^* \times \Sigma^*$. The componentwise product in M exactly corresponds to the diagonal concatenation in Σ^{**} (see Definition 1 and Remark 1). It is known that a non-empty rational subset of a monoid can be expressed, starting with singletons, by a finite number of unions, products and stars. Thus $L \in \mathcal{L}(D)$ if and only if $L = L(T)$ for some rational relation $T \subseteq M$. On the other hand, it is well known that $T \subseteq M$ is a rational relation iff it is accepted by a transducer, that is an automaton over M. Further this automaton can be viewed as a one-way automaton with two tapes (cf. [15]). Then, in analogy to Remark 2, one-way two-tape automata are equivalent to 4NFA that move only right and down.

Example 2. Let $L_{n,n}$ be the language of squares, as in Example 1. It can be easily shown that $L_{n,n}$ is denoted by the following D-RE: $E_{n,n} = (\lambda_{0,1} \oslash \lambda_{1,0})^{*\bigcirc}$. We have $L_{n,n} = L(T)$, where T is the rational relation $T = \{(a^n, a^n) \mid n \geq 1\}$. Further $L = \mathcal{L}(A)$ where A is the 4NFA that starting in the top-left corner moves along the main diagonal until it eventually reaches the bottom-right corner and accepts. More generally, the languages $L_{n,n+i} = \{p \mid l_1(p) = n, l_2(p) = n+i, n \geq 1\}$, for some $i \geq 0$, are denoted by the D-RE: $E_{n,n+i} = E_{n,n} \oplus (((\lambda_{0,1}^{i\bigcirc} \oslash \lambda_{1,0}))^{*\ominus})$.

The following example shows that also for a one-letter alphabet, four-way automata that move only right and down are strictly less powerful than 3DFA.

Example 3. Language $L = \{(kn, n) | \ k, n \geq 1\}$ is recognized by a 3DFA that, starting in the top-left corner moves along the main diagonal until it reaches the right boundary and then moves along the secondary diagonal until it reaches the left boundary and so on until it eventually reaches some corner and accepts. By Proposition 2, language L cannot be recognized by a four-way automaton that moves only right and down, since $\{(a^{kn}, a^n) | \ k, n \geq 1\}$ is not a rational relation (see [3]).

3.2   CRD-Regular Expressions

In this section we consider regular expressions involving all three concatenations and corresponding stars, as defined in previous sections. We show that the class of denoted languages, denoted by $\mathcal{L}(\mathrm{CRD})$, is strictly included in the family REC, and strictly contains $\mathcal{L}(\mathrm{3DFA})^{(1)}$. Further we show that $\mathcal{L}(\mathrm{CRD})$ is not comparable with $\mathcal{L}(\mathrm{4NFA})^{(1)}$.

Let $\mathrm{CRD} = \{\cup, \oplus, \ominus, \oslash, *\oplus, *\ominus, *\oslash\}$, where C, R, D stand for "column", "row" and "diagonal". The elements of CRD are called *CRD-regular operations*.

Definition 4. *A CRD-regular expression (CRD-RE), is defined recursively as:*
1. $\emptyset$, $\{\lambda_{0,1}\}$, $\{\lambda_{1,0}\}$ are CRD-RE
2. if α, β are CRD-RE then $(\alpha) \cup (\beta)$, $(\alpha) \oplus (\beta)$, $(\alpha)^{*\oplus}$, $(\alpha) \ominus (\beta)$, $(\alpha)^{*\ominus}$, $(\alpha) \oslash (\beta)$, $(\alpha)^{*\oslash}$ are CRD-RE.

Every CRD-RE denotes a language using the standard notation. Languages denoted by CRD-RE are called *CRD-regular languages*. The family of CRD-regular languages (over a one-letter alphabet) will be denoted by $\mathcal{L}(\mathrm{CRD})$.

Example 4. Let $L = \{(n, k_1(n+1) + k_2(n+2) + k_3(n+3) \mid n \geq 1, k_1, k_2, k_3 \geq 0\}$. Consider the languages $L_{n,n+i}$ denoted by: $E_{n,n+i} = E_{n,n} \oplus ((E_{1,i})^{*\ominus})$, as in Example 2. Language L can be denoted by the following CRD-RE: $E = E_{n,n+1}^{*\oplus} \oplus E_{n,n+2}^{*\oplus} \oplus E_{n,n+3}^{*\oplus}$.

Example 5. Let $L = \{(hn, hkn + n) \mid n, h, k \geq 1 \ \}$. We have $L = L_1 \oplus L_2$, where $L_1 = \{(n, kn) \mid n, k \geq 1 \ \}$ and $L_2 = \{(hm, m) \mid m, h \geq 1 \ \}$. If $E_{n,n}$ is a D-RE for the languages of squares (see Example 2) a CRD-RE for L is $E = (E_{n,n}^{*\oplus}) \oplus (E_{n,n}^{*\ominus})$.

We now introduce the notion of *extension* of a picture p in a language L in order to formulate a necessary condition for CRD-regular languages. The condition concerns some regularity on infinite sequences of extensions of any picture in the language.

Definition 5. *Let $\Sigma = \{a\}$, $L \subseteq \Sigma^{**}$, $p = (n, m)$, $p' = (n', m') \in L$. Picture p' is a column-extension of p in L if $n = n'$ and $m' > m$. Picture p' is a row-extension of p in L if $n' > n$ and $m = m'$. Picture p' is a diagonal-extension of p in L if $n' > n$ and $m' > m$. In any of these cases, p' is an extension of p in L.*

Definition 6. *A sequence $(p_1, p_2, \ldots)$ of column- (row-, diagonal-, respectively) extensions of p in L is strictly increasing if, for any $i \geq 2$, p_i is a column- (row-diagonal-, respectively) extension of p_{i-1} in L. A strictly increasing sequence $(p_1, p_2, \ldots)$ of column- (row-, diagonal-, respectively) extensions of $p = (n, m)$ in L with $p_i = (n_i, m_i)$ is looping if there exist some $a, b, h, k \geq 0$ such that $n_i = n + a + ih$ and $m_i = m + b + ik$ for any $i \geq 1$.*

Observe that for a strictly increasing looping sequence of column- (row-, respectively) extensions, we necessarily have $a = h = 0$ ($b = k = 0$, respectively).

Proposition 3. *If $L \in \mathcal{L}(CRD)$ then for every $p \in L$, either p has only a finite number of extensions in L or any strictly increasing sequence of extensions of p in L has a looping subsequence.*

Proof. Let r be CRD-regular expression for L. The proof is by induction on the number of operators in r. The basis is obvious. Assume proposition is true for languages denoted by CRD-regular expression with less than i operators, $i \geq 1$, and let r have i operators. There are seven cases depending on the form of r: $r = r_1 \cup r_2$, $r = r_1 \oplus r_2$, $r = r_1 \ominus r_2$, $r = r_1 \oslash r_2$, $r = r_1^{*\ominus}$, $r = r_1^{*\oplus}$, or $r = r_1^{*\oslash}$. In any case, r_1 and r_2 denote some language L_1 and L_2, respectively, that satisfies the condition. In the first case ($r = r_1 \cup r_2$) the result easily follows. We sketch the proof of the second case ($r = r_1 \oplus r_2$); the other cases can be handled using similar techniques and reasonings. If $r = r_1 \oplus r_2$ then $L = L_1 \oplus L_2$. By inductive hypothesis both L_1 and L_2 satisfy the condition. Let $p \in L$. Clearly, $p = p_1 \oplus p_2$ for some $p_1 \in L_1$ and $p_2 \in L_2$. If p has only a finite number of extensions in L, we are done. Suppose, instead, that p has an infinite number of extensions in L: so we have to prove that any strictly increasing sequence of column- (row-, diagonal-, respectively) extensions has a looping subsequence. We give some details only for diagonal extensions; the other cases can be proved analogously. Then, suppose there exists a strictly increasing sequence, s, of diagonal-extensions of p in L. Without loss of generality, we can suppose that, from s, we can extract a subsequence s' obtained by column concatenation of pictures of s'_1, a strictly increasing sequence of diagonal-extensions of p_1 in L_1, with pictures of s'_2, a sequence of pictures in L_2 with the same number of columns or a strictly increasing sequence of diagonal-extensions of p_2 in L_2. Then, using the looping subsequences of s'_1 and s'_2, we can obtain a looping subsequence of s.

Proposition 3 can be used to prove that some picture languages are not in $\mathcal{L}(CRD)$, as shown in the following example. Analogous motivations can be used to show that also the language $L = \{(n, n^2) \mid n \geq 1\}$ is not in $\mathcal{L}(CRD)$.

Example 6. Let $L = \{(2^n, 2^n) \mid n \geq 1\}$. Language L does not satisfy the condition in Proposition 3 and thus $L \notin \mathcal{L}(CRD)$. Indeed, every picture in L has an infinite number of extensions and any extension is a diagonal-extension because, for any $n \geq 1$, there is only one picture with 2^n rows and one picture with 2^n

columns. Suppose by the contrary that $p \in L$ and that some strictly increasing sequence of extensions has a subsequence $\{p_i\}_{i \geq 0}$ where $p_i = (n+a+ih, m+b+ik)$, for some $a, b \geq 0$, $h, k \geq 1$. Since $p_0 = (n+a, m+b) \in L$, we have that $m+b = n+a = 2^x$; and since $p_1 = (n+a+h, m+b+k) \in L$, then $m+b+k = n+a+k = 2^{x+y}$ with $y > 0$ and thus $k = 2^{x+y} - 2^x$. Consider now $p_2 = (n+a+2h, m+b+2k)$; we have that $m+b+2k = 2^x + 2(2^{x+y} - 2^x) = 2^x(1 + 2^{y+1} - 2) = 2^x(2^{y+1} - 1)$. Therefore $m+b+2k$ is the product of a power of 2 times an odd number and cannot be a power of 2, against $p_2 \in L$.

We now show that the family of CRD-regular languages lies between the class $\mathcal{L}(3DFA)^{(1)}$ and $REC^{(1)}$. On the other hand, it is not comparable with $\mathcal{L}(4NFA)^{(1)}$.

Proposition 4. *$\mathcal{L}(3DFA)^{(1)} \subset \mathcal{L}(CRD) \subset REC^{(1)}$, with strict inclusions.*

Proof. Let $L \in \mathcal{L}(3DFA)^{(1)}$. Following [13], we have that L is a finite union of languages R whose elements are $(a_0 + a_1 n, h(b_0 + b_1 n) + b_2 n + b_3 k + b_4)$ with $a_0, a_1, b_0, b_1, b_2, b_3, b_4$ positive integers and n, h, k positive integer variables. Any such language R is in $\mathcal{L}(CRD)$. Indeed let $E_{n,n}$ denote the language of squares (Example 2). The language $\{(a_0 + a_1 n, b_0 + b_1 n) \mid a_0, a_1, b_0, b_1, n \in N\}$ is denoted by $E_{a_0, a_1, b_0, b_1} = ((a_0, b_0) \oslash ((E_{n,n})^{a_1} \ominus)^{b_1} \oplus)$. Therefore a CRD-RE for R is $E = (E_{a_0, a_1, b_0, b_1})^{*\oplus} \oplus ((a_0, 1)^{*\oplus} \ominus E_{0, a_1, 0, b_2}) \oplus ((1, b_3)^{*\ominus})^{*\oplus}) \oplus ((1, b_4)^{*\ominus})$. Moreover the inclusion $\mathcal{L}(3DFA)^{(1)} \subset \mathcal{L}(CRD)$ is strict: the language $L = \{(n, k_1(n+1) + k_2(n+2) + k_3(n+3))\}$ in Example 4 is in $\mathcal{L}(CRD)$, but $L \notin \mathcal{L}(3DFA)^{(1)}$ (cf. [13]). Further we have $\mathcal{L}(CRD) \subseteq REC^{(1)}$ because $REC^{(1)}$ is closed under operations in CRD (cf. [5] and Proposition 1). An example of languages in $REC^{(1)} \setminus \mathcal{L}(CRD)$ is $\{(2^n, 2^n) \mid n \geq 1\}$ (see Example 6).

Proposition 5. *There exist $L \in \mathcal{L}(CRD) \setminus L(4NFA)^{(1)}$ and $L' \in \mathcal{L}(4NFA)^{(1)} \setminus \mathcal{L}(CRD)$.*

Proof. Let $L = \{(hn, hkn + n) \mid n, h, k \geq 1\}$; $L \in \mathcal{L}(CRD)$ (Example 5), but it can be shown that $L \notin \mathcal{L}(4NFA)^{(1)}$. Consider now the language $L' = \{(2^n, 2^n) \mid n \geq 1\}$; $L' \notin \mathcal{L}(CRD)$ (Example 6), but $L' \in \mathcal{L}(4DFA)^{(1)}$ ([12]).

3.3   A Collection of Examples

In this section, we give a collection of examples of two-dimensional languages and classify them with respect to their machine-type and their regular expression-type. Languages are given by their representative element, where $n, m, h, k \geq 1$ are integer variables and $c \geq 1$ is an integer constant. Moreover $f_1(n) = a_1 + \cdots + a_n$, where $a_1, \cdots, a_n$ are all chosen in a finite subset of N, and $f_2(n) = k_1(n+1) + k_2(n+2) + k_3(n+3)$, where $k_1, k_2, k_3 \geq 1$ are integer variables.

Element	2DFA	2NFA	3DFA	3NFA	4DFA	4NFA	D-RE	CRD-RE	REC
(n, n)	Y	Y	Y	Y	Y	Y	Y	Y	Y
$(2, 2n)$	Y	Y	Y	Y	Y	Y	Y	Y	Y
$(2n, 2n)$	Y	Y	Y	Y	Y	Y	Y	Y	Y
$(2n, 2m)$	Y	Y	Y	Y	Y	Y	Y	Y	Y
(n, cn)	Y	Y	Y	Y	Y	Y	Y	Y	Y
$(n, f_1(n))$	N	Y	Y	Y	Y	Y	Y	Y	Y
(kn, n)	N	N	Y	Y	Y	Y	N	Y	Y
$(n, f_2(n))$	N	N	N	Y	Y	Y	N	Y	Y
(n, kn)	N	N	N	N	Y	Y	N	Y	Y
$(2^n, 2^n)$	N	N	N	N	Y	Y	N	N	Y
$(hn, hkn + n)$	N	N	N	N	N	N	N	Y	Y
(n, n^2)	N	N	N	N	N	N	N	N	Y
(n^2, n)	N	N	N	N	N	N	N	N	Y
(n^2, n^2)	N	N	N	N	N	N	N	N	Y
$(n, 2^n)$	N	N	N	N	N	N	N	N	Y
$(n, n!)$	N	N	N	N	N	N	N	N	N

3.4 Advanced Star Operations

Using three types of concatenation operation (row, column and diagonal) and the three corresponding stars we get regular expressions describing a quite large family of two-dimensional languages. Nevertheless, all those operations together seem not enough to describe the whole family REC[(1)] because REC contains very "complex" languages even in the case of a one-letter (see for example [4]).

The peculiarity of the "classical" star operation (along which such column, row or diagonal stars are defined) is that it corresponds to an iterative process that at each step adds (concatenates) always the same set. We can say that it corresponds to the idea of an iteration for some recursive H defined like $H(1) = S$ and $H(n + 1) = H(n) \cdot S$ where S is a given set. This seems to cause the main difficulty in defining expressions for languages of pictures (n, n^2) or (n^2, n^2) or $(2^n, 2^n)$, or $(f(n), f(n))$ with $f(n)$ polynomial or exponential function, for $n > 0$.

In this section we propose a new type of iteration operation, an *advanced star*, that results much more powerful because somehow "implements" the idea of iteration for a recursive H defined like $H(1) = S$ and $H(n+1) = H(n) \cdot K(n)$ where S is a given set and $K(n)$ is another recursive function. That is at each step of the iteration we add something that depends on that step.

Definition 7. *Let L, L_r, L_d be two-dimensional languages. The star of L with respect to (L_r, L_d) is defined as: $L^{(L_r, L_d)*} = \bigcup_i L^{(L_r, L_d)i}$ where $L^{(L_r, L_d)0} = L$ and $L^{(L_r, L_d)i+1} = \left\{ p' = \begin{array}{|c|c|} \hline p & p_r \\ \hline p_d & q \\ \hline \end{array} \mid p \in L^{(L_r, L_d)i}, \; p_r \in L_r, p_d \in L_d, q \in \Sigma^{**} \right\}.$*

Remark that the operation we defined cannot be simulated by a sequence of ① and ⊖ operations because to get p' we first concatenate $p ① p_r$ and $p ⊖ p_d$, then we overlay them and then we fill the hole with a picture $q \in \Sigma^{**}$. Notice that this advanced star is based on a reverse principle with respect to the diagonal star: we "decide" what to concatenate to the right and down to the given picture and then fill the hole in the bottom-right corner. Moreover, observe that, at $(i+1)$st step of the iteration, we are forced to select pictures $p_r \in L_r$ and $p_d \in L_d$ that have the same number of rows and the same number of columns, respectively, of pictures generated at the ith step. Therefore, we exploit the fact that column and row concatenations are partial operations to synchronize the steps of the iterations with the pictures we concatenate.

As a simple example, we can define the language $L' = \{(n, n^2) \,|\, n > 0\}$ as $L^{(L_r, L_d)*}$ where $L = \{(1,1)\}$, $L_r = \{(n, 2n+1) \,|\, n > 0\}$ and $L_d = \{(1, n) \,|\, n > 0\}$. Moreover we can define the language $L'' = \{(n^2, n^2) \,|\, n > 0\}$ as $M^{(M_r, M_d)*}$ where $M = \{(1,1)\}$, $M_r = \{(n^2, 2n+1) \,|\, n > 0\}$ and $M_d = \{(2n+1, n^2) \,|\, n > 0\}$.

We state the following proposition without proof for like of space. It can be proved using techniques similar to main proof in [4], despite the details are more involved.

Proposition 6. *If L, L_r, L_d are languages in REC, then $L^{(L_r, L_d)*}$ is in REC.*

Remark that it seems not possible to define the language of pictures of size $(n, 2^n)$ using such advanced star. In fact the definition should involve an iteration like: $L^{(L_d)i+1} = \left\{ p' = \boxed{\begin{array}{c|c} p & p \\ \hline p_d & q \end{array}} \,\Big|\, p \in L^{(L_d)i} , \, p_d \in L_d, q \in \Sigma^{**} \right\}$. That is, we would need to use as L_r the language itself.

4   Concluding Remarks and Further Research

We have proposed new operations so that a quite wide class of two-dimensional languages inside REC could be described in terms of regular expressions. Nevertheless we still have not gain a complete description of class REC, even for the case of one-letter. Further steps are surely to define and refine other "advanced" star operations in the aim of proving a two-dimensional Kleene's Theorem.

References

1. Berstel, J.: Transductions and Context-free Languages. Teubner (1979)
2. Blum, M., Hewitt, C.: Automata on a two-dimensional tape. IEEE Symposium on Switching and Automata Theory. (1967) 155-160
3. De Prophetis, L., Varricchio, S.: Recognizability of rectangular pictures by wang systems. Journal of Automata, Languages, Combinatorics. **2** (1997) 269-288
4. Giammarresi, D.: Two-dimensional languages and recognizable functions. (Proc., Developments in language theory, Finland, 1993), Rozenberg, G., Salomaa A. (Eds), World Scientific Publishing Co. (1994)
5. Giammarresi, D., Restivo, A.: Two-dimensional finite state recognizability. Fundamenta Informaticae **25:3, 4** (1996) 399-422
6. Giammarresi, D., Restivo, A.: Two-dimensional languages. In: G. Rozenberg *et al* Eds: Handbook of Formal Languages Vol. III. Springer Verlag (1997) 215-268

7. Hopcroft, J. E., Ullman, J. D.: Introduction to Automata Theory, Languages and Computation. Addison-Wesley (1979)
8. Inoue, K., Nakamura, A.: Some properties of two-dimensional on-line tessellation acceptors. Information Sciences. **13** (1997) 95-121
9. Inoue, K., Nakamura, A.: Two-dimensional finite automata and unacceptable functions. Intern. J. Comput. Math. Sec. A **7** (1979) 207–213
10. Inoue, K., Takanami, I.: A Characterization of recognizable picture languages. (Proc., Second International Colloquium on Parallel Image Processing) LNCS **654** Springer-Verlag, Berlin (1993)
11. Inoue, K., Takanami, I., Nakamura, A.: A note on two-dimensional finite automata. Information Processing Letters **7:1** (1978) 49-52
12. Kari, J., Moore, C.: Rectangles and squares recogized by two-dimensional automata. `http://www.santafe.edu/~moore/pubs/picture.html`
13. Kinber, E. B.: Three-way Automata on Rectangular Tapes over a One-Letter Alphabet. Information Sciences **35** Elsevier Sc. Publ. (1985) 61-77
14. Matz, O.: Regular expressions and Context-free Grammars for picture languages. (Proc. STACS'97) LNCS **1200** Springer Verlag (1997) 283-294
15. Rabin, M., Scott, T.: Finite automata and their decision problems. IBM Journal Res. and Dev. **3** (1959) 114-125
16. Simplot, D.: A characterization of recognizable picture languages by tilings by finite sets. Theoretical Computer Science **218:2** (1999) 297-323
17. Wilke, T.: Star-free picture expressions are strictly weaker than first-order logic. (Proc. ICALP'97) LNCS **1256** Springer-Verlag (1997) 347-357

On Competence in CD Grammar Systems[*]

Maurice H. ter Beek[1],[**], Erzsébet Csuhaj-Varjú[2],
Markus Holzer[3], and György Vaszil[2]

[1] Istituto di Scienza e Tecnologie dell'Informazione, Pisa, Italy
maurice.terbeek@isti.cnr.it
[2] Computer and Automation Research Institute,
Hungarian Academy of Sciences, Budapest, Hungary
{csuhaj, vaszil}@sztaki.hu
[3] Institut für Informatik, Technische Universität München
München, Germany
holzer@in.tum.de

Abstract. We investigate the generative power of cooperating distributed grammar systems (CDGSs), if the cooperation protocol is based on the level of competence on the underlying sentential form. A component is said to be $=k$-competent ($\leq k$-, $\geq k$-competent, resp.) on a sentential form if it is able to rewrite exactly k (at most k, at least k, resp.) different nonterminals appearing in that string. In most cases CDGSs working according to the above described cooperation strategy turn out to give new characterizations of the language families based on random context conditions, namely random context (context-free) languages and the biologically motivated family of languages generated by ET0L systems with random context. Thus, the results presented in this paper can shed new light on some longstanding open problems in the theory of regulated rewriting.

1 Introduction

A grammar system is a set of grammars that under a specific cooperation protocol generates one language. The idea to consider—contrary to the "one grammar generating one language" paradigm of classical formal language theory—a set of cooperating grammars generating one language first appeared in [9]. An intensive exploration of the potential of grammar systems was not undertaken until [3] established a link between cooperating distributed grammar systems (CDGSs) and blackboard systems as known from artificial intelligence. A blackboard system consists of several autonomous agents, a blackboard, and a control

---

[*] The first author was supported by an ERCIM postdoctoral fellowship and the third author was supported by project Centre of Excellence in Information Technology, Computer Science and Control, ICA1-CT-2000-70025, HUN-TING project, WP 5.
[**] This author's work for this paper was fully carried out during his stay at the Computer and Automation Research Institute of the Hungarian Academy of Sciences.

C.S. Calude, E. Calude and M.J. Dinneen (Eds.): DLT 2004, LNCS 3340, pp. 76–88, 2004.

mechanism. The control mechanism dictates some rules which the agents must respect during their joint effort to solve a problem stated on the blackboard. The only way in which the agents may communicate is via the blackboard, which represents the current state of the problem solving. If the problem solving is successful, the solution appears on the blackboard. CDGSs form a language-theoretic framework for modelling blackboard systems. Agents are represented by grammars, the blackboard is represented by the sentential form, control is regulated by a cooperation protocol of the grammars, and the solution is represented by a terminal word. By now, grammar systems form a well-established and well-recognized area within the theory of formal languages. The interested reader is referred to [6] for more information.

In this paper we examine some variants of cooperation protocols for CDGSs based on the level of competence that a component has on a sentential form. Competence-based cooperation protocols have already been studied in the literature, e.g., [1, 3–5, 9] We consider cooperation protocols that allow a component to start rewriting when such a competence condition is satisfied, and that require it to do so as long as the grammar satisfies this condition. Intuitively, a component is $=k$-competent ($\leq k$-competent, $\geq k$-competent, resp.) on a sentential form if it is able to rewrite exactly k (at most k, at least k, resp.) different nonterminals appearing in the sentential form. In the sequel we will call these cooperation protocols the $=k$-comp.-mode ($\leq k$-comp.-mode, $\geq k$-comp.-mode, resp.) of derivation. Hence the more different nonterminals of a sentential form a component is able to rewrite, the higher its (level of) competence on that string. By restricting the rewriting of the sentential form to components having a certain (level of) competence, we provide a formal interpretation of the requirement that agents must be competent enough before being able to participate in the problem solving taking place on the blackboard.

We demonstrate that these competence-based cooperation protocols are very powerful and closely related to rewriting mechanisms based on random context conditions. To be more precise, it is shown that CDGSs working in the $=1$-comp.- or ≤ 1-comp.-mode of derivation are at least as powerful as the family of languages generated by forbidding random context grammars, while CDGSs working according to the ≥ 1-comp.-mode of derivation characterize the family of ET0L languages. A slight increase in the level of competence gives a significant increase in generative power, namely already CDGSs working in the $=2$-comp.- or ≤ 2-comp.-mode of derivation characterize the family of random context languages or, equivalently, that of the recursively enumerable languages, while the ≥ 2-comp.-mode leads to the biologically motivated family of languages generated by ET0L systems with random context [11]. This is yet another characterization of the family of random context ET0L languages, which recently appeared several times in relation with CDGSs and non-standard derivation modes—see, e.g., [2]. The family of random context ET0L languages is of interest because it coincides with the family of recurrent programmed context-free languages and forms an intermediate class between the families of context-free random context languages and programmed context-free languages generated by grammars without appearance checking [8]. In fact we show that rather simple component grammars suffice to simulate random context grammars or ET0L systems with random context, thus showing that it is indeed the cooperation protocol that is very powerful. So we hope that one can gain a deeper insight into the nature of (recurrent) programmed versus random context grammars without appearance checking such that new light is shed on some longstanding open questions.

2   Definitions

We assume the reader to be familiar with the basic notions of formal languages as, e.g., contained in [7]. In general, we have the following conventions. Set difference is denoted by $\setminus$, set inclusion by $\subseteq$, and strict set inclusion by $\subset$. The cardinality of a set M is denoted by $|M|$. The empty word is denoted by λ.

A *random context grammar* is a quadruple $G = (N, T, P, S)$, where N, T, and $S \in N$ are the set of nonterminals, the set of terminals, and the start symbol, respectively. Moreover, P is a finite set of random context rules, i.e., triples of the form $(\alpha \rightarrow \beta, Q, R)$, where $\alpha \rightarrow \beta$ is a context-free production and $Q, R \subseteq N$ are its permitting and forbidding context, respectively. For $x, y \in (N \cup T)^*$ we write $x \Rightarrow y$ if and only if $x = x_1 \alpha x_2$, $y = x_1 \beta x_2$, all symbols of Q appear in $x_1 x_2$, and no symbol of R appears in $x_1 x_2$. If either Q and/or R is empty, then the corresponding context check is omitted. The language generated by G is defined as $L(G) = \{ w \in T^* \mid S \stackrel{*}{\Rightarrow} w \}$, where $\stackrel{*}{\Rightarrow}$ is the reflexive transitive closure of $\Rightarrow$. The family of languages generated by random context grammars is denoted by $\mathcal{L}(\mathrm{RC, CF})$. It is known—see, e.g., [7]—that $\mathcal{L}(\mathrm{RC, CF}) = \mathcal{L}(\mathrm{RE})$, where $\mathcal{L}(\mathrm{RE})$ denotes the class of recursively enumerable languages.

Random context grammars where all permitting contexts are empty are called *forbidding random context grammars*. In this case we are led to the family $\mathcal{L}(\mathrm{fRC, CF})$ of forbidding random context languages. It is known—see, e.g., [7]— that $\mathcal{L}(\mathrm{ET0L}) \subset \mathcal{L}(\mathrm{fRC, CF}) \subseteq \mathcal{L}(\mathrm{RC, CF})$, where $\mathcal{L}(\mathrm{ET0L})$ denotes the family of languages generated by ET0L systems.

A *random context ET0L system* is a sixtuple $G = (\Sigma, H, \omega, \Delta, oc, noc)$, where the four tuple $(\Sigma, H, \omega, \Delta)$ is an ordinary ET0L system, with Σ as its total alphabet, $\Delta \subseteq \Sigma$ as its terminal alphabet, H as its set of tables (finite substitutions from Σ into Σ^*), $\omega \in \Sigma^+$ as its axiom, and oc, noc as functions from H to the subsets of Σ. For two strings $x, y \in \Sigma^*$, the relation $x \Rightarrow y$ holds if and only if there is an $h \in H$, such that all letters in $oc(h)$ occur in x, no letter of $noc(h)$ occurs in x, and $y \in h(x)$. Let $\stackrel{*}{\Rightarrow}$ denote the reflexive and transitive closure of $\Rightarrow$. The language generated by G is defined as $L(G) = \{ w \in \Delta^* \mid \omega \stackrel{*}{\Rightarrow} w \}$. The family of languages generated by random context ET0L systems is denoted by $\mathcal{L}(\mathrm{RC, ET0L})$. It is known—see, e.g., [7]— that $\mathcal{L}(\mathrm{ET0L}) \subset \mathcal{L}(\mathrm{RC, ET0L}) \subseteq \mathcal{L}(\mathrm{RE})$, but it is an open problem whether the latter inclusion is strict.

3 Competence in CD Grammar Systems

A *cooperating distributed grammar system* (CDGS) of degree n, with $n \geq 1$, is an $(n+3)$-tuple $G = (N, T, \alpha, P_1, \ldots, P_n)$, in which N and T are its disjoint alphabets of nonterminals and terminals, respectively, $\alpha \in (N \cup T)^*$ is its axiom, and $P_1, \ldots, P_n$ are finite sets of context-free productions over $N \times (N \cup T)^*$ that are called its components. The given definition of CDGSs differs from the usual one since arbitrary words from $(N \cup T)^*$ may serve as its axioms. For $x, y \in (N \cup T)^*$ and $1 \leq i \leq n$, we define a single rewriting step as $x \Rightarrow_i y$ if and only if $x = x_1 A x_2$ and $y = x_1 z x_2$, for some $A \to z \in P_i$. The subscript i thus refers to the component being used.

Next we recall from [4] the notion of competence that components of a CDGS have on a particular sentential form. First we define the domain of a component as $\mathrm{dom}(P_i) = \{ A \in N \mid A \to z \in P_i \}$. Consequently, component P_i, with $1 \leq i \leq n$, is said to be k-*competent* on a sentential form x in $(N \cup T)^*$ if and only if $|\mathrm{alph}_N(x) \cap \mathrm{dom}(P_i)| = k$, where $\mathrm{alph}_N(x) = \{ A \in N \mid x \in (N \cup T)^* A (N \cup T)^* \}$, i.e., it denotes the set of all nonterminals occurring in x. We abbreviate the (level of) competence of component P_i on x by $\mathrm{clev}_i(x)$.

Based on the (level of) competence that the components have on a sentential form, we define the following cooperation protocols for CDGSs:

1. $x \Rightarrow_i^{\leq k\text{-comp.}} y$ if and only if there is a derivation $x = x_0 \Rightarrow_i x_1 \Rightarrow_i \cdots \Rightarrow_i x_{m-1} \Rightarrow_i x_m = y$ and it satisfies
   (a) $\mathrm{clev}_i(x_j) \leq k$ for $0 \leq j < m$ and (i) $\mathrm{clev}_i(x_m) = 0$ or (ii) $y \in T^*$, or
   (b) $\mathrm{clev}_i(x_j) \leq k$ for $0 \leq j < m$ and $\mathrm{clev}_i(x_m) > k$,
2. $x \Rightarrow_i^{= k\text{-comp.}} y$ if and only if there is a derivation $x = x_0 \Rightarrow_i x_1 \Rightarrow_i \cdots \Rightarrow_i x_{m-1} \Rightarrow_i x_m = y$ and it satisfies
   (a) $\mathrm{clev}_i(x_j) = k$ for $0 \leq j < m$ and $\mathrm{clev}_i(x_m) \neq k$, or
   (b) $\mathrm{clev}_i(x_0) = k$, $\mathrm{clev}_i(x_j) \leq k$ for $1 \leq j \leq m$, and $y \in T^*$.
3. $x \Rightarrow_i^{\geq k\text{-comp.}} y$ if and only if there is a derivation $x = x_0 \Rightarrow_i x_1 \Rightarrow_i \cdots \Rightarrow_i x_{m-1} \Rightarrow_i x_m = y$ and it satisfies
   (a) $\mathrm{clev}_i(x_j) \geq k$ for $0 \leq j < m$ and $\mathrm{clev}_i(x_m) < k$, or
   (b) $\mathrm{clev}_i(x_0) \geq k$ and $y \in T^*$.

Let $D = \{ \leq k\text{-comp.}, = k\text{-comp.}, \geq k\text{-comp.} \mid k \geq 1 \}$ and let $\Rightarrow^f$ denote $\Rightarrow_i^f$ for some i, with $1 \leq i \leq n$, and $f \in D$. The reflexive transitive closure of $\Rightarrow^f$ is denoted by $\stackrel{*}{\Rightarrow}^f$. The language generated by G in the f-mode of derivation, with $f \in D$, is $L_f(G) = \{ w \in T^* \mid \alpha \stackrel{*}{\Rightarrow}^f w \}$. The family of languages generated by CDGSs working in the f-mode of derivation is denoted by $\mathcal{L}(\mathrm{CD}, \mathrm{CF}, f)$.

Example 1. Let $G = (N, T, \alpha, P_1, \ldots, P_8)$ be a CDGS with set of nonterminals $N = \{ A, A', B, B', C, D \}$, terminals $T = \{ a, b, c \}$, axiom AB, and components

$$P_1 = \{ A \to a A' b, B' \to B', C \to C \}, \qquad P_5 = \{ A' \to C, B \to B \},$$
$$P_2 = \{ A \to A, B \to B'c, C \to C \}, \qquad P_6 = \{ A \to A, A' \to A', B' \to D \},$$
$$P_3 = \{ A' \to A, B \to B, C \to C \}, \qquad P_7 = \{ B' \to B', C \to \lambda \}, \text{ and}$$
$$P_4 = \{ A' \to A', B' \to B, C \to C \}, \qquad P_8 = \{ D \to \lambda \}.$$

When working in the $\leq$1-comp.-mode or $=$1-comp.-mode of derivation, G generates the language $L(G) = \{\, a^n b^n c^n \mid n \geq 1 \}$. This can be seen as follows.

Starting from the axiom, the components P_1, P_3, P_5, and P_6 are all 1-competent. However, except for P_1, their application does not alter the axiom and hence these components remain 1-competent forever. In those cases the derivation thus enters a loop. From the axiom, the only two-step derivation that does not loop is thus $AB \Rightarrow_1^{=1\text{-comp.}} aA'bB \Rightarrow_2^{=1\text{-comp.}} aA'bB'c$. Consequently, a choice must be made. First we can apply P_5 to obtain the derivation $aA'bB'c \Rightarrow_5^{=1\text{-comp.}} aCbB'c \Rightarrow_6^{=1\text{-comp.}} aCbDc$, after which the derivation can be finished by $aCbDc \Rightarrow_7^{=1\text{-comp.}} abDc \Rightarrow_8^{=1\text{-comp.}} abc$ or instead by applying P_8 before P_7. Secondly, we can apply P_3 to obtain $aA'bB'c \Rightarrow_3^{=1\text{-comp.}} aAbB'c \Rightarrow_4^{=1\text{-comp.}} aAbBc$, after which this sequence of applications of P_1, P_2, P_3, and P_4 can be repeated $n-1$ times, for some $n \geq 1$, to obtain $a^n Ab^n Bc^n$. Subsequently, the derivation can be finished by $a^n Ab^n Bc^n \Rightarrow_1^{=1\text{-comp.}} a^n A'b^n Bc^n \Rightarrow_2^{=1\text{-comp.}} a^n A'b^n B'c^n \Rightarrow_5^{=1\text{-comp.}} a^n Cb^n B'c^n \Rightarrow_6^{=1\text{-comp.}} a^n Cb^n Dc^n \Rightarrow_7^{=1\text{-comp.}} a^n b^n Dc^n \Rightarrow_8^{=1\text{-comp.}} a^n b^n c^n$ or, instead, by interchanging the application of P_7 and P_8. Clearly, indeed the language $L_f(G) = \{\, a^n b^n c^n \mid n \geq 1 \,\}$, with $f \in \{\leq$1-comp., $=$1-comp.$\}$, is generated.

4 The Power of $\leq$ k- and $=$ k-Competence in CDGSs

It turns out that CDGSs working in the $\leq$1-comp.-mode or in the $=$1-comp.-mode are at least as powerful as forbidding random context grammars, but it remains an open problem to establish their exact computational power. Due to the lack of space the proof of the following theorem is left to the reader.

Theorem 1. *For $f \in \{\leq 1, =1\}$, $\mathcal{L}(\text{fRC}, \text{CF}) \subseteq \mathcal{L}(\text{CD}, \text{CF}, f\text{-comp.})$.*   □

Next we consider CDGSs working in the $=k$-comp.-mode, with $k \geq 2$. It turns out that already for $k = 2$, such CDGSs characterize the class of random context languages (and thus the class of recursively enumerable languages).

Theorem 2. *For $f \in \{\leq k, =k\}$ and $k \geq 2$, $\mathcal{L}(\text{CD}, \text{CF}, f\text{-comp.}) = \mathcal{L}(\text{RC}, \text{CF})$.*

Proof. The inclusions from left to right are rather obvious, since CDGSs working in any of the above competence modes can be simulated by a random context grammar. As a formal proof would be quite tedious, we leave the technical details to the reader. We now prove the inclusion from right to left for the $=k$ case, for $k \geq 2$. The $\leq k$ case can be proved in a similar way, but is left to the reader.

We first prove the case that $k = 2$, and then sketch the necessary modifications for the cases with $k > 2$. Let $G = (N, T, P, S)$ be a random context grammar in normal form[1] with rules $p : (A \rightarrow z, Q, R) \in P$, where Q is the per-

---

[1] A (forbidding) random context grammar $G = (N, T, P, S)$ is in *normal form* if for every (forbidding) random context rule $(A \rightarrow z, Q, R) \in P$, we have $A \notin Q \cup R$. It is easy to show that for every (forbidding) random context grammar G, there exists a (forbidding) random context grammar G' in normal form that generates the same language, i.e., $L(G') = L(G)$.

mitting context and R is the forbidding context of p. Note that the fact that G is in normal form implies that $A \notin Q \cup R$. Obviously, we can assume that $Q \cap R = \emptyset$.

To simulate G we construct a CDGS G' with nonterminals $N' = M \cup N \cup \{F, X, Y, Z\}$, where $M = \{\, [p],\ [p, B],\ [p, \neg C] \mid p : (A \rightarrow z, Q, R) \in P,\ B \in Q,\ C \in R \,\}$, such that the unions above are disjoint, the set of terminals T is disjoint from N', the axiom is SZ, and the components are as defined below.

For each random context rule $p : (A \rightarrow z, Q, R)$, we construct the components $\{P_{p,start}\} \cup \{\, P_{p,B}^{check} \mid B \in Q \,\} \cup \{\, P_{p,\neg C}^{check} \mid C \in R \,\} \cup \{P_{p,apply}\}$ described below. At any moment, we can see from the subscripts of these components which step of the simulation is performed: After we *start* simulating the application of the rule $p \in P$, we check the (non-)presence of the permitting (forbidding) symbols from the permitting context, Q (forbidding context, R) of this rule, or we *apply* the context-free production $A \rightarrow z$ of this rule. Next to these components we introduce below two more components, P_X and P_{YZ}. For now we assume that both Q and R are nonempty. The other cases will later be dealt with separately.

The idea of the simulation is the following. Before the simulation of the application of a random context rule, a marker from the set M is introduced in the sentential form. With the aid of this marker, we check the presence of the permitting and the non-presence of the forbidding symbols. First we check the presence of the permitting symbols starting with the first such symbol. The only component that is able to rewrite the leading marker is 1-competent whenever this permitting symbol is not present in the sentential form, and 2-competent whenever this symbol is present. The moment in which this component is applied it introduces the symbol X and by doing so becomes ≥ 3-competent. This procedure is repeated for all the remaining permitting symbols. Secondly, we check that no forbidding symbol is present. Again, we start with the first such forbidding symbol. This time the only component that is able to rewrite the leading marker is 3-competent whenever this forbidding symbol is present, in which case no successful derivation exists.

We now present more details of the construction. The simulation starts with the application of the component

$$P_{p,start} = \{A \rightarrow [p]XY\} \cup \{X \rightarrow F,\ Y \rightarrow F,\ Z \rightarrow Z\} \cup \{\, L \rightarrow F \mid L \in M \,\},$$

for some rule $p : (A \rightarrow z, Q, R) \in P$. This component introduces the leading marker $[p]$ indicating that we start to simulate p, which thus requires the presence of A. The moment in which one occurrence of A is rewritten, this component moreover becomes ≥ 3-competent, which thus guarantees that only one such an occurrence is rewritten. Since during the whole simulation the leading marker as well as the symbols Y and Z are present in the sentential form, no simulation of a rule different from p can be started once the simulation of p was started.

Before testing the presence of the permitting context Q of p, we have to remove X (or the derivation eventually is blocked) with the component

$$P_X = \{X \rightarrow \lambda,\ Z \rightarrow Z\}.$$

This component is 2-competent whenever X is present in the sentential form, and becomes 1-competent after it has erased X.

Let the permitting context of p be $Q = \{B_1, \ldots, B_{\ell_p}\}$. The simulation continues by applying, for all $2 \leq j \leq \ell_p$, the components

$$P_{p,B_1}^{check} = \{[p] \to [p, B_1]\} \cup \{B_1 \to B_1 X\} \cup \{X \to F\} \cup \{L \to F \mid L \in M\} \text{ and}$$

$$P_{p,B_j}^{check} = \{[p, B_{j-1}] \to [p, B_j]\} \cup \{B_j \to B_j X\} \cup \{X \to F\} \cup \{L \to F \mid L \in M\},$$

alternated with the component P_X erasing the X's inbetween.

The sequence of components thus applied is $P_{p,B_1}^{check}, P_X, \ldots, P_{p,B_{\ell_p}}^{check}, P_X$. Each such a component P_{p,B_j}^{check}, $1 \leq j \leq |Q|$, is 2-competent once it is applied due to the presence of the symbol B_j, which is in accordance with the fact that rule p can only be applied when this symbol from its permitting context is present. If this symbol is not present, then such a component is 1-competent due to the presence of the leading marker $[p, B_{j-1}]$ and thus not applicable. Note, moreover, that if the X was not removed and the symbol B_j is not present, then such a component would be 2-competent. In that case it could either replace the X by an F or replace the leading marker $[p, B_{j-1}]$ by $[p, B_j]$, remain 2-competent, and replace either the X or $[p, B_j]$ by an F. Also all these cases are in accordance with the permitting context of p.

Now the moment in which the production $B_j \to B_j X$ is applied, this component becomes ≥ 3-competent, and it has successfully tested the presence of B_j. Note that no derivation can be successful in case the leading marker $[p, B_{j-1}]$ is not replaced by the application of $[p, B_{j-1}] \to [p, B_j]$ *before* the application of $B_j \to B_j X$. We also note that in case there were more occurrences of B_j in the sentential form, then still only one occurrence is rewritten.

When we arrive at this point, we have thus successfully tested the presence of all the symbols from the permitting context of rule p. Hence we are ready to test the non-presence of all the symbols from its forbidding context. Let $p : (A \to z, Q, R)$, and let $R = \{C_1, \ldots, C_{m_p}\}$. The simulation continues with the application of the component

$$P_{p,\neg C_1}^{check} = \{[p, B_{\ell_p}] \to [p, \neg C_1]\} \cup \{C_1 \to F\}$$
$$\cup \{X \to F, \ Z \to Z\} \cup \{L \to F \mid L \in M \setminus \{[p, \neg C_1]\}\}.$$

Given the current sentential form, this component is 2-competent or 3-competent, depending on the presence of C_1 in the sentential form. If C_1 is present, then no successful derivation exists. This is in accordance with the fact that in that case the rule p cannot be applied due to the fact that the symbol C_1 from its forbidding context is present in the sentential form.

Subsequently we apply, for all $2 \leq k \leq m_p$, the components

$$P_{p,\neg C_k}^{check} = \{[p, \neg C_{k-1}] \to [p, \neg C_k]\} \cup \{C_k \to F\}$$
$$\cup \{X \to F, \ Z \to Z\} \cup \{L \to F \mid L \in M \setminus \{[p, \neg C_k]\}\}.$$

The sequence of components thus applied is $P_{p,\neg C_2}^{check}, \ldots, P_{p,\neg C_{m_p}}^{check}$. If along the way a symbol C_k, for $2 \leq k \leq |R|$, from the forbidding context of the rule p is present, then no successful derivation exists.

When we arrive at this point without having introduced an F, we have thus successfully tested also the non-presence of all the symbols from the forbidding context of rule p. Hence we are ready to actually simulate the application of the context-free production $A \to z$ of the rule p. Obviously, we need to do so only once, but this is guaranteed by the fact that there is only one occurrence of the leading marker $[p, \neg C_{m_p}]$. To this aim, we apply the 2-competent component

$$P_{p,apply} = \{[p, \neg C_{m_p}] \to z\} \cup \{Z \to Z\} \cup \{L \to F \mid L \in M\},$$

which becomes 1-competent as soon as $[p, C_{m_p}] \to z$ is applied, in which case we have successfully simulated the application of the rule p.

All that remains is to bring the sentential form back to a form from which the simulation of another rule from G can be started or to finish the derivation. This is done by removing Y. To this aim we apply the 2-competent component

$$P_{YZ} = \{Y \to \lambda, \ Z \to \lambda\}.$$

At this point it is important to note that eventually it is this component P_{YZ} that can finish the derivation by removing not only Y, but also Z. However, it can be seen that no successful derivation exists if Z is removed rather than Y.

If component P_{YZ} is applied earlier on in the derivation, then such an application would remove either Y or Z, but not both. For the same reason as above, no successful derivation exists if Z is removed. Now assume that Y is removed. Since the only use of Y is to guarantee that component P_{YZ} is 2-competent when we want to finish the derivation by removing both Y and Z, an earlier application of component P_{YZ} is harmless as long as it occurs *before* the final application of a rule from G. If, on the contrary, Y is removed after the final application of a rule from G, then component P_{YZ} can never become 2-competent, thus Z can never be removed, and no successful derivation exists.

Let us now describe how to adapt the construction for the cases dealing with a random context rule $p : (A \to z, Q, R)$ in which Q and/or R is empty. We distinguish three cases and describe only the components that must be changed:

(1) In case $R \neq Q = \emptyset$, we construct no components of the form P_{p,B_j}^{check} and replace production $[p, B_{\ell_p}] \to [p, \neg C_1]$ by $[p] \to [p, \neg C_1]$ in component $P_{p,\neg C_1}^{check}$.

(2) In case $Q \neq R = \emptyset$, we remove all components of the form $P_{p,\neg C_k}^{check}$ and replace production $[p, \neg C_{m_p}] \to z$ by $[p, B_{\ell_p}] \to z$ in component $P_{p,apply}$.

(3) In case $Q = R = \emptyset$, we construct no components of the form P_{p,B_j}^{check} or $P_{p,\neg C_k}^{check}$ and replace production $[p, \neg C_{m_p}] \to z$ by $[p] \to z$ in component $P_{p,apply}$.

The CDGS G' constructed above correctly simulates the random context grammar G and generates the language $L(G)$, when working in the $=2$-comp.-mode. This proves the statement of this theorem for the case $k = 2$.

Let us now briefly discuss the proof of the more general case. Let $k > 2$. We now sketch how to adapt G' such that the resulting CDGS G'' correctly simulates the random context grammar G and generates the language $L(G)$, when working in the $=k$-comp.-mode. We do not specify all the resulting modifications, but rather take one such component and show how it is adapted for inclusion in G''.

Recall that the component $P_{p,start}$ becomes ≥ 3-competent (and can thus no longer be applied) the moment in which production $A \to [p]XY$ is applied. The reason for this is as follows. The application of this production introduces each of the symbols $[p]$, X, and Y to the sentential form. Since the component moreover contains a production for each of these symbols, this immediately makes the component ≥ 3-competent. To make the simulation work in the $=k$-comp. mode for $k \geq 3$, we replace the component $P_{p,start}$ in G' by the component $P''_{p,start} = \{A \to [p]XYZ_1 \cdots Z_{k-2}\} \cup \{X \to F, \; Y \to F, \; Z \to Z\} \cup \{Z_1 \to F, \; \ldots, \; Z_{k-2} \to F\} \cup \{L \to F \mid L \in M\}$, where $Z_1, \ldots, Z_{k-2}$ are new symbols different from $N' \cup T$. We leave the other modifications to the reader. $\square$

5   The Power of $\geq$ k-Competence in CDGSs

We start our investigations with CDGSs working in the ≥ 1-comp.-mode. Since the ≥ 1-comp.-mode by definition equals the t-mode of derivation, as introduced in [3], we immediately obtain the following result, which is due to [3].

Theorem 3. $\mathcal{L}(\mathrm{CD}, \mathrm{CF}, \geq 1\text{-comp.}) = \mathcal{L}(\mathrm{ET0L})$. $\square$

Next we consider CDGSs working in the $\geq k$-comp.-mode, with $k \geq 2$. It turns out that already for $k = 2$, such CDGSs characterize the class of random context ET0L languages.

Theorem 4. *For* $k \geq 2$, $\mathcal{L}(\mathrm{CD}, \mathrm{CF}, \geq k\text{-comp.}) = \mathcal{L}(\mathrm{RC}, \mathrm{ET0L})$.

Proof. Here we prove the inclusion from right to left. To prove the reverse inclusion, for any CGDS working in the $\geq k$-comp.-mode of derivation a recurrent programmed grammar can be constructed that simulates it. The quite tedious details are left to the reader.

The construction we use is strongly based on the one used in the proof of Theorem 2, except that the test for forbidding symbols is now incorporated in the component applying the simulated productions. Again, we first prove the case that $k = 2$, and then sketch the necessary modifications for the cases with $k > 2$. Let $G = (\Sigma, H, \omega, \Delta, oc, noc)$ be a random context ET0L system in normal form[2]. Without loss of generality we can assume $oc(h) \cap noc(h) = \emptyset$ for every table $h \in H$. To simulate G we construct a CDGS G' with nonterminals $N' = M \cup N \cup \{B' \mid B \in oc(h), \; h \in H\} \cup \{F, X, Y\}$, where $M = \{[h,B]_1, [h,B]_2, [h,B]_3 \mid h \in$

---

[2] A random context ET0L system $G = (\Sigma, H, \omega, \Delta, oc, noc)$ is in *normal form* if every table $h \in H$ is of the form $\{B \to B \mid B \in \Sigma \setminus \{A\}\} \cup h_A$, where $h_A = \{A \to z, \; A \to A \mid A \in \Sigma, z \in \Sigma^*, z \neq A\}$ or $h_A = \{A \to z \mid A \in \Sigma, z \in \Sigma^*, z \neq A\}$, and $A \notin oc(h) \cup noc(h)$. If table h is of the form $\{B \to B \mid B \in \Sigma \setminus \{A\}\} \cup h_A$ for some $A \in \Sigma$, then A is called the *active symbol* of h and $A \to z$ in h_A the *active production* of h. By standard constructions one can show that for every random context ET0L system G, there exists a random context ET0L system G' in normal form that generates the same language, i.e., $L(G') = L(G)$.

H, $B \in oc(h)$}, such that the unions are disjoint, terminals T disjoint from N', axiom XYS, and the components defined below.

For each table $h \in H$, with $oc(h) = \{B_1, \ldots, B_{\ell_h}\}$, we construct the components $\{P_{h,B_j,1}, P_{h,B_j,2}, P_{h,B_j,3} \mid 1 \leq j \leq \ell_h\} \cup \{P_{h,apply}\}$ described below. We also introduce one more component, P_{finish}. For now we assume that both $oc(h)$ and $noc(h)$ are nonempty. The other cases will later be dealt with separately.

The idea of the simulation is similar to that of the proof of Theorem 2. We now describe more details of the construction. Let $h \in H$ be a table of G and let $oc(h) = \{B_1, \ldots, B_{\ell_h}\}$. The simulation of h starts by applying the component

$$P_{h,B_1,1} = \{A \to [h, B_1]_1 \mid A \text{ is the active symbol of } h\}$$
$$\cup \{[h, B_1]_1 \to [h, B_1]_1, \ B_1 \to B_1'\} \cup \{L \to F \mid L \in M\}.$$

This component can be applied if and only if both A, the active symbol of h, and $B_1 \in oc(h)$ are present, after which it remains ≥ 2-competent until all occurrences of B_1 have been primed. Moreover, we shall shortly see that no successful derivation exists unless all occurrences of A have been replaced by $[h, B_1]_1$.

Consequently, the component

$$P_{h,B_1,2} = \{[h, B_1]_1 \to [h, B_1]_2\} \cup \{B_1' \to B_1, \ B_1 \to B_1\} \cup \{A \to F \mid$$
$$A \text{ is the active symbol of } h\} \cup \{L \to F \mid L \in M \setminus \{[h, B_1]_2\}\}$$

remains ≥ 2-competent until all occurrences of $[A_i, p_i, 1]_1$ have been replaced by $[h, B_1]_2$. Moreover, we shall shortly see that no successful derivation exists unless at least one occurrence of B_1' is unprimed. However, due to the presence of $B_1 \to B_1$ this means that this component remains ≥ 2-competent until all occurrences of B_1' are unprimed. Since this is the only component capable of unpriming B_1', it is this component that guarantees that no successful derivation exists if component $P_{h,B_1,1}$ has not replaced all occurrences of A by $[h, B_1]_1$.

The component which guarantees that no successful derivation exists if component $P_{h,B_1,2}$ has not unprimed all occurrences of B_1' is

$$P_{h,B_1,3} = \{[h, B_1]_2 \to [h, B_1]_3\} \cup \{B_1 \to B_1\} \cup \{L \to F \mid L \in M \setminus \{[h, B_1]_3\}\}.$$

This component is ≥ 2-competent if and only if B_1 is present. Since this is the only component replacing $[h, B_1]_2$ by $[h, B_1]_3$, no successful derivation exists if this component is not applied.

Now that we have successfully tested the presence of the first permitting symbol, the simulation continues by doing the same for the remaining permitting symbols, i.e., by applying, for all $2 \leq j \leq \ell_h$, the components

$$P_{h,B_j,1} = \{[h, B_{j-1}]_3 \to [h, B_j]_1, \ [h, B_j]_1 \to [h, B_j]_1\} \cup \{B_j \to B_j'\}$$
$$\cup \{L \to F \mid L \in M\},$$

$$P_{h,B_j,2} = \{[h, B_j]_1 \to [h, B_j]_2\} \cup \{B_j' \to B_j, \ B_j \to B_j\} \cup \{A \to F \mid$$
$$A \text{ is the active symbol of } h\} \cup \{L \to F \mid L \in M \setminus \{[h, B_j]_2\}\}, \text{ and}$$

$$P_{h,B_j,3} = \{[h, B_j]_2 \to [h, B_j]_3\} \cup \{B_j \to B_j\} \cup \{L \to F \mid L \in M \setminus \{[h, B_j]_3\}\}.$$

Hence, the sequence of components applied is $P_{h,B_2,1}$, $P_{h,B_2,2}$, $P_{h,B_2,3}$, $\ldots$, $P_{h,B_{\ell_h},1}$, $P_{h,B_{\ell_h},2}$, $P_{h,B_{\ell_h},3}$. If along the way a permitting symbol B_j, for $2 \leq j \leq \ell_h$, is not present, then the derivation is blocked due to the fact that in that case component $P_{h,B_j,1}$ is 1-competent and thus cannot be applied.

When we arrive at this point, we have thus successfully tested the presence of all the symbols from the permitting context of table h. Hence we are ready to test the non-presence of all the symbols from its forbidding context and to subsequently simulate the application of its active production $A \to z$ by replacing some of the occurrences of $[h, B_{\ell_h}]_3$ by z (and the remaining occurrences by A). The simulation continues with the application of the component

$$P_{h,apply} = \{[h, B_{\ell_h}]_3 \to z, \, [h, B_{\ell_h}]_3 \to A \mid A \to z \text{ is the active production of } h\}$$
$$\cup \{X \to X\} \cup \{C \to F \mid C \in noc(h)\} \cup \{L \to F \mid L \in M\}.$$

In this component we use the extra marker X to guarantee its ≥ 2-competence whenever all the permitting symbols are present. In case any forbidding symbol $C \in noc(h)$ is present, then a failure symbol F must be introduced or else $P_{h,apply}$ remains ≥ 2-competent. This is in accordance with the fact that in that case no active production from table h can be applied due to the fact that a symbol from its forbidding context is present in the sentential form. Hence we have successfully applied table $h \in H$ and the sentential form is in a form from which the simulation of another table from G can be started.

It remains to erase the symbols X and Y from the sentential form as soon as a successful derivation of a terminal word in G has been simulated, i.e., when the sentential form is XYw, for some $w \in T^*$. This is achieved by the component

$$P_{finish} = \{X \to \lambda, \, Y \to \lambda\}.$$

Note that both X and Y are erased by this component if and only if it is applied to a sentential form XYw, for some $w \in T^*$. In all other cases, only one of these symbols is erased because this component becomes 1-competent the moment this happens. Since neither of these symbols can be rewritten by any component other than P_{finish}, no successful derivation exists if this component is applied to a sentential form that is not of the form XYw with $w \in T^*$.

Similar to the way we did this in the proof of Theorem 2, our construction can easily be adapted for tables $h \in H$, where $oc(h)$ and/or $noc(h)$ is empty.

This completes the description of the CDGS G'. It is left to the reader to verify that whenever components are applied in an order different from the one prescribed above by the leading marker, then no successful derivation exists. This is achieved by the inclusion of productions in components which guarantee—where necessary—that a failure symbol F (which can never be rewritten) is introduced, or that the derivation is blocked because no more component can be applied. Both of these cases clearly block the derivation. The CDGS G' constructed above correctly simulates the random context ET0L system G and generates the language $L(G)$, when working in the ≥ 2-comp.-mode. This proves the statement of this theorem for the case $k = 2$. The proof of the more general case is rather straightforward and it is thus left to the reader.   □

6   Conclusion

In this paper we have introduced the $\leq k$-comp.-, $= k$-comp.-, and $\geq k$-comp.-mode of derivation, with $k \geq 1$, as cooperation protocols for CDGSs. They enable a component of a CDGS to rewrite a sentential form only if it is at most, exactly, or at least k-competent, resp., on that string. CDGSs working in the ≤ 2-comp.-or $= 2$-comp.-mode of derivation characterize the class of recursively enumerable languages, while those working in the ≥ 2-comp.-mode of derivation character-ize the class of random context ET0L languages, which in turn equals the class of recurrent programmed languages with appearance checking [11], that is ob-viously included in $\mathcal{L}(\text{RE})$, but it is not known whether it is strictly included or not. In Theorem 4 we provide yet another alternative characterization of this language class, which thus might shed new light on this longstanding open problem.

Finally, the components of the CDGSs used in the proofs in this paper are very simple grammars, with only a limited number of productions. The results of this paper thus demonstrate that cooperating agents with a rather restricted level of competence are able to solve arbitrarily complicated problems. Furthermore, if these agents are represented by context-free grammars, then there is no difference between the cases in which each agent has an exact level of competence and those in which it has a bounded level of competence—compare with [9].

References

1. H. Bordihn and E. Csuhaj-Varjú, On competence and completeness in CD grammar systems. *Acta Cybernetica* 12, 4 (1996), 347–361.
2. H. Bordihn and M. Holzer. Grammar systems with negated conditions in their cooperation protocols. *JUCS* 6 (2000), 1165–1184.
3. E. Csuhaj-Varjú and J. Dassow, On cooperating distributed grammar systems. *EIK* 26 (1990), 49–63.
4. E. Csuhaj-Varjú, J. Dassow, and M. Holzer, On a Competence-based Cooper-ation Strategy in CD Grammar Systems. Technical Report 2004/3, Theoretical Computer Science Research Group, Computer and Automation Research Insti-tute, Hungarian Academy of Sciences, 2004.
5. E. Csuhaj-Varjú, J. Dassow, and M. Holzer. CD Grammar Systems with Com-petence Based Entry Conditions in Their Cooperation Protocols. Technical Re-port 2004/4, Theoretical Computer Science Research Group, Computer and Au-tomation Research Institute, Hungarian Academy of Sciences, 2004.
6. E. Csuhaj-Varjú, J. Dassow, J. Kelemen, and Gh. Păun, *Grammar Systems— A Grammatical Approach to Distribution and Cooperation*, Gordon and Breach, 1994.

7. J. Dassow and Gh. Păun, *Regulated Rewriting in Formal Language Theory*, *EATCS Monographs on Theoretical Computer Science* 18, Springer-Verlag, Berlin, 1989.
8. H. Fernau and D. Wätjen, Remarks on regulated limited ET0L systems and regulated context-free grammars. *TCS* 194 (1998), 35–55.
9. R. Meersman and G. Rozenberg, Cooperating grammar systems. In *Proceedings MFCS'78*, *LNCS* 64, Springer-Verlag, Berlin, 1978, 364–374.
10. Gh. Păun and G. Rozenberg, Prescribed teams of grammars. *Acta Informatica* 31 (1994), 525–537.
11. S.H. von Solms. Some notes on ET0L-languages. *IJCM* 5 (1976), 285–296.

The Dot-Depth and the Polynomial Hierarchy Correspond on the Delta Levels

Bernd Borchert[1], Klaus-Jörn Lange[1], Frank Stephan[2], and Pascal Tesson[1], and Denis Thérien[3]

[1] Universität Tübingen, Germany
{borchert, lange, tesson}@informatik.uni-tuebingen.de
[2] National University of Singapore
fstephan@comp.nus.edu.sg
[3] McGill University, Montréal, Canada
denis@cs.mcgill.ca

Abstract. The leaf-language mechanism associates a complexity class to a class of regular languages. It is well-known that the Σ_k- and Π_k-levels of the dot-depth hierarchy and the polynomial hierarchy correspond in this formalism. We extend this correspondence to the Δ_k-levels of these hierarchies: $\text{Leaf}^P(\Delta_k^L) = \Delta_k^p$. These results are obtained in part by relating operators on varieties of languages to operators on the corresponding complexity classes.

1 Introduction

The leaf-language mechanism associates a complexity class to any class of languages. It is well-known that the Σ_k- and Π_k-levels of the dot-depth hierarchy and the polynomial hierarchy correspond via leaf languages, i.e. for all $k \geq 1$ it holds:

$$\text{Leaf}^P(\Sigma_k^L) = \Sigma_k^p \quad \text{and} \quad \text{Leaf}^P(\Pi_k^L) = \Pi_k^p.$$

This was shown by Burtschick & Vollmer [BV98]. As an immediate consequence the class of all starfree regular languages $\mathcal{SF}$ and the polynomial hierarchy correspond via leaf languages: $\text{Leaf}^P(\mathcal{SF}) = \text{PH}$. Furthermore, the k-th full level $\mathcal{DD}_k$ of the dot-depth hierarchy (the Boolean closure of Σ_k^L) and the Boolean closure of Σ_k^p (for $k = 1$ called the *Boolean hierarchy over* NP) correspond via leaf languages, i.e. $\text{Leaf}^P(\mathcal{DD}_k) = \text{BC}(\Sigma_k^p)$ – results due originally to Hertrampf et al. [HL*93]. Schmitz, Wagner and Selivanov [ScW98, Sel02] further obtained correspondences between the classes of the Boolean hierarchies defined over the respective Σ_k classes.

In this paper, we extend the correspondence of the dot-depth hierarchy and the polynomial hierarchy to the Δ_k-levels of the two hierarchies. For the dot-depth hierarchy, they are the intersections of the corresponding Σ_k- and Π_k-levels and for the polynomial hierarchy they are the polynomial-time Turing reducibility closure of the Σ_{k-1}-class. For level 2 the correspondence $\text{Leaf}^P(\Delta_2^L) = \Delta_2^p$ was already shown in the unpublished manuscript [BSS99].

C.S. Calude, E. Calude and M.J. Dinneen (Eds.): DLT 2004, LNCS 3340, pp. 89–101, 2004.
© Springer-Verlag Berlin Heidelberg 2004

The proof of this result used in one direction Schützenberger's characterization of Δ_L^2 as unambiguous products [Sch76] and, for the other containment, a method of Wagner [Wa90] showing the Δ_2^p-completeness of ODD MAX SAT.

The main result of the present paper is more general, showing that for all $k \geq 2$:

$$\mathrm{Leaf}^{\mathrm{P}}(\Delta_k^L) = \Delta_k^p.$$

A key step is to show that the complexity class captured via leaf languages by the unambiguous polynomial closure of a variety $\mathcal{V}$ of languages is contained in the Turing-reducibility closure of the class captured by $\mathcal{V}$, i.e. $\mathrm{Leaf}^{\mathrm{P}}(\mathrm{UPol}(\mathcal{V})) \subseteq \mathrm{T} \cdot \mathrm{Leaf}^{\mathrm{P}}(\mathcal{V})$. This makes use of the characterization of the UPol operator of Pin, Straubing and Thérien [PST88].

In fact, we argue that our methods yield a much more general result, namely that under some technical assumptions we have:

$$\mathrm{Leaf}^{\mathrm{P}}(\mathrm{UPol}(\mathrm{BPol}(\mathcal{V}))) = \mathrm{T} \cdot \exists \cdot \mathrm{Leaf}^{\mathrm{P}}(\mathcal{V})$$

where BPol denotes the Boolean polynomial closure operator. In other words, we can relate purely language theoretic operators to operators on complexity classes. This is particularly significant in light of the links between algebraic automata theory and computational complexity first uncovered by Barrington and Thérien [BT88]. The generality of our result also allows us to shed new light on a conjecture of Straubing and Thérien concerning regular languages K such that $\mathrm{Leaf}^{\mathrm{P}}(K)$ contains the complexity class BPP [ST03].

The paper is organized as follows. In Section 2 we recall the definition of the dot-depth hierarchy, and of the related operators Pol and UPol. In Section 3 we review the notion of leaf languages and the polynomial hierarchy. We establish the main result about the Δ-classes in Section 4. Finally, we present in Section 5 two general theorems relating operators on varieties with operators on complexity classes and discuss their main consequences. An extended version of this paper is available on one of the authors' web page: `www.cs.mcgill.ca/~ptesso`.

2   The Dot-Depth Hierarchy

A *class of languages* $\mathcal{L}$ is a mapping which assigns to each alphabet Σ a set of languages $\Sigma^*\mathcal{L}$ over Σ. We will write $L \in \mathcal{L}$ as an abbreviation for $L \in \Sigma^*\mathcal{L}$ for some alphabet Σ.

Let a class of languages $\mathcal{L}$ be given. The *polynomial closure*[1] $\mathrm{Pol}(\mathcal{L})$ is the class of languages consisting, for every alphabet Σ, of the finite unions of *marked products* of languages from $\Sigma^*\mathcal{L}$, i.e. languages $L = L_0 a_1 L_1 \cdots a_n L_n$ such that the L_i are languages from $\Sigma^*\mathcal{L}$ and the a_i are letters from Σ. We further say that the product $L = L_0 a_1 L_1 \cdots a_n L_n$ is *unambiguous* if for every word w in L there is

---

[1] The terminology stems from the interpretation of concatenation and union as multiplication and addition respectively in the semiring Σ^*.

a unique factorization $w = w_0 a_1 w_1 \cdots a_n w_n$ with $w_i \in L_i$. We denote as $\mathrm{UPol}(\mathcal{L})$ the class of finite disjoint unions of unambiguous marked products of languages of $\mathcal{L}$. The classes of languages Co-$\mathcal{L}$ and B$\mathcal{L}$ are defined, for every alphabet Σ, as the set of complements of languages from $\Sigma^* \mathcal{L}$ and as the Boolean closure of $\Sigma^* \mathcal{L}$, respectively. Co-Pol and BPol are defined as the combined operators Co- $\circ$ Pol and B $\circ$ Pol, respectively.

Let $\mathcal{I}$ be set the class of languages which consists for every alphabet Σ only of the two languages $\emptyset$ and Σ^*. The classes of the dot-depth hierarchy are defined as the following classes of regular languages:

Definition 1 (Dot-Depth Hierarchy).

(a) $\Sigma_0^L := \Pi_0^L := \Delta_0^L := \mathcal{DD}_0 := \mathcal{I}$ (d) $\mathcal{DD}_{k+1} := \mathrm{BPol}(\mathcal{DD}_k)$

(b) $\Sigma_{k+1}^L := \mathrm{Pol}(\mathcal{DD}_k)$ (e) $\Delta_{k+1}^L := \Sigma_{k+1}^L \cap \Pi_{k+1}^L$

(c) $\Pi_{k+1}^L := \mathrm{Co\text{-}Pol}(\mathcal{DD}_k)$ (f) $\mathcal{SF} := \bigcup_{i \geq 0} \Sigma_i^L$

It should be noted that the dot-depth hierarchy presented here is sometimes known as the Straubing-Thérien hierarchy. Other papers use the term referring to the closely related Cohen-Brzozowski hierarchy which is defined analogously: level 0 of the Cohen-Brzozowski hierarchy further includes the so-called generalized definite languages. Although the hierarchies thus defined do not coincide, our results can also be obtained for the levels of the Cohen-Brzozowski hierarchy.

Results of Pin and Weil [PW97] show that $\Sigma_{k+1}^L \cap \Pi_{k+1}^L = \mathrm{UPol}(\mathcal{DD}_k)$ for all $k \geq 0$. Therefore the line (e) in the above definition could be equivalently given in terms of the UPol operator. It is known that the dot-depth hierarchy is infinite, i.e. all classes $\Sigma_k^L, \Pi_k^L, \mathcal{DD}_k, \Delta_k^L$ for $k \geq 1$ are all different [St94]. Their union $\mathcal{SF}$ is the class of *star-free* languages.

A *variety of languages* $\mathcal{V}$ is a class of regular languages closed under Boolean operations (i.e. each $\Sigma^* \mathcal{V}$ is closed under finite Boolean operations), left and right quotients (for any u in Σ^* and L in $\Sigma^* \mathcal{V}$ the languages $u^{-1}L = \{w \mid uw \in L\}$ and $Lu^{-1} = \{w \mid wu \in L\}$ are also in $\Sigma^* \mathcal{V}$) and inverse homomorphic images (if Γ, Σ are alphabets and h is a homomorphism from Γ^* to Σ^* then L in $\Sigma^* \mathcal{V}$ implies $h^{-1}(L) = \{x \in \Gamma^* \mid h(x) \in L\}$ in $\Gamma * \mathcal{V}$). A positive variety is defined similarly but the closure under complement is not required. Varieties have been established as a most natural unit of classification of regular languages. Examples of varieties of languages include $\mathcal{DD}_k$ and Δ_k^L for $k \geq 0$, and $\mathcal{SF}$ while Σ_k^L and Π_k^L form only positive varieties.

The class $\mathcal{I}$ defined above is called the *trivial* variety and is contained in every variety (and every positive variety) of languages.

3   Leaf Languages and the Polynomial Hierarchy

Let P (NP) be the set of languages computable by a Turing machine in deterministic (nondeterministic) polynomial time. For a complexity class $\mathcal{C}$ let $\mathrm{T} \cdot \mathcal{C}$ be the set of languages computable in deterministic polynomial time via an oracle

Turing machine (see e.g. [Pa94]) which uses a language from $\mathcal{C}$ as an oracle[2]. Let $\exists \cdot \mathcal{C}$ be the set of all languages L such that

$$L = \{x \mid \text{ there exists } y \text{ with } |y| \leq q(|x|) \text{ such that } \langle x, y \rangle \in A\}$$

for some language $A \in \mathcal{C}$ and some polynomial q. Similarly, for an integer p, the class $\text{Mod}_p \cdot \mathcal{C}$ denotes the set of languages L such that

$$L = \{x \mid \text{ there exists } 0 \pmod{p} \ y \text{ with } |y| \leq q(|x|) \text{ and such that } \langle x, y \rangle \in A\}$$

for some language $A \in \mathcal{C}$ and some polynomial q. Let co $\cdot \mathcal{C}$ be the set of complements of $\mathcal{C}$ and $\text{BC} \cdot \mathcal{C}$ be the Boolean closure of $\mathcal{C}$. Using this operator notation one can write for example $P = T \cdot \emptyset$, $NP = \exists \cdot P$ and $\Delta_2^p = T \cdot NP$. The classes of the polynomial hierarchy are defined as follows [Pa94]:

Definition 2 (Polynomial Hierarchy). *Let $k \geq 1$.*

(a) $\Sigma_0^p := \Pi_0^p := \Delta_0^p := P$

(b) $\Sigma_{k+1}^p := \exists \cdot \Pi_k^p$

(c) $\Pi_{k+1}^p := \text{co} \cdot \Sigma_{k+1}^p$

(d) $\Delta_{k+1}^p := T \cdot \Sigma_k^p$

(e) $PH := \bigcup_{i \geq 0} \Sigma_{k+1}^p$

Note that $NP = \Sigma_1^p$ and co-$NP = \Pi_1^p$.

It is not known whether the polynomial hierarchy is in fact infinite. Nevertheless, there exists an oracle relative to which all classes Σ_k^p, Π_k^p and Δ_k for $k \geq 1$ are different from one another [Yao85]. Moreover, in this relativized world, Δ_k^p is different from $\text{BC}(\Sigma_{k-1}^p)$ because Δ_k^p still has a $\leq_m^p$-complete language while by the (relativizable) results of Kadin [Ka88] the class $\text{BC}(\Sigma_{k-1}^p)$ does not. For $k = 1$ and $k = 2$ oracles separating Δ_k^p from its superset $\Sigma_k^p \cap \Pi_k^p$ were constructed in [BGS75] and [He84], respectively, but for larger k the authors could not find a construction in the literature. On the other hand, any PSPACE-complete oracle A collapses all these classes to P.

Let some language L over some alphabet Σ be given. The *leaf language* approach [BCS92, HL*93, BKS99] assigns to the language L a class $\text{Leaf}^P(L)$ of languages on the alphabet $\{0, 1\}$ the following way. Let some nondeterministic polynomial-time Turing machine N be given. Assume that it not only accepts or rejects on every computation path but outputs a letter from the alphabet Σ on each computation path when it terminates. N produces for every input $x \in \{0, 1\}^*$ a computation tree (not necessary balanced) whose paths are ordered in the natural way and whose leaves are labeled by letters from Σ. Therefore the letters on the leaves form a word over the alphabet Σ which we call the leafstring(N, x) or the yield of the computation tree. Let for each N the language $\text{Leaf}^N(L) \in \{0, 1\}^*$ be the set of inputs x such that leafstring(N, x) is in L, and let $\text{Leaf}^P(L)$ be the set of languages $\text{Leaf}^N(L)$ for some N. As an example note that if S_1 is the language $\{a, b\}^* a \{a, b\}^*$ over alphabet $\{a, b\}$ then

---

[2] This class is often denoted $P(\mathcal{C})$, $P^{\mathcal{C}}$ or $\leq_T^p(\mathcal{C})$.

$\mathrm{Leaf}^{\mathrm{P}}(S_1) = \mathrm{NP}$. For a class $\mathcal{V}$ of languages let $\mathrm{Leaf}^{\mathrm{P}}(\mathcal{V})$ be[3] the union of the classes $\mathrm{Leaf}^{\mathrm{P}}(L)$ for $L \in \mathcal{V}$. Note that all classes $\mathrm{Leaf}^{\mathrm{P}}(L)$ and $\mathrm{Leaf}^{\mathrm{P}}(\mathcal{V})$ are by definition subsets of the set of languages over the alphabet $\{0,1\}$. We will call such classes *complexity classes*. Throughout this paper, we use $\mathcal{C}$ to denote some complexity class and $\mathcal{V}$ to denote a variety.

Proposition 1. *Let $\mathcal{C}$ be a complexity class and $\mathcal{V}$ be a (positive) variety of languages.*

(a) $\mathrm{T} \cdot \mathrm{T} \cdot \mathcal{C} = \mathrm{T} \cdot \mathcal{C}$.
(b) *If $\mathcal{C} \subseteq \mathrm{T} \cdot \mathrm{Leaf}^{\mathrm{P}}(\mathcal{V})$ then $\mathrm{BC} \cdot \mathcal{C} \subseteq \mathrm{T} \cdot \mathrm{Leaf}^{\mathrm{P}}(\mathcal{V})$.*

4   Correspondence of the Δ Levels of the Hierarchies

In this section, we establish our main theorem and show that the Δ_k-level of the dot-depth hierarchy corresponds via leaf languages to the Δ_k-level of the polynomial hierarchy.

We will use the following characterization of the UPol operator. Let a variety of languages $\mathcal{V}$ be given: $\ell_1 \square \mathcal{V}$ is the Boolean closure of languages[4] of *unambiguous* products $L_0 a L_1$ where L_0, L_1 are in $\Sigma^* \mathcal{V}$ and $a \in \Sigma$. By [PST88], $\ell_1 \square \mathcal{V}$ is itself a variety of languages and this operator characterizes $\mathrm{UPol}(\mathcal{V})$:

Theorem 1 ([PST88]). *Let $\mathcal{V}$ be a variety of languages. $\mathrm{UPol}(\mathcal{V})$ is the class of languages obtained from $\mathcal{V}$ by finitely many applications of the operator $\ell_1 \square$.*

We first show that one application of this $\ell_1 \square$ can basically be simulated by a polynomial-time Turing reduction.

Lemma 1. *Let $\mathcal{V}$ be a variety of languages:*

$$\mathrm{Leaf}^{\mathrm{P}}(\ell_1 \square \mathcal{V}) \subseteq \mathrm{T} \cdot \mathrm{Leaf}^{\mathrm{P}}(\mathcal{V}).$$

Proof (Sketch). We recall the notion of syntactic congruence of a language $L \subseteq \Sigma^*$: Let $x \equiv_L y$ if for any $u, v \in \Sigma^*$ we have $uxv \in L$ iff $uyv \in L$. It is well-known that the syntactic congruence of L has finite index if and only if L is regular. Assume now that $L_0 a L_1$ is an unambiguous concatenation with L_0, L_1 in $\Sigma^* \mathcal{V}$.

For a word $x \in \Sigma^*$, let $[x]_{L_0}$ and $[x]_{L_1}$ denote the $\equiv_{L_0}$ and $\equiv_{L_1}$ equivalence classes of x respectively. Following [PST88], we let G be the graph containing:

- For any $x, y \in \Sigma^*$, a vertex labeled with the pair $([x]_{L_0}, [y]_{L_1})$: since L_0 and L_1 are regular, there are only finitely many vertices.
- For any letter $b \in \Sigma$ edges from $([x]_{L_0}, [by]_{L_1})$ to $([xb]_{L_0}, [y]_{L_1})$. We label these edges with triples $\langle [x]_{L_0}, b, [y]_{L_1} \rangle$.

---

[3] For the trivial variety, however, we define for technical reasons $\mathrm{Leaf}^{\mathrm{P}}(\mathcal{I}) = \mathrm{P}$.
[4] The symbols $\square$ and ℓ_1 actually have a meaning as block product and the set of locally trivial categories, respectively, see [PST88].

Note that every $x \in \Sigma^*$ traces a path in this graph from vertex $([\epsilon]_{L_0}, [x]_{L_1})$ to vertex $([x]_{L_0}, [\epsilon]_{L_1})$. Furthermore, if $x = x_0 b x_1$ where b is the jth letter of b then the jth edge of this path is $\langle [x_0]_{L_0}, b, [x_1]_{L_1} \rangle$. Call an edge *critical* if it is of the form $\langle [x_0]_{L_0}, a, [x_1]_{L_1} \rangle$ with $x_0 \in L_0$ and $x_1 \in L_1$: we have $x \in L_0 a L_1$ if and only if the path traced out by x in G contains such a critical edge. Because $L_0 a L_1$ is unambigous however, any path in G contains at most one critical edge (as first observed by [PST88]). In particular, if $\langle [x]_{L_0}, b, [y]_{L_1} \rangle$ is some edge in G then either no path with a critical edge ends at vertex $([x]_{L_0}, [by]_{L_1})$ or no path with a critical edge starts at vertex $([xb]_{L_0}, [y]_{L_1})$.

Let N be a polynomial-time NDTM: we exhibit a polynomial-time algorithm D which, using queries to a $\text{Leaf}^P(\mathcal{V})$ oracle, can test whether the leafstring x produced by N on a given input t lies in $L_0 a L_1$. The algorithm will check if the path traced by x in G contains a critical edge. Suppose for simplicity that on any input t the computation tree of N is complete and of depth $q(|t|)$ for some polynomial q. Thus, x has length $2^{q(|t|)}$. For any $1 \leq k \leq 2^{q(|t|)}$ let b_k denote the kth letter in x and p_{k-1} and s_{k+1} the words such that $x = p_{k-1} b_k s_{k+1}$.

Lemma 2. *For any polynomial-time NDTM N and any $L_0, L_1 \in \Sigma^* \mathcal{V}$ for some variety $\mathcal{V}$, there exists a polynomial-time DTM querying a $\text{Leaf}^P(\mathcal{V})$ oracle which given $k \leq 2^{q(|t|)}$ computes the triple $\langle [p_{k-1}]_{L_0}, b_k, [s_{k+1}]_{L_1} \rangle$ defined above.*

The proof of this lemma which can be found in the full report relies on the simple observation that computing the $\equiv_{L_0}$-class of x is equivalent to checking membership of x in a finite number of languages of the form $u^{-1} L_0 v^{-1}$, all of which also lie in $\mathcal{V}$. We also need to argue that DTM working with m different oracles in $\text{Leaf}^P(\mathcal{V})$ can be simulated by a DTM having access to a single oracle in $\text{Leaf}^P(\mathcal{V})$.

To check if the path traced out by x contains a critical edge, our algorithm uses binary search to zoom in on a segment r of the path where that edge might lie if it exists at all. Initially, we set $x = r$. Now let $k \leq 2^{q(|t|)}$ be the middle position of the segment r. Using Lemma 2, we compute $\langle [p_{k-1}]_{L_0}, b_k, [s_{k+1}]_{L_1} \rangle$: If this edge is critical then we know that x is in $L_0 a L_1$ and we can stop. Otherwise, we consider two cases:

- if no path with a critical edge ends at vertex $([p_{k-1}]_{L_0}, [b_k s_{k+1}]_{L_1})$ then any critical edge in the path traced out by x must occur at some later $k > 2^{q(|t|)-1}$ and we continue our search on the rightmost half of r.
- symmetrically if no path with a critical edge starts from $([p_{k-1} b_k]_{L_0}, [s_{k+1}]_{L_1})$ we must continue our search for a critical edge in the first half of r.

As we noted above the unambiguity of $L_0 a L_1$ ensures that at least one of the above holds for any k. If *both* of the above hold then x cannot lie in $L_0 a L_1$ and the algorithm rejects. After $q(|t|)$ steps of this binary search the segment r in which the critical edge might lie consists of a single edge and we can check if it is indeed critical and reject if it is not. So we have shown $\text{Leaf}^P(\mathcal{U}) \subseteq \text{T} \cdot \text{Leaf}^P(\mathcal{V})$ for the set $\mathcal{U}$ of unambiguous concatenations of the form $L = L_0 a L_1$ with $L_0, L_1 \in \mathcal{V}$.

Therefore:

$$\text{Leaf}^P(\ell_1 \Box \mathcal{V}) = \text{Leaf}^P(B\mathcal{U}) \subseteq BC \cdot \text{Leaf}^P(\mathcal{U}) \subseteq T \cdot \text{Leaf}^P(\mathcal{V})$$

by Proposition 1 (b).

Now combining Theorem 1 with the idempotency of the Turing closure operator T· stated in Proposition 1(a), we can obtain:

Corollary 1. *Let $\mathcal{V}$ be a variety of languages.*
$\text{Leaf}^P(\text{UPol}(\mathcal{V})) \subseteq T \cdot \text{Leaf}^P(\mathcal{V})$.

The authors do not know whether the reverse inclusion in the above Corollary 1 holds in general, i.e. whether the UPol operator on varieties corresponds to the Turing reducibility closure on the complexity classes. Nevertheless, Theorem 3 below states that, under certain conditions, a correspondence holds at least for the combined operator UPol ∘ BPol on language classes and the combined operator T · ∃· on complexity classes.

We say that a language A is *polynomial-time conjunctively reducible* to a language B if there is a polynomial-time oracle Turing machine for A which on input x asks membership questions to B and accepts if and only if all questions are answered positively. Note that the questions can be assumed to be non-adaptive, i.e. the oracle Turing machine produces a (polynomially long) list of questions which are asked all at once. A complexity class $\mathcal{C}$ is *closed under polynomial-time conjunctive reductions* if every language which is polynomial-time conjunctively reducible to a language in $\mathcal{C}$ is also in $\mathcal{C}$. Of course every complexity class closed under polynomial-time Turing reductions is also closed under polynomial-time conjunctive reductions. Therefore, the Δ_k^p classes have this property. Other classes have this property and in particular Σ_k^p and Π_k^p (via two different arguments).

For $i \geq 2$, we define inductively a sequence of languages $D_i \subseteq \Sigma_i^*$ where $|\Sigma_i| = 2i-1$. First $D_2 = 0^* a_2 \{0, a_2, b_2\}^*$. For $k \geq 3$, let first $\Sigma_k = \Sigma_{k-1} \cup \{a_k, b_k\}$ (where a_k, b_k are two new "marker" symbols) and define D_k as the set of words $w_1 m_1 w_2 m_2 \cdots w_n m_n w_{n+1}$ such that the $m_i \in \{a_k, b_k\}$ are markers, the w_i are words in Σ_{k-1}^*, and

- there exists $i \leq n$ such that $w_i \in D_{k-1}$.
- if $i_{\min}$ is the smallest i with $w_i \in D_{k-1}$, then $m_{i_{\min}} = a_k$.

Note that $i_{\min}$ is the *only* position holding a marker and whose prefix lies in $\Sigma_k^*\{a_k, b_k\}D_{k-1}$ but not in $\Sigma_k^*\{a_k, b_k\}D_{k-1}\{a_k, b_k\}\Sigma_k^*$. This can be used to formally show:

Lemma 3. *For $k \geq 2$ the language D_k lies in Δ_k^L.*

We can now prove our main theorem:

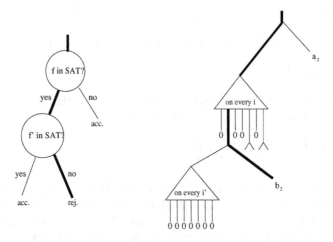

Fig. 1. Building a computation tree for the leaf language $0^* a_2 \{0, a_2, b_2\}^*$

Theorem 2 (Main). *For every $k \geq 2$:*
$$\text{Leaf}^P(D_k) = \text{Leaf}^P(\Delta_k^L) = \Delta_k^p.$$

Proof. By Corollary 1 we have for each k

$$\text{Leaf}^P(\Delta_k^L) = \text{Leaf}^P(\text{UPol}(\mathcal{DD}_{k-1})) \subseteq \text{T} \cdot \text{Leaf}^P(\mathcal{DD}_{k-1}) \subseteq \text{T} \cdot \text{BC} \cdot \Sigma_{k-1}^p = \Delta_k^p.$$

Of course, $\text{Leaf}^P(D_k) \subseteq \text{Leaf}^P(\Delta_k^L)$ so it remains to show that $\Delta_k^p \subseteq \text{Leaf}^P(D_k)$. We proceed by induction on k: our argument generalizes an original idea of Wagner [Wa90] whose result can be interpreted as showing that $\Delta_2^p \subseteq \text{Leaf}^P(D_2)$. We sketch this proof as our induction base.

We need to show that there exists a $\text{Leaf}^P(D_2)$ machine N that can simulate any deterministic polynomial time machine D querying, say, a SAT oracle. N begins by simulating the deterministic behavior of D until D asks a first query Q. At this point N simulates the query by non-deterministically branching in two computations (see Figure 1):

- On the left branch N attempts to verify that the oracle answers positively to the question Q: it produces a path for each possible witness of membership of Q in SAT. On every such path, if first checks whether the candidate-witness is correct. If it is not, N terminates on this computation path, writing a 0 on this leaf, otherwise, N resumes the deterministic simulation of D assuming a positive answer to the query Q.
- On the right branch N continues the deterministic simulation of D assuming a negative answer to the query Q.

We proceed in the same fashion for each query of D. When the simulation of D is complete, N terminates and writes an a_2 on the leaf if D accepts, and a b_2

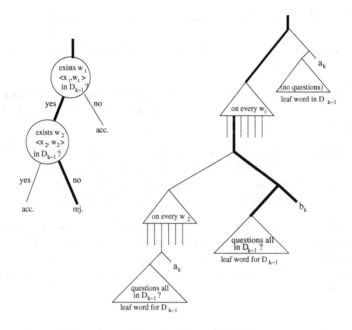

Fig. 2. Building a computation tree for the leaf language D_k

otherwise. The key observation is that the leftmost path of N with a non-0 on its leaf corresponds to a correct simulation of D because if a query Q is answered negatively, all candidate witnesses are rejected and the left subtree thus created has all its leaves labeled with 0.

Therefore the first non-0 in the leafstring is an a_2 if and only if D accepts its input x. In other words, leafstring$(N, x) \in D_2$ iff x is accepted by D. This shows $\Delta_2^p = \mathrm{P}^{\mathrm{SAT}} \subseteq \mathrm{Leaf}^{\mathrm{P}}(D_2)$.

More generally, we now want to show that a deterministic polynomial-time D querying a Σ_k^p oracle can be simulated by a $\mathrm{Leaf}^{\mathrm{P}}(D_{k+1})$ machine N. It is convenient here to think of Σ_k^p as $\exists \cdot \Delta_k^p$ because the induction hypothesis guarantees that $\Delta_k^p \subseteq \mathrm{Leaf}^{\mathrm{P}}(D_k)$. The oracle thus answers positively on query x if there exists a witness w such that $\langle x, w \rangle$ belongs to some language K in Δ_k^p. Since this class is closed under polynomial-time conjunctive reductions there exists a NDTM M such that on input $(\langle x_1, w_1 \rangle, \ldots, \langle x_n, w_n \rangle)$ M produces a leafstring from D_k if and only if each pair $\langle x_i, w_i \rangle$ lies in K.

We can now simulate D as follows: whenever D queries the oracle $\exists \cdot K$ with question t, N branches into two computations:

- On the left branch, N produces a path for each possible witness w_i of membership of t. However, instead of immediately verifying that the pair $\langle t, w_i \rangle$ lies in K, our simulation simply assumes this hypothesis, postpones its check and resumes the simulation of D assuming a positive answer to the query.

– On the right branch, N continues the simulation of D assuming a negative answer to the query.

Once a branch of N's simulation of D terminates, we need to check that all input/witness pairs along that branch *do* lie in K. For this, N again branches into two paths:

– On the left path it produces a computation tree whose leafstring lies in D_k if and only if all input/witness pairs on that branch lie in K.

– On the right branch it writes a marker a_{k+1} if D has accepted on this path of assumed oracle answers, and a marker b_{k+1} in the other case.

Once again, if we consider the leafstring of N, we observe that the first subword y_i sitting between two markers and lying in L indicates the valid computation of D. Accordingly, the marker which lies at the right of y_i is a_{k+1} if and only if D accepts its input and, by definition, this happens if and only if the leafstring lies in D_{k+1}.

5   Operators on Varieties *Versus* Operators on Complexity Classes

In fact, in the extended version of this report, we obtain through similar arguments the following much more general result:

Theorem 3. *Let $\mathcal{V}$ be a variety of languages such that* $\mathrm{Leaf}^{\mathrm{P}}(\mathcal{V})$ *is closed under polynomial-time conjunctive reductions. Then* $\mathrm{Leaf}^{\mathrm{P}}(\mathrm{UPol}(\mathrm{BPol}(\mathcal{V}))) = \mathrm{T} \cdot \exists \cdot \mathrm{Leaf}^{\mathrm{P}}(\mathcal{V})$.

We also give in the full report an explicit proof of a result hinted at but not proved formally in [HL*93] and based on the ideas of [BT88].

Theorem 4. *For any variety of languages $\mathcal{V}$:*

$$\mathrm{Leaf}^{\mathrm{P}}(\mathrm{Pol}(\mathcal{V})) = \exists \cdot \mathrm{Leaf}^{\mathrm{P}}(\mathcal{V}).$$

All these results relativize and this allows us to shed new light on the problem of finding a leaf-language upper bound for BPP: the classical result of Lautemann and Sipser (see [Pa94]) shows that BPP is contained in $\Sigma_2^p \cap \Pi_2^p$ and it is natural to ask whether there is a language L in $\Sigma_2^L \cap \Pi_2^L$ with BPP $\subseteq \mathrm{Leaf}^{\mathrm{P}}(L)$. This we now know cannot be the case with respect to all oracles since we will have $\mathrm{Leaf}^{\mathrm{P}}(L) \subseteq \Delta_2^p$ whereas relativized worlds exist in which Δ_2^p is strictly contained in BPP (see e.g. [BT00]).

In [ST03], it was shown that the class of regular languages which can be defined by two-variable sentences using ordinary and modular quantifiers is exactly the class $\mathcal{DA} * \mathcal{G}_{sol} = \mathrm{UPol}(\mathrm{BPol}(\mathcal{G}_{\mathrm{sol}}))$ where $\mathcal{G}_{sol}$ denotes the class of languages whose syntactic monoid is a solvable group. Theorem 3 can be combined with a result of [HL*93] to show

$$\mathrm{Leaf}^{\mathrm{P}}(\mathcal{DA} * \mathcal{G}_{\mathrm{sol}}) = \mathrm{T} \cdot \exists \cdot \mathrm{Leaf}^{\mathrm{P}}(\mathcal{G}_{\mathrm{sol}}) = \mathrm{T} \cdot \exists \cdot \mathrm{MOD}^{*}\mathrm{P}$$

where $\mathrm{MOD}^{*}\mathrm{P}$ denotes the closure of P under the Mod_q operators. Similarly, for any prime q, let $\mathcal{G}_q$ be the class of languages whose syntactic monoids are q-groups: we can show

$$\mathrm{Leaf}^{\mathrm{P}}(\mathrm{UPol}(\mathrm{BPol}(\mathcal{G}_q)))) = \mathrm{T} \cdot \exists \cdot \mathrm{Leaf}^{\mathrm{P}}(\mathcal{G}_q) = \mathrm{T} \cdot \exists \cdot \mathrm{MOD}_q\mathrm{P}.$$

Straubing and Thérien have conjectured that BPP and indeed the whole polynomial hierarchy is contained in $\mathrm{Leaf}^{\mathrm{P}}(\mathrm{UPol}(\mathrm{BPol}(\mathcal{G}_{\mathrm{sol}})))$ but there exist relativized worlds in which $\mathrm{T} \cdot \exists \cdot \mathrm{Mod}_2 \cdot \mathrm{P}$ does not contain BPP [BT00] and it is perhaps possible to further show that in this world even $\mathrm{T} \cdot \exists \cdot \mathrm{MOD}^{*}\mathrm{P}$ does not contain BPP. Such a result would rule out any relativizable proof of the aforementioned conjecture.

6 Conclusion

We have shown that the Delta classes of the dot-depth hierarchy and of the polynomial hierarchy correspond via leaf-languages. This result also holds if we consider the Cohen-Brzozowski definition of the dot-depth hierarchy. This extension of our result can be obtained using the "bridging" method outlined e.g. in [Pin98] which relates the two hierarchies in a straightforward way. It is also clear from our proofs that this correspondence still holds if we consider only balanced computation trees in the definition of the leaf language mechanism (see e.g. [BCS92]).

We know by Theorem 2 that there are languages in Δ_2^L, for example D_2, that capture the class Δ_2^{P}. Using algebraic methods, one can prove that if L is in Δ_2^L then either $\mathrm{Leaf}^{\mathrm{P}}(L) \subseteq \mathrm{BC}(\mathrm{NP})$ or $\mathrm{Leaf}^{\mathrm{P}}(L) = \Delta_2^{\mathrm{P}}$ but we do not know if a similar phenomenon occurs for $k \geq 3$. A related question is whether the D_k languages which we defined are complete for the Δ_k^L-classes with respect to the reductions defined in [SeW04].

For all varieties considered in this paper, we do have that $\mathrm{Leaf}^{\mathrm{P}}(\mathcal{V})$ is closed under polynomial-time conjunctive reductions and this can perhaps be concluded simply using the closure properties of the varieties. This would make Theorem 3 "cleaner" because it would shift the technical requirements to the class of languages $\mathcal{V}$, with no requirements left for the complexity class $\mathrm{Leaf}^{\mathrm{P}}(\mathcal{V})$.

Acknowledgments. We thank Jean-Eric Pin and Klaus Wagner for useful discussions. The fourth author was supported by the Alexander von Humboldt foundation and the fifth author by the Humboldt foundation and NSERC and FQRNT grants.

References

[BGS75] T. P. BAKER, J. GILL, R. SOLOVAY: *Relativizations of the P =? NP Question*, SIAM J. Computing **4**, 1975, pp. 431-442

[BT88] D. BARRINGTON, D. THÉRIEN: *Finite Monoids and the Fine Structure of NC^1*, J. of the ACM **35(4)**, 1988, pp. 941–952.

[BKS99] B. BORCHERT, D. KUSKE, F. STEPHAN: *On existentially first-order definable languages and their relation to NP*, Theoret. Informatics Appl. **33**, 1999, pp. 259–269

[BSS99] B. BORCHERT, H. SCHMITZ, F. STEPHAN: $\text{Leaf}^P(\Delta_2^B) = \Delta_2^p$, unpublished manuscript, 1999

[BCS92] D. P. BOVET, P. CRESCENZI, R. SILVESTRI: *A uniform approach to define complexity classes*, Theoretical Computer Science **104**, 1992, pp. 263–283.

[BT00] H. BUHRMAN, L. TORENVLIET, *Randomness is Hard*. SIAM J. Computing **30(5)**: 1485-1501 (2000)

[BV98] H.-J. BURTSCHICK, H. VOLLMER: *Lindström Quantifiers and Leaf Language Definability*, International J. of Foundations of Computer Science **9**, 1998, pp. 277-294.

[He84] H. HELLER: *Relativized Polynomial Hierarchies Extending Two Levels* Mathematical Systems Theory **17**, 1984, pp. 71-84.

[HL*93] U. HERTRAMPF, C. LAUTEMANN, T. SCHWENTICK, H. VOLLMER, K. WAGNER: *On the power of polynomial-time bit-computations*, Proc. 8th Structure in Complexity Theory Conference, 1993, pp. 200–207.

[Ka88] J. KADIN: *The Polynomial Time Hierarchy collapses if the Boolean Hierarchy collapses*, SIAM J. Computing **17**, 1988, pp. 1263–1282.

[Pa94] C. PAPADIMITRIOU: *Computational Complexity*, Addison Wesley, 1990.

[Pin98] J.-E. PIN: *Bridges for Concatenation Hierarchies*, Proc. Int. Conf. Automata, Languages and Programming, 1998, pp. 431–442.

[PST88] J.-E. PIN, H. STRAUBING, D. THÉRIEN: *Locally trivial categories and unambiguous concatenation*, J. of Pure and Applied Algebra **52**, 1988, pp. 297–311.

[PW97] J.-E. PIN, P. WEIL: *Polynomial closure and unambiguous product*, Theory of Computing Systems **30**, 1997, pp. 383–422.

[Sch76] M. P. SCHÜTZENBERGER: *Sur le produit de concatenation non ambigu*, Semigroup Forum **13**, 1976, pp. 47–75.

[ScW98] H. SCHMITZ, K. W. WAGNER, *The Boolean hierarchy over level 1/2 of the Straubing-Thérien hierarchy*, Technical report 201, Inst. für Informatik, Uni. Würzburg, 1998.

[Sel02] V. L. SELIVANOV: *Relating Automata-Theoretic Hierarchies to Complexity-Theoretic Hierarchies*, RAIRO–Theoretical Informatics & Applications **36(1)**, 2002, pp. 29–42.

[SeW04] V. L. SELIVANOV, K. W. WAGNER, *A Reducibility for the Dot-Depth Hierarchy*, Proc. MFCS, 2004, pp. 783–793.

[St94] H. STRAUBING: *Finite Automata, Formal Logic, and Circuit Complexity*, Birkhäuser, Boston, 1994.

[ST03] H. STRAUBING, D. THÉRIEN, *Regular Languages Defined by Generalized First-Order Formulas with a Bounded Number of Bound Variables*, Theory Comput. Syst. **36(1)**, 2003, pp. 29-69.

[TT02]   P. TESSON, D. THÉRIEN: *Diamonds are forever: the Variety DA*, in Semi-groups, Algorithms, Automata and Languages, WSP, 2002, 475–499.

[Wa90]   K. W. WAGNER: *Bounded Query Classes*, SIAM J. Computing **19**, 1990, pp. 833–846

[Yao85]   A. C.C. YAO: *Separating the Polynomial Hierarchy by oracles*, Proc. 26th IEEE Symp. on the Foundations of Computer Science, 1985, pp. 1–10

Input Reversals and Iterated Pushdown Automata: A New Characterization of Khabbaz Geometric Hierarchy of Languages

Henning Bordihn[1], Markus Holzer[2], and Martin Kutrib[3]

[1] Institut für Informatik, Universität Potsdam,
August-Bebel-Straße 89, D-14482 Potsdam, Germany
henning@cs.uni-potsdam.de
[2] Institut für Informatik, Technische Universität München,
Boltzmannstraße 3, D-85748 Garching bei München, Germany
holzer@informatik.tu-muenchen.de
[3] Institut für Informatik, Universität Gießen,
Arndtstraße 2, D-35392 Gießen, Germany
kutrib@informatik.uni-giessen.de

Abstract. Input-reversal pushdown automata are pushdown automata with the additional power to reverse the unread part of the input. We show that these machines characterize the family of linear context-free indexed languages, and that $k + 1$ input reversals are better than k for both deterministic and nondeterministic input-reversal pushdown automata, i.e., there are languages which can be recognized by a deterministic input-reversal pushdown automaton with $k + 1$ input reversals but which cannot be recognized with k input reversals (deterministic or nondeterministic). In passing, input-reversal finite automata are investigated. Moreover, an inherent relation between input-reversal pushdown automata and controlled linear context-free languages are shown, leading to an alternative description of Khabbaz geometric hierarchy of languages by input-reversal iterated pushdown automata. Finally, some computational complexity problems for the investigated language families are considered.

1 Introduction

A pushdown automaton is a one-way finite automaton with a separate pushdown storage (PD), that is a last-in first-out (LIFO) storage structure, which is manipulated by pushing and popping. Probably, such machines are best known for capturing the family of context-free languages $\mathcal{L}$(CFL). Pushdown automata have been extended or restricted in various ways, and the results obtained for these classes of machines hold for a large variety of formal language classes, when appropriately abstracted. This led to the rich theory of abstract families of automata (AFA), which is the equivalent of the theory of abstract families of languages (AFL); for the general treatment of machines and languages we refer to Ginsburg [8].

C.S. Calude, E. Calude and M.J. Dinneen (Eds.): DLT 2004, LNCS 3340, pp. 102–113, 2004.

In this paper, we consider an extension of pushdown automata, so called input-reversal pushdown automata. These machines were inspired by the recently introduced flip-pushdown automata [17] and the "flip-pushdown input-reversal" theorem of [11]. Basically, an input-reversal pushdown automaton is an ordinary pushdown automaton with the additional ability to reverse the unread part of the input during the computation. This allows the machine to read from both ends of the input. Thus, we show that input-reversal pushdown automata become equally powerful as linear context-free indexed languages as defined in [7]. On the other hand, if the number of input reversals is zero, obviously the family of context-free languages is characterized. Thus it remains to investigate the number of input reversals as a natural computational resource. Obviously, since by a single input reversal one can accept the non-context-free language $\{ ww \mid w \in \{a,b\}^* \}$, the base level of that hierarchy is already separated. But what about the other levels? We show that $k + 1$ input reversals are better than k for both deterministic and nondeterministic input-reversal pushdown automata. To this end, we develop a technique to decrease the number of input reversals, whose immediate consequence is that every input-reversal pushdown language can be transformed into a context-free language. Then by pumping arguments, the hierarchy induced by the number of input reversals can be separated.

A closer look at the characterization of linear context-free indexed languages in terms of input-reversal pushdown automata reveals an inherent deep relation between input-reversal automata (in general) and controlled linear context-free grammars [9]. Loosely speaking, while the control language is verified by the underlying basic device, the input-reversal feature mimics the actual linear context-free derivation. Intuitively, for a family of languages $\mathcal{A}$, an $\mathcal{A}$-controlled linear context-free grammar consists of a linear context-free grammar G and a control language L in $\mathcal{A}$, where the terminals of L are interpreted as labels of rules of G. Observe, that the control of linear context-free grammars can be iterated by starting with $\mathcal{A}$ and by taking the result of the kth step as class of control languages for the $(k + 1)$st step. When starting this iteration process from the context-free languages we obtain the so called geometric hierarchy of languages [14, 15], which has its name from the geometric series involved in the pumping lemmata for these language families. Moreover, the levels of the geometric hierarchy of languages are characterized by, e.g., context-free based finite-reversal checking-stack automata [10] or alternatively by iterated one-turn pushdown automata where the innermost pushdown is unrestricted [21]. By the above mentioned relation of input-reversal automata and controlled linear context-free languages we can show that a $(k + 1)$-iterated one-turn pushdown automaton (where the innermost pushdown is unrestricted, respectively) can be simulated by an input-reversal k-iterated one-turn pushdown automaton (where the innermost pushdown is unrestricted, respectively) and *vice versa*, thus trading one-turn pushdown iteration by input-reversal.

Finally, computational complexity aspects of languages accepted by iterated one-turn pushdown automata (where the innermost pushdown is unrestricted) with input-reversal are considered. It is well known that all these languages are context-sensitive [14, 15]. We show that input-reversal iterated one-turn push-down automata languages are contained within nondeterministic logarithmic space, and moreover, whenever the innermost pushdown is unrestricted, it belongs to the important complexity class LOG(CFL) ⊆ P. This nicely resembles a simulation technique of [18] for linear context-free languages. Based on these results we prove that linear context-free (restricted) indexed languages are complete for LOG(CFL) (NL, respectively). This generalizes the results for the based classes, namely for linear context-free and context-free languages in general as given in [18] and [19].

2   Definitions

We assume the reader to be familiar with the basics of complexity theory as contained in the book of Balcázar *et al.* [4]. In particular we consider the following well-known chain of inclusions: NL ⊆ LOG(CFL) ⊆ P ⊆ NP. Here NL is the class of problems accepted by nondeterministic logspace bounded Turing machines, LOG(CFL) is the class of problems logspace many-one reducible to a context-free language, and P (NP, respectively) is the set of problems accepted by deterministic (nondeterministic, respectively) polynomially time bounded Turing machines. Completeness and hardness are always meant with respect to deterministic log-space many-one reducibilities.

For the details in formal language theory we refer the reader to the book of Hopcroft and Ullman [12]. Concerning our notations, for any set Σ, let Σ^+ be the free semigroup and Σ^* the free monoid with identity λ generated by Σ. Set inclusion and strict set inclusion are denoted by $\subseteq$ and $\subset$, respectively. The Chomsky hierarchy is the strict chain of inclusions $\mathcal{L}(\text{REG}) \subset \mathcal{L}(\text{LIN}) \subset \mathcal{L}(\text{CFL}) \subset \mathcal{L}(\text{CS}) \subset \mathcal{L}(\text{RE})$, where $\mathcal{L}(\text{REG})$ denotes the family of regular languages, $\mathcal{L}(\text{LIN})$ the family of linear context-free languages, $\mathcal{L}(\text{CFL})$ the family of context-free languages, $\mathcal{L}(\text{CS})$ the family of context-sensitive languages, and $\mathcal{L}(\text{RE})$ the family of recursively enumerable languages.

In the following we consider variants of pushdown automata with the ability to reverse part of the unread input.

Definition 1. *A* (nondeterministic) input-reversal pushdown automaton, *is a tuple* $A = (Q, \Sigma, \Gamma, \delta, \Delta, q_0, Z_0, F)$, *where* Q *is a finite set of states,* Σ *is a finite input alphabet,* Γ *is a finite pushdown alphabet,* δ *is a mapping from* $Q \times (\Sigma \cup \{\lambda\}) \times \Gamma$ *to finite subsets of* $Q \times \Gamma^*$ *called the transition function,* Δ *is a mapping from* Q *to* 2^Q, $q_0 \in Q$ *is a initial state,* $Z_0 \in \Gamma$ *is a particular pushdown symbol, called the bottom-of-pushdown symbol, which initially appears on the pushdown storage, and* $F \subseteq Q$ *is a set of final states.*

A *configuration* or *instantaneous description* of an input-reversal pushdown automaton is a triple (q, w, γ), where q is a state in Q, w is a string of input symbols, and γ is a string of pushdown symbols. An input-reversal pushdown automaton A is said to be in configuration (q, w, γ) if A is in state q with w as remaining input, and γ on the pushdown storage, the rightmost symbol of γ

being the top symbol on the pushdown storage. If a is in $\Sigma \cup \{\lambda\}$, w in Σ^*, γ and β in Γ^*, and Z is in Γ, then we write $(q, aw, \gamma Z) \vdash_A (p, w, \gamma\beta)$, if the pair (p, β) is in $\delta(q, a, Z)$, for "ordinary" pushdown transitions, and $(q, aw, \gamma) \vdash_A (p, w^R a, \gamma)$, if p is in $\Delta(q)$, for input-reversal transitions. Whenever there is a choice between an ordinary pushdown transition or an input-reversal one, then the automaton nondeterministically chooses the next move. As usual, the reflexive transitive closure of $\vdash_A$ is denoted by $\vdash_A^*$. The subscript A will be dropped from $\vdash_A$ and $\vdash_A^*$ whenever the meaning remains clear.

Let k be a natural number. For an input-reversal pushdown automaton A we define $T_k(A)$, the language *accepted by final state and exactly k input reversals*[1], to be

$$T_k(A) = \{\, w \in \Sigma^* \mid (q_0, w, Z_0) \vdash_A^* (q, \lambda, \gamma) \text{ with exactly } k$$
$$\text{input reversals, for any } \gamma \in \Gamma^* \text{ and } q \in F \,\}.$$

Furthermore, we define $N_k(A)$, the language *accepted by empty pushdown and exactly k input reversals*, to be

$$N_k(A) = \{\, w \in \Sigma^* \mid (q_0, w, Z_0) \vdash_A^* (q, \lambda, \lambda) \text{ with exactly } k$$
$$\text{input reversals, for any } q \in Q \,\}.$$

If the number of input reversals is not limited, the language accepted by final state (empty pushdown, respectively) is analogously defined as above and denoted by $T(A)$ ($N(A)$, respectively). When accepting by empty pushdown, the set of final states is irrelevant. Thus, in this case, we usually let the set of final states be the empty set. The special case of *input-reversal finite automata* and languages is defined in the obvious way and will be investigated in the succeeding section.

In order to clarify our notation we give an example.

Example 2. Let $A = (\{q_0, q_1\}, \{a, b\}, \{X, Y, Z_0\}, \delta, \Delta, q_0, Z_0, \emptyset)$ be an input-reversal pushdown automaton where

1. $\delta(q_0, a, Z_0) = \{(q_0, Z_0 X)\}$
2. $\delta(q_0, b, Z_0) = \{(q_0, Z_0 Y)\}$
3. $\delta(q_0, a, X) = \{(q_0, XX)\}$
4. $\delta(q_0, b, X) = \{(q_0, XY)\}$
5. $\delta(q_0, a, Y) = \{(q_0, YX)\}$
6. $\delta(q_0, b, Y) = \{(q_0, YY)\}$
7. $\delta(q_1, a, X) = \{(q_1, \lambda)\}$
8. $\delta(q_1, b, Y) = \{(q_1, \lambda)\}$
9. $\delta(q_1, \lambda, Z_0) = \{(q_1, \lambda)\}$

and $\Delta(q_0) = \{q_1\}$ that accepts by empty pushdown the non-context-free language $L = \{\, ww \mid w \in \{a, b\}^* \,\}$. This is seen as follows.

---

[1] One may define language acceptance of input-reversal pushdown automata with *at most* k input reversals. Since an input-reversal pushdown automaton can count the number of reversals performed during its computation in its finite control and it can perform additional input-reversals at the end of the nondeterministic computation, it is an easy exercise to show that these two language acceptance mechanisms are equally powerful.

The transitions (1) through (6) allow A to store the input on the pushdown. If A decides that the middle of the input string has been reached, then the input reversal operation specified by $\Delta(q_0) = \{q_1\}$ is selected and A goes to the state q_1 and tries to match the remaining input symbols with reversed input. This is done with the transitions (7) and (8). Thus, A will empty its pushdown with transition (9) and therefore accept the input string (by empty pushdown) if and only if the guess of A was correct and the input is of the form ww.

The next theorem can be shown with a simple adaption of the proof for ordinary pushdown automata—see, e.g., [12]. Thus, we omit the proof.

Theorem 3. *Let k be some natural number. Then language L is accepted by some input-reversal pushdown automaton A_1 with empty pushdown making exactly k input reversals, i.e., $L = N_k(A_1)$, if and only if language L is accepted by some input-reversal pushdown automaton A_2 by final state making exactly k input reversals, i.e., $L = T_k(A_2)$. The statement remains valid for input-reversal pushdown automata with an unbounded number of input reversals.*   □

Moreover, we need the notion of context-free indexed grammars and languages as contained in Aho [1, 2]. A *context-free indexed grammar* is a five-tuple $G = (N, T, I, P, S)$, where N, T, and I are the finite pairwise disjoint alphabets of nonterminals, terminals, and indexed symbols, respectively, $S \in N$ is the axiom, and P is a finite set of productions of the form $A \to \alpha$ or $Af \to \alpha$ with $A \in N$, $f \in I$, and $\alpha \in (NI^* \cup T)^*$. A sentential form x in $(NI^* \cup T)^*$ directly derives y, for short $x \Rightarrow y$, if and only if

1. $x = x_1 A \xi x_2$, for $x_1, x_2 \in (NI^* \cup T)^*$, $A \in N$, and $\xi \in I^*$, production $A \to X_1\eta_1 X_2\eta_2 \ldots X_k\eta_k$ is in P with $X_i \in N \cup T$, $\eta_i \in I^*$, for $1 \le i \le k$, and $y = x_1 X_1\theta_1 X_2\theta_2 \ldots X_k\theta_k x_2$, where, for $1 \le i \le k$, $\theta_i = \eta_i\xi$, if X_i is in N, or $\theta_i = \lambda$, if X_i is in T;
2. $x = x_1 A f \xi x_2$, for $x_1, x_2 \in (NI^* \cup T)^*$, $A \in N$, $f \in I$, and $\xi \in I^*$, production $Af \to X_1\eta_1 X_2\eta_2 \ldots X_k\eta_k$ is a production in P with $X_i \in N \cup T$, $\eta_i \in I^*$, for $1 \le i \le k$, and $y = x_1 X_1\theta_1 X_2\theta_2 \ldots X_k\theta_k x_2$, where, for $1 \le i \le k$, $\theta_i = \eta_i\xi$, if X_i is in N, or $\theta_i = \lambda$, if X_i is in T.

Then the language generated by a context-free indexed grammar G with terminal alphabet T and axiom S is defined as $L(G) = \{ w \in T^* \mid S \Rightarrow^* w \}$, where $\Rightarrow^*$ denotes the reflexive transitive closure of the relation $\Rightarrow$. A language is said to be a context-free indexed language, if it is generated by a context-free indexed grammar.

A context-free indexed grammar $G = (N, T, I, P, S)$ is a *context-free restricted indexed grammar* if $N = N_1 \cup N_2$ with $N_1 \cap N_2 = \emptyset$ and P contains productions of the two forms: (1) $A \to \alpha$ with $A \in N_1$ and $\alpha \in (N_1 I^* \cup N_2 I^* \cup T)^*$, or (2) $Af \to \alpha$ with $A \in N_2$, $f \in I$, and $\alpha \in (N_2 I^* \cup T)^*$. A context-free (restricted) indexed grammar is called *linear context-free (restricted) indexed grammar* if every right-hand side of each production is in $T^* N I^* T^* \cup T^*$. A language generated by a linear context-free (restricted) indexed grammar is referred to as a linear context-free (restricted) indexed language.

3   Finite Automata with Input Reversals

Before we consider input-reversal pushdown automata in more detail, we investigate input-reversal finite automata. First we show that a constant number of input reversals does not increase the computational power.

Theorem 4. *Let k be some natural number. Then language L is accepted by some input-reversal finite automaton $A = (Q, \Sigma, \delta, \Delta, q_0, F)$ with exactly k (at most k) input-reversals, i.e., $L = T_k(A)$ if and only if L is regular.*

Proof. The implication from right to left is immediate. Conversely, we argue as follows: First define the ordinary finite automata $A_p = (Q, \Sigma, \delta, q_0, \{p\})$, for every p in Q. Moreover, let $B_q = (Q, \Sigma, \delta, \Delta, q, F)$, for every q in Q , be the input-reversal finite automaton, which is build from A by replacing initial state q_0 by q. Then by induction we show that the language $T_k(A)$ is regular. For $k = 0$ the statement is obvious. By induction hypothesis assume that every language accepted by some input-reversal finite automaton making exactly $k - 1$ input reversals, for $k \geq 1$, is regular. Obviously, the language $T_k(A)$ can be written as

$$T_k(A) = \bigcup_{\substack{p \in Q \\ q \in \Delta(p)}} L(A_p) \cdot (T_{k-1}(B_q))^R,$$

by cutting at the first input reversal that appears during a computation. Then by induction hypothesis $T_{k-1}(B_q)$ is regular, and so is L by the closure properties of regular languages under concatenation and reversal. This proves the stated claim.   □

Whenever the number of input reversals is not restricted to be constant, then we find the following situation:

Theorem 5. *The language L is accepted by some input-reversal finite automaton (with an unbounded number of input-reversals) if and only if L is a linear context-free language.*

Proof. Let L be accepted by some input-reversal nondeterministic finite automaton $A = (Q, \Sigma, \delta, \Delta, q_0, F)$. We construct a linear context-free grammar as follows: Let $G = (N, \Sigma, P, q_0)$, where $N = Q \cup \{p^R \mid p \in Q\}$, the union being disjoint, and the production set $P = P_1 \cup P_2 \cup P_3$ with

$$P_1 = \{p \to aq \mid q \in \delta(p, a)\} \cup \{p^R \to q^R a \mid q \in \delta(p, a)\},$$
$$P_2 = \{p \to q^R \mid q \in \Delta(p)\} \cup \{p^R \to q \mid q \in \Delta(p)\},$$

and

$$P_3 = \{p \to \lambda \mid p \in F\} \cup \{p^R \to \lambda \mid p \in F\}.$$

Then by induction one can verify that

$$(p, w) \vdash_A^* (q, \lambda) \quad \text{if and only if} \quad p \Rightarrow^* wq \text{ and } p^R \Rightarrow q^R w^R \text{ in } G,$$

if the computation of A on word w is without any input reversal. Moreover, if $q \in \Delta(p)$, then

$$(p, \lambda) \vdash_A (q, \lambda) \quad \text{if and only if} \quad p \Rightarrow q^R \text{ and } p^R \Rightarrow q \text{ in } G.$$

Combining these statements one immediately deduces that the language accepted by A equals the language generated by G, i.e, $T(A) = L(G)$. To this end rules from the production set P_3 have to be applied in order to terminate the derivation. The tedious details are left to the reader.

For the converse implication let $G = (N, T, P, S)$ be a linear context-free grammar in normal form, that is, all productions are of the forms $A \to aB$, $A \to Ba$, or $A \to a$—see, e.g., [12]. We construct an input-reversal finite automaton from G as follows: Let $A = (Q, T, \delta, \Delta, S, F)$, where $Q = N \cup \{ A^R \mid A \in N \} \cup \{\Lambda\}$ the union being disjoint, $F = \{\Lambda\}$, and δ and Δ are specified as follows: For every A in N and a in T let

$$\delta(A, a) = \{ B \in N \mid A \to aB \in P, \} \cup \{ \Lambda \mid A \to a \in P \}$$

and

$$\delta(A^R, a) = \{ B^R \in N \mid A \to Ba \in P, \} \cup \{ \Lambda \mid A \to a \in P \}.$$

Moreover, for every A in N define $\Delta(A) = \{A^R\}$ and $\Delta(A^R) = \{\Lambda\}$. By easy means, one observes that $A \Rightarrow^* wB$ if and only if $(A, w) \vdash_A^* (B, \lambda)$ and $A \Rightarrow^* Bw$ if and only if $(A^R, w^R) \vdash_A^* (B, \lambda)$, for $A, B \in N$ and $w \in T^*$. Similar statements are valid in case of termination. This immediately implies, together with the input-reversal relation Δ, that every word generated by the linear context-free grammar G is acceptable by the input-reversal finite automaton A and *vice versa*. Thus, $L(G) = T(A)$. $\qquad\square$

4   Pushdown Automata with Input Reversals

In this section we will see that Theorem 5 from the previous section nicely generalizes to an infinite hierarchy of language classes within the context-sensitive languages—more precisely, we will obtain an alternative characterization of Khabbaz geometric hierarchy of languages [15]. Before we consider input-reversal pushdown automata in general, we have to introduce $\mathcal{A}$-controlled linear context-free grammars. Intuitively, for some family of languages $\mathcal{A}$, an $\mathcal{A}$-controlled linear context-free grammar consists of a linear context-free grammar $G = (N, T, P, S)$ and a control language L in $\mathcal{A}$, where the terminals of L are interpreted as labels of rules of G. More formally, this reads as follows:

Let $G = (N, T, P, S)$ be a linear context-free grammar, where every rule in P has some label r. Let Λ denote the set of all labels. Take a language $L \subseteq \Lambda^*$. A derivation $S \Rightarrow \alpha_1 \Rightarrow \alpha_2 \Rightarrow \cdots \Rightarrow \alpha_n = w \in T^*$ in G using the rules with labels $r_1, r_2, \ldots, r_n$ in Λ is *valid* if and only if the word $r_1 r_2 \ldots r_n$ belongs to L. Then the language generated by G under L-control is the set of all terminal words that can be generated by a valid derivation. If we take the language L from a formal language class $\mathcal{A}$, then we call the grammar an $\mathcal{A}$-controlled linear context-free grammar, the generated language an $\mathcal{A}$-controlled linear context-free language. The control of linear context-free grammars can be iterated by starting with $\mathcal{A}$ and by taking the result of the kth step as class of control languages for the $(k + 1)$st step. For $k \geq 1$, let $\mathrm{CTRL}_k(\mathcal{A})$ refer to the kth level of this hierarchy and define $\mathrm{CTRL}_0(\mathcal{A}) = \mathcal{A}$. In this way, we obtain two hierarchies, namely $\mathrm{CTRL}_k(\mathrm{LIN})$ and $\mathrm{CTRL}_k(\mathrm{CFL})$, respectively. Observe, that $\mathrm{CTRL}_1(\mathrm{CFL})$ ($\mathrm{CTRL}_1(\mathrm{LIN})$, respectively) is equal to the family of linear context-free (restricted, respectively) indexed languages [7]. Now we are ready to state our next theorem.

Theorem 6. *The language L is accepted by some input-reversal pushdown automaton with empty pushdown (by final state) number of input-reversals if and only if L is a linear context-free indexed language.*

Proof. Recall the proof of Theorem 5. When constructing a grammar from a given input-reversal pushdown automaton we end up with a linear context-free grammar whose derivation is controlled by a context-free language. Since the family of context-free-controlled linear context-free languages is equal to the family of a linear context-free indexed languages the implication from left to right follows.

On the other hand, when starting with a linear context-free indexed language or equivalently with a context-free-controlled linear context-free language, again the proof of Theorem 5 can be adapted, such that now an input-reversal pushdown automaton instead of an input-reversal finite automaton is constructed. As the reader may see, the control language is verified by the underlying basic device, while the input-reversal feature mimics the actual linear derivation. Since the construction is quite straight-forward we omit the technical details. This shows the implication from right to left.   □

An immediate consequence of Theorem 6 is that *unary* linear context-free indexed languages, i.e., languages over a singleton letter alphabet, are regular. Therefore, we obtain the following statement.

Corollary 7. *If L is a unary language accepted by some input-reversal pushdown automaton, then L is a regular language.*   □

In fact the relation between controlled linear context-free languages and input-reversal pushdown automata generalizes further. First, if one restricts to one-turn pushdown automata, we find the following corollary. Since by the proof of the above given theorem we have argued, that the control language of a controlled linear context-free grammar is verified by the underlying basic device, the input-reversal feature mimics the actual linear context-free derivation, the "one-turn" restriction of the indexed strings translates to the appropriate restriction of the pushdown storage. Due to the lack of space we omit the proof of the following corollary, which reads as follows:

Corollary 8. *The language L is accepted by some input-reversal one-turn push-down automaton with empty pushdown (by final state) if and only if L is a linear context-free restricted indexed language.* □

Moreover, one can give also an alternative characterization of Khabbaz geometric hierarchy of languages. To this end, one has to introduce k-iterated one-turn pushdown automata—see, e.g., [21]. Note, that a language L belongs to $\text{CTRL}_k(\text{LIN})$ ($\text{CTRL}_k(\text{CFL})$, respectively), for $k \geq 1$, if and only if L is accepted by a k-iterated one-turn pushdown automaton (where the innermost pushdown is unrestricted, respectively). The below given theorem shows that a $(k+1)$-iterated one-turn pushdown automaton can be simulated by an input-reversal k-iterated one-turn pushdown automaton and *vice versa*, thus trading pushdown iteration by input-reversal. Here input-reversal k-iterated one-turn pushdown automata are defined as k-iterated one-turn pushdown automata, with the additional ability of input reversals. Since the proof follows similar lines as that of Theorem 6 and is based on the relation between controlled linear-context free languages and iterated one-turn pushdown automata (where the innermost storage is an unrestricted pushdown) we omit it.

Theorem 9. *Let k be some natural number. Then language L is accepted by some input-reversal k-iterated one-turn pushdown automaton with empty push-down (by final state) if and only if L is accepted by some $(k+1)$-iterated one-turn pushdown automaton with empty pushdown (by final state). The statement remains valid in case the first storage is an unrestricted pushdown.* □

In the remainder of this section we restrict ourself to input-reversal (one-turn) pushdown automata, considering the hierarchy of languages induced by the number of input reversals. First we state an essential technique for input-reversal pushdown automata, which reads as follows—we omit the proof and leave the details to the reader:

Theorem 10. *Let k be a natural number. Language L is accepted by an input-reversal pushdown automaton $A_1 = (Q, \Sigma, \Gamma, \delta, \Delta, q_0, Z_0, \emptyset)$ by empty pushdown with $k+1$ input reversals, i.e., $L = N_{k+1}(A_1)$, if and only if the language*

$$L_R = \{\, w\$v^R \mid (q_0, w, Z_0) \vdash^*_{A_1} (q_1, \lambda, Z_0\gamma) \text{ without any reversals, } q_2 \in \Delta(q_1),$$
$$\text{and } (q_2, v^R, Z_0\gamma) \vdash^*_{A_1} (q_3, \lambda, \lambda) \text{ with exactly } k \text{ input reversals} \,\}$$

is accepted by an input-reversal pushdown automaton A_2 by empty pushdown with k input reversals, i.e., $L_R = N_k(A_2)$, where $\$$ is a new symbol not contained in Σ. The statement remains valid if state acceptance is considered. □

Now we are ready for the separation of the levels of the input-reversal push-down hierarchy. The proof of the hierarchy follows closely the proof of a hierarchy on the number of pushdown flips in [11]. Nevertheless, there are some slight differences, which make our statement even harder to prove. For convenience we omit most of the technical details.

Theorem 11. *Let k be a natural number. Then there is a language L, which is accepted by a $(k+1)$-input-reversal (deterministic) pushdown automaton, but cannot be accepted by any k-input-reversal pushdown automaton.*

Sketch of Proof. We define the language

$$L_k = \{\, \#w_1\#w_2\#\ldots\#w_k\#w_k\#\ldots\#w_2\#w_1\# \mid w_i \in \{a,b\}^* \text{ for } 1 \le i \le k \,\}.$$

Language L_{k+1} is accepted by a (deterministic) input-reversal pushdown automaton A making $k+1$ input reversals as follows. Automaton A starts to read and to store subword w_1. On the symbol $\#$ between w_1 and w_2 it reverses the remaining input. So, $(q_0, \#w_1\#\ldots\#w_1\#, Z_0) \vdash^*_A (q_1, \#w_1^R\#\ldots\#w_2^R\#, Z_0w_1)$ is computed. Next, automaton A compares the pushdown content w_1 with the next subword of the current input, i.e., with w_1^R. The computed configuration is $(q_2, \#w_2^R\#\ldots\#w_2^R\#, Z_0)$. Now the process repeats. The next stage is $(q_3, \#w_3^R\#\ldots\#w_2^R\#, Z_0w_2^R) \vdash^*_A (q_4, \#w_2\#\ldots\#w_3\#, Z_0w_2^R)$, which leads to $(q_5, \#w_3\#\ldots\#w_3\#, Z_0)$. Finally, when all pairs of subwords match, the input is accepted.

Next we give evidence that L_{k+1} cannot be accepted by any k-input-reversal pushdown automaton. Assume to the contrary, that language L_{k+1} is accepted by some k-input-reversal pushdown automaton A with exactly k input reversals. Then applying Theorem 10 exactly k times (from left to right), results in a context-free language L. Now the idea is to pump an appropriate word from L and to undo the input-reversals, in order to obtain a word that must be in L_{k+1}, but is not due to its form. From this contradiction the assertion follows. If the pumping is done such that no input reversal boundaries in the word are pumped, then the input-reversals can be undone. In order to prevent the boundaries from being pumped, we use the generalization of Ogden's lemma, which is due to Bader and Moura [3] and incorporates excluded positions. Though little is known about the context-free language, it is easy to see that the words must have subwords of the forms $\Sigma^* v\Sigma^* v\Sigma^*$ or $\Sigma^* u\Sigma^* v\Sigma^* u\Sigma^* v\Sigma^*$, where $\Sigma = \{a, b, \#\}$. On these subwords we can apply the generalized pumping lemma in order to obtain words in L_{k+1} that are not of the appropriate form.   □

5   On the Complexity of Input-Reversal Languages

We consider some computational complexity problems of input-reversal languages in more detail.

Theorem 12. *Let $k \ge 1$ be some natural number. Then every language L accepted by some k-iterated one-turn pushdown automaton (where the first storage is an unrestricted pushdown, respectively) is contained in NL (LOG(CFL), respectively).*

Proof. We prove the statement by induction on k. If $k = 1$ the the statement follows in both cases from Sudborough [18, 19]. Now assume $k \ge 1$. Instead of iterated pushdowns we use the characterization in terms of controlled linear context-free grammars. So let a linear context-free grammar $G = (N, T, P, S, \Lambda)$ with control language from level $k - 1$ be given. Without loss of generality we may assume that the linear context-free grammar is in normal form. Assume $w = a_1 a_2 \ldots a_n$ to be the input of length n. Define the triples $[A, i, j]$ with $A \in N$ and $0 \leq i, j \leq n$. To each triple $[A, i, j]$ associate the word $a_1 a_2 \ldots a_i A a_{j+1} a_{j+2} \ldots a_n$ if $i < j$ and λ otherwise. Then the following algorithm checks membership for the language under consideration: Start the algorithm with triple $[S, 0, n]$, where S is the axiom of G. (1) Guess a rule $r = (A \to \alpha)$ from P, where A is the nonterminal of the actual triple. (2) Check whether $A \to \alpha$ is applicable to the associated sentential form and replace the triple accordingly. More precisely, replace $[A, i, j]$ with (i) $[B, i + 1, j]$, if $\alpha = aB$ and $a_{i+1} = a$, (ii) with $[B, i, j - 1]$, if $\alpha = Ba$ and $a_j = a$, (iii) with the empty word if $\alpha = a$, $i + 1 = j$, and $a_{i+1} = a$. If this is not the case, then halt and reject. (3) We continue the simulation starting with (1) until we have reached the empty word. (4) In passing the algorithm writes down the labels of the guessed rules. If the written word belongs to the control language, then we halt and accept. Otherwise, we halt and reject.

The interested reader may verify the correctness of the described algorithm. We implement the algorithm on an oracle auxiliary one-turn pushdown automaton, where the oracle tape is written deterministically [16], in order to simulate step (4). To be more precise, the one-turn pushdown storage is used to protocol the labels of the guessed rules, and to written them on the oracle tape at the very ending of the simulation. The space of the machine (as usual the oracle tape is not taken into consideration) is bounded by the size of the triples $[A, i, j]$, which is clearly logarithmic.

Hence membership for k-iterated one-turn pushdown languages can be verified in NL, because the space of the oracle auxiliary one-turn pushdown automaton is logarithmically bounded, while the time is polynomial, the oracle set is in NL by induction hypothesis, and NAux1-tPDA-SpaceTime$(\log n, \text{pol } n)^{\langle \text{NL} \rangle} \subseteq$ NL [13, 20]. Analogously we estimate the complexity of k-iterated one-turn pushdown languages, where the first storage is an unrestricted pushdown to be contained within LOG(CFL), since the oracle set is in LOG(CFL) be induction hypothesis, and NAuxPDA-SpaceTime$(\log n, \text{pol } n)^{\langle \text{LOG(CFL)} \rangle} \subseteq$ LOG(CFL), which is due to Borodin *et al.* [6, 5]. □

Since the lowest levels, i.e., the linear context-free and context-free languages, respectively, of the above mentioned hierarchies are NL- and LOG(CFL)-complete, respectively, we immediately obtain the following completeness result:

Theorem 13. *The following problems are complete with respect to deterministic logspace many-one reductions: Let $k \geq 1$ be some natural number.*

1. *The fixed membership problem for k-input-reversal one-turn pushdown languages, where the first storage is an unrestricted pushdown, is LOG(CFL)-complete and*
2. *the fixed membership problem for k-input-reversal one-turn pushdown automata languages is NL-complete.* □

References

1. A. V. Aho. Indexed grammars—an extension of context-free grammars. *Journal of the ACM*, 15(4):647–671, October 1968.
2. A. V. Aho. Nested stack automata. *Journal of the ACM*, 16(3):383–406, July 1969.
3. Ch. Bader and A. Moura. A generalization of Ogden's lemma. *Journal of the ACM*, 29(2):404–407, April 1982.
4. J. L. Balcázar, J. Díaz, and J. Gabarró. *Structural Complexity I*, volume 11 of *EATCS Monographs on Theoretical Computer Science*. Springer, 1988.
5. A. Borodin, S. A. Cook, P. W. Dymond, W. L. Ruzzo, and M. Tompa. Erratum: Two applications of inductive counting for complementation problems. *SIAM Journal on Computing*, 18(6):1283, December 1989.
6. A. Borodin, S. A. Cook, P. W. Dymond, W. L. Ruzzo, and M. Tompa. Two applications of inductive counting for complementation problems. *SIAM Journal on Computing*, 18(3):559–578, June 1989.
7. J. Duske and R. Parchmann. Linear indexed languages. *Theoretical Computer Science*, 32(1–2):47–60, July 1984.
8. S. Ginsburg. *Algebraic and Automata-Theoretic Properties of Formal Languages*. North-Holland, Amsterdam, 1975.
9. S. Ginsburg and E. H. Spanier. Cotrol sets on grammars. *Mathematical Systems Theory*, 2(2):159–177, 1968.
10. S. A. Greibach. One way finite visit automata. *Theoretical Computer Science*, 6(2):175–221, April 1978.
11. M. Holzer and M. Kutrib. Flip-pushdown automata: $k + 1$ pushdown reversals are better than k. In J. C. M. Baeten, J. K. Lenstra, J. Parrow, and G. J. Woeginger, editors, *Proceedings of the 30th International Colloqium on Automata, Languages and Propgramming*, number 2719 in LNCS, pages 490–501, Eindhoven, The Netherlands, June–July 2003. Springer.
12. J. E. Hopcroft and J. D. Ullman. *Introduction to Automata Theory, Languages and Computation*. Addison-Wesley, 1979.
13. N. Immerman. Nondeterministic space is closed under complementation. *SIAM Journal on Computing*, 17(5):935–938, October 1988.
14. N. A. Khabbaz. Control sets and linear grammars. *Information and Control*, 25(3):206–221, July 1974.
15. N. A. Khabbaz. A geometric hierarchy of languages. *Journal of Computer and System Sciences*, 8(2):142–157, April 1974.
16. W. L. Ruzzo, J. Simon, and M. Tompa. Space-bounded hierarchies and probabilistic computations. *Journal of Computer and System Sciences*, 28(2):216–230, April 1984.
17. P. Sarkar. Pushdown automaton with the ability to flip its stack. Report TR01-081, Electronic Colloquium on Computational Complexity (ECCC), November 2001.
18. I. H. Sudborough. A note on tape-bounded complexity classes and linear context-free languages. *Journal of the ACM*, 22(4):499–500, October 1975.
19. I. H. Sudborough. On the tape complexity of deterministic context-free languages. *Journal of the ACM*, 25(3):405–414, July 1978.
20. R. Szelepcsényi. The method of forced enumeration for nondeterministic automata. *Acta Informatica*, 26(3):279–284, November 1988.
21. H. Vogler. Iterated linear control and iterated one-turn pushdowns. *Mathematical Systems Theory*, 19(2):117–133, 1986.

On the Maximum Coefficients of Rational Formal Series in Commuting Variables

Christian Choffrut[1], Massimiliano Goldwurm[2], and Violetta Lonati[2]

[1] L.I.A.F.A., Université Paris VII, Paris, France
Christian.Choffrut@liafa.jussieu.fr
[2] Dipartimento di Scienze dell'Informazione, Universit degli Studi di Milano
Milano, Italy
{goldwurm, lonati}@dsi.unimi.it

Abstract. We study the *maximum function* of any $\mathbb{R}_+$-rational formal series S in two commuting variables, which assigns to every integer $n \in \mathbb{N}$, the maximum coefficient of the monomials of degree n. We show that if S is a power of any primitive rational formal series, then its maximum function is of the order $\Theta(n^{k/2}\lambda^n)$ for some integer $k \geq -1$ and some positive real λ. Our analysis is related to the study of limit distributions in pattern statistics. In particular, we prove a general criterion for establishing Gaussian local limit laws for sequences of discrete positive random variables.

1   Introduction

The general observation motivating this paper is the following. Consider a rational fraction $\frac{p(x)}{q(x)}$ where $p(x)$ and $q(x)$ are two polynomials with coefficients in the field of real numbers (with $q(0) \neq 0$). It is well-known that the coefficient of the term x^n of its Taylor expansion is asymptotically equivalent to a linear combination of expressions of the form $n^{k-1}\lambda^n$ where λ is a root of $q(x)$ and k its multiplicity, cf. [10, Theorem 6.8] or [16, Lemma II.9.7]. It is natural to ask whether a similar evaluation holds for formal series in two variables both in the commutative and in the noncommutative case.

The purpose of this work is to pose the problem in more general terms. Indeed, assume we are given a semiring $\mathbb{K}$ whose underlying set is contained in the reals, e.g., $\mathbb{Z}$, $\mathbb{N}$, $\mathbb{R}_+$, etc. Assume further we are given a monoid $\mathcal{M}$ that is either the free monoid or the free commutative monoid over a finite alphabet. Consider next a rational formal series r over $\mathcal{M}$ with coefficients in $\mathbb{K}$ and denote by (r, x) the corresponding *coefficient* of the element $x \in \mathcal{M}$. Denoting by $|x|$ the length of x, we want to investigate the *maximum function* $g_r(n) = \max\{|(r, x)| : x \in \mathcal{M}, |x| = n\}$, where $|(r, x)|$ denotes the absolute value of (r, x). We note that the same definition can be extended to the formal series over trace monoids [6].

For rational formal series over a free monoid with integer coefficients, the growth of the coefficients was investigated in [17] (see also [14]), where it is

C.S. Calude, E. Calude and M.J. Dinneen (Eds.): DLT 2004, LNCS 3340, pp. 114–126, 2004.
© Springer-Verlag Berlin Heidelberg 2004

proved that for such a series r either there exists $k \in \mathbb{N}$ such that $g_r(n) = O(n^k)$ or $|(r, \omega_j)| \geq c^{|\omega_j|}$ for a sequence of words $\{\omega_j\}$ of increasing length and for some constant $c > 1$. In the first case, the series is the sum of products of at most $k + 1$ characteristic series of regular languages over the free monoid (see also [2, Corollary 2.11]). When the semiring of coefficients is $\mathbb{N}$ the problem is related to the analysis of ambiguity of formal grammar (or finite automata) generating (recognizing, resp.) the support of the series; a wide literature has been devoted to this problem (see for instance [12, 20] and [21, 22] for a similar analysis in the algebraic case).

As far as the growth of coefficients is concerned, another case can be found in the literature. It is related to the tropical semiring $\mathcal{T}$ whose support is the set $\mathbb{N} \cup \{\infty\}$ and whose operations are the min for the addition and the + for the multiplication. In [19], Imre Simon proves that for all $\mathcal{T}$-rational series s over the free monoid $\{a, b\}^*$, there exists an integer k such that $(s, w) = O(|w|^{1/k})$ holds for all $w \in \{a, b\}^*$. Moreover it is proven that for each positive integer k, there exists a $\mathcal{T}$-rational series s_k such that $g_{s_k}(n) = \Theta(n^{1/k})$. Thus, the hierarchy is strict though it is not proven that all series have an asymptotic growth of this kind.

In the present work, we study the maximum function of rational formal series in two commuting variables a, b with coefficients in $\mathbb{R}_+$. As far as we know, the general problem of characterizing the order of magnitude of $g_S(n)$ for such a series S is still open, though some contributions can be found in the literature [3]. We prove the following

Theorem. *For any positive $k \in \mathbb{N}$ and any primitive $\mathbb{R}_+$-rational formal series r in the noncommutative variables a, b, if S is the commutative image of r^k then, for some $\lambda > 0$, its maximum function satisfies the relation*

$$g_S(n) = \begin{cases} \Theta(n^{k-(3/2)} \lambda^n) & \text{if } r \text{ is not degenerate} \\ \Theta(n^{k-1} \lambda^n) & \text{otherwise} \end{cases}$$

This result is obtained by studying the limit distribution of the discrete random variables $\{X_n\}$ naturally associated with the series S: each X_n takes on values only in $\{0, 1, \ldots, n\}$ and $\Pr\{X_n = i\}$ is proportional to the coefficient $(S, a^i b^{n-i})$. We prove that X_n has a Gaussian limit distribution whenever S is a power of a primitive formal series r as described above. Under the same hypothesis we also give a local limit theorem for $\{X_n\}$ based on the notion of symbol-periodicity introduced in [4]. When the symbol-periodicity associated with r is 1 we just obtain a local limit theorem for $\{X_n\}$ in the sense of DeMoivre-Laplace [9]. The material we present also includes a general criterion for local limit laws that is of interest in its own right and could be useful in other contexts. It holds for sequences of discrete random variables with values in a linear progression of a fixed period d included in $\{0, 1, \ldots, n\}$.

The paper is organized as follows: after some preliminaries on rational series and probability theory, in Section 4 we present the new criterion for local limit laws. Then, we recall basic definitions and properties of the stochastic models defined via rational formal series. In Section 6 we give global and local limit theorems for the pattern statistics associated with powers of primitive rational formal series (*power model*). In the last section, we illustrate our main result on the maximum function of rational formal series in two commutative variables.

Due to space constraints, the proofs of Section 4 are omitted and can be found in Appendix for referees' convenience.

2   Preliminaries on Formal Series

Given a monoid $\mathcal{M}$ and a semiring $\mathbb{K}$ we call *formal series* over $\mathcal{M}$ any application $r : \mathcal{M} \longrightarrow \mathbb{K}$, usually denoted as an infinite formal sum $r = \sum_{w \in \mathcal{M}} (r, w)\, w$, which associates with each $w \in \mathcal{M}$ its *coefficient* $(r, w) \in \mathbb{K}$. The set of all series over $\mathcal{M}$ with coefficients in $\mathbb{K}$ is a monoid algebra, provided with the usual operations of sum, product and star restricted to the elements r such that $(r, 1_{\mathcal{M}}) = 0_{\mathbb{K}}$. These operations are called rational operations. A series r is called *rational* if it belongs to the smallest set closed under rational operations, containing the series 0 and all the series χ_w, for $w \in \mathcal{M}$, such that $(\chi_w, w) = 1$ and $(\chi_w, x) = 0$ for each $x \neq w$.

In particular, in this work we fix the alphabet $\{a, b\}$ and consider the formal series over the free monoid $\{a, b\}^*$ or the free commutative monoid $\{a, b\}^{\otimes}$ with coefficients in the semiring $\mathbb{R}_+$ of nonnegative real numbers. In the former case, the family of all formal series is denoted by $\mathbb{R}_+\langle\langle a, b \rangle\rangle$. By Kleene's Theorem [16, 15], we know that every rational $r \in \mathbb{R}_+\langle\langle a, b \rangle\rangle$ admits a *linear representation* over $\{a, b\}$, i.e. a triple (ξ, μ, η) such that, for some integer $m > 0$, ξ and η are (non-null column) vectors in $\mathbb{R}_+^m$ and $\mu : \{a, b\}^* \longrightarrow \mathbb{R}_+^{m \times m}$ is a monoid morphism, satisfying $(r, w) = \xi^T \mu(w)\, \eta$ for each $w \in \{a, b\}^*$. We say that m is the *size* of the representation.

Observe that considering such a triple is equivalent to defining a (weighted) nondeterministic finite automaton over the alphabet $\{a, b\}$, where the state set is given by $\{1, 2, \ldots, m\}$ and the transitions, the initial and the final states are assigned weights in $\mathbb{R}_+$ by μ, ξ and η respectively.

Analogously, the family of all formal series over $\{a, b\}^{\otimes}$ with coefficients in $\mathbb{R}_+$ is denoted by $\mathbb{R}_+[[a, b]]$. In this case, any element of $\{a, b\}^{\otimes}$ is represented in the form $a^i b^j$. The canonical morphism $\varphi : \{a, b\}^* \to \{a, b\}^{\otimes}$, associating with each $w \in \{a, b\}^*$ the monomial $a^i b^j$ where $i = |w|_a$ and $j = |w|_b$, extends to the semiring of formal series: for every $r \in \mathbb{R}_+\langle\langle a, b \rangle\rangle$

$$(\varphi(r), a^i b^j) = \sum_{|x|_a = i, |x|_b = j} (r, x)\ .$$

Notice that, since φ is a morphism, for every rational series $r \in \mathbb{R}_+\langle\langle a, b \rangle\rangle$, the commutative image $\varphi(r)$ is rational in $\mathbb{R}_+[[a, b]]$.

For our purpose, a subset of rational series of particular interest is defined by the so-called primitive linear representations. We recall that a matrix $M \in \mathbb{R}_+^{m \times m}$ is *primitive* if for some $k \in \mathbb{N}$ all entries of M^k are strictly positive. The main property of these matrices is given by the Perron–Frobenius Theorem, stating that any primitive matrix M has a unique eigenvalue λ of largest modulus, which is real positive (see for instance [18, Chapter 1]). Such a λ is called the Perron-Frobenius eigenvalue of M.

Thus, a linear representation (ξ, μ, η) defined over a set of generators $\{a, b\}$ is called primitive if $\mu(a) + \mu(b)$ is a primitive matrix. We also say that a series $r \in \mathbb{R}_+ \langle\langle a, b \rangle\rangle$ is *primitive* if it is rational and admits a primitive linear representation. Moreover, we say that (ξ, μ, η) is *degenerate* if $\mu(\sigma) = 0$ for some $\sigma \in \{a, b\}$. Analogously, we say that $r \in R_+ \langle\langle a, b \rangle\rangle$ is degenerate if, for some $\sigma \in \{a, b\}$, $(r, w) \neq 0$ implies $w \in \{\sigma\}^*$.

Given a monoid $\mathcal{M}$, let us assume that a length function $|\cdot| : \mathcal{M} \longrightarrow \mathbb{N}$ is well-defined for $\mathcal{M}$ such that the set $\{x \in \mathcal{M} : |x| = n\}$ is finite for each $n \in \mathbb{N}$. Then, for any formal series $s : \mathcal{M} \longrightarrow \mathbb{C}$ we define the *maximum function* $g_s : \mathbb{N} \longrightarrow \mathbb{R}_+$ as

$$g_s(n) = \max\{|(s, x)| \; : \; x \in \mathcal{M}, |x| = n\} \qquad \text{(for every } n \in \mathbb{N}).$$

In the following $|\cdot|$ will denote both the modulus of a complex number and the length of a word: the meaning will be clear from the context.

Our main results concern the order of magnitude of $g_r(n)$ for some $r \in \mathbb{R}_+[[a, b]]$. To state them precisely we use the symbol Θ with the standard meaning: given two sequences $\{f_n\}, \{g_n\} \subseteq \mathbb{R}_+$, the equality $g_n = \Theta(f_n)$ means that for some pair of positive constant c_1, c_2, the relation $c_1 f_n \leq g_n \leq c_2 f_n$ holds for any n large enough.

3   Convergence in Distribution

Let X be a random variable (r.v.) with values in a set $\{x_0, x_1, \ldots, x_k, \ldots\}$ of real numbers and set $p_k = \Pr\{X = x_k\}$, for every $k \in \mathbb{N}$. We denote by F_X its distribution function, i.e. $F_X(\tau) = \Pr\{X \leq \tau\}$ for every $\tau \in \mathbb{R}$. If the set of indices $\{k \mid p_k \neq 0\}$ is finite we can consider the moment generating function of X, given by

$$\Psi_X(z) = \mathbb{E}(e^{zX}) = \sum_{k \in \mathbb{N}} p_k e^{zx_k} \; ,$$

which is well-defined for every $z \in \mathbb{C}$. This function can be used to compute the first two moments of X, since $\mathbb{E}(X) = \Psi_X'(0)$ and $\mathbb{E}(X^2) = \Psi_X''(0)$, and to prove convergence in distribution. We recall that, given a sequence of random variables $\{X_n\}_n$ and a random variable X, X_n converges to X *in distribution* (or *in law*) if $\lim_{n \to \infty} F_{X_n}(\tau) = F_X(\tau)$ for every point $\tau \in \mathbb{R}$ of continuity for F_X. It is well-known that if Ψ_{X_n} and Ψ_X are defined all over $\mathbb{C}$ and $\Psi_{X_n}(z)$ tends to $\Psi_X(z)$ for every $z \in \mathbb{C}$, then X_n converges to X in distribution [9].

A convenient approach to prove the convergence in law to a Gaussian random variable relies on the so called "quasi-power" theorems introduced in [11] and implicitly used in the previous literature [1] (see also [7]). For our purpose we present the following simple variant of such a theorem.

Theorem 1. *Let $\{X_n\}$ be a sequence of random variables, where each X_n takes values in $\{0, 1, \ldots, n\}$ and let us assume the following conditions:*

C1 *There exist two functions $r(z)$, $y(z)$, both analytic at $z = 0$ where they take the value $r(0) = y(0) = 1$, and a positive constant c, such that for every $|z| < c$*

$$\Psi_{X_n}(z) = r(z) \cdot y(z)^n \left(1 + O(n^{-1})\right);\tag{1}$$

C2 *The constant $\sigma = y''(0) - (y'(0))^2$ is strictly positive (variability condition).*

Also set $\mu = y'(0)$. Then $\frac{X_n - \mu n}{\sqrt{\sigma n}}$ converges in distribution to a normal random variable of mean 0 and variance 1, i.e. for every $x \in \mathbb{R}$

$$\lim_{n \longrightarrow +\infty} Pr\left\{\frac{X_n - \mu n}{\sqrt{\sigma n}} \leq x\right\} = \frac{1}{\sqrt{2\pi}} \int_{-\infty}^{x} e^{-\frac{t^2}{2}} dt.$$

The main advantage of this theorem, with respect to other classical statements of this kind, is that it does not require any condition of independence concerning the random variables X_n. For instance, the standard central limit theorems assume that each X_n is a partial sum of the form $X_n = \sum_{j \leq n} U_j$, where the U_j's are independent random variables [9].

4   A General Criterion for Local Convergence Laws

Convergence in law of a sequence of r.v.'s $\{X_n\}$ does not yield an approximation of the probability that X_n has a specific value. Theorems concerning approximations for expressions of the form $Pr\{X_n = x\}$ are usually called *local limit theorems* and often give a stronger property than a traditional convergence in distribution[1]. A typical example is given by the so-called de Moivre-Laplace Local Limit Theorem [9], which intuitively states that, for certain sequences of binomial random variables X_n, up to a factor $\Theta(\sqrt{n})$ the probability that X_n takes on a value x approximates a Gaussian density at x.

In this section we present a general criterion that guarantees, for a sequence of discrete random variables, the existence of a local convergence property of a Gaussian type more general than DeMoivre-Laplace's Theorem mentioned above. In the subsequent section, using such criterion, we show that the same local convergence property holds for certain pattern statistics.

Theorem 2 (Local Limit Criterion). *Let $\{X_n\}$ be a sequence of random variables such that, for some integer $d \geq 1$ and every $n \geq d$, X_n takes on values only in the set*

$$\{x \in \mathbb{N} \mid 0 \leq x \leq n, x \equiv \rho \,(mod\ d)\}\tag{2}$$

for some integer $0 \leq \rho < d$. Assume that conditions C1 and C2 of Theorem 1 hold true and let μ and σ be the positive constants defined in the same theorem. Moreover assume the following property:

---

[1] For this reason, theorems showing convergence in distribution of a sequence of r.v.'s are sometimes called global or integral limit theorems.

C3 *For all* $0 < \theta_0 < \pi/d$ $\lim\limits_{n \to +\infty} \left\{ \sqrt{n} \sup\limits_{|\theta| \in [\theta_0, \pi/d]} |\Psi_{X_n}(i\theta)| \right\} = 0$

Then, as n grows to $+\infty$ the following relation holds uniformly for every $j = 0, 1, \ldots, n$:

$$Pr\{X_n = j\} = \begin{cases} \dfrac{de^{-\frac{(j-\mu n)^2}{2\sigma n}}}{\sqrt{2\pi \sigma n}} \cdot (1 + o(1)) & \text{if } j \equiv \rho \ (mod \ d) \\ \\ 0 & \text{otherwise} \end{cases} \tag{3}$$

Due to space constraints, the proof is omitted. We only note that it extends some ideas used in [3, Section 5] and [4, Theorem 3]. However, the present approach is much more general: we drop any rationality hypothesis on the distribution of the r.v.'s X_n and only rely on conditions C1, C2, C3 together with the assumption that each X_n takes on values only in the set (2).

Observe that $\Psi_{X_n}(i\theta)$ is the so-called characteristic function of X_n, it is periodic of period 2π and it assumes the value 1 at $\theta = 0$. Condition C3 states that, for every constant $0 < \theta_0 < \pi/d$, as n grows to $+\infty$, the value $\Psi_{X_n}(i\theta)$ is of the order $o(n^{-1/2})$ uniformly with respect to $\theta \in [-\pi/d, -\theta_0] \cup [\theta_0, \pi/d]$. Note that ρ may depend on n even if $\rho = \Theta(1)$.

One can easily show that any sequence $\{X_n\}$ of binomial r.v.'s of parameters n and p, where $0 < p < 1$ (i.e. representing the number of successes over n independent trials of probability p), satisfies the hypothesis of the theorem with $d = 1$. In this case (3) corresponds to DeMoivre-Laplace Local Limit Theorem. Thus our general criterion includes the same theorem as a special case.

Relations of the form (3) already appeared in the literature. In particular in [9, Section 43], (3) is proved when X_n is the sum of n independent *lattice* r.v.'s of equal distribution and maximum span d. Note that our theorem is quite general since it does not require any condition of independence for the X_n's.

We also note that, for $d = 1$ a similar criterion for local limit laws has been proposed in [7, Theorem 9.10] where, however, a different condition is assumed, i.e. one requires that the probability generating function $p_n(u)$ of X_n has a certain expansion, for $u \in \mathbb{C}$ belonging to an annulus $1 - \varepsilon \leq |u| \leq 1 + \varepsilon$ ($\varepsilon > 0$), that corresponds to assume an equation of the form (1) for $z \in \mathbb{C}$ such that $|\Re e(z)| \leq \delta$ (for some $\delta > 0$).

5   Pattern Statistics Over Rational Series

In this section we turn our attention to sequences of random variables defined by means of rational formal series in two non-commuting variables. We recall definitions and properties introduced in [3, 4].

Let us consider the binary alphabet $\{a, b\}$ and, for $n \in \mathbb{N}$, let $\{a, b\}^n$ denote the set of all words of length n in $\{a, b\}^*$. Given a formal series $r \in \mathbb{R}_+\langle\langle a, b \rangle\rangle$, let n be a positive integer such that $(r, x) \neq 0$ for some $x \in \{a, b\}^n$. Consider the probability space of all words in $\{a,b\}^n$ equipped with the probability measure given by

$$\Pr\{\omega\} = \frac{(r,\omega)}{\sum_{x \in \{a,b\}^n}(r,x)} \qquad (\omega \in \{a,b\}^n). \tag{4}$$

In particular, if r is the characteristic series χ_L of a language $L \subseteq \{a,b\}^*$, then $\Pr$ is just the uniform distribution over the set of words on length n in L: $\Pr\{\omega\} = \sharp(L \cap \{a,b\}^n)^{-1}$ if $\omega \in L$, while $\Pr\{\omega\} = 0$ otherwise. We define the random variable $Y_n : \{a,b\}^n \rightarrow \{0,1,\ldots,n\}$ such that $Y_n(\omega) = |\omega|_a$ for every $\omega \in \{a,b\}^n$. Then, for every $j = 0,1,\ldots,n$, we have

$$\Pr\{Y_n = j\} = \frac{\sum_{|\omega|=n,|\omega|_a=j}(r,\omega)}{\sum_{x \in \{a,b\}^n}(r,x)}. \tag{5}$$

For sake of brevity, we say that Y_n counts the occurrences of a in the stochastic model defined by r. If $r = \chi_L$ for some $L \subseteq \{a,b\}^*$, then Y_n represents the number of occurrences of a in a word chosen at random in $L \cap \{a,b\}^n$ under uniform distribution.

A useful tool to study the distribution of the pattern statistics Y_n is given by certain generating functions associated with formal series. Given $r \in \mathbb{R}_+\langle\langle a,b\rangle\rangle$, for every $n,j \in \mathbb{N}$ let $r_{n,j}$ be the coefficient of $a^j b^{n-j}$ in the commutative image $\varphi(r)$ of r, i.e.

$$r_{n,j} = (\varphi(r),a^j b^{n-j}) = \sum_{|x|=n,|x|_a=j}(r,x).$$

Then, we define the function $r_n(z)$ and the generating function $\mathbf{r}(z,w)$ by

$$r_n(z) = \sum_{j=0}^n r_{n,j}\, e^{jz} \qquad \text{and} \qquad \mathbf{r}(z,w) = \sum_{n=0}^{+\infty} r_n(z)\, w^n = \sum_{n=0}^{+\infty}\sum_{j=0}^n r_{n,j}\, e^{jz}\, w^n$$

where z and w are complex variables. Thus, from the definition of $r_{n,j}$ and from equation (5) we have

$$\Pr\{Y_n = j\} = \frac{r_{n,j}}{r_n(0)}, \qquad \Psi_{Y_n}(z) = \frac{r_n(z)}{r_n(0)}. \tag{6}$$

Moreover, we remark that the relation between a series r and its generating function $\mathbf{r}(z,w)$ can be expressed in terms of a semiring morphism. Denoting by $\Sigma^\oplus$ the free commutative monoid over the alphabet Σ, consider the monoid morphism $\mathcal{H} : \{a,b\}^* \longrightarrow \{e^z,w\}^\oplus$ defined by setting $\mathcal{H}(a) = e^z w$ and $\mathcal{H}(b) = w$. Then, such a map extends to a semiring morphism from $\mathbb{R}_+\langle\langle a,b\rangle\rangle$ to $\mathbb{R}_+[[e^z,w]]$ so that

$$\mathcal{H}(r) = \mathbf{r}(z,w) \tag{7}$$

for every $r \in \mathbb{R}_+\langle\langle a,b\rangle\rangle$. This property translates arithmetic relations among formal series into analogous relations among the corresponding generating functions.

When r is rational, the probability spaces given by (4) define a stochastic model (called *rational* stochastic model) of interest for the analysis of pattern statistics. A typical goal in that context is to estimate the limit distribution of the number of occurrences of patterns in a word of length n generated at random according to a given probabilistic model (usually a Markovian process [13]). In the rational model, the pattern is reduced to a single letter a. However, the analysis of Y_n in such a model includes as a particular case the study of the frequency of occurrences of regular patterns in words generated at random by a Markovian process [3, Section 2.1].

The limit distribution of Y_n is studied in [3] in the global sense and in [4] in the local sense, assuming that r admits a primitive linear representation (ξ, μ, η). Set $A = \mu(a)$ and $B = \mu(b)$. Then it is easy to see that in this case

$$\mathbf{r}(z, w) = \xi^T \left(I - w(Ae^z + B)\right)^{-1} \eta . \tag{8}$$

It turns out that Y_n has a Gaussian limit distribution [3, Theorem 4], and this extends a similar result, earlier presented in [13] for pattern statistics in a Markovian model. A local limit property of the form (3) also holds, where d is the so-called x-period of $Ax + B$ [4, Theorem 4].

We recall (see [4]) that given a polynomial $f = \sum_k f_k x^k \in \mathbb{R}_+[x]$, the x-period of f is defined as the value $D(f) = \mathrm{GCD}\{|h - k| : f_h \neq 0 \neq f_k\}$, where we assume $\mathrm{GCD}(\{0\}) = \mathrm{GCD}(\emptyset) = +\infty$. For any matrix $M \in \mathbb{R}_+[x]^{m \times m}$ and any index $q \in \{1, 2, \ldots, m\}$, the x-period of q is the value $d(q) = \mathrm{GCD}\{D((M^n)_{qq}) \mid n \geq 0\}$, assuming that every non-zero element in $\mathbb{N} \cup \{+\infty\}$ divides $+\infty$. It turns out that, for every matrix $M(x) \in \mathbb{R}_+[x]^{m \times m}$ such that $M(1)$ is primitive, all indices have the same x-period, which is called the x-period of M.

We conclude this section presenting some results proved in [4] we use in the following section. They give interesting properties of the x-period of a matrix of the form $M_1 x + M_2$, where $M_1 \neq 0 \neq M_2$ and $M_1 + M_2$ is primitive.

Proposition 3. *Consider two non-null matrices $M_1, M_2 \in \mathbb{R}_+^{m \times m}$, assume that $M_1 + M_2$ is primitive and let d be the x-period of $M_1 x + M_2$. Then d is finite and there exists an integer $0 \leq \gamma < d$ such that for any pair of indices $p, q \in \{1, 2, \cdots, m\}$ and for any integer n large enough, we have $D((M_1 x + M_2)^n_{pq}) \equiv \gamma n + \delta_{pq} \pmod{d}$ for a suitable integer $0 \leq \delta_{pq} < d$ independent of n.*

Proposition 4. *Consider two non-null matrices $M_1, M_2 \in \mathbb{R}_+^{m \times m}$, assume that $M_1 + M_2$ is primitive and denote by λ its Perron-Frobenius eigenvalue. Moreover let d be the x-period of $M_1 x + M_2$. Then, for any real number $\theta \neq 2k\pi/d$ $(k \in \mathbb{Z})$, all eigenvalues of $M_1 e^{i\theta} + M_2$ are in modulus smaller than λ.*

6   Pattern Statistics in the Power Model

In this section, we consider a stochastic model defined by the *power* of any primitive rational series (note that in this case the model is not primitive anymore) and we study the central and local behaviour of the associated pattern statistics Y_n. The results we obtain extend the analysis developed in [3] and [4] concerning the primitive rational stochastic models. They also extend some results presented in [5], where the (global) limit distribution of Y_n is determined whenever r is the product of two primitive formal series.

Theorem 5. *For any positive integer k and any primitive nondegenerate $r \in \mathbb{R}_+\langle\langle a,b \rangle\rangle$, let s be defined by $s = r^k$ and let Y_n be the random variables counting the occurrences of a in the model defined by s. Then the following properties hold true.*

T1 *There exist two constants α and β, satisfying $0 < \alpha$ and $0 < \beta < 1$, such that $\frac{Y_n - \beta n}{\sqrt{\alpha n}}$ converges in distribution to a normal random variable of mean value 0 and variance 1.*

T2 *If (ξ, μ, η) is a primitive linear representation for r and d is the x-period of $\mu(a)x + \mu(b)$, then there exist d many functions $C_i : \mathbb{N} \longrightarrow \mathbb{R}_+$, $i = 0, 1, \ldots, d-1$, such that $\sum_i C_i(n) = 1$ for every $n \in \mathbb{N}$ and further, as n grows to $+\infty$, the relation*

$$Pr\{Y_n = j\} = \frac{d \, C_{\langle j \rangle_d}(n)}{\sqrt{2\pi\alpha n}} \, e^{-\frac{(j - \beta n)^2}{2\alpha n}} \cdot (1 + o(1)) \tag{9}$$

holds uniformly for every $j = 0, 1, \ldots, n$ (here $\langle j \rangle_d = j - \lfloor j/d \rfloor d$).

We note that in case $k = 1$ statement T1 coincides with [3, Theorem 4] and statement T2 corresponds to [4, Theorem 4].

We split the proof of the previous theorem in two separate parts and we use the criteria presented in Theorem 1 and in Theorem 2. We still use the notation introduced in the previous section: set $A = \mu(a)$, $B = \mu(b)$, $M = A + B$ and denote by λ the Perron Frobenius eigenvalue of the primitive matrix M.

Proof of T1. Since $s = r^k$, by applying the morphism $\mathcal{H}$ defined in (7) we get

$$\mathbf{s}(z, w) = \mathbf{r}(z, w)^k .$$

From equation (8), since $A + B$ is primitive and both A and B are non-null, one can show [3, Section 4] that near the point $(0, \lambda^{-1})$ the function $\mathbf{r}(z, w)$ admits a Laurent expansion of the form

$$\mathbf{r}(z, w) = \frac{R(z)}{1 - u(z)w} + O(1)$$

where $R(z)$ and $u(z)$ are complex functions, non-null and analytic at $z = 0$. Moreover, the constants $\alpha = (u''(0) - \beta^2)/\lambda$ and $\beta = u'(0)/\lambda$ are strictly positive. We also recall that α and β can be expressed as function of the matrix M and in particular of its eigenvectors.

As a consequence, in a neighbourhood of $(0, \lambda^{-1})$ we have

$$\mathbf{s}(z, w) = \left(\frac{R(z)}{1 - u(z)w}\right)^k + O\left(\frac{1}{1 - u(z)w}\right)^{k-1}$$

and hence the associated sequence is of the form

$$s_n(z) = R(z)^k \binom{n + k - 1}{k - 1} u(z)^n + O\left(n^{k-2} u(z)^n\right).$$

Now, by the definition of our stochastic model, the characteristic function of Y_n is given by $\Psi_{Y_n}(z) = s_n(z)/s_n(0)$ and hence, in a neighbourhood of $z = 0$, it has an expansion of the form

$$\Psi_{Y_n}(z) = \frac{s_n(z)}{s_n(0)} = \left(\frac{R(z)}{R(0)}\right)^k \cdot \left(\frac{u(z)}{\lambda}\right)^n \cdot (1 + \mathrm{O}(n^{-1})) .$$

As a consequence, both conditions of Theorem 1 hold with $\mu = \beta$ and $\sigma = \alpha$ and this proves the result.   □

Proof of T2 (Outline). For every $p, q \in \{1, 2, \ldots, m\}$, let $r^{(pq)}$ be the series defined by the linear representation $(\xi_p e_p,\ \mu,\ \eta_q e_q)$, where e_i is the characteristic array of entry i. Then $r = \sum_{p,q} r^{(pq)}$. Thus, since $s = r^k$, we have

$$s = \sum_* r^{(p_1 q_1)} \cdot r^{(p_2 q_2)} \cdots r^{(p_k q_k)} \tag{10}$$

where the sum is over all sequences $\ell = p_1 q_1 p_2 q_2 \cdots p_k q_k \in \{1, 2, \ldots, m\}^{2k}$. For sake of brevity, for every such ℓ, let $r^{(\ell)}$ be the series

$$r^{(\ell)} = r^{(p_1 q_1)} \cdot r^{(p_2 q_2)} \cdots r^{(p_k q_k)} \tag{11}$$

and let $Y_n^{(\ell)}$ denote the r.v. counting the occurrences of a in the model defined by $r^{(\ell)}$. Then, the primitivity hypothesis allows one to prove that the relation

$$\Pr\{Y_n = j\} = \sum_* C_\ell \Pr\{Y_n^{(\ell)} = j\} + \mathrm{O}(n^{-1})$$

holds for every $j \in \{0, 1, \ldots, n\}$, where C_ℓ is a non-negative constant for each ℓ and $\sum_\ell C_\ell = 1$.

Thus, to determine the local behaviour of $\{Y_n\}$, we first study $\{Y_n^{(\ell)}\}$ for any ℓ such that $r^{(\ell)} \neq 0$. Indeed, by the previous relation, it is sufficient to prove that the equation

$$\Pr\{Y_n^{(\ell)} = j\} = \begin{cases} \dfrac{d\, e^{-\frac{(j - \beta n)^2}{2\alpha n}}}{\sqrt{2\pi\alpha n}} \cdot (1 + o(1)) & \text{if } j \equiv \rho_\ell \ (\mathrm{mod}\ d) \\[4mm] 0 & \text{otherwise} \end{cases}$$

holds uniformly for every $j = 0, 1, \ldots, n$, where α and β are defined as in T1, while ρ_ℓ is an integer (possibly depending on n) such that $0 \leq \rho_\ell < d$ (in particular $C_i(n) = \sum_{\rho_\ell = i} C_\ell$ for each i). To this aim, we simply have to show that, for every $n \in \mathbb{N}$, $Y_n^{(\ell)}$ satisfies the hypotheses of Theorem 2.

First, one can prove that $Y_n^{(\ell)}$ takes on values only in a set of the form (2). This is a consequence of the fact that the values $r_{n,j}^{(\ell)}$ are given by the convolutions

$$r_{n,j}^{(\ell)} = \sum_{\substack{n_1 + n_2 + \cdots + n_k = n \\ j_1 + j_2 + \cdots + j_k = j}} \prod r_{n_i, j_i}^{(p_i q_i)}$$

together with Proposition 3.

As far as condition C1 and C2 are concerned, we can argue (with obvious changes) as in the proof of T1 and observe that the two constants α and β are the same for all series $r^{(\ell)} \neq 0$, since they depend on the matrices A and B (not on the initial and final arrays).

To prove condition C3 let us consider the generating function of $\{r_n^{(\ell)}(z)\}$, obtained by applying the morphism $\mathcal{H}$ to (11):

$$\mathbf{r}^{(\ell)}(z, w) = \prod_{j=1}^{k} \xi_{p_j} (I - w(Ae^z + B))^{-1}{}_{p_j q_j} \eta_{q_j} .$$

Applying Proposition 4, one can prove that all singularities of $\mathbf{r}^{(\ell)}(i\theta, w)$ are in modulus greater than λ^{-1}. Hence, by Cauchy's integral formula, for any arbitrary $\theta_0 \in (0, \pi/d)$ we can choose $0 < \tau < \lambda$ such that the associated sequence $\{r_n^{(\ell)}(i\theta)\}$ is bounded by $O(\tau^n)$ for every $|\theta| \in [\theta_0, \pi/d]$. Analogously, one gets $r_n^{(\ell)}(0) = \Theta(n^{k-1}\lambda^n)$ and this also implies $\Psi_{Y_n^{(\ell)}}(i\theta) = r_n^{(\ell)}(i\theta)/r_n^{(\ell)}(0) = O(\epsilon^n)$ for some $0 < \epsilon < 1$. This yields condition C3 and concludes the proof. $\square$

As a final remark, we note that Theorem 5 cannot be extended to all rational models because the "quasi-power" condition C1 does not hold for $\Psi_{Y_n}(z)$ in the general case. In fact, a large variety of limit distributions for Y_n are obtained in rational models that have two primitive components [5] and more complicated behaviours occur in multicomponent models [8].

7   Estimate of the Maximum Coefficients

The result proved in the last section provides us an asymptotic evaluation for the maximum coefficients of formal series in commuting variables that are commutative image of powers of primitive rational formal series.

Corollary 6. *For any $k \in \mathbb{N}$, $k \neq 0$ and any primitive series $r \in \mathbb{R}_+\langle\langle a, b\rangle\rangle$, let $s = r^k$ and consider its commutative image $S = \varphi(s) \in \mathbb{R}_+[[a, b]]$. Then, for some $\lambda > 0$, the maximum function of S satisfies the relation*

$$g_S(n) = \begin{cases} \Theta(n^{k-(3/2)}\lambda^n) & \text{if } r \text{ is not degenerate} \\ \Theta(n^{k-1}\lambda^n) & \text{otherwise} \end{cases}$$

Proof. Let (ξ, μ, η) be a primitive linear representation of r and let λ be the Perron-Frobenius eigenvalue of $\mu(a) + \mu(b)$. To determine $g_S(n)$ we have to compute the maximum of the values $s_{n,j} = (S, a^j b^{n-j})$ for $j = 0, 1, \ldots, n$.

First consider the case when r is not degenerate. Then, let Y_n count the occurrences of a in the model defined by $s = r^k$ and recall that $\Pr(Y_n = j) = s_{n,j}/s_n(0)$. Now, reasoning as above, we have $s_n(0) = \Theta(n^{k-1}\lambda^n)$ and by Theorem 5, the set of probabilities $\{\Pr(Y_n = j) \mid j = 0, 1, \ldots, n\}$ has the maximum at some integer $j \in [\beta n - d, \beta n + d]$, where it takes on a value of the order $\Theta(n^{-1/2})$ and this proves the first equation.

On the other hand, if r is degenerate, then either $\mu(a) = 0$ or $\mu(b) = 0$. In the first case, all $r_{n,j}$ vanish except $r_{n,0}$ which is of the order $\Theta(\lambda^n)$. Hence for every n, the value $\max_j\{s_{n,j}\} = s_n(0)$ is given by the k-th convolution of $r_{n,0}$, which is of the order $\Theta(n^{k-1}\lambda^n)$. The case $\mu(b) = 0$ is similar.   □

Example. Consider the rational function $(1 - a - b)^{-k}$. Its Taylor expansion near the origin yields the series

$$S = \sum_{n=0}^{+\infty} \binom{n+k-1}{k-1} \sum_{j=0}^{n} \binom{n}{j} a^j b^{n-j}$$

By direct computation, one can verify that

$$g_S(n) = \binom{n+k-1}{k-1}\binom{n}{\lfloor n/2 \rfloor} = \Theta(n^{k-3/2}2^n) \, .$$

In fact, it turns out that $S = \varphi(r^k)$ where $r = \chi_{\{a,b\}^*} \in \mathbb{R}_+\langle\langle a, b\rangle\rangle$.   □

Even though the statement of Theorem 5 cannot be extended to all rational models, we believe that the property given in Corollary 6 well represents the asymptotic behaviour of maximum coefficients of all rational formal series in two commutative variables. We actually think that a similar result holds for all rational formal series in commutative variables. More precisely, let us introduce the symbol $\widehat{\Theta}$ with the following meaning: for any pair of sequences $\{f_n\}, \{g_n\} \subseteq \mathbb{R}_+$, we have $g_n = \widehat{\Theta}(f_n)$ if $g_n = O(f_n)$ and $g_{n_j} = \Theta(f_{n_j})$ for some monotone strictly increasing sequence $\{n_j\} \subseteq \mathbb{N}$. Then we conjecture that the asymptotic behaviour of the maximum function of every rational formal series $t \in \mathbb{R}_+[[\sigma_1, \cdots, \sigma_\ell]]$, is of the form

$$g_t(n) = \widehat{\Theta}\left(n^{k/2}\lambda^n\right)$$

for some integer $k \geq 1 - \ell$ and some $\lambda \in \mathbb{R}_+$.

References

1. E. A. Bender. Central and local limit theorems applied to asymptotic enumeration. *Journal of Combinatorial Theory*, 15:91–111, 1973.
2. J. Berstel and C. Reutenauer. *Rational series and their languages*, Springer-Verlag, New York - Heidelberg - Berlin, 1988.
3. A. Bertoni, C. Choffrut, M. Goldwurm, and V. Lonati. On the number of occurrences of a symbol in words of regular languages. *Theoretical Computer Science*, 302(1-3):431–456, 2003.
4. A. Bertoni, C. Choffrut, M. Goldwurm, and V. Lonati. Local limit distributions in pattern statistics: beyond the Markovian models. *Proceedings 21st S.T.A.C.S.*, V. Diekert and M. Habib editors, Lecture Notes in Computer Science, vol. n. 2996, Springer, 2004, 117–128.

5. D. de Falco, M. Goldwurm, and V. Lonati. Frequency of symbol occurrences in bicomponent stochastic models. Rapporto Interno n. 299-04, D.S.I. Università degli Studi di Milano, April 2004, to appear in *Theoretical Computer Science*. Preliminary version in *Proceedings of the 7th D.L.T. Conference*, Z. Esik and Z. Fülop editors, Lecture Notes in Computer Science, vol. n. 2710, Springer, 2003, 242–253.
6. V. Diekert and G. Rozenberg (editors). *The book of traces*. World Scientific, 1995.
7. P. Flajolet and R. Sedgewick. The average case analysis of algorithms: multivariate asymptotics and limit distributions. *Rapport de recherche* n. 3162, INRIA Rocquencourt, May 1997.
8. M. Goldwurm, and V. Lonati. Pattern occurrences in multicomponent models. Manuscript, September 2004, submitted for pubblication.
9. B.V. Gnedenko. *The theory of probability* (translated by G. Yankovsky). Mir Publishers - Moscow, 1976.
10. P. Henrici. *Elements of numerical analysis*, John Wiley, 1964.
11. H.K. Hwang. *Théorèmes limites pour les structures combinatoires et les fonctions arithmétiques*. Ph.D. Dissertation, École polytechnique, Palaiseau, France, 1994.
12. W. Kuich. Finite automata and ambiguity. Technical Report n.253, I.I.G. University of Graz, 1988.
13. P. Nicodeme, B. Salvy, and P. Flajolet. Motif statistics. *Theoretical Computer Science*, 287(2):593–617, 2002.
14. C. Reutenauer. *Propriétés arithmétiques et topologiques de séries rationnelles en variables non commutatives*, These Sc. Maths, Doctorat troisieme cycle, Université Paris VI, 1977.
15. J. Sakarovitch. *Eléments de théorie des automates*, Vuibert Informatique, 2003.
16. A. Salomaa and M. Soittola. *Automata-Theoretic Aspects of Formal Power Series*, Springer-Verlag, 1978.
17. M.-P. Schützenberger. Finite counting automata. *Information and Control*, 5:91–107, 1962.
18. E. Seneta. *Non-negative matrices and Markov chains*, Springer–Verlag, New York Heidelberg Berlin, 1981.
19. I. Simon. The nondeterministic complexity of a finite automaton. In *Mots*, M. Lothaire editor, Lecture Notes in Mathematics, Hermes, Paris, 1990, 384–400.
20. A. Weber and H. Seidl. On the degree of ambiguity of finite automata. *Theoretical Computer Science*, 88:325–349, 1991.
21. K. Wich. Exponential ambiguity of context-free grammars. In *Proceedings of the 4th DLT*, G. Rozenberg and W. Thomas editors. World Scientific, Singapore, 2000, 125–138.
22. K. Wich. Sublinear ambiguity. In *Proceedings of the 25th MFCS*, M. Nielsen and B. Rovan editors. Lecture Notes in Computer Science, vol. n.1893, Springer-Verlag, 2000, 690–698.

On Codes Defined by Bio-operations

Mark Daley[1] and Michael Domaratzki[2],[*]

[1] Department of Computer Science, University of Western Ontario,
London, ON, Canada
daley@csd.uwo.ca
[2] Jodrey School of Computer Science, Acadia University,
Wolfville, NS, Canada
mike.domaratzki@acadiau.ca

Abstract. We consider the classes of $\oplus$-codes and $\otimes$-codes, which are superclasses of outfix and hypercodes, respectively. These restrictions are based on the synchronized insertion operation, which serves as a model for the gene rearrangement function in certain unicellular organisms. We investigate the classes of $\oplus$-codes and $\otimes$-codes from a theoretical perspective, examine their relationships with traditional code classes and consider related decidability problems.

1 Introduction

The theory of codes is a fundamental area in formal language theory, and is crucial to several applied areas, such as data compression and error detection. Several interesting classes of codes can be defined by use of binary word operations via a fixed language equation, namely $L \cap (L \diamond \Sigma^+) = \emptyset$. These classes include, for example, the classes of prefix, outfix and hypercodes. Recently, languages which are defined by shuffle on trajectories and the same language equation have been studied [5].

In this paper, we examine the classes of languages defined by the same fixed equation and using the synchronized insertion operation, recently defined and studied by Daley *et al.* [2–4]. The synchronized insertion operation models part of the DNA unscrambling process of the stichotrichous ciliates [3]. We note that the synchronized insertion operation falls outside the class of language operations defined by shuffle on trajectories.

These classes of languages have many interesting theoretical properties, and their study is inspired by the biology of gene scrambling in ciliates. One of the hypotheses for the existence of scrambled genes in ciliates is the notion that errors introduced during the descrambling process may speed up the process of evolution. We can view languages in this class as a set of genomes which are maximally evolved in the sense that further applications of the synchronized insertion operation will not allow us to obtain a genome which is a member of

---

[*] Research supported in part by an NSERC PGS-B graduate scholarship.

C.S. Calude, E. Calude and M.J. Dinneen (Eds.): DLT 2004, LNCS 3340, pp. 127–138, 2004.

our starting population. An alternate view is that these languages may represent genomes which are "evolutionary dead-ends".

The concept of codes in relation to DNA has previously received attention in the literature. Kari *et al.* [12, 13] have studied DNA languages which avoid undesirable bonding properties and also posses certain coding properties (such as being code or a solid code). We also note the more recent work of Kari *et al.* [14] which recasts the previous work in a more general framework (the article also contains a list of references for works relating to DNA codewords and the theory of codes).

Most of the previous work on DNA languages and codes has focused on the bonding properties of DNA strands, and the applications of bond-free strands to DNA computing. This work takes an alternate view, by considering languages which satisfy code-like properties under operations implied by nature.

In this paper, we are primarily interested with the theoretical properties of languages which are the analogs of outfix and hyper-codes under the equivalent operations on DNA strands (i.e., those derived from synchronized insertion). We find that synchronized operations yield codes which have interesting differences from the usual ("unsynchronized") operations. For example, every maximal synchronized outfix code must be infinite, a fact which is not seen in the related unsynchronized classes of codes.

2   Preliminaries

Let Σ be an *alphabet*, i.e, a finite set of elements called *letters*. Finite sequences of letters are called *words*. The empty word (the word consisting of no letters) is denoted by ϵ. Let Σ^* be the set of finite all words over Σ, and let $\Sigma^+ = \Sigma^* - \epsilon$. A *language* is any subset of Σ^*.

Given a word $w = w_1 \ldots w_n \in \Sigma^*$, where $w_i \in \Sigma$ for all $1 \leq i \leq n$, its length, denoted $|w|$, is n. For all $w \in \Sigma^*$ and $a \in \Sigma$, $|w|_a$ is the number of occurrences of a in w. If $L \subseteq \Sigma^*$ is a language, let $alph(L) \subseteq \Sigma$ be the set of all letters appearing in some word in L: $alph(L) = \{a \in \Sigma : \Sigma^* a \Sigma^* \cap L \neq \emptyset\}$. Given alphabets Σ, Δ, a *morphism* is any function $h : \Sigma^* \to \Delta^*$ satisfying $h(xy) = h(x)h(y)$ for all $x, y \in \Sigma^*$. Recall that a morphism $h : \Delta^* \to \Sigma^*$ is a *weak coding* if $h(b) \in \Sigma \cup \{\epsilon\}$ for all $b \in \Delta$. For additional background of formal languages, we refer the reader to Rozenberg and Salomaa [17].

We now turn to the *synchronized insertion* operation, originally defined by Daley and Kari [3]. The synchronized insertion operation, denoted $\oplus$, is defined on words $x, y \in \Sigma^*$ as follows: $x \oplus y = \{x_1 \alpha y_1 \alpha x_2 : x_1, x_2, y_1 \in \Sigma^*, \alpha \in \Sigma^+, x = x_1 \alpha x_2, y = y_1 \alpha\}$. It is known that we may take the word α to be of length one without loss of generality [2], that is, we have that

$$x \oplus y = \{x_1 a y_1 a x_2 : x_1, x_2, y_1 \in \Sigma^*, a \in \Sigma, x = x_1 a x_2, y = y_1 a\}.$$

The *synchronized deletion* operation, denoted $\ominus$, is defined on words $x, y \in \Sigma^*$ as follows: $x \ominus y = \{x_1 \alpha x_2 : x_1, x_2, y_1 \in \Sigma^*, \alpha \in \Sigma^+, x = x_1 \alpha y_1 \alpha x_2, y = y_1 \alpha\}$.

Once again, α may be taken to be of length one without loss of generality:

$$x \ominus y = \{x_1 a x_2 \ : \ x_1, x_2, y_1 \in \Sigma^*, a \in \Sigma^+, x = x_1 a y_1 a x_2, y = y_1 a\}.$$

The following result [2], which states that $\ominus$ and $\oplus$ are mutual *left-inverses* (in the sense defined by Kari [11]), will prove useful:

Lemma 2.1. *For all* $x, y, z \in \Sigma^*$, $x \in y \oplus z \iff y \in x \ominus z$.

Define the *shuffle* of two words $x, y \in \Sigma^*$ as follows:

$$x \sqcup y = \{\prod_{i=1}^{n} x_i y_i \ : \ x = \prod_{i=1}^{n} x_i, y = \prod_{i=1}^{n} y_i; \ x_i, y_i \in \Sigma^* (1 \le i \le n)\}.$$

Define the insertion operation as $x \leftarrow y = \{x_1 y x_2 \ : \ x_1, x_2 \in \Sigma^*, x_1 x_2 = x\}$.

We introduce the operation of *synchronized scattered insertion*[1], denoted $\otimes$, on words $x, y \in \Sigma^*$ as follows:

$$x \otimes y = \left\{ \begin{array}{l} (\prod_{i=1}^{n} x_i a_i y_i a_i) x_{n+1} \ : \ x = (\prod_{i=1}^{n} x_i a_i) x_{n+1}, \\[2mm] y = (\prod_{i=1}^{n} y_i a_i), x_j, y_i \in \Sigma^*, a_i \in \Sigma, (1 \le j \le n+1, 1 \le i \le n) \end{array} \right\}.$$

Further, we introduce the analogous *synchronized scattered deletion* operation, denoted $\odot$, on words $x, y \in \Sigma^*$ as follows:

$$x \odot y = \left\{ \begin{array}{l} (\prod_{i=1}^{n} x_i a_i) x_{n+1} \ : \ x = (\prod_{i=1}^{n} x_i a_i y_i a_i) x_{n+1}, \\[2mm] y = (\prod_{i=1}^{n} y_i a_i), x_j, y_i \in \Sigma^*, a_i \in \Sigma, (1 \le j \le n+1, 1 \le i \le n) \end{array} \right\}.$$

We note that the contexts a_i in the previous two definitions could be allowed to be of arbitrary length, however, this does not affect the definitions of $\otimes$ or $\odot$.

We note that for all words $x, y \in \Sigma^*$, $x \oplus y \subseteq x \otimes y$ (resp., $x \ominus y \subseteq x \odot y$). However, equality does not hold, as is easily observed. We note that $\otimes$ (resp., $\odot$) serves as the *transitive closure* of $\oplus$ (resp., $\ominus$), in the same sense that shuffle $\sqcup$ is the transitive closure of insertion $\leftarrow$ (see Domaratzki [5] for definitions relating trajectory-based operations to binary relations and transitivity). This is our motivation for introducing $\otimes$ and $\odot$; we note that these operations themselves are not designed to model a single step in any biological process.

We extend the definition of these operations from words to languages as follows: for all $\diamond \in \{\ominus, \oplus, \leftarrow, \sqcup, \otimes, \odot\}$, and all $L_1, L_2 \subseteq \Sigma^*$,

$$L_1 \diamond L_2 = \bigcup_{\substack{x \in L_1 \\ y \in L_2}} x \diamond y.$$

---

[1] We avoid the term *synchronized shuffle*, which has been used in the literature to denote a different operation, see, e.g., Latteux and Roos [15].

A language L is a *code* (or *-*code*, if there is confusion) if each word $w \in L^*$ has a unique decomposition in L, i.e., for all $n, m \geq 0$, $w_1, \ldots, w_n, u_1, \ldots, u_m \in L$, $u_1 u_2 \cdots u_m = w_1 w_2 \cdots w_n$ implies that $n = m$ and $u_i = w_i$ for all $1 \leq i \leq n$.

A language $L \subseteq \Sigma^+$ is said to be a *prefix code* (resp., *suffix, outfix, hyper* code) if $L\Sigma^+ \cap L = \emptyset$ (resp. $\Sigma^+ L \cap L = \emptyset$, $(L \leftarrow \Sigma^+) \cap L = \emptyset$, $(L \sqcup \Sigma^+) \cap L = \emptyset$). A language $L \subseteq \Sigma^+$ is said to be a *biprefix code* if L is both a prefix and suffix code. For more background on codes, see Berstel and Perrin [1], or Jürgensen and Konstantinidis [10].

We note that the prefix, suffix, outfix and hypercodes are defined by an equation of the form $L \cap (L \diamond \Sigma^+) = \emptyset$ for some $\diamond$. This motivates our main definitions, which relate the notions of codes to bio-operations. For all non-empty languages $L \subseteq \Sigma^+$, say that L is a

1. *synchronized outfix code* (or, briefly, $\oplus$-code) if $(L \oplus \Sigma^+) \cap L = \emptyset$;
2. *synchronized infix code* (or, briefly, $\oplus^R$-code) if $(\Sigma^+ \oplus L) \cap L = \emptyset$;
3. *synchronized hypercode* (or, briefly, $\otimes$-code) if $(L \otimes \Sigma^+) \cap L = \emptyset$.

Informally, we collectively refer to synchronized infix, synchronized outfix and hypercodes as *bio-codes*. We note some examples:

Example 2.1. Let $L = \{a^n b^n : n \geq 1\} \subseteq \{a, b\}^+$. Then L is a $\oplus$-code. To see this, let $x \in a^i b^i \oplus y$ for some $y \in \{a, b\}^+$. If $y \in \{a, b\}^* a$, then $x \in \{a^j y a^{i-j} b^i : 1 \leq j \leq i\}$. But certainly then $x \notin \{a^n b^n : n \geq 1\}$. The case where $y \in \{a, b\}^* b$ is similar. Thus, L is a $\oplus$-code. L is not a $\otimes$-code as $ab \otimes ab \ni aabb$.

Example 2.2. Let $L_{ab} = \{x \in \{a, b\}^+ : |x|_a = |x|_b\}$. Then as $abab \in ab \oplus ba$, $L \cap (L \oplus \{a, b\}^+) \neq \emptyset$. Thus, L_{ab} is not a $\oplus$-code.

Example 2.3. The language $L = \{a, ab, ba, b\}$ is a $\otimes$-code. Note that L is not a hypercode, as $ab \in a \sqcup b$.

We briefly note the motivation for the definitions of $\oplus$- and $\otimes$-codes. As the operation $\oplus$ is modelled to represent the modification of DNA by the insertion of a strand of DNA between marked positions, languages which are $\oplus$-codes are sets of DNA strands which are immune to further insertion of genetic material via the $\oplus$ operation. As for $\otimes$-codes, these represents sets of DNA strands which are immune not only to one insertion via the $\oplus$ operation, but several of these insertions.

3   The Synchronized Scattered Insertion and Deletion Operations

As the operations $\otimes, \odot$ have not been investigated before, we begin by noting some of their properties. The following result states that $\otimes$ and $\odot$ are mutual left-inverses of each other. The proof is straight-forward, and is left to the reader.

Lemma 3.1. *For all $x, y, z \in \Sigma^*$, $x \in y \otimes z \iff y \in x \odot z$.*

We will also need the following technical lemma:

Lemma 3.2. *Let Σ be an alphabet. Let $u, \beta_1, \beta_2 \in \Sigma^*$ $z \in \Sigma^+$ and $a \in \Sigma$. If $ua\beta_1 \in ua\beta_2 \otimes z$, then there exists $z' \in \Sigma^*$ such that $ua\beta_1 \in ua\beta_2 \otimes z'$ and no portion of z' is inserted into ua (i.e., $a\beta_1 \in a\beta_2 \otimes z'$).*

Next, we briefly investigate the closure properties of $\otimes$ and $\odot$.

Lemma 3.3. *Let Σ be an alphabet. There exist an alphabet $\Delta \supseteq \Sigma$, a regular language $R \subseteq \Delta^*$ (resp., $R' \subseteq \Delta^*$) and weak codings $\rho_1, \rho_2, \varphi : \Delta^* \to \Sigma^*$ (resp., $\pi_1, \pi_2, \psi : \Delta^* \to \Sigma^*$) such that for all $L_1, L_2 \subseteq \Sigma^*$, $L_1 \otimes L_2 = \varphi(\rho_1^{-1}(L_1) \cap \rho_2^{-1}(L_2) \cap R)$. (resp., $L_1 \odot L_2 = \psi(\pi_1^{-1}(L_1) \cap \pi_2^{-1}(L_2) \cap R')$).*

Recall that a *cone* (or *full trio*) is a class of languages closed under morphism, inverse morphism and intersection with regular languages. The regular, context-free and recursively enumerable languages are all examples of cones. The following positive closure properties follow immediately:

Corollary 3.1. *Let $\mathcal{C}$ be any cone. Then for all languages L_1, L_2 such that one of L_1, L_2 is in $\mathcal{C}$ and the other is regular, the languages $L_1 \otimes L_2, L_1 \odot L_2 \in \mathcal{C}$.*

Thus, for instance, the regular languages are closed under $\otimes$ and $\odot$.

4   Bio-codes and Other Code Classes

We first note that the use of the word 'code' for $\oplus$-code (as well as $\oplus^R$- and $\otimes$-codes) is somewhat of an abuse of terminology, as $\oplus$-codes are not necessarily $*$-codes. Indeed, if we consider $L = \{a, ab, b\}$, we can verify that L is a $\oplus$-code, but $a \cdot b = ab$, so that L is not a $*$-code. Similarly, L is also a $\otimes$-code.

Further, even if we replace concatenation with synchronized insertion, $\oplus$-codes do not satisfy the corresponding $*$-code property. In particular, say that a language L is a $*$-$\oplus$-code if, for all $n, m \geq 1$ and all $x_i, y_j \in L$ (for $1 \leq i \leq n$ and $1 \leq j \leq m$), $((\cdots(x_1 \oplus x_2) \oplus x_3) \cdots \oplus x_n) = ((\cdots(y_1 \oplus y_2) \oplus y_3) \cdots \oplus y_m)$ implies that $n = m$ and $x_i = y_i$ for all $1 \leq i \leq n$. (A definition which requires not equality but a non-trivial intersection is also possible; it does not affect the observation below.) Note that $L = \{ab, ba\}$ is a $\oplus$-code. However, L is not a $\oplus^*$-code, since $ab \oplus ab = \{abab\} = ab \oplus ba$.

Next, we note that the classes of $*$-codes and $\oplus$-codes are incomparable. As we have noted, not every $\oplus$-code is a $*$-code. Further, the language ab^* is a prefix code, and hence a $*$-code, but it is not a $\oplus$-code. The following result gives relations between classical code classes and their synchronized counterparts.

Lemma 4.1. *Let $L \subseteq \Sigma^+$ be an outfix code (resp., infix code, hypercode). Then L is a $\oplus$-code (resp., $\oplus^R$,-code, $\otimes$-code).*

Further, we note that every $\otimes$-code is a $\oplus$-code. Other relationships between bio-codes and classical code classes are easily verified. We summarize the relationships between codes, $\oplus$-codes (SO), $\otimes$-codes (SH) outfix codes (O), prefix codes (P), suffix codes (S), and hypercodes (H) in Figure 1. The languages referred to in Figure 1 by number are as follows: $L_1 = \{aab, a, ab\}$, $L_2 = \{ab, aabb, a, b\}$, $L_3 = \{a, b, ab\}$, $L_4 = ab^* + d^*c$, $L_5 = \{a, abcd, aabccd, d\}$, $L_6 = \{a, ab, c, dc\}$, $L_7 = ba^*$, $L_8 = \{a^n b^n : n \geq 1\} \cup \{a\}$, $L_9 = \{b, ba\}$, $L_{10} = a^*b$, $L_{11} = \{a^n b^n : n \geq 1\} \cup \{b\}$, $L_{12} = \{b, ab\}$, $L_{13} = ba^*b$, $L_{14} = \{a^n b^n : n \geq 1\}$ $L_{15} = \{ac, abc\}$, $L_{16} = \{a^n ba^n : n \geq 1\}$ and $L_{17} = \{ac, abcd\}$.

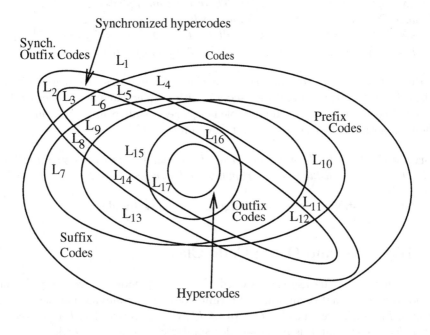

Fig. 1. Relationships between bio-codes and various classes of codes

5   Properties of Bio-codes

We now investigate some general properties of the classes of bio-codes under consideration. The following property holds for outfix codes as well as synchronized outfix codes (see, e.g., Ito *et al.* [9]):

Lemma 5.1. *Let $L \subseteq \Sigma^+$ be a regular language. If L is a $\oplus$-code then L is finite.*

Our main result in this section will be an extension of Higman's Theorem [8], which can interpreted as follows:

Theorem 5.1. *Let Σ be a finite alphabet. Every hypercode over Σ is finite.*

We note also the work of Haines [6], which gives some interesting formal language theoretic consequences of Higman's Theorem. We require the following easily proven lemma:

Lemma 5.2. *Let Σ be a finite alphabet. Let S be an infinite language over Σ. Then there exist $u, v \in \Sigma^*$, $a \in \Sigma$ and $S' \subseteq S$ such that S' is infinite and $S' \subseteq uava\Sigma^*$.*

We now prove our main result. Our proof is based on the proof of Higman's Theorem given by Lothaire [16].

Theorem 5.2. *Let Σ be a finite alphabet. Every synchronized hypercode over Σ is finite.*

Proof. Assume, contrary to what we want to prove, that there is an infinite synchronized hypercode over Σ. Then there exists an infinite sequence $\{x_i\}_{i \geq 1}$ of words over Σ such that $i < j$ implies $x_j \notin x_i \otimes \Sigma^+$. By abuse of terminology, call such a sequence *division-free*.

Choose (using the axiom of choice) a minimal such division-free sequence as follows: let y_1 be the shortest word starting an infinite division-free sequence. Let y_2 be the shortest word such that y_1, y_2 begins an infinite division-free sequence. Continuing in this way, let $\{y_i\}_{i \geq 1}$ be the resulting division-free sequence.

By Lemma 5.2, the infinite set $\{y_i\}_{i \geq 1}$ contains an infinite subset which is contained in $uava\Sigma^*$ for some $a \in \Sigma$ and $u, v \in \Sigma^*$. Let $\{i_j\}_{j \geq 1} \subseteq \mathbb{N}$ be defined such that $y_{i_j} = uava\beta_j$ for all $j \geq 1$. For all $j \geq 1$, define $z_j = ua\beta_j$.

Consider the sequence $\{y_1, y_2, \ldots, y_{i_1-1}, z_1, z_2, \ldots\}$. Clearly, it is smaller than our minimal sequence $\{y_i\}_{i \geq 1}$. Thus, it is not division-free. We have that $y_j \notin y_i \otimes \Sigma^*$ for all $1 \leq i < j < i_1$, by our choice of $\{y_i\}_{i \geq 1}$. Thus, there are two cases:

(a) There exists $1 \leq k < k'$ such that $z_{k'} \in z_k \otimes \Sigma^+$. Let $\alpha \in \Sigma^+$ be such that $z_{k'} \in z_k \otimes \alpha$. Consider $z_{k'} = ua\beta_{k'}$ and $z_k = ua\beta_k$. But now it is clear that there exists α' such that $y_{i_{k'}} = uava\beta_{k'} \in uava\beta_k \otimes \alpha' = y_{i_k} \otimes \alpha'$ by Lemma 3.2. This contradicts that $\{y_i\}_{i=1}^k$ is division-free.

(b) There exist $1 \leq k < i_1$ and $k' \geq 1$ such that $z_{k'} \in y_k \otimes \Sigma^+$. Note that $i_{k'} > k$. Let $\alpha \in \Sigma^+$ be chosen so that $z_{k'} = ua\beta_{k'} \in y_k \otimes \alpha$. It is not hard to show that there exist $\alpha_1, \alpha_2 \in \Sigma^*$ such that $\alpha = \alpha_1\alpha_2$ and $uava\beta_{k'} \in y_k \otimes \alpha_1 va\alpha_2$. This is a contradiction, since $y_{i_{k'}} = uava\beta_{k'}$, and thus $y_{i_{k'}} \in y_k \otimes \Sigma^+$, contradicting that $\{y_i\}_{i \geq 1}$ is division-free.

Thus, we have arrived at a contradiction. $\square$

6   Maximal Bio-codes

Call a $\oplus$-code $L \subseteq \Sigma^+$ *maximal* if L' is not a $\oplus$-code for all $L \subset L' \subseteq \Sigma^+$, where $\subset$ denotes proper inclusion. Similarly, we define maximal $\otimes$-codes and maximal $\oplus^R$-codes.

By Zorn's Lemma, it is easy to see that if $L \subseteq \Sigma^+$ is a $\oplus$-code ($\otimes$-, $\oplus^R$-code), there exists some $L_m \subseteq \Sigma^+$ such that $L \subseteq L_m$ and L_m is a maximal $\oplus$-code ($\otimes$-, $\oplus^R$-code). We can also appeal to dependency theory [10].

We now explicitly demonstrate a maximal $\oplus$-code.

Example 6.1. The language $L \subseteq \{a, b\}^+$ given by

$$L = \{a^n b^n \; : \; n \geq 1\} \cup \{b^n a^n \; : \; n \geq 1\} \cup \{a, b\}$$

is a maximal $\oplus$-code. To see this, consider an arbitrary $x \in \{a, b\}^+$ such that $x \notin L$. We show that $L \cup \{x\}$ is not a $\oplus$-code. We assume that x begins with a; the case where x begins with a b is completely symmetrical. There are three cases:

(a) $x \in a(a + b)^* a$: Let $x = aua$ for some $u \in \{a, b\}^*$. Note that $x \in a \oplus ua \subseteq L \oplus \{a, b\}^+$.
(b) $x \in a^* b^*$: Let $x = a^i b^j$ with $i \neq j$. If $i > j$ then $x \in a^j b^j \oplus a^{i-j}$. If $i < j$ then $x \in a^i b^i \oplus b^{j-i}$. In either case $x \in L \oplus \{a, b\}^+$.
(c) $x \in a^+(b\{a, b\}^* a)b^+$. Let $x = a^i buab^j$ for some $i, j \geq 1$ and $u \in \Sigma^*$. There are two subcases: If $i \geq j$, then note that $x = a^j(a^{i-j}bua)b^j \in a^j b^j \oplus a^{i-j}bua$. If $i < j$, then $x = a^i b(uab^{j-i+1})b^{i-1} \in a^i b^i \oplus uab^{j-i+1}$. In any case, $x \in L \oplus \{a, b\}^+$. Thus, L is a maximal $\oplus$-code.

Example 6.2. The language $L = \{a, ab, ba, b\}$ is a maximal $\otimes$-code.

We now turn to finite maximal $\oplus$-codes. It is easy to see that for $\Sigma = \{a\}$, the language $L = \{a^i\}$ is a maximal $\oplus$-code for any $i \geq 1$. We now show that unary alphabets are the only alphabets for which finite maximal $\oplus$-codes exist. This is in contrast to the case of outfix codes, where there exist finite maximal outfix codes, e.g., $L = \{a^3, ab, ba, b^3\} \subseteq \{a, b\}^+$ [9]. We require the following observation, which is easily established:

Observation 6.1. *Let Σ be an alphabet with $|\Sigma| \geq 2$. Let $a, b \in \Sigma$ with $a \neq b$. For all $x \in \Sigma^+$, $|(x \oplus \Sigma^+) \cap \{a^n b^n \; : \; n \geq 1\}| \leq 1$.*

Lemma 6.1. *Let Σ be an alphabet with $|\Sigma| \geq 2$. If $L \subseteq \Sigma^+$ is a maximal $\oplus$-code, then L is infinite.*

Proof. Let $L \subseteq \Sigma^+$ be a maximal $\oplus$-code. Let $a, b \in \Sigma$ with $a \neq b$. Assume, contrary to what we want to prove, that L is finite. Let $m = |L|$ and $\ell = \max\{|x| \; : \; x \in L\}$.

Consider the set $S = \{a^{\ell+1+j} b^{\ell+1+j} \; : \; 0 \leq j \leq m\}$. Note that $S \cap L = \emptyset$. As L is a maximal $\oplus$-code, $L \cup \{x\}$ is not a $\oplus$-code for all $x \in S$. In particular, since any $x \in S$ is longer than any word in L, for all $x \in S$ there exists some $y_x \in L$ such that $x \in y_x \oplus \Sigma^+$. By Observation 6.1, if $x, x' \in S$ with $x \neq x'$, then $y_x \neq y_{x'}$. But as $|S| > |L|$, this is a contradiction. Thus, L is not a maximal $\oplus$-code, as there is some $x \in S$ such that $x \notin L \oplus \Sigma^+$, and thus $L \cup \{x\}$ is a $\oplus$-code. □

Thus, we note that singleton languages over a unary alphabet are the only regular languages which are maximal $\oplus$-codes.

Corollary 6.1. *If $L \subseteq \Sigma^+$ is a regular maximal $\oplus$-code, then $\Sigma = \{a\}$ and $L = \{a^i\}$ for some $i \geq 1$.*

As a consequence, we also note that, though every outfix code is a $\oplus$-code, not every maximal outfix code is a maximal $\oplus$-code. Indeed, we have previously noted that there exist finite maximal outfix codes over a two letter alphabet. The following characterization of maximal $\oplus$-codes will prove useful:

Lemma 6.2. *Let $L \subseteq \Sigma^+$ be a $\oplus$-code (resp., $\oplus^R$-code, $\otimes$-code). Then L is a maximal $\oplus$-code (resp., maximal $\oplus^R$-code, maximal $\otimes$-code) iff $L \cup (L \oplus \Sigma^+) \cup (L \ominus \Sigma^+) = \Sigma^+$. (resp., $L \cup (\Sigma^+ \oplus L) \cup (\Sigma^+ \ominus L) = \Sigma^+$, $L \cup (L \otimes \Sigma^+) \cup (L \odot \Sigma^+) = \Sigma^+$).*

7   Closure Properties

We now consider the closure properties of the classes of $\oplus$- and $\otimes$-codes. It is clear that every subset of a $\oplus$-code is an $\oplus$-code, and thus the non-empty intersection of an $\oplus$-code with an arbitrary language is an $\oplus$-code.

Further, it is clear that the class of $\oplus$-codes are not closed under union with a singleton language, which is shown by the existence of a maximal $\oplus$-code (Example 6.1).

We now state the positive closure properties:

Theorem 7.1. *The class of $\oplus$-codes is closed under (a) ϵ-free inverse morphism, (b) reversal and (c) quotient with a single word (modulo the empty word).*

The following can also be established in the same manner as Theorem 7.1.

Theorem 7.2. *The class of $\otimes$-codes is closed under (a) ϵ-free inverse morphism, (b) reversal and (c) quotient with a single word (modulo the empty word).*

We now turn to non-closure properties:

Theorem 7.3. *The class of $\otimes$-codes (resp., $\oplus$-codes) is not closed under (a) concatenation (with a single letter), (b) Kleene closure, (c) (1-uniform) morphism, (d) arbitrary inverse morphism, (e) quotient (with an arbitrary finite language), (f) synchronized insertion (with an arbitrary finite language), (g) synchronized scattered insertion (with an arbitrary finite language), and (h) synchronized deletion of a single letter.*

We now demonstrate some more specialized closure properties.

Lemma 7.1. *Let $L_1, L_2 \subseteq \Sigma^+$ be $\oplus$-codes such that L_1 is a prefix code and L_2 is a suffix code. Then $L_1 L_2$ is an $\oplus$-code.*

The following lemma is analogous to the corresponding result for outfix codes [9–Lemma 3.13]:

Lemma 7.2. *Let $L_1, L_2 \subseteq \Sigma^+$. If $L_1 L_2$ is a non-empty $\oplus$-code, then L_1, L_2 are $\oplus$-codes.*

8   Decidability Problems

We now investigate the decidability questions related to bio-codes.

Lemma 8.1. *Let $\diamond \in \{\oplus, \oplus^R\}$. Given a regular language R, it is decidable whether R is a $\diamond$-code.*

For $\otimes$-codes, a better decidability result is possible, due to the fact that all $\otimes$-codes are finite.

Lemma 8.2. *Given a CFL L, it is decidable whether L is a $\otimes$-code.*

We now turn to undecidability:

Lemma 8.3. *Let $L \subseteq \Sigma^+$ be a linear CFL. Then it is undecidable whether L is a $\oplus$-code.*

For $\otimes$-codes, we have seen that it is decidable whether a CFL L is a $\otimes$-code. It is not surprising that it is undecidable whether a given CSL is a $\otimes$-code:

Lemma 8.4. *Given a context-sensitive language $L \subseteq \Sigma^+$, it is undecidable whether L is a $\otimes$-code.*

We now turn to the questions of deciding whether a particular language is a maximal $\oplus$-code.

Lemma 8.5. *Let $\diamond \in \{\oplus, \oplus^R, \otimes\}$. Then given a regular language $R \subseteq \Sigma^+$, it is decidable whether R is a maximal $\oplus$-code over Σ.*

Proof. By Lemma 8.1, given $R \subseteq \Sigma^+$, we can decide if R is a $\diamond$-code. Let $\star \in \{\ominus, \ominus^R, \odot\}$ be chosen so that $\star$ is the left-inverse of $\diamond$. If R is a $\diamond$-code, by Lemma 6.2, R is a maximal $\diamond$-code iff $\Sigma^+ = R \cup (R \diamond \Sigma^+) \cup (R \star \Sigma^+)$. Since all the involved languages are regular, this equality is decidable.   □

Lemma 8.6. *Let $\mathcal{C}$ be any class of languages such that*

(a) $\{a^n b^n : n \geq 0\} \cup \{b^n a^n : n \geq 0\} \cup \{a, b\} \in \mathcal{C}$;
(b) given a language $L \subseteq \Sigma^+$ in $\mathcal{C}$, it is undecidable whether $L = \emptyset$, for $|\Sigma| = 2$;
(c) $\mathcal{C}$ is closed under union and concatenation with a regular language.

Then given a language $L \subseteq \Sigma^+$, it is undecidable whether L is a maximal $\oplus$-code.

Corollary 8.1. *Given $L \in \mathrm{CS}$, it is undecidable whether L is a maximal $\oplus$-code.*

Lemma 8.7. *Let $\mathcal{C}$ be a class of languages such that*

(a) $\mathcal{C}$ is closed under difference, and quotient with a regular language;
(b) $\{a^n b^n : n \geq 0\} \cup \{b^n a^n : n \geq 0\} \cup \{a, b\} \in \mathcal{C}$;
(c) given a language $L \subseteq \Sigma^+$ in $\mathcal{C}$, it is undecidable whether $L = \emptyset$, for $|\Sigma| = 2$.

Then given a $\oplus$-code $L \subseteq \Sigma^+$, it is undecidable whether L is a maximal $\oplus$-code.

Corollary 8.2. *Let Σ be an alphabet with $|\Sigma| \geq 2$. Let $L \subseteq \Sigma^+$ be a recursive $\oplus$-code. Then it is undecidable whether L is a maximal $\oplus$-code.*

9   Synchronized-Insertion Closed Languages and $\oplus$-Codes

Let $L \subseteq \Sigma^+$. Recall [4] the following definition:

$$sins(L) = \{x \in \Sigma^* : L \oplus x \subseteq L\}.$$

We define the class of languages SINS $= \{L \ : \ L \subseteq sins(L)\}$. A language $L \in$ SINS is said to be *sins-closed*. For results on SINS, see Daley *et al.* [4]. We now consider $sins(L)$ when L is a $\oplus$-code.

Lemma 9.1. *Let $L \subseteq \Sigma^+$ be a $\oplus$-code. Then*

$$sins(L) = \{\epsilon\} \cup \{x \in \Sigma^+ \ : \ L \oplus x = \emptyset\} = \{\epsilon\} \cup \Sigma^*(\Sigma - alph(L)).$$

Corollary 9.1. *The following equality holds:*

$$\{L \ : \ L \text{ is a } \oplus \text{-code and } L \cup \{\epsilon\} \in \text{SINS}\} = \{\emptyset\}.$$

Let $L = \{a, aa, b\} \subseteq \{a, b\}^+$. Note that $sins(L) = \{\epsilon\} = \{\epsilon\} \cup \{x \in \Sigma^+ \ : \ L \oplus x = \emptyset\}$. However, L is not a $\oplus$-code. Thus, the converse of Lemma 9.1 does not hold.

10   Concluding Remarks

We have introduced and studied here the $\oplus$-codes and the $\otimes$-codes which are based on an operation inspired by the gene descrambling process found in stichotrichous ciliates.

We have shown that the classes of $\oplus$-codes and $\otimes$-codes are disjoint from the traditionally studied classes of $*$-codes and that all regular $\oplus$-codes and all $\otimes$-codes must be finite. We then considered maximal $\oplus$-codes and demonstrated that, for alphabets of size at least two, all maximal $\oplus$-codes must be infinite. We gave also an effective characterization of maximal $\oplus$-codes.

The classes of $\oplus$-codes and $\otimes$-codes were shown to be closed under nonerasing inverse morphism, reversal and quotient with a singleton (modulo the empty word). The same classes were shown not to be closed under concatenation (with a single letter), Kleene closure, (1-uniform) morphism, arbitrary inverse morphism, quotient (with a finite language), synchronized insertion and synchronized scattered insertion (with a finite language), and synchronized deletion of a single letter.

Turning to problems of decidability, we demonstrated that it is decidable if a regular language is an $\oplus$-code while the same property is undecidable for linear context-free languages. In contrast, we have shown that it is decidable if an arbitrary context-free language is an $\otimes$-code while the same property is, unsurprisingly, undecidable for context-sensitive languages.

Finally, we considered the effect of the *sins* operation [4] on $\oplus$-codes and showed an exclusive relationship between the class of $\oplus$-codes and the class of synchronized insertion closed languages.

The work presented here represents a theoretical investigation of classes of codes defined by an operation inspired by a biological process. While it is our hope that this investigation will prove to be biologically relevant we also feel that it has generated some theoretically compelling results and represents well the rich variety of interesting abstract constructs which may be inferred from, and inspired by, biological systems.

References

1. J. Berstel and D. Perrin. *Theory of Codes*. Available at http://www-igm.univ-mlv.fr/%7Eberstel/LivreCodes/Codes.html, 1996.
2. M. Daley, O. Ibarra, and L. Kari. Closure properties and decision questions of some language classes under ciliate bio-operations. *Theor. Comp. Sci.*, 306(1):19–38, 2003.
3. M. Daley and L. Kari. Some properties of ciliate bio-operations. In M. Ito and M. Toyama, editors, *DLT 2002: Developments in Language Theory, Sixth International Conference*, volume 2450 of *LNCS*, pages 116–127. Springer, 2003.
4. M. Daley, L. Kari, and I. McQuillan. Families of languages defined by ciliate bio-operations. *Theor. Comput. Sci.*, 320(1):51–69, 2004.
5. M. Domaratzki. Trajectory-Based Codes. *Acta Inf.* 40(6–7):491–527, 2004.
6. L. Haines. On free monoids partially ordered by embedding. *J. Comb. Theory*, 6:94–98, 1969.
7. T. Harju and J. Karhumäki. Morphisms. pages 439–510. In [17].
8. G. Higman. Ordering by divisibility in abstract algebras. *Proc. London Math. Soc.*, 2(3):326–336, 1952.
9. M. Ito, H. Jürgensen, H. Shyr, and G. Thierrin. Outfix and infix codes and related classes of languages. *J. Comp. Sys. Sci.*, 43:484–508, 1991.
10. H. Jürgensen and S. Konstantinidis. Codes. pages 511–600. In [17].
11. L. Kari. On language equations with invertible operations. *Theor. Comput. Sci.*, 132:129–150, 1994.
12. L. Kari, R. Kitto, and G. Thierrin. Codes, involutions and DNA encoding. In volume 2300 of *LNCS*, pages 376–393. Springer, 2002.
13. L. Kari, S. Konstantinidis, E. Losseva, and G. Wozniak. Sticky-free and overhang-free DNA languages. *Acta. Inf.*, 40(2):119–157, 2003.
14. L. Kari, S. Konstantinidis, and P. Sosík. On properties of bond-free DNA languages. Technical Report 609, Computer Science Department, University of Western Ontario, 2003. Submitted for publication.
15. M. Latteux and Y. Roos. Synchronized shuffle and regular languages. In J. Karhumäki *et al.* editors, *Jewels are Forever: Contributions on Theoretical Computer Science in Honour of Arto Salomaa*, pages 35–44. Springer, 1999.
16. M. Lothaire. *Combinatorics on Words*. Addison-Wesley, 1983.
17. G. Rozenberg and A. Salomaa, editors. *Handbook of Formal Languages, Vol. I*. Springer-Verlag, 1997.

Avoidable Sets and Well Quasi-Orders[*]

Flavio D'Alessandro[1] and Stefano Varricchio[2]

[1] Dipartimento di Matematica, Università di Roma "La Sapienza", Roma, Italy
dalessan@mat.uniroma1.it
[2] Dipartimento di Matematica, Università di Roma "Tor Vergata", Roma, Italy
varricch@mat.uniroma2.it

Abstract. Let I be a finite set of words and $\Rightarrow_I^*$ be the derivation relation generated by the set of productions $\{\epsilon \rightarrow u \mid u \in I\}$. Let L_I^ϵ be the set of words u such that $\epsilon \Rightarrow_I^* u$. We prove that the set I is unavoidable if and only if the relation $\Rightarrow_I^*$ is a well quasi-order on the set L_I^ϵ. This result generalizes a theorem of [7]. Further generalizations are investigated.

1 Introduction

A *quasi-order* on a set S is called a *well quasi-order* (*wqo*) if every non-empty subset X of S has at least one minimal element in X but no more than a finite number of (non-equivalent) minimal elements.

A set of words I is called *unavoidable* if there exists an integer $k > 0$ such that any word $w \in A^+$, with $A = \text{alph}(I)$ and $|w| \geq k$, contains as a factor a word of I. A finite set I is called *avoidable* if it is not unavoidable.

Well quasi-orders have been widely investigated in the past. We recall the celebrated Higman and Kruskal results [10, 15]. Higman gives a very general theorem on division orders in abstract algebras from which one derives that the *subsequence ordering* in free monoids is a wqo. Kruskal extends Higman's result, proving that certain embeddings on finite trees are well quasi-orders. Some remarkable extensions of the Kruskal theorem are given in [12, 16].

In the last years many papers have been devoted to the applications of wqo's to formal language theory [1–8, 11].

In [7], a remarkable class of grammars, called *unitary grammars*, has been introduced in order to study the relationships between the classes of context-free and regular languages. If I is a finite set of words then we can consider the set of productions

$$\{\epsilon \rightarrow u, \; u \in I\}$$

and the derivation relation $\Rightarrow_I^*$ of the semi-Thue system associated with I. Moreover the language generated by the unitary grammar associated with I

---

[*] This work was partially supported by MIUR project "Linguaggi formali e automi: teoria e applicazioni".

C.S. Calude, E. Calude and M.J. Dinneen (Eds.): DLT 2004, LNCS 3340, pp. 139–150, 2004.
© Springer-Verlag Berlin Heidelberg 2004

is $L_I^\epsilon = \{w \in A^* \mid \epsilon \Rightarrow_I^* w\}$. Unavoidable sets of words are characterized in terms of the wqo property of the unitary grammars. Precisely it is proved that I is unavoidable if and only if the derivation relation $\Rightarrow_I^*$ is a wqo.

In this paper we give the following improvement of the previous result of [7]: *A finite set of words I is unavoidable if and only if the relation $\Rightarrow_I^*$ is a well quasi-order on the language L_I^ϵ.* The crucial step of our main result is the construction of a *bad sequence* of elements of L_I^ϵ, when I is avoidable. As a consequence of our theorem and of some results of [7] one obtains the equivalence of the following conditions:

- I is unavoidable;
- L_I^ϵ is regular;
- $\Rightarrow_I^*$ is a well quasi-order on L_I^ϵ.

It is worth noticing that the problems we have discussed above, may be considered with respect to other quasi-orders. In [9], Haussler investigated the relation $\vdash_I^*$ defined as the transitive and reflexive closure of $\vdash_I$ where $v \vdash_I w$ if

$$v = v_1 v_2 \cdots v_{n+1},$$

$$w = v_1 a_1 v_2 a_2 \cdots v_n a_n v_{n+1},$$

where the a_i's are letters, and $a_1 a_2 \cdots a_n \in I$. In particular, a characterization of the wqo property of $\vdash_I^*$ in terms of subsequence unavoidable sets of words was given in [9]. In the last part of the paper, we focus our attention on a possible extension of our main result with respect to $\vdash_I^*$.

2   Preliminaries

The main notions and results concerning quasi-orders and languages are shortly recalled in this section.

Let A be a finite *alphabet* and let A^* be the free monoid generated by A. The elements of A are usually called *letters* and those of A^* *words*. The identity of A^* is denoted ϵ and called the *empty word*.

A nonempty word $w \in A^*$ can be written uniquely as a sequence of letters as $w = a_1 a_2 \cdots a_n$, with $a_i \in A$, $1 \leq i \leq n$, $n > 0$. The integer n is called the *length* of w and denoted $|w|$. For all $a \in A$, $|w|_a$ denotes the number of occurrences of the letter a in w. Let $w \in A^*$. The word $u \in A^*$ is a *factor* of w if there exist $p, q \in A^*$ such that $w = puq$. If $w = uq$, for some $q \in A^*$ (resp. $w = pu$, for some $p \in A^*$), then u is called a *prefix* (resp. a *suffix*) of w.

The set of all prefixes (resp. suffixes, factors) of w is denoted $\mathrm{Pref}(w)$ (resp. $\mathrm{Suff}(w)$, $\mathrm{Fact}(w)$). A word u is a *subsequence* of a word v if $u = a_1 a_2 \cdots a_n$, $v = v_1 a_1 v_2 a_2 \cdots v_n a_n v_{n+1}$ with $a_i \in A$, $v_i \in A^*$. A subset L of A^* is called a *language*. If L is a language of A^*, then $\mathrm{alph}(L)$ is the smallest subset B of A such that $L \subseteq B^*$. Moreover, $\mathrm{Pref}(L)$ denotes the set of prefixes of all words of L. A language of A^* is called *recognizable* if it is accepted by a finite automaton or, equivalently, *via* the well known characterization of Myhill and Nerode, if it is saturated by a finite index congruence of A^*. The family of recognizable languages of A^* is denoted $Rec(A^*)$. A binary relation $\leq$ on a set S is a *quasi-order* (qo) if $\leq$ is reflexive and transitive. Moreover, if $\leq$ is symmetric, then $\leq$ is an equivalence relation. The meet $\leq \cap \leq^{-1}$ is an equivalence relation $\sim$ and the quotient of S by $\sim$ is a *poset* (partially ordered set). A quasi-order $\leq$ in a semigroup S is *monotone on the right (resp. on the left)* if for all $x_1, x_2, y \in S$

$$x_1 \leq x_2 \text{ implies } x_1 y \leq x_2 y \text{ (resp. } y x_1 \leq y x_2).$$

A quasi-order is *monotone* if it is monotone on the right and on the left.

An element $s \in X \subseteq S$ is *minimal* in X with respect to $\leq$ if, for every $x \in X$, $x \leq s$ implies $x \sim s$. For $s, t \in S$ if $s \leq t$ and s is not equivalent to t mod $\sim$, then we set $s < t$.

A quasi-order in S is called a *well quasi-order* (wqo) if every non-empty subset X of S has at least one minimal element but no more than a finite number of (non-equivalent) minimal elements. We say that a set S is *well quasi-ordered* (wqo) by $\leq$, if $\leq$ is a well quasi-order on S.

There exist several conditions which characterize the concept of well quasi-order and that can be assumed as equivalent definitions (cf. [6]).

Theorem 1. *Let S be a set quasi-ordered by $\leq$. The following conditions are equivalent:*

i. $\leq$ *is a well quasi-order;*
ii. *every infinite sequence of elements of S has an infinite ascending subsequence;*
iii. *if $s_1, s_2, \ldots, s_n, \ldots$ is an infinite sequence of elements of S, then there exist integers i, j such that $i < j$ and $s_i \leq s_j$;*
iv. *there exists neither an infinite strictly descending sequence in S (i.e. $\leq$ is well founded), nor an infinity of mutually incomparable elements of S.*

A partial order satisfying the wqo property is also called a *well partial order*. The quasi-orders considered in this paper are actually partial orders. However, according to the current terminology, we refer to them as quasi-orders.

Let $\sigma = \{s_i\}_{i \geq 1}$ be an infinite sequence of elements of S. Then σ is called *good* if it satisfies condition (iii) of Theorem 1 and it is called *bad* otherwise, that is, for all integers i, j such that $i < j$, $s_i \not\leq s_j$.

It is worth noting that, by condition (iii) above, a useful technique to prove that $\leq$ is a wqo on S is to prove that no bad sequence exists in S.

If ρ and σ are two relations on sets S and T respectively, then the direct product $\rho \otimes \sigma$ is the relation on $S \times T$ defined as

$$(a, b) \, \rho \otimes \sigma \, (c, d) \iff a \, \rho \, c \text{ and } b \, \sigma \, d.$$

The following lemma is well known (*see* [6], Ch. 6).

Lemma 1. *The following conditions hold:*

i. *Every subset of a wqo set is wqo.*
ii. *If S and T are wqo by $\leq_S$ and $\leq_T$ respectively, then $S \times T$ is wqo by $\leq_S \otimes \leq_T$.*

Following [6], we recall that a *rewriting system*, or *semi-Thue system* on an alphabet A is a pair (A, π) where π is a binary relation on A^*. Any pair of words $(p, q) \in \pi$ is called a *production* and denoted by $p \to q$. Let us denote by $\Rightarrow_\pi$ the derivation relation of π, that is, for $u, v \in A^*$, $u \Rightarrow_\pi v$ if

$$\exists\, (p, q) \in \pi \text{ and } \exists\, h,\, k \in A^* \text{ such that } u = hpk,\ v = hqk.$$

The *derivation relation* $\Rightarrow_\pi^*$ is the transitive and reflexive closure of $\Rightarrow_\pi$. One easily verifies that $\Rightarrow_\pi^*$ is a monotone quasi-order on A^*.

A semi-Thue system is called *unitary* if π is a finite set of productions of the kind

$$\epsilon \to u,\ u \in I,\ I \subseteq A^+.$$

Such a system, also called *unitary grammar*, is then determined by the finite set $I \subseteq A^+$. Its derivation relation and its transitive and reflexive closure are denoted by $\Rightarrow_I$ (or, simply, $\Rightarrow$) and $\Rightarrow_I^*$ (or, simply, $\Rightarrow^*$), respectively. We set $L_I^\epsilon = \{u \in A^* \mid \epsilon \Rightarrow^* u\}$.

Unitary grammars have been introduced in [7], where the following theorem is proved.

Theorem 2. *Let I be a finite set of A^+ and assume that $A = \mathrm{alph}(I)$. The following conditions are equivalent:*

i. the derivation relation $\Rightarrow_I^*$ is a wqo on A^*;
ii. the set I is unavoidable;
iii. the language L_I^ϵ is regular.

3   Main Result

The main result of this section will be stated in Corollary 4 which is an improvement of Theorem 2 where Condition i. is substituted by the weaker condition that L_I^ϵ is well quasi ordered by the relation $\Rightarrow_I^*$. In order to achieve this result we have first to prove the following non-trivial theorem.

Theorem 3. *Let I be a finite set of words. If I is avoidable then $\Rightarrow_I^*$ is not a wqo on the language L_I^ϵ.*

The proof of Theorem 3 is divided into the following three cases.

3.1   First Case

We suppose that $\mathrm{Card}(I) = 1$ so that $I = \{w\}$. Set $A = \mathrm{alph}(I)$. Let us first observe that $\mathrm{Card}(A) \geq 2$. Indeed, if $\mathrm{Card}(A) = 1$ then $w = a^k$, $k \geq 1$ so that I is an unavoidable set of A^* which contradicts the assumption on the set I. Hence, w may be factorized as $w = w'ab^k$, where $a, b \in A, a \neq b, w' \in A^*$ and $k > 0$.

Now we construct the bad sequence of L_I^ϵ. For any $n > 0$, let x_n be the word defined as

$$x_n = (w'a)^{n-1}w(w'a)b^{kn}.$$

The following lemma states some useful properties of the words of the sequence $\{x_n\}$.

Lemma 2. *The following conditions hold:*
i. For any $n > 0$, $x_n \in L_I^\epsilon$.
ii. For any $n > 0$, $|x_n| = (n+1)|w|$.

Corollary 1. *Let n, m be positive integers. If $x_n \Rightarrow_I^\ell x_{n+m}$ then $\ell = m$.*

Proof. By condition (ii) of the previous lemma, $|x_{n+m}| = (n+m+1)|w| = |x_n| + m|w|$, which implies that the length of the derivation $x_n \Rightarrow_I^\ell x_{n+m}$ is $\ell = m$. □

Lemma 3. *Let y be a word and let n, ℓ be positive integers. If $x_n \Rightarrow_I^\ell y$ then*
i. $y = y'b^h$ where $y' \notin A^*b$ and $1 \le h \le k(n+\ell)$;
ii. if $h = k(n+\ell)$ then $y = (w'a)^{n-1}w(w'a)^{\ell+1}b^h$.

Proposition 1. $\Rightarrow_I^*$ *is not a wqo on L_I^ϵ.*

Proof. The proof is by contradiction. Suppose that $\Rightarrow_I^*$ is a wqo on L_I^ϵ. Therefore, every sequence of words of L_I^ϵ is good. Hence there exist positive integers n, ℓ such that $x_n \Rightarrow_I^* x_{n+\ell}$. By Corollary 1, one has

$$x_n \Rightarrow_I^\ell y \tag{1}$$

with $y = x_{n+\ell}$. By the fact that $b^{k(n+\ell)}$ is a suffix of y, condition (ii) of Lemma 3 yields

$$y = x_{n+\ell} = (w'a)^{n+\ell-1}w(w'a)b^{k(n+\ell)} = (w'a)^{n-1}w(w'a)^{\ell+1}b^{k(n+\ell)}. \tag{2}$$

Finally Equality (2) yields

$$(w'a)^\ell w = (w'a)^\ell w'ab^k = w(w'a)^\ell,$$

so that $a = b$ which is a contradiction. Hence $\Rightarrow_I^*$ is not a wqo on L_I^ϵ. □

3.2   Second Case

We suppose that $\mathrm{Card}(I) \ge 2$ and, for every letter a of $\mathrm{alph}(I)$, there exists a word of I which begins with a. Set $\mathrm{alph}(I) = A$.

Lemma 4. *Let I be a finite avoidable set of A^+. Then there exists a word $w \in A^+$ such that, for any $n \ge 0$, $\mathrm{Fact}(w^n) \cap I = \emptyset$.*

Proof. Let $X = A^* \setminus A^*IA^*$. Since I is finite, $X \in \mathrm{Rec}(A^*)$. Moreover, since I is avoidable in A^*, X is infinite. By the latter two conditions and by using the well known pumping lemma, one has that there exists a word $v = fwg \in X$ with $f, g, w \in A^*$ such that $w \ne \epsilon$ and, for any $n \ge 0$, $fw^ng \in X$. Since X is closed by factors, we have that, for any $n \ge 0$, $w^n \in X$ and, thus, $\mathrm{Fact}(w^n) \cap I = \emptyset$. □

From now on, w denotes the word defined in the statement of Lemma 4.

Lemma 5. *Let $a \in A$ such that $w \notin aA^*$. Then there exist words $ax, ay \in L_I^\epsilon$ such that $x \notin \mathrm{Suff}(y)$ and $|x| < |y|$.*

Proof. First suppose that there exists a word u of I of period at least two. Hence $u = u'cd^k$ with $u' \in A^*$, $c, d \in A$, $c \neq d$ and $k > 0$. Then $\epsilon \Rightarrow_I^2 (u'c)^2 d^{2k}$. Let $x = avu$ and $y = av(u'b)^2 a^{2k}$ with $av \in I$. Thus, x and y satisfy the claim.

If every word of I has period 1, then there exist words $a^i, b^j \in I$, $a \neq b$. Hence take $x = a^i$ and $y = a^i b^j$. $\square$

Now it is convenient to notice that, by hypothesis, for every $a \in A$, $I \cap aA^* \neq \emptyset$. Hence there exists a word $z \in A^+$ such that $\epsilon \Rightarrow_I^l wz$ with $l > 0$. Therefore the sequence of words $\{z^n\}$ is such that

$$\forall\, n \geq 1, \ \epsilon \Rightarrow_I^{ln} w^n z^n. \tag{3}$$

Let us denote $\{z_n\}$ a sequence of words of A^+ such that, for any $n > 0$, condition (3) holds if one replaces z^n with z_n and such that z_n is of minimal length.

Lemma 6. *The sequence $\{|z_n|\}$ is not upper bounded.*

Proof. By contradiction, suppose that our sequence is upper bounded. Thus there exists a positive integer M such that, for any $n > 0$, $|z_n| < M$. For any $n > 0$, let l_n be the length of the derivation $\epsilon \Rightarrow_I^{l_n} w^n z_n$. Since, for any $n > 0$, $\mathrm{Fact}(w^n) \cap I = \emptyset$, then $l_n < M$ and, hence, $|w^n| < MN$, where N is the maximal length of a word of I. The latter inequality is not possible if $n > MN$. Hence the sequence $\{|z_n|\}$ is not upper bounded. $\square$

By possibly replacing the sequence $\{z_n\}$ with one of its subsequence, Lemma 6 yields the following corollary.

Corollary 2. *The sequence of words $\{z_n\}$ is such that, for any $n, m > 0$, $|z_n| + |y| < |z_{n+m}|$ where y is the word defined in Lemma 5.*

Now, starting from the words ax, ay, w and those of the sequence $\{z_n\}$ previously defined, we consider the following two sequences $\{x_n\}, \{y_n\}$ of words: for any $n > 0$,

$$x_n = w^n ax z_n, \quad y_n = w^n ay z_n.$$

The condition that, for any $n > 0$, $x_n, y_n \in L_I^\epsilon$ immediately follows from the definition of the sequences $\{x_n\}$ and $\{y_n\}$. The following Lemma is used in the sequel. Its proof is an ease consequence of the definition of the relation $\Rightarrow_I^*$.

Lemma 7. *Let $f, g, v \in A^*$ and let $a \in A$. If $fag \Rightarrow_I^* v$ then $v = f'ag'$ where f', g' are words of A^* such that*

$$f \Rightarrow_I^* f', \quad g \Rightarrow_I^* g', \quad fag \Rightarrow_I^* f'ag \Rightarrow_I^* f'ag' = v.$$

Lemma 8. *Let n, k be positive integers. If $x_n \Rightarrow_I^* x_{n+k}$ then $z_{n+k} = z' x z_n$, $z' \in A^*$. Similarly, if $y_n \Rightarrow_I^* y_{n+k}$ then $z_{n+k} = z'' y z_n$, $z'' \in A^*$.*

Proof. We deal with the case when $w^n a x z_n \Rightarrow_I^* w^{n+k} a x z_{n+k}$, the other case being completely analogous. By applying Lemma 7 to $f = w^n$ and $g = x z_n$ one obtains words $f', g' \in A^*$ such that

1. $f' a g' = w^{n+k} a x z_{n+k}$,
2. $w^n \Rightarrow_I^* f'$,
3. $x z_n \Rightarrow_I^* g'$.

First we remark that if $f' = w^n$ then by (1) $w \in a A^*$ which is not possible since w does not begin with the letter a. Hence, by (2), $w^n \Rightarrow_I^+ f'$. This implies that there exists at least a word $u \in I$ such that $u \in \mathrm{Fact}(f')$. If $|f'| < |w^{n+k}|$ then, by condition (1), $f' \in \mathrm{Pref}(w^{n+k})$ so that $u \in \mathrm{Fact}(w^{n+k})$. By Lemma 4 the latter condition is not possible. Hence $|f'| \geq |w^{n+k}|$ so that, by condition (1), $f' = w^{n+k} \zeta$, where $\zeta \in A^*$. Since $\epsilon \Rightarrow_I^* w^n z_n$, the previous condition and condition (2) yield $\epsilon \Rightarrow_I^* w^{n+k} \zeta z_n$. Hence, by the definition of z_{n+k}, $|\zeta z_n| \geq |z_{n+k}|$, so that $|w^{n+k} \zeta a x z_n| \geq |w^{n+k} a x z_{n+k}|$. Since, by Lemma 7,

$$w^n a x z_n \Rightarrow_I^* w^{n+k} \zeta a x z_n \Rightarrow_I^* w^{n+k} a x z_{n+k},$$

the latter condition implies that $w^{n+k} \zeta a x z_n = w^{n+k} a x z_{n+k}$. Hence, by Corollary 2, $z_{n+k} = z' x z_n$, with $x' \in A^*$. □

Proposition 2. *The relation $\Rightarrow_I^*$ is not a wqo on L_I^ϵ.*

Proof. The proof is by contradiction. Set $L = L_I^\epsilon$ and denote $\leq$ the relation $\Rightarrow_I^*$. Suppose that L is well quasi ordered by $\leq$. Then, by Lemma 1, the set $L \times L$ is well quasi ordered by the canonical relation defined by $\leq$ on $L \times L$. Hence every sequence of elements of $L \times L$ is good with respect to that quasi order. Now consider the sequence $\{(x_n, y_n)\}$. Hence there exist integers $n, k > 0$ such that $x_n \leq x_{n+k}$ and $y_n \leq y_{n+k}$. By Lemma 8, $z_{n+k} = z' x z_n = z'' y z_n$ with $z', z'' \in A^*$. On the other hand, by Lemma 5, $|x| < |y|$ and therefore x is a suffix of y which is a contradiction. □

Remark 1. The same result of Proposition 2 may be obtained under the assumption that, for every letter $a \in A$, there exists a word of the set I that ends with a. In this case, the proof is completely analogous.

3.3   Third Case

Now we suppose that the set I has at least two words and it does not satisfy the hypothesis of the previous case. Set $\mathrm{alph}(I) = A$. Therefore, according to Remark 1, we assume that there exists a letter c of the set A such that, for every $f \in I$, $f \notin A^* c$. In order to study this case, it is useful to introduce some preliminary definitions and results. For any $f \in A^*$, we set

$$\nu_c(f) = \frac{|f|_c}{|f|}.$$

We adopt the following conventions. The word u denotes a prefix of a word of the set I such that $\nu_c(u)$ is maximal. Moreover w denotes a word of the set I with $u \in \mathrm{Pref}(w)$ and we set $w = uv, v \in A^*$. We state two lemmas whose proof is omitted for the sake of brevity.

Lemma 9. *The following conditions hold.*
i) The word u ends with the letter c and $v \neq \epsilon$.
ii) Let f be a word of I such that, for any $g \in I$, $\nu_c(g) \leq \nu_c(f)$. Then, for any $g \in L_I^\epsilon$, $\nu_c(g) \leq \nu_c(f)$. Moreover $\nu_c(f) < \nu_c(u)$.
iii) Let $n > 0$ and let f be a word of A^* such that $u^n \Rightarrow_I^* f$. Then $\nu_c(f) \leq \nu_c(u)$.
iv) Let $v_0, \ldots, v_i$ be words of the set $\mathrm{Pref}(L_I^\epsilon)$. Then $\nu_c(v_0 \cdots v_i) \leq \nu_c(u)$.

Now it is convenient to notice that, by hypothesis, $\epsilon \Rightarrow_I w = uv$ and therefore, for any $n > 0$,

$$\epsilon \Rightarrow_I^n u^n v^n. \tag{4}$$

Let us denote $\{z_n\}$ a sequence of words of A^+ such that, for any $n > 0$, condition (4) holds if one replaces v^n with z_n and such that z_n is of minimal length.

Lemma 10. *The sequence $\{|z_n|\}$ is not upper bounded.*

The following result is useful. Its proof is similar to that of Lemma 5.

Lemma 11. *Let $a \in A$ with $a \neq c$ and let $H = |w| + 1$. Then there exist words $a^H x, a^H y \in L_I^\epsilon$ such that $x \notin \mathrm{Suff}(y)$ and $|x| < |y|$.*

By possibly replacing the sequence $\{z_n\}$ with one of its subsequence, Lemma 10 yields the following corollary.

Corollary 3. *The sequence of words $\{z_n\}$ is such that, for any $n, m > 0$, $|z_n| + |y| < |z_{n+m}|$ where y is the word defined in Lemma 11.*

Now, starting from the words $a^H x, a^H y, w = uv$ and those of the sequence $\{z_n\}$, we consider the following two sequences $\{x_n\}, \{y_n\}$ of words: for any $n > 0$,

$$x_n = u^n a^H x z_n, \quad y_n = u^n a^H y z_n.$$

The condition that, for any $n > 0$, $x_n, y_n \in L_I^\epsilon$ immediately follows from the definition of the sequences $\{x_n\}$ and $\{y_n\}$.

Lemma 12. *Let n, k be positive integers. If $x_n \Rightarrow_I^* x_{n+k}$ then $z_{n+k} = z' x z_n$, $z' \in A^+$. Similarly, if $y_n \Rightarrow_I^* y_{n+k}$ then $z_{n+k} = z'' y z_n$, $z'' \in A^+$.*

Proof. We deal with the case when $u^n a^H x z_n \Rightarrow_I^* u^{n+k} a^H x z_{n+k}$, the other being completely analogous. By applying Lemma 7 to $f = u^n$ and $g = a^{H-1} x z_n$ one obtains words $f', g' \in A^*$ such that

1. $f' a g' = u^{n+k} a^H x z_{n+k}$,
2. $u^n \Rightarrow_I^* f'$,
3. $a^{H-1} x z_n \Rightarrow_I^* g'$.

Let us prove that $u^{n+k} \in \text{Pref}(f')$. By (1), it suffices to show that $|f'| \geq |u^{n+k}|$. By contradiction, suppose that $|f'| < |u^{n+k}|$. First we notice that in the derivation process

$$g = a^{H-1}xz_n \Rightarrow_I^* g'$$

at least a word of I must be inserted in the prefix a^{H-1} of g. Indeed, otherwise, we have $g' = a^{H-1}g''$, $g'' \in A^*$ and therefore

$$f'ag' = f'a^H g'' = u^{n+k}a^H xz_{n+k}.$$

Since $|f'| < |u^{n+k}|$ and $|u| < H$ we obtain $u \in A^*a$, with $a \neq c$ which contradicts condition (i) of Lemma 9. Therefore, the prefix of g' of length $H - 1$ is of the form

$$p = av_1 \cdots av_i,$$

where $1 \leq i \leq H - 1$, $v_1, \ldots, v_i \in \text{Pref}(L_I^\epsilon)$. Again, the equality $f'ag' = u^{n+k}a^H xz_{n+k}$ and the condition $|f'| < |u^{n+k}|$, $|u| < H$ yield the existence of a power u^j of u such that $j \leq n + k$ and

$$u^j = f'q,$$

where q is a proper prefix of p. Set $q = av_1 \cdots av_k'$. Then, by Lemma 9 – (iv), we have $\nu_c(q) = \nu_c(av_1 \cdots av_k') < \nu_c(v_1 \cdots v_k') \leq \nu_c(u)$ and by Lemma 9 – (iii) $\nu_c(f') \leq \nu_c(u)$ whence

$$\nu_c(u) = \nu_c(u^j) = \nu_c(f'q) < \nu_c(u),$$

which is a contradiction. Hence $|f'| \geq |u^{n+k}|$ and thus by (1), $f' = u^{n+k}\zeta$, $\zeta \in A^*$. Therefore we have $f = u^n \Rightarrow_I^+ f' = u^{n+k}\zeta$. On the other hand, we have $\epsilon \Rightarrow_I^* u^n z_n$ which thus gives

$$\epsilon \Rightarrow_I^* u^{n+k}\zeta z_n.$$

By the definition of z_{n+k}, we have $|\zeta z_n| \geq |z_{n+k}|$ and thus $|u^{n+k}\zeta axz_n| \geq |u^{n+k}axz_{n+k}|$. Now, by Lemma 7,

$$fag = u^n axz_n \Rightarrow_I^+ f'ag = u^{n+k}\zeta axz_n \Rightarrow_I^* u^{n+k}axz_{n+k} = f'ag',$$

which gives $u^{n+k}\zeta axz_n = u^{n+k}axz_{n+k}$. By Corollary 3, $|xz_n| < |z_{n+k}|$ which gives $z_{n+k} = z'xz_n$, with $z' \in A^+$. $\qquad \square$

The proof of the following proposition follows *verbatim* the argument of that of Proposition 2.

Proposition 3. *The relation $\Rightarrow_I^*$ is not a wqo on L_I^ϵ.*

By Theorem 3, we have that $\Rightarrow_I^*$ is a well quasi-order on A^* if and only if $\Rightarrow_I^*$ is a well quasi-order on L_I^ϵ. Hence Theorem 2 gives the following corollary.

Corollary 4. *Let I be a finite set of words over the alphabet A. Then the following conditions are equivalent:*

- *I is unavoidable;*
- *L_I^ϵ is regular;*
- *$\Rightarrow_I^*$ is a well quasi-order on L_I^ϵ.*

4   Open Problems and Perspectives

As announced in the introduction of this paper, one can consider a possible extension of the previous results with respect to other significant quasi orders and, in particular, in the case of the relation $\vdash_I^*$ we now introduce. Let I be a finite subset of A^+. Then we denote by $\vdash_I$ the binary relation of A^* defined as: for every $u, v \in A^*$, $u \vdash_I v$ if

$$u = u_1 u_2 \cdots u_{n+1},$$

$$v = u_1 a_1 u_2 a_2 \cdots u_n a_n u_{n+1},$$

with $u_i \in A^*$, $a_i \in A$, and $a_1 \cdots a_n \in I$.

The relation $\vdash_I^*$ is the transitive and reflexive closure of $\vdash_I$. One easily verifies that $\vdash_I^*$ is a monotone quasi-order on A^*. Moreover $L_{\vdash_I}^\epsilon$ denotes the set of all words derived from the empty word by applying $\vdash_I^*$, that is

$$L_{\vdash_I}^\epsilon = \{u \in A^* \mid \epsilon \vdash_I^* u\}.$$

The relation $\vdash_I^*$ has been first considered in [9] where the following theorem has been proved.

Theorem 4. *Let $I \subseteq A^+$ and assume that $A = \mathrm{alph}(I)$. The following conditions are equivalent:*

  i. *the derivation relation $\vdash_I^*$ is a wqo on A^*;*
 ii. *the set I is subsequence unavoidable in A^*, that is there exists a positive integer k such that any word $u \in A^*$, with $|u| \geq k$, contains as a subsequence a word of I;*
iii. *the language $L_{\vdash_I}^\epsilon$ is regular.*

In [9] it is also proved that I is subsequence unavoidable if and only if, for every $a \in A$, $I \cap \{a\}^+ \neq \emptyset$. It is also worth noticing that the relationships between the quasi-orders $\vdash_I^*$ and $\Rightarrow_I^*$ have been deeply investigated in [2], [3] where, as a consequence of a more general result, the following result is proved:

Theorem 5. *For any finite set I, $\vdash_I^*$ is a wqo on L_I^ϵ.*

In this theoretical setting, it is natural to ask whether Theorem 4 may be extended by replacing condition (i) with the weaker condition that the derivation relation $\vdash_I^*$ is a wqo on $L_{\vdash_I}^\epsilon$. Unfortunately this is not true as shown by the following example. Consider the set $I = \{ab\}$. It is easily verified that $L_{\vdash_I}^\epsilon = L_I^\epsilon$ and therefore, by a well known construction, $L_{\vdash_I}^\epsilon$ is generated by a context-free grammar with only one variable. Precisely, $L_{\vdash_I}^\epsilon$ is the language of all *semi-Dyck words* over the alphabet $\{a, b\}$. By Theorem 5, $\vdash_I^*$ is a well quasi order on $L_{\vdash_I}^\epsilon = L_I^\epsilon$ while this language is not regular. This example lead us to raise the following conjecture.

Conjecture 1. The following two conditions are equivalent:

1. $\vdash_I^*$ is wqo on $L_{\vdash_I}^\epsilon$;
2. $L_{\vdash_I}^\epsilon$ is context-free.

At the present we are not able to solve this conjecture. However it seems that a significant step of a possible solution of our problem is the combinatorial characterization of finite sets I such that $L_{\vdash_I}^\epsilon$ is context-free. In the literature, the language $L_{\vdash_I}^\epsilon$ is also called the *iterated shuffle of I* or *the shuffle closure of I* [13]. Many papers have been devoted to the studying of the shuffle closure of finite languages (see for instance [13, 14]) but, as far as we know, no characterization has been given for the context-freeness property of them. Here, we give such a characterization when I is a singleton. Precisely, we prove.

Theorem 6. *Let $I = \{w\}$. Then $L_{\vdash_I}^\epsilon$ is context-free if and only if $w = a^k b^h$ where a and b are distinct letters and h, k are non negative integers.*

Proof. **(Sketch)** Let us prove the necessary condition. Let

$$I = \{w\} = \{a_1^{i_1} a_2^{i_2} \cdots a_k^{i_k}\},$$

where $k \geq 3$, $i_1, \ldots, i_k \geq 1$ and the $a_i's$ are letters such that, for every $i = 1, \ldots, k-1$, $a_i \neq a_{i+1}$. By contradiction, suppose that $L_{\vdash_I}^\epsilon$ is context-free. Let

$$X = \{a_1^{i_1 n} a_2^{i_2 n} \cdots a_k^{i_k n} \mid n \geq 1\}.$$

By applying the Pumping Lemma for context-free languages to X, one proves that X is not context-free. On the other hand, one may prove that

$$X = L_{\vdash_I}^\epsilon \cap (a_1^{i_1})^* (a_2^{i_2})^* \cdots (a_k^{i_k})^*.$$

Since the family of context-free languages is closed under intersection with regular languages, one has that X is context-free which is a contradiction. This proves the necessary condition.

Let us now prove the sufficient condition. Let $I = \{w\} = a^h b^k$, where a and b are distinct letters and h, k are non negative integers. If $k = 0$ (resp. $h = 0$), then $L_{\vdash_I}^\epsilon = (a^h)^*$ (resp. $= (b^k)^*$) and, hence, it is regular. Suppose that $h, k > 0$. For any word u over the alphabet $\{a, b\}$, one can consider the following integer parameters

$$q_a^u = |u|_a / h, \quad q_b^u = |u|_b / k, \quad \text{and}$$
$$r_a^u = |u|_a \bmod h, \quad r_b^u = |u|_b \bmod k.$$

Then, one can prove that, for any word w,

$$w \in L_{\vdash_I}^\epsilon$$

if and only if the following condition holds: $q_a^w = q_b^w$, $r_a^w = r_b^w = 0$ and for any prefix u of w, either $q_a^u > q_b^u$ or $q_a^u = q_b^u$ and $r_b^u = 0$.

By using the characterization above, one may construct a push-down automaton that accepts $L_{\vdash_I}^\epsilon$. $\square$

References

1. D. P. Bovet and S. Varricchio, On the regularity of languages on a binary alphabet generated by copying systems. *Information Processing Letters* **44**, 119–123 (1992).
2. F. D'Alessandro and S. Varricchio, On Well Quasi-orders On Languages. *Lecture Notes in Computer Science*, Vol. 2710, pp. 230–241, Springer-Verlag, Berlin, 2003.
3. F. D'Alessandro and S. Varricchio, Well quasi-orders and context-free grammars, to appear on *Theoretical Computer Science*.
4. A. de Luca and S. Varricchio, Some regularity conditions based on well quasi-orders. *Lecture Notes in Computer Science*, Vol. 583, pp. 356–371, Springer-Verlag, Berlin, 1992.
5. A. de Luca and S. Varricchio, Well quasi-orders and regular languages. *Acta Informatica* **31**, 539–557 (1994).
6. A. de Luca and S. Varricchio, *Finiteness and regularity in semigroups and formal languages*. EATCS Monographs on Theoretical Computer Science. Springer, Berlin, 1999.
7. A. Ehrenfeucht, D. Haussler, and G. Rozenberg, On regularity of context-free languages. *Theoretical Computer Science* **27**, 311–332 (1983).
8. T. Harju and L. Ilie, On well quasi orders of words and the confluence property. *Theoretical Computer Science* **200**, 205–224 (1998).
9. D. Haussler, Another generalization of Higman's well quasi-order result on Σ^*. *Discrete Mathematics* **57**, 237–243 (1985).
10. G. H. Higman, Ordering by divisibility in abstract algebras. *Proc. London Math. Soc.* **3**, 326–336 (1952).
11. L. Ilie and A. Salomaa, On well quasi orders of free monoids. *Theoretical Computer Science* **204**, 131–152 (1998).
12. B. Intrigila and S. Varricchio, On the generalization of Higman and Kruskal's theorems to regular languages and rational trees. *Acta Informatica* **36**, 817–835 (2000).
13. M. Ito, L. Kari, and G. Thierrin, Shuffle and scattered deletion closure of languages. *Theoretical Computer Science* **245 (1)**, 115–133 (2000).
14. M. Jantzen, Extending regular expressions with iterated shuffle. *Theoretical Computer Science* **38**, 223–247 (1985).
15. J. Kruskal, The theory of well-quasi-ordering: a frequently discovered concept. *J. Combin. Theory, Ser. A* **13**, 297–305 (1972).
16. L. Puel, Using unavoidable sets of trees to generalize Kruskal's theorem, *J. Symbolic Comput.* **8 (4)**, 335–382 (1989).

A Ciliate Bio-operation and Language Families

Jürgen Dassow

Otto-von-Guericke-Universität Magdeburg, Fakultät für Informatik,
Magdeburg, Germany
dassow@iws.cs.uni-magdeburg.de

Abstract. We formalize the hairpin inverted repeat operation, which
is known in ciliate genetics as an operation on words and languages by
defining $\mathcal{HI}(w, P)$ as the set of all words $x\alpha y^R\alpha^R z$ where $w = x\alpha y\alpha^R z$
and the pointer α is in P. We extend this concept to language families
which results in families $\mathcal{HI}(\mathcal{L}_1, \mathcal{L}_2)$. For $\mathcal{L}_1$ and $\mathcal{L}_2$ being the families
of finite, regular, context-free, context-sensitive or recursively enumer-
able language, respectively, we determine the hierarchy of the families
$\mathcal{HI}(\mathcal{L}_1, \mathcal{L}_2)$ and compare these families with those of the Chomsky hierarchy. Furthermore, we give some results on the decidability of the
membership problem, emptiness problem and finiteness problem for the
families $\mathcal{HI}(\mathcal{L}_1, \mathcal{L}_2)$.

1 Introduction and Definitions

DNA molecules can be described as words over the alphabet $\{A, C, G, T\}$ where
the letters stand for Adenine, Cytosine, Guanine and Thymine or over the alphabet of the pairs of the Watson-Crick complementary letters. Thus operations
on DNA structures, genes and chromosomes can be interpreted as operations on
words. They can be extended to operations on languages (as sets of words) and
language families.

For instance, the splicing as a basic recombination operation on DNA molecules has been modelled as a language-theoretic operation by T. Head in [8]. A
splicing rule r can be given as a word $r = u_1\#u_2\$v_1\#v_2$. Its application to two
words x and y is only possible if $x = x_1u_1u_2x_2$ and $y = y_1v_1v_2y_2$ and results in
$z = x_1u_1v_2y_2$. We write

$$z = Spl_r(x, y).$$

This can be extended to languages and language families by

$$Spl(L, R) = \{z \mid z = Spl_r(x, y),\ x, y \in L,\ r \in R\},$$
$$Spl(\mathcal{L}_1, \mathcal{L}_2) = \{K \mid K = Spl(L, R),\ L \in \mathcal{L}_1,\ R \in \mathcal{L}_2\}.$$

In [10] Gh. Păun has compared the language families $Spl(\mathcal{L}_1, \mathcal{L}_2)$, where $\mathcal{L}_1$
and $\mathcal{L}_2$ are families of the Chomsky hierarchy, with the families of the Chomsky
hierarchy. These results are summarized in [9] and [11], too.

In [14], D. B. Searls introduced formal language-theoretic counterparts of
some large scale rearrangements in DNA molecules, genes and chromosomes

C.S. Calude, E. Calude and M.J. Dinneen (Eds.): DLT 2004, LNCS 3340, pp. 151–162, 2004.
© Springer-Verlag Berlin Heidelberg 2004

as inversion, translocation and duplication. In [5] and [4] the effect of these operations applied to language families from the Chomsky hierarchy has been studied.

In the last years the loop direct repeat excision and insertion, the hairpin inverted repeat operation and the double loop alternating direct repeat excision, which are well-known operations occurring in the descrambling of ciliates, have been formulated as language-theoretic operations (see [6]). In order to model these operation one uses some pointers which determine the places where the operations can be applied. For the hairpin inverted repeat operation we obtain the following definition

$$\mathcal{HI}(w) = \{xay^R\alpha^R z \mid w = x\alpha y\alpha^R z \text{ and } \alpha \text{ is a pointer}\}$$

where α, x, y and z are words over some alphabet V and y^R denotes the mirror word obtained from y (i.e., $\lambda^R = \lambda$ for the empty word, $x^R = x$ for any letter x, and $(w_1 w_2)^R = w_2^R w_1^R$ for words w_1 and w_2). In the paper [1], the authors generalized this operation to languages by

$$\mathcal{HI}(L) = \{xay^R\alpha^R z \mid x\alpha y\alpha^R z \in L, \ x, z \in V^*, \ y \in V^+, \ \alpha \in V^+\}$$

(i.e., $\mathcal{HI}(L)$ consist of all words which can be obtained from words of L by the hairpin inverted repeat operation where any non-empty word can be used as a pointer). In [1], [2] and [3] the closure of some language families under this operation has been studied.

In this paper we continue all these investigations. In accordance with molecular biology, where mostly only a finite set of pointers can be recognized, instead of allowing any word as a pointer, we require that the pointers have to belong to a certain language P. This leads to the following concepts.

Let V be an alphabet. For $w \in V^+$, $L \subseteq V^+$ and $P \subseteq V^+$, we set

$$\mathcal{HI}(w, P) = \{xay^R\alpha^R z \mid w = x\alpha y\alpha^R z, \ x, y \in V^*, \ y \in V^+, \ \alpha \in P\}$$

and

$$\mathcal{HI}(L, P) = \bigcup_{w \in L} \mathcal{HI}(w, P).$$

Moreover, we extend the definition to language families $\mathcal{L}_1$ and $\mathcal{L}_2$ by

$$\mathcal{HI}(\mathcal{L}_1, \mathcal{L}_2) = \{\mathcal{HI}(L, P) \mid L \in \mathcal{L}_1, \ P \in \mathcal{L}_2\}.$$

In this paper we study the language families $\mathcal{HI}(\mathcal{L}_1, \mathcal{L}_2)$, where $\mathcal{L}_1$ and $\mathcal{L}_2$ are families of the Chomsky hierarchy or the family of finite languages. We give some results on the place of these families within the Chomsky hierarchy. Furthermore, we compare the families $\mathcal{HI}(\mathcal{L}_1, \mathcal{L}_2)$ with each other and present some results on the decidability of the membership problem, emptiness problem and finiteness problem for the families $\mathcal{HI}(\mathcal{L}_1, \mathcal{L}_2)$.

Let us mention that in [7] the authors studied related families using a modified concept of hairpin operation and finite sets of pointers (and using methods for proofs which differ from our ideas) and that [12] contains statements analogous to our results for an operation modelling the double loop alternating direct repeat excision.

Throughout the paper we assume that the reader is familiar with the basic notions of the theory of formal languages which can be found in [13].

For a language L, $alph(L)$ denotes the smallest alphabet V such that $L \subseteq V^*$. For a word w, $sub(w)$ denotes the set of subwords of w. We extend this notion to languages by $sub(L) = \{z \mid z \in sub(w), w \in L\}$.

By FIN, REG, CF, CS and RE we denote the families of all finite, regular, context-free, context-sensitive and recursively enumerable languages, respectively, and by $\mathcal{H}$ we denote the set of these five families.

Given $\mathcal{L}_1, \mathcal{L}_2$ and $\mathcal{L}$ in $\mathcal{H}$, we say that $\mathcal{L}$ is an optimal (upper) bound for $\mathcal{HI}(\mathcal{L}_1, \mathcal{L}_2)$, if $\mathcal{HI}(\mathcal{L}_1, \mathcal{L}_2) \subseteq \mathcal{L}$ and $\mathcal{HI}(\mathcal{L}_1, \mathcal{L}_2)$ is not contained in any family $\mathcal{L}' \in \mathcal{H}$ with $\mathcal{L}' \subseteq \mathcal{L}$.

For a language family $\mathcal{L}$, $U(\mathcal{L})$ denotes the family of unary languages of $\mathcal{L}$.

2   Upper Bounds for Hairpin Families

The following lemma immediately follows from the definitions.

Lemma 1. *For any language families $\mathcal{L}_1$, $\mathcal{L}_2$, $\mathcal{L}_3$ and $\mathcal{L}_4$ such that $\mathcal{L}_1 \subseteq \mathcal{L}_2$ and $\mathcal{L}_3 \subseteq \mathcal{L}_4$, $\mathcal{HI}(\mathcal{L}_1, \mathcal{L}_3) \subseteq \mathcal{HI}(\mathcal{L}_2, \mathcal{L}_4)$.*   □

By definition, any word belonging to $\mathcal{HI}(L, P)$ for some languages L and P has the form $x_1 \alpha x_2 \alpha^R x_3$ for some words α, x_1, x_2, x_3. Obviously, there are words which do not have this structure, e.g. the word abc over $\{a, b, c\}$. Thus the finite language only consisting of abc cannot be in $\mathcal{HI}(\mathcal{L}_1, \mathcal{L}_2)$ for all language families $\mathcal{L}_1$ and $\mathcal{L}_2$. Therefore we have the following result.

Lemma 2. *For any two language families X and Y, FIN is not contained in $\mathcal{HI}(X, Y)$.*   □

By Lemma 2, we do not have lower estimations $\mathcal{L} \subseteq \mathcal{HI}(\mathcal{L}_1, \mathcal{L}_2)$ with a language family $\mathcal{L}$ which contains the family of finite languages. Especially, we do not have lower estimations by the families of the Chomsky hierarchy.

We now give some upper estimations.

Theorem 1. $\mathcal{HI}(FIN, FIN) = \mathcal{HI}(FIN, REG) = \mathcal{HI}(FIN, CF)$
$$= \mathcal{HI}(FIN, CS) = \mathcal{HI}(FIN, RE) \subset FIN.$$

Proof. Obviously,

$$\mathcal{HI}(L, P) = \bigcup_{w \in P} \mathcal{HI}(L, \{w\}).$$

Moreover, if $\mathcal{HI}(L, \{w\})$ is non-empty for some pointer w, then w has to be a subword of some word of L, i.e., $w \in sub(L)$. Thus we have

$$\mathcal{HI}(L, P) = \bigcup_{w \in P \cap sub(L)} \mathcal{HI}(L, \{w\}) = \mathcal{HI}(L, P \cap sub(L)). \tag{1}$$

If L is a finite language, then $sub(L)$ and $P \cap sub(L)$ are finite, too. Thus, by (1), any language $\mathcal{HI}(L, P)$ with a finite language L belongs to $\mathcal{HI}(FIN, FIN)$. Therefore $\mathcal{HI}(FIN, X) \subseteq \mathcal{HI}(FIN, FIN)$ for all language families. The opposite inclusions follow from Lemma 1.

In order to finish the proof we now show that $\mathcal{HI}(FIN, FIN) \subset FIN$. Let L and P be two finite languages. We set

$$m = \#(L), \quad n = \max\{|w| \mid w \in L\} \text{ and } m' = \#(P).$$

By definition, if $z = x_1 \alpha x_2 \alpha^R x_3 \in \mathcal{HI}(L, P)$, then L contains the word $y = x_1 \alpha x_2^R \alpha^R x_3$. For any word $y \in L$ and any $\alpha \in P$, we have less than n^2 possible compositions $y = x_1 \alpha x_2^R \alpha^R x_3$ since we can use at most n positions where α and α^R start. Thus the number of words in $\mathcal{HI}(L, P)$ is bounded by $m \cdot m' \cdot n^2$, and therefore $\mathcal{HI}(L, P)$ is finite. The inclusion is strict by Lemma 2.

Theorem 2. $\mathcal{HI}(REG, FIN) \subset REG$.

Proof. Let L be a regular language which is accepted by the deterministic finite automaton $\mathcal{A} = (V, Z, z_0, F, \delta)$. Further let $P = \{w_1, w_2, \ldots, w_n\}$ be a finite set where $w_i = a_{i,1} a_{i,2} \ldots a_{i,r_i}$ with $a_{i,j} \in V$ for $1 \leq i \leq n$, $1 \leq j \leq r_i$. Then we consider the nondeterministic finite automaton

$$\mathcal{A}' = (V, \{u_1, u_2, \ldots, u_6\} \times K \times Z \times (Z \cup \{q\}) \times (Z \cup \{q\}), (u_1, t_1, z_0, q, q), F', \delta')$$

where $q, u_1, u_2, u_3, u_4, u_5$ are new symbols,

$$K = \{z_{i,j} \mid 1 \leq i \leq n, \ 1 \leq j \leq r_i\} \cup \{t_1, t_2\},$$
$$F' = \{(u_6, t_2, z, z, z') \mid z \in Z, z' \in F\}$$

and δ' is defined as follows:

$$(u_1, t_1, \delta(z, a), q, q) \in \delta'((u_1, t_1, z, q, q), a) \text{ for } z \in Z, a \in V$$

(starting in the initial state of $\mathcal{A}$ and reading v_1 we obtain $(u_1, t_1, \delta(z_0, v_1), q, q)$),

$$(u_2, z_{i,1}, \delta(z, a_{i,1}), q, q) \in \delta'((u_1, t_1, z, q, q), a_{i,1}) \text{ for } z \in Z,$$
$$(u_2, z_{i,j}, \delta(z, a_{i,j}), q, q) \in \delta'((u_2, z_{i,j-1}, z, q, q), a_{i,j}) \text{ for } z \in Z, 1 \leq j \leq r_i - 1,$$
$$(u_3, z_{i,r_i}, \delta(z, a_{i,r_i}), z', z') \in \delta'((u_2, z_{i,r_i-1}, z, q, q), a_{i,r_i}) \text{ for } z \in Z, z' \in Z$$

(we check whether v_1 is followed by a word $w_i \in P$; in the third component we continue the simulation of $\mathcal{A}$; finally we get $(u_3, z_{i,r_i}, \delta(z_0, v_1 w_i), z', z')$ for some $z' \in Z$),

$$(u_4, z_{i,r_i}, z, z_2, z') \in \delta'((u_3, z_{i,r_i}, z, z_1, z'), a) \text{ for } z, z_1 \in Z, z_1 = \delta(z_2, a),$$
$$(u_4, z_{i,r_i}, z, z_2, z') \in \delta'((u_4, z_{i,r_i}, z, z_1, z'), a) \text{ for } z, z_1 \in Z, z_1 = \delta(z_2, a)$$

(we remember z_{i,r_i}, z and z' and simulate $\mathcal{A}$ nondeterministically backwards in the fourth component; thus after reading the non-empty word v_2 which follows $u_1 w_i$ we obtain the state $(u_4, z_{i,r_i}, \delta(z_0, v_1 w_i), z'', z')$ with $\delta(z'', v_2^R) = z'$),

$(u_5, z_{i,r_i-1}, z, z, \delta(z', a_{i,r_i})) \in \delta'((u_4, z_{i,r_i}, z, z, z'), a_{i,r_i})$ for $z \in Z$,
$(u_5, z_{i,j-1}, z, z, \delta(z'', a_{i,j})) \in \delta'((u_5, z_{i,j}, z, z, z''), a_{i,j})$ for $z, z'' \in Z, 1 < j < r_i$,
$(u_6, t_2, z, z, \delta(z'', a_{i,1})) \in \delta'((u_5, z_{i,1}, z, z, z''), a_{i,1})$ for $z, z'' \in Z$

(we check whether $v_1 w_i v_2$ is followed by w_i^R; this process can only be started if $\underline{z} = \delta(z_0, v_1 w_i)$ and $z' = \delta(\underline{z}, v_2^R)$; during this phase we simulate $\mathcal{A}$ in the fifth component; t_2 remembers that the check was successful; finishing this phase we get the state $(u_6, t_2, \delta(z_0, u_1 w_i), \delta(z_0, u_1 w_i), \delta(z', w_i^R))$),

$$(u_6, t_2, z, z, \delta(z'', a)) \in \delta'((u_6, t_2, z, z, z''), a) \text{ for } z'' \in Z, a \in V$$

(we obtain $(u_6, t_2, \delta(z_0, v_1 w_i), \delta(z_0, v_1 w_i), \delta(z', w_i^R v_3))$ if we read the remaining part v_3 of the input word).

By these explanations and the definition of F, the language accepted by $\mathcal{A}'$ consists of all words of the form $v_1 w_i v_2 w_i^R v_3$ with $v_1, v_3 \in V^*$, $v_2 \in V^+$, $w_i \in P$ and

$$\delta(z_0, v_1 w_i v_2^R w_i^R v_3) = \delta(\underline{z}, v_2^R w_i^R v_3) = \delta(z', w_i^R v_3) \in F,$$

i.e., $v_1 w_i v_2^R w_i^R v_3 \in L$. Hence $\mathcal{A}'$ accepts $\mathcal{HI}(L, R)$ which proves the regularity of $\mathcal{HI}(L, R)$.

The strictness of the inclusion follows by Lemma 2.

We mention that the upper bound given in Theorem 2 is optimal with respect to $\mathcal{H}$ since $\mathcal{HI}(REG, FIN)$ contains the infinite language

$$\{ca^n c \mid n \geq 1\} = \mathcal{HI}(\{ca^n c \mid n \geq 1\}, \{c\}).$$

Theorem 3. $\mathcal{HI}(REG, REG) \subset CF$.

Proof. The proof can be given analogous to that of Theorem 2 using an pushdown automaton instead of the finite automaton and the fact that we can store the reversal of a word on the pushdown tape.

The following result shows that the upper bound given in Theorem 3 is also optimal.

Lemma 3. $\mathcal{HI}(REG, REG)$ *contains a non-regular language.*

Proof. Let $V = \{a, b, c\}$. We set

$$L = \{ba^n ca^m ca^p b \mid n, m, p \geq 1\} \text{ and } P = \{ba^n c \mid n \geq 1\}.$$

Then

$$\mathcal{HI}(L, P) = \{ba^n ca^m ca^n b \mid n, m \geq 1\} \in \mathcal{HI}(REG, REG).$$

Using closure properties of REG it is easy to prove that $\mathcal{HI}(L, P)$ is not regular.

Theorem 4. $\mathcal{HI}(CS, CS) \subset CS$.

Proof. Let L and P be two context-sensitive languages, and let $\mathcal{A}_L$ and $\mathcal{A}_P$ be the linearly bounded automata accepting L and P, respectively. We now construct the Turing machine $\mathcal{A}$ which is able to perform the following steps. $\mathcal{A}$ divides the input word x into five parts $x_1' x_2 x_3' x_4 x_5'$ by priming the original letters of the subwords x_1, x_3 and x_5. $\mathcal{A}$ copies the subword x_2 which results in $x_1' x_2 x_3' x_4 x_5' \$ x_2$ (where $\$$ is a marker) and checks by a simulation of $\mathcal{A}_P$ on x_2 whether or not x_2 belongs to P. If the answer is negative, then $\mathcal{A}$ rejects x. Otherwise $\mathcal{A}$ deletes the marker and the word obtained from x_2 which gives $x_1' x_2 x_3' x_4 x_5'$, again. $\mathcal{A}$ copies x_2 and x_4 which results in $x_1' x_2 x_3' x_4 x_5' \$ x_2 \$ x_4$ and checks whether $x_2 = x_4^R$ holds. If the answer is negative, then $\mathcal{A}$ rejects the input. Otherwise $\mathcal{A}$ deletes all letters besides $x_1' x_2 x_3' x_4 x_5'$. $\mathcal{A}$ transforms $x_1' x_2 x_3' x_4 x_5'$ into $x_1 x_2 x_3^R x_4 x_5$ and checks by a simulation of $\mathcal{A}_L$ whether or not $x_1 x_2 x_3^R x_4 x_5 \in L$. If the answer is negative, then $\mathcal{A}$ rejects, otherwise $\mathcal{A}$ accepts the input. Obviously, $\mathcal{A}$ accepts $\mathcal{HI}(L, P)$.

It is easy to see that the space complexity function of $\mathcal{A}$ is linearly bounded. Therefore $\mathcal{HI}(L, P) \in CS$.

The strictness follows by Lemma 2, again.

By Lemma 1, we obtain immediately the following consequences from Theorem 4.

Corollary 1.

  i) $\mathcal{HI}(REG, CF) \subseteq \mathcal{HI}(REG, CS) \subset CS$.
  ii) $\mathcal{HI}(CF, FIN) \subseteq \mathcal{HI}(CF, REG) \subseteq \mathcal{HI}(CF, CF) \subseteq \mathcal{HI}(CF, CS) \subset CS$.
  iii) $\mathcal{HI}(CS, FIN) \subseteq \mathcal{HI}(CS, REG) \subseteq \mathcal{HI}(CS, CF) \subset CS$.   □

We now show that we cannot improve the relations of Theorem 4 and Corollary 1 within the families of the Chomsky hierarchy, i.e., we show that the family of context-free languages is not an upper bound. To prove this it is sufficient to show that the smallest families $\mathcal{HI}(REG, CF)$ and $\mathcal{HI}(CF, FIN)$ occurring in Corollary 1 contain non-context-free languages.

Lemma 4.

  i) $\mathcal{HI}(CF, FIN)$ *contains a non-context-free language.*
  ii) $\mathcal{HI}(REG, CF)$ *contains a non-context-free language.*

Proof. i) Let $V = \{a, b, c\}$. We set

$$L = \{xcx^R c \mid x \in \{a, b\}^+\} \text{ and } P = \{c\}.$$

Then

$$\mathcal{HI}(L, P) = \{xcxc \mid x \in \{a, b\}^+\} \in \mathcal{HI}(CF, FIN)$$

is not context-free as easily can be shown.

ii) can be shown by an analogous proof.

Theorem 5. $\mathcal{HI}(RE, RE) \subset RE$.

Proof. The proof follows by the construction given in the proof of Theorem 4.

Corollary 2.

  i) $\mathcal{HI}(REG, RE) \subseteq \mathcal{HI}(CF, RE) \subseteq \mathcal{HI}(CS, RE) \subset RE$.

  ii) $\mathcal{HI}(RE, FIN) \subseteq \mathcal{HI}(RE, REG) \subseteq \mathcal{HI}(RE, CF) \subseteq \mathcal{HI}(RE, CS) \subset RE$.   □

We now present two lemmas which imply that the bounds given in Corollaries 2 are optimal, too. Furthermore, these lemmas show the optimality of the given bounds (for $\mathcal{HI}(\mathcal{L}_1, \mathcal{L}_2)$ and $\mathcal{HI}(\mathcal{L}_2, \mathcal{L}'_1)$ where $\mathcal{L}_1 \in \mathcal{H}$, $\mathcal{L}_2 \in \mathcal{H}$, $\mathcal{L}_2 \in \{CS, RE\}$, $\mathcal{L}_1 \subseteq \mathcal{L}_2$, $\mathcal{L}_1 \neq FIN$ and $\mathcal{L}'_1 \subseteq \mathcal{L}_2$) within language families satisfying certain (weak) closure properties.

Lemma 5. *Let $\mathcal{L}_1$ and $\mathcal{L}_2$ be two language families such that $REG \subseteq \mathcal{L}_1$, $\mathcal{L}_2$ is closed under concatenation with letters and $\mathcal{HI}(\mathcal{L}_1, \mathcal{L}_2) \subseteq \mathcal{L}_2$. Then there is no language family $\mathcal{L}$ closed under non-erasing gsm-mapppings and linear erasings such that $\mathcal{HI}(\mathcal{L}_1, \mathcal{L}_2) \subseteq \mathcal{L} \subset \mathcal{L}_2$.*

Proof. Assume that $\mathcal{HI}(\mathcal{L}_1, \mathcal{L}_2) \subseteq \mathcal{L} \subset \mathcal{L}_2$. Let $K \subseteq V^*$ be an arbitrary language of $\mathcal{L}_2$. Further let c and d be two additional letters not in V. We define the languages

$$L = \{cx_1 dx_2 dx_3 c \mid x_1, x_3 \in V^*, \ x_2 \in V^+\} \text{ and } P = \{c\}K\{d\}.$$

By the suppositions, $L \in \mathcal{L}_1$ and $P \in \mathcal{L}_2$ which implies that

$$\mathcal{HI}(L, P) = \{cx_1 dx_2^R dx_1^R c \mid x_1 \in K, \ x_2 \in V^+\} \in \mathcal{HI}(\mathcal{L}_1, \mathcal{L}_2) \subseteq \mathcal{L}.$$

By the closure of $\mathcal{L}$ under non-erasing gsm mappings and linear erasings, we get

$$\{cx_1 d(x'_2)^R) d(x'_1)^R c \mid x_1 \in K, \ x_2 \in V^+\} \in \mathcal{L}$$

($x' \in (V')^*$ denotes a primed version of the word $x \in V^*$) and then $K \in \mathcal{L}$. Thus $\mathcal{L}_2 \subseteq \mathcal{L}$ in contrast to our assumption.

Analogously, one can show the following statement.

Lemma 6. *Let $\mathcal{L}_1$ and $\mathcal{L}_2$ be two language families such that $\mathcal{L}_1$ is closed under reversal and concatenation with letters, $\mathcal{L}_2$ contains a language consisting of one letter only and $\mathcal{HI}(\mathcal{L}_1, \mathcal{L}_2) \subseteq \mathcal{L}_1$. Then there is no language family $\mathcal{L}$ closed under quotients by letters such that $\mathcal{HI}(\mathcal{L}_1, \mathcal{L}_2) \subseteq \mathcal{L} \subset \mathcal{L}_1$.*   □

3   Comparison of the Hairpin Families

Let $\mathcal{L}_1$ and $\mathcal{L}_2$ be language families with $\mathcal{L}_1 \subset \mathcal{L}_2$. By Lemma 1, we have the inclusions $\mathcal{HI}(\mathcal{L}_1, \mathcal{L}) \subseteq \mathcal{HI}(\mathcal{L}_2, \mathcal{L})$ and $\mathcal{HI}(\mathcal{L}, \mathcal{L}_1) \subseteq \mathcal{HI}(\mathcal{L}, \mathcal{L}_2)$ for any language family $\mathcal{L}$. In this section we study the strictness of these inclusions.

Lemma 7. *Let $\mathcal{L}_1$ and $\mathcal{L}_2$ are language families which are closed under union with finite sets and intersections with regular sets and satisfy $\mathcal{L}_1 \subset \mathcal{L}_2$ and $U(\mathcal{L}_1) \subset U(\mathcal{L}_2)$. Then $\mathcal{HI}(\mathcal{L}_1, \mathcal{L}) \subset \mathcal{HI}(\mathcal{L}_2, \mathcal{L})$ for any language family $\mathcal{L}$ which contains a language consisting of one letter only and is closed under intersection with regular sets.*

Proof. The inclusion $\mathcal{HI}(\mathcal{L}_1, \mathcal{L}) \subseteq \mathcal{HI}(\mathcal{L}_2, \mathcal{L})$ holds by Lemma 1. We now prove its strictness.

Let $L \in U(\mathcal{L}_2) \backslash U(\mathcal{L}_1)$ for some language $L \subseteq \{a\}^*$. By the closure properties supposed, $L' = L \cap \{a^n \mid n \geq 3\}$ is in $\mathcal{L}_2 \setminus \mathcal{L}_1$. Moreover, $L' = \mathcal{HI}(L, \{a\}) \in \mathcal{HI}(\mathcal{L}_2, \mathcal{L})$.

Assume that $L' \in \mathcal{HI}(\mathcal{L}_1, \mathcal{L})$. Then $L' = \mathcal{HI}(L'', P)$ for some $L'' \in \mathcal{L}_1$ and some $P \in \mathcal{L}$. By the closure under intersections by regular sets, we can assume that P and L'' are subsets of $\{a\}^+$. This implies $\mathcal{HI}(L'', P) = L''$. Thus $L'' = L'$ in contrast to the fact $L' \notin \mathcal{L}_1$ shown above.

Theorem 6. *The following diagram holds: if two families are connected by a (double) arrow, then the upper or right family includes (strictly) the lower or left family; the double bars in the first column denote equality.*

$$
\begin{array}{ccccccccc}
\mathcal{HI}(FIN, RE) & \Longrightarrow & \mathcal{HI}(REG, RE) & \Longrightarrow & \mathcal{HI}(CF, RE) & \Longrightarrow & \mathcal{HI}(CS, RE) & \Longrightarrow & \mathcal{HI}(RE, RE) \\
\| & & \Uparrow & & \Uparrow & & \Uparrow & & \uparrow \\
\mathcal{HI}(FIN, CS) & \Longrightarrow & \mathcal{HI}(REG, CS) & \Longrightarrow & \mathcal{HI}(CF, CS) & \Longrightarrow & \mathcal{HI}(CS, CS) & \Longrightarrow & \mathcal{HI}(RE, CS) \\
\| & & \Uparrow & & \Uparrow & & \Uparrow & & \uparrow \\
\mathcal{HI}(FIN, CF) & \Longrightarrow & \mathcal{HI}(REG, CF) & \Longrightarrow & \mathcal{HI}(CF, CF) & \Longrightarrow & \mathcal{HI}(CS, CF) & \Longrightarrow & \mathcal{HI}(RE, CF) \\
\| & & \Uparrow & & \Uparrow & & \Uparrow & & \uparrow \\
\mathcal{HI}(FIN, REG) & \Rightarrow & \mathcal{HI}(REG, REG) & \Rightarrow & \mathcal{HI}(CF, REG) & \Rightarrow & \mathcal{HI}(CS, REG) & \Rightarrow & \mathcal{HI}(RE, REG) \\
\| & & \Uparrow & & \Uparrow & & \Uparrow & & \uparrow \\
\mathcal{HI}(FIN, FIN) & \Rightarrow & \mathcal{HI}(REG, FIN) & \Rightarrow & \mathcal{HI}(CF, FIN) & \Rightarrow & \mathcal{HI}(CS, FIN) & \Rightarrow & \mathcal{HI}(RE, FIN)
\end{array}
$$

Proof. The inclusions follow by Lemma 1 and the equalities by Lemma 1. We do not prove all strictnesses of inclusions. We only present a proof for some cases; the remaining relations can be proved by analogous considerations.

i) $\mathcal{HI}(REG, X) \subset \mathcal{HI}(CF, X)$ for $X \in \{FIN, REG, CF, CS, RE\}$

By Lemma 1, we are done if there is a language U in $\mathcal{HI}(CF, FIN)$ which is not contained in the family $\mathcal{HI}(REG, RE)$.

Let $L = \{cad\}\{a^r b^r \mid r \geq 1\}\{dac\}$ and $P = \{cad\}$, then

$$U = \mathcal{HI}(L, P) = \{cad\}\{b^r a^r \mid r \geq 1\}\{dac\} \in \mathcal{HI}(CF, FIN).$$

Let us assume that $U \in \mathcal{HI}(REG, RE)$, i.e., $U = \mathcal{HI}(K, Q)$ for some regular language K and some recursively enumerable language Q. Q can only contain words α such that $w = x\alpha y\alpha^R z \in U$. This implies

$$Q \subseteq \{cad, ad, ca\} \cup \{a^n \mid n \in I_1\} \cup \{b^m \mid m \in I_2\}$$

where I_1 and I_2 are some sets of positive integers.

If the pointer α is cad, then $x = z = \lambda$ and $y = b^r a^r$ for some r. Then $cada^r b^r dac \in K$. The same situation holds if the pointer is ca or ad. If $\alpha = a^n$ or $\alpha = b^m$, then $w = xay\alpha^R z = xay^R \alpha^R z$, i.e., $w \in K$. Thus we have

$$K \subseteq \{cada^r b^r dac \mid r \geq 1\} \cup \{cadb^r a^r dac \mid r \geq 1\}.$$

Moreover, $K_1 = K \cap \{cada^r b^s dac \mid r \geq 1, s \geq 1\}$ or $K_2 = K \cap \{cadb^r a^s dac \mid r \geq 1, s \geq 1\}$ are regular languages and at least one of them has to be infinite. Assume that K_1 is infinite (the other case can be handled analogously). Then $K_1 = \{cada^r b^r dac \mid r \in I\}$ is regular, where I is an infinite set of positive integers. Using the pumping lemma it is easy to prove a contradiction.

ii) $\mathcal{HI}(X, CS) \subset \mathcal{HI}(X, RE)$ for $X \in \{REG, CF, CS\}$.

Let $P' \subset \{a\}^+$ be a non-recursive language and $P = \{c\}P'\{d\}$. Further we consider the regular language $L = \{c\}\{a\}^+\{dad\}\{a\}^+\{c\}$. Then

$$U = \mathcal{HI}(L, P) = \{ca^n dada^n c \mid a^n \in P'\} \in \mathcal{HI}(REG, RE).$$

By Theorem 1, $U \in \mathcal{HI}(X, RE)$.

Let us assume that $U \in \mathcal{HI}(X, CS)$. By Lemma 4, $U \in CS$. Then the language $U' = \{a^{2n+5} \mid a^n \in P'\}$ is context-sensitive. Let $\mathcal{A}$ be the following algorithm working on input n (or a^n). We construct $ca^n dada^n c$ and decide whether or not $ca^n dada^n c \in U$. If the answer is negative, then $\mathcal{A}$ says "no", too. Otherwise, $\mathcal{A}$ answers "yes". Obviously, $\mathcal{A}$ decides whether or not $a^n \in P'$. This contradicts the choice of P' as a non-recursive language. Therefore $U \notin \mathcal{HI}(X, CS)$.

iii) $\mathcal{HI}(CF, FIN) \subset \mathcal{HI}(CF, REG)$.

Let $L = \{ca^r da^r da^k c \mid r \geq 1, k \geq 1\}$ and $P = \{c\}\{a\}^+\{d\}$. Then

$$U = \mathcal{HI}(L, P) = \{ca^n da^n da^n c \mid n \geq 1\} \in \mathcal{HI}(CF, REG).$$

We now prove that $U \notin \mathcal{HI}(CF, FIN)$. Assume the contrary. Then $U = \mathcal{HI}(K, Q)$ for some context-free language K and some finite set Q. Moreover, let

$$K' = K \cap \{c\}\{a\}^+\{d\}\{a\}^+\{d\}\{a\}^+\{c\}.$$

Then K' is a context-free language. Moreover, by the structure of words in U, it is easy to see that $\mathcal{HI}(K', Q) = \mathcal{HI}(K, Q)$. Now let m be the maximal length of words in Q, m' the constant of the pumping lemma by Bar-Hillel/Perles/Shamir for K' and $n = \max\{m, m'\}$. We consider the word $z = ca^n da^n da^n c \in U$. Assume that $ca^r \in Q$ for some $r \leq n-1$ and $z = \mathcal{HI}(z', ca^r)$ for some $z' \in K'$. We obtain $z' = z$. Let $z = z_1 z_2 z_3 z_4 z_5$ be the decomposition of z such that $z_1 z_2^i z_3 z_4^i z_5 \in K'$ for any $i \geq 0$. Obviously, $\#_c(z_2) = \#_d(z_2) = \#_c(z_4) = \#_d(z_4) = 0$. Thus $z_2 = a^s$ and $z_4 = a^t$ with $s+t > 0$. Then $z_1 z_2^i z_3 z_4^i z_5$ is obtained by increasing the number of occurrences of a in at most two blocks of a's. Assume that the increase occurs in the first two blocks (the other cases can be handled analogously). Then we get $ca^{n+s} da^{n+t} da^n c \in K'$. Using the pointer ca^r we obtain $ca^{n+s} da^{n+t} da^n c \in \mathcal{HI}(K', Q) = U$ in contrast to the structure of words in U. Analogously, we can derive a contradiction for the other possible cases for words in Q.

4   Decidability Results for Hairpin Families

Throughout this section we assume that a language $L \in \mathcal{HI}(\mathcal{L}_1, \mathcal{L}_2)$ is given by two devices A_1 and A_2 which describe languages $L(A_1) \in \mathcal{L}_1$ and $L(A_2) \in \mathcal{L}_2$, respectively, such that $L = \mathcal{HI}(L(A_1), L(A_2))$. The device can be a grammar generating a language or an automaton accepting the language or a (regular) expression. Since there are algorithms to transform a generating device into an accepting device and vice versa, the concrete type of the describing device is not of importance.

We say that a property E is (un)decidable for $\mathcal{HI}(\mathcal{L}_1, \mathcal{L}_2)$ if, given two devices A_1 and A_2 describing languages $L(A_1) \in \mathcal{L}_1$ and $L(A_2) \in \mathcal{L}_2$, it is (un)decidable whether or not $\mathcal{HI}(L(A_1), L(A_2))$ has the property E.

The following theorem states the (un)decidabilities of the membership problem, emptiness problem and finiteness problem for language families $\mathcal{HI}(\mathcal{L}_1, \mathcal{L}_2)$.

Theorem 7. *The following table holds. (The table has to be read as follows: In the meet of the column associated with $\mathcal{L}_1$ and the row associated with $\mathcal{L}_2$ we give a triple (a, b, c), where a, b and c are the status of decidability of the membership problem, emptiness problem and finiteness problem for $\mathcal{HI}(\mathcal{L}_1, \mathcal{L}_2)$, respectively. The decidability and undecidability of a problem are denoted by $+$ and $-$, respectively; a question mark denotes that the status of decidability is presently unknown; T denotes the case that the property holds for all languages of the family.)*

	FIN	REG	CF	CS	RE
RE	$(?,?,T)$	$(-,-,-)$	$(-,-,-)$	$(-,-,-)$	$(-,-,-)$
CS	$(+,+,T)$	$(+,-,-)$	$(+,-,-)$	$(+,-,-)$	$(-,-,-)$
CF	$(+,+,T)$	$(+,?,?)$	$(+,-,-)$	$(+,-,-)$	$(-,-,-)$
REG	$(+,+,T)$	$(+,+,+)$	$(+,-,-)$	$(+,-,-)$	$(-,-,-)$
FIN	$(+,+,T)$	$(+,+,+)$	$(+,+,+)$	$(+,-,-)$	$(-,-,-)$

Proof. We do not prove all relations. We only present the proof for some cases; the remaining relations can be proved by analogous considerations.

i) *The membership problem is undecidable for $\mathcal{HI}(REG, RE)$.*

Let A_2 be an arbitrary device for a recursively enumerable language and $V = alph(L(A_2))$. Let c, d and e be symbols not contained in V. We construct a device A_2' describing $\{c\}L(A_2)\{d\}$. Further let A_1 be a device which describes the regular language $L(A_1) = \{c\}V^*\{ded\}V^*\{c\}$. Then

$$\{cwdedw^R c \mid w \in L(A_2)\} = \mathcal{HI}(L(A_1), L(A_2')) \in \mathcal{HI}(REG, RE).$$

Let us assume that there is an algorithm to decide the membership problem for languages in $\mathcal{HI}(REG, RE)$. Then, given w, we can construct $cwdedw^R c$ and decide whether or not $cwdedw^R c \in \mathcal{HI}(L(A_1), L(A_2'))$. Since the answer is positive if and only if $w \in L(A_2)$ holds, we can decide whether or not $w \in L(A_2)$ for an arbitrary device A_2 which is impossible.

ii) *The emptiness problem is decidable for $\mathcal{HI}(FIN, CS)$.*

Given A_1 and A_2 describing/generating the finite language L_1 and the context-sensitive language L_2, we can construct the finite language $L_2 \cap sub(L_1)$ and then the finite language $\mathcal{HI}(L_1, L_2) = \mathcal{HI}(L_1, L_2 \cap sub(L_1))$ (see the proof of Theorem 1) whose emptiness is decidable.

iii) *The emptiness problem is undecidable for $\mathcal{HI}(REG, CS)$.*

We take the construction of part i) starting with an arbitrary device A_2 describing a context-sensitive language. Then we have

$$\{cwdedw^R c \mid w \in L(A_2)\} = \mathcal{HI}(L(A_1), L(A_2')) \in \mathcal{HI}(REG, CS).$$

Obviously, $\mathcal{HI}(L(A_1), L(A_2')) \neq \emptyset$ if and only if $L(A_2) \neq \emptyset$. The undecidability of the emptiness of $L(A_2)$ implies the undecidability of the emptiness of $\mathcal{HI}(L(A_1), L(A_2'))$.

iv) *The emptiness problem is undecidable for $\mathcal{HI}(CF, REG)$.*

Let $P = \{(u_1, v_1), (u_2, v_2), \ldots, (u_n, v_n)\}$ be an instance of the Post Correspondence Problem over some alphabet V. Let c, d and e be letters not contained in V. We construct the context-free grammar

$$G_1 = (\{S, A\}, V \cup \{c, d, e\}, P, S)$$

with

$$P = \{S \to cAc, A \to ded\} \cup \{A \to u_i A v_i^R \mid 1 \leq i \leq n\},$$

which generates the language

$$L(G_1) = \{cu_{i_1} u_{i_2} \ldots u_{i_k} dedv_{i_k}^R v_{i_{k-1}}^R \ldots v_{i_1}^R c \mid 1 \leq i_j \leq n, 1 \leq j \leq k\},$$

and a regular grammar G_2 with $L(G_2) = \{c\}V^+\{d\}$. Then

$$\mathcal{HI}(L(G_1), L(G_2)) = \{cu_{i_1} u_{i_2} \ldots u_{i_k} dedv_{i_k}^R v_{i_{k-1}}^R \ldots v_{i_1}^R c \mid$$
$$1 \leq i_j \leq n, 1 \leq j \leq k, u_{i_1} u_{i_2} \ldots u_{i_k} = v_{i_1} v_{i_2} \ldots v_{i_k}\}.$$

Hence $\mathcal{HI}(L(G_1), L(G_2))$ is non-empty if and only if P has a solution. Thus the undecidability of the existence of a solution for a Post Correspondence Problem implies the undecidability of the emptiness of $\mathcal{HI}(L(A_1), L(A_2'))$.

We have left open the status of the membership problem and emptiness problem for the family $\mathcal{HI}(FIN, RE)$ and of the emptiness problem and finiteness problem for $\mathcal{HI}(REG, CF)$.

References

1. M. Daley and L. Kari, Some properties of ciliated bio-operations. In: M. Ito and M. Toyama (eds.), *Developments of Language Theory 2002*, LNCS 2450, Springer-Verlag, Berlin, 2003, 116–127.
2. M. Daley, O. H. Ibarra and L. Kari, Closure and decidability properties of some language classes with respect to ciliate bio-operations. *Theor. Comp. Sci.* **306** (2003) 19–38.

3. M. Daley, O. H. Ibarra, L. Kari, M. McQuillan and K. Nakano, The ld- and dlad bio-operation on formal languages. *J. Automata, Languages and Combinatorics* **8** (2004) 477–498.

4. J. Dassow, V. Mitrana and A. Salomaa, Operations and languages generating devices suggested by genome evolution. *Theor. Comp. Sci.* **270** (2002) 701–738.

5. J. Dassow and Gh. Paun, Remarks on operations suggested by mutations in genomes. *Fundamenta Informaticae* **36** (1998) 183–200.

6. A. Ehrenfeucht, T. Harju, I. Petre, D. M. Prescott and G. Rozenberg, Formal systems for gene assembly in ciliates. *Theor. Comp. Sci.* **292** (2003) 199–219.

7. R. Freund, C. Martin-Vide and V. Mitrana, On some operations on strings suggested by gene assembly in ciliates. *New Generation Computing* **20** (2002) 279–293.

8. T. Head, Formal language theory and DNA: an analysis of the generative capacity of specific recombinant behaviours. *Bull. Math. Biology* **49** (1987) 737–759.

9. T. Head, Gh. Păun and D. Pixton, Language Theory and molecular genetics. In: [13], Vol. II, Springer-Verlag, 1997, 295–360.

10. Gh. Păun, On the splicing operation. *Discrete Appl. Math.* **70** (1996) 57–79.

11. Gh. Păun, G. Rozenberg and A. Salomaa, *DNA Computing - New Computing Paradigms.* EATCS Monographs, Springer-Verlag, Berlin, 1998

12. I. Rössling, Über eine Ciliaten-Operation auf formalen Sprachen. Diploma Thesis, University of Magdeburg, 2004.

13. G. Rozenberg and A. Salomaa, *Handbook of Formal Languages*, Vol. I–III. Springer-Verlag, 1997.

14. D. B. Searls, The computational linguistics of biological sequences. In *Artificial Intelligence and Molecular Biology* (L. Hunter ed.), AAAI Press, The MIT Press, 1993, 47–120.

Semantic Shuffle on and Deletion Along Trajectories

Michael Domaratzki*

Jodrey School of Computer Science, Acadia University
Wolfville, NS, Canada
mike.domaratzki@acadiau.ca

Abstract. We introduce semantic shuffle on trajectories (SST) and semantic deletion along trajectories (SDT). These operations generalize the notion of shuffle on trajectories, but add sufficient power to encompass various formal language operations used in applied areas. However, the added power given to SST and SDT does not destroy many desirable properties of shuffle on trajectories, especially with respect to solving language equations involving SST. We also investigate closure properties and decidability questions related to SST and SDT.

1 Introduction and Motivation

Shuffle on trajectories, introduced by Mateescu et al. [24], is a powerful tool for generalizing operations on formal languages which act by inserting the letters of one word into another. There has been much research into this formalism, see, e.g., Harju et al. [13], Mateescu and Salomaa [26], Mateescu et al. [27] and others. Mateescu [23] also introduced the related concept of splicing along routes, which is an extension of shuffle on trajectories designed to model splicing operations on DNA. Recently, both the author [6] and Kari and Sosík [18] have independently introduced the notion of *deletion along trajectories*, which is an analogue of shuffle on trajectories for operations which delete letters of one word from another. This has led to even more research on trajectory-based operations [7–10, 15, 16]. Kari et al. [17] also introduce the notion of substitution and difference on trajectories, related concepts which have applications to modelling noisy channels.

In the paper which introduced shuffle on trajectories, Mateescu et al. make a distinction between *syntactic* and *semantic* operations on words:

> [Shuffle on trajectories is] based on syntactic constraints on the shuffle operations. The constraints are referred to as syntactic constraints since they do not concern properties of the words that are shuffled, or properties of the letters that occur in these words.

---

* Research supported in part by an NSERC PGS-B graduate scholarship.

C.S. Calude, E. Calude and M.J. Dinneen (Eds.): DLT 2004, LNCS 3340, pp. 163–174, 2004.

Instead, the constraints involve the general strategy to switch from one word to another word. Once such a strategy is defined, the structure of the words that are shuffled does not play any role.

However, constraints that take into consideration the inner structure of the words that are shuffled together are referred to as semantic constraints. [24, p. 2].

In this paper, we introduce a semantic variant of shuffle on trajectories, and investigate the properties of the operation. We naturally call the semantic variant *semantic shuffle on trajectories* (SST). It is a proper extension of the notion of shuffle on trajectories, and can simulate many more operations than shuffle on trajectories, especially operations of interest in applied areas of formal language theory. We also introduce the corresponding notion for deletion on trajectories, which we call *semantic deletion on trajectories* (SDT). The advantages of SST and SDT are that they preserves many of the desirable properties of the usual, syntactic shuffle on trajectories, while being capable of simulating more operations of interest. However, SST and SDT have some fundamental differences from the syntactic case. For instance, the problem of determining whether two sets of trajectories define the same operation is trivial in the syntactic case. In the semantic case, we employ the theory of trace languages to solve this equivalence problem.

We show how SST and SDT can be used to simulate operations in bioinformatics and DNA computing. These operations include *synchronized insertion* and *synchronized deletion*, introduced by Daley *et al.* [2, 3] in the study of the DNA operations of certain ciliates, and *contextual insertion and deletion*, introduced by Kari and Thierrin [19].

We further demonstrate the power of SST and SDT by giving many examples of other semantic operations simulated by these formalisms, including operations in concurrency theory, formal methods of software engineering and discrete event systems (DES). For example, one of the examples given by Mateescu *et al.* [24] of a semantic operation is *distributed concatenation*, defined by Kudlek and Mateescu [21, 20]. We observe that SST can simulate distributed concatenation, as well as other mix operations defined by Kudlek and Mateescu [21]. We can also simulate Latin product [25], infiltration product [28] and usual set intersection.

The two semantic constructs we introduce are *synchronization* and *content restriction*. Synchronization allows for only one letter to be output for two corresponding, identical symbols in the input words. Content restriction allows a trajectory to specify that a particular letter must appear at a specific point. This is inspired by bio-informatical operations, where operations occur only in the context of certain subsequences of the DNA strand.

2   Definitions

Let Σ be a finite set of symbols, called *letters*. Then Σ^* is the set of all finite sequences of letters from Σ, which are called *words*. The empty word ϵ is the empty sequence of letters. The *length* of a word $w = w_1 w_2 \cdots w_n \in \Sigma^*$, where

$w_i \in \Sigma$, is n, and is denoted $|w|$. A *language* L is any subset of Σ^*. By $\overline{L}$, we mean $\Sigma^* - L$, the complement of L.

A *morphism* $h : \Delta^* \to \Sigma^*$ is any function satisfying $h(xy) = h(x)h(y)$ for all $x, y \in \Delta^*$. A *substitution* $h : \Delta^* \to 2^{\Sigma^*}$ is any function satisfying $h(xy) = h(x)h(y)$ for all $x, y \in \Delta^*$. If $h(a)$ is regular (resp., finite) for all $a \in \Delta$, we say that h is a regular (resp., finite) substitution. Recall that a morphism $h : \Delta^* \to \Sigma^*$ is a *weak coding* if $h(a) \in \Sigma \cup \{\epsilon\}$ for all $a \in \Delta$. For additional background in formal languages and automata theory, please see Yu [30].

Before we define SST, we define the trajectory alphabet. Let $\Gamma = \{0, 1, \sigma\}$. For any alphabet Σ, let $\Gamma_\Sigma = \Gamma \cup (\Gamma \times \Sigma)$. For ease of readability, we denote $[c, a]$ by $\overset{a}{c}$ for all $a \in \Sigma$ and $c \in \Gamma$.

We can now define the SST operation. Let Σ be an alphabet, $t \in \Gamma_\Sigma^*$ and $x, y \in \Sigma^*$. Then the SST of x and y along t, denoted $x \sqcup\!\sqcup_t y$, is defined as follows: If $x = ax'$, $y = by'$ (where $a, b \in \Sigma$, $x', y' \in \Sigma^*$), and $t = ct'$, where $c \in \Gamma_\Sigma$ and $t' \in \Gamma_\Sigma^*$, then

$$
x \sqcup\!\sqcup_t y = \begin{cases} a(x' \sqcup\!\sqcup_{t'} y) & \text{if } c \in \{0, \overset{a}{0}\}, \\ b(x \sqcup\!\sqcup_{t'} y') & \text{if } c \in \{1, \overset{b}{1}\}, \\ a(x' \sqcup\!\sqcup_{t'} y') & \text{if } a = b \text{ and } c \in \{\sigma, \overset{a}{\sigma}\}, \\ \emptyset & \text{otherwise.} \end{cases}
$$

If $x = ax'$, $y = \epsilon$ and $t = ct'$ then

$$
x \sqcup\!\sqcup_t \epsilon = \begin{cases} a(x' \sqcup\!\sqcup_{t'} \epsilon) & \text{if } c \in \{0, \overset{a}{0}\}, \\ \emptyset & \text{otherwise.} \end{cases}
$$

If $x = \epsilon$, $y = by'$ and $t = ct'$ then

$$
\epsilon \sqcup\!\sqcup_t y = \begin{cases} b(\epsilon \sqcup\!\sqcup_{t'} y') & \text{if } c \in \{1, \overset{b}{1}\}, \\ \emptyset & \text{otherwise.} \end{cases}
$$

If $x = y = \epsilon$, then $x \sqcup\!\sqcup_t y = \epsilon$ if $t = \epsilon$ and $\emptyset$ otherwise. Finally, if $\{x, y\} \neq \{\epsilon\}$, then $x \sqcup\!\sqcup_\epsilon y = \emptyset$. If $x, y \in \Sigma^*$ and $T \subseteq \Gamma_\Sigma^*$, then $x \sqcup\!\sqcup_T y = \cup_{t \in T} x \sqcup\!\sqcup_T y$. If $L_1, L_2 \subseteq \Sigma^*$ and $T \subseteq \Gamma_\Sigma^*$, then $L_1 \sqcup\!\sqcup_T L_2 = \cup_{x \in L_1, y \in L_2} x \sqcup\!\sqcup_T y$.

Intuitively, we can consider the set of trajectories as consisting of instructions $\{0, 1, \sigma\}$, as well as an instruction on what letter can be present while that instruction is performed. The letters of Γ indicate a 'don't care'–this represents our syntactic shuffle on trajectories[1]. Note that if $T \subseteq \{0, 1\}^*$, then $L_1 \sqcup\!\sqcup_T L_2$ is the syntactic shuffle on trajectories operation.

We also introduce a corresponding deletion operation. Let $\Delta = \{i, d, \sigma\}$. For any alphabet Σ, let $\Delta_\Sigma = \Delta \cup (\Delta \times \Sigma)$. If $x = ax'$, $y = by'$ (where $a, b \in \Sigma$), and $t = ct'$ (where $c \in \Delta_\Sigma$), then

---

[1] Technically, the alphabet Γ is not needed, since we can replace, e.g., 0 with the expression $\cup_{a \in \Sigma} \overset{a}{0}$, however, its presence will make our expressions more readable and understandable.

$$
x \rightsquigarrow_t y = \begin{cases} a(x' \rightsquigarrow_{t'} y) & \text{if } c \in \{i, \overset{a}{i}\} \\ (x' \rightsquigarrow_{t'} y') & \text{if } a = b \text{ and } c \in \{d, \overset{a}{d}\} \\ a(x' \rightsquigarrow_{t'} y') & \text{if } a = b \text{ and } c \in \{\sigma, \overset{a}{\sigma}\} \\ \emptyset & \text{otherwise.} \end{cases}
$$

We also have the following base cases: if $x = ax'$ $(a \in \Sigma)$ $y = \epsilon$, and $t = ct'$ (where $c \in \Delta_\Sigma$), then

$$
x \rightsquigarrow_t \epsilon = \begin{cases} a(x' \rightsquigarrow_{t'} y) & \text{if } c \in \{i, \overset{a}{i}\} \\ \emptyset & \text{otherwise.} \end{cases}
$$

If $x = \epsilon$, then $\epsilon \rightsquigarrow_t y = \epsilon$ if $t = y = \epsilon$, and $\epsilon \rightsquigarrow_t y = \emptyset$ otherwise. Finally, $x \rightsquigarrow_\epsilon y = \emptyset$ if $x \neq \epsilon$. If $x, y \in \Sigma^*$ and $T \subseteq \Delta_\Sigma^*$, then $x \rightsquigarrow_T y = \bigcup_{t \in T} x \rightsquigarrow_T y$. If $L_1, L_2 \subseteq \Sigma^*$ and $T \subseteq \Delta_\Sigma^*$, then $L_1 \rightsquigarrow_T L_2 = \bigcup_{x \in L_1, y \in L_2} x \rightsquigarrow_T y$.

3   Equivalence of Sets of Trajectories

We now consider, given Σ and two sets of trajectories $T_1, T_2 \subseteq \Gamma_\Sigma^*$, whether the operations $\sqcap_{T_1}, \sqcap_{T_2}$ coincide, that is, whether $L_1 \sqcap_{T_1} L_2 = L_1 \sqcap_{T_2} L_2$ for all languages $L_1, L_2 \subseteq \Sigma^*$. If $\sqcap_{T_1}, \sqcap_{T_2}$ represent, in this sense, the same operation, we say that T_1, T_2 are *equivalent* sets of trajectories. We can also consider the same problem for $T_1, T_2 \subseteq \Delta_\Sigma^*$, however, the results we obtain can be easily transferred to that setting.

We note that for $T_1, T_2 \subseteq \{0, 1\}^*$ or $T_1, T_2 \subseteq \{i, d\}^*$, it is known that T_1 and T_2 are equivalent if and only if they are equal. We note that this is not the case for $T_1, T_2 \subseteq \Gamma_\Sigma^*$. As a simple example, consider $T_1 = \{\overset{a}{0}\overset{a}{1}\}$ and $T_2 = \{\overset{a}{1}\overset{a}{0}\}$. Note that for $i = 1, 2$,

$$
L_1 \sqcap_{T_i} L_2 = \begin{cases} \{aa\} & \text{if } L_1 \cap L_2 \supseteq \{a\}; \\ \emptyset & \text{otherwise.} \end{cases}
$$

Thus, T_1, T_2 are equivalent, but not equal.

Our first step is to consider only those $T \subseteq (\Gamma \times \Sigma)^*$. This is accomplished through the finite substitution $\text{sem}_\Sigma : \Gamma_\Sigma^* \to 2^{(\Gamma \times \Sigma)^*}$:

$$
\text{sem}_\Sigma(\overset{a}{c}) = \{\overset{a}{c}\} \quad \forall c \in \Gamma, a \in \Sigma;
$$
$$
\text{sem}_\Sigma(c) = \bigcup_{a \in \Sigma} \overset{a}{c} \quad \forall c \in \Gamma.
$$

If Σ is understood, then we denote sem_Σ by sem.

Note that $\text{sem}_\Sigma(T)$ is a regular set of trajectories if T is a regular set of trajectories.

Lemma 3.1. *Let $T \subseteq \Gamma_\Sigma^*$. Then T and $\text{sem}_\Sigma(T)$ are equivalent.*

Thus, it suffices to consider $T_1, T_2 \subseteq (\Gamma \times \Sigma)^*$. We now define an equivalence relation on words as follows: Let $I \subseteq (\Gamma \times \Sigma)^2$ be the following subset

$$I = \{(\overset{a}{c_1}, \overset{a}{c_2}) \ : \ a \in \Sigma, c_1, c_2 \in \Gamma\}.$$

For two words $z_1, z_2 \in (\Gamma \times \Sigma)^*$, say that $z_1 \leftrightarrow_I z_2$ if $z_1 = \alpha\beta_1\beta_2\gamma$ and $z_2 = \alpha\beta_2\beta_1\gamma$, where $(\beta_1, \beta_2) \in I$. We define the equivalence relation $\sim_I$ on $(\Gamma \times \Sigma)^*$ as the reflexive and transitive closure of $\leftrightarrow_I$. That is, for all $u, v \in (\Gamma \times \Sigma)^*$, $u \sim_I v$ if there exist $k \geq 1$ and $z_1, z_2, \cdots, z_k \in (\Gamma \times \Sigma)^*$ such that $u = z_1, v = z_k$ and for all $1 \leq i \leq k - 1$, $z_i \leftrightarrow_I z_{i+1}$.

This construction is a special case of partial commutation and trace languages, see Diekert and Métivier [5]. We note that I defines an *independence relation*, and the quotient monoid $\mathbb{M} = \mathbb{M}(\Gamma \times \Sigma, I) = (\Gamma \times \Sigma)^*/ \sim_I$ is called the *free partially commutative monoid*. Elements of $\mathbb{M}$ are called *traces*. We denote the equivalence class of a word $x \in (\Gamma \times \Sigma)^*$ by $[x]$. For a language $L \subseteq (\Gamma \times \Sigma)^*$, we denote $[L] = \cup_{x \in L}[x]$.

Lemma 3.2. *For all $t_1, t_2 \in (\Gamma \times \Sigma)^*$, $t_1 \sim_I t_2$, if and only if, for all $x, y \in \Sigma^*$, $x \sqcap_{t_1} y = x \sqcap_{t_2} y$.*

The above lemmata imply the following result:

Theorem 3.1. *Let Σ be an alphabet and $T_1, T_2 \subseteq \Gamma_\Sigma^*$. Then T_1, T_2 are equivalent if and only if $[\mathrm{sem}(T_1)] = [\mathrm{sem}(T_2)]$.*

We now address decidability. We say that a language $R \subseteq \mathbb{M}$ is a rational trace language if there exists a regular expression over $\mathbb{M}$ whose language equals R. The regular expressions are defined in the natural way, see Diekert and Métivier [5, Sect. 4.1] for details.

Since I is transitive (i.e., $(\beta_1, \beta_2), (\beta_2, \beta_3) \in I$ imply $(\beta_1, \beta_3) \in I$), we immediately conclude the following result (see Diekert and Métivier [5, Thm. 5.2]):

Theorem 3.2. *Let Σ be an alphabet and $T_1, T_2 \subseteq \Gamma_\Sigma^*$. If $[\mathrm{sem}(T_1)], [\mathrm{sem}(T_2)]$ are rational trace languages, it is decidable whether T_1 and T_2 are equivalent.*

We note that since the mapping $x \rightarrow [x]$ defines a morphism between the monoids Σ^* and $\mathbb{M}(\Sigma, I)$, every regular language $R \subseteq \Sigma^*$ yields a rational trace language $[R] \subseteq \mathbb{M}(\Sigma, I)$. Thus, we have the following corollary:

Corollary 3.1. *Let Σ be an alphabet and $T_1, T_2 \subseteq \Gamma_\Sigma^*$. If T_1, T_2 are regular, it is decidable whether T_1 and T_2 are equivalent.*

4   Closure Properties

Theorem 4.1. *Let Σ be an alphabet. There exist weak codings $\rho_1, \rho_2, \tau, \varphi$ and a regular language R such that for all $L_1, L_2 \subseteq \Sigma^*$ and $T \subseteq \Gamma_\Sigma^*$.*

$$L_1 \sqcap_T L_2 = \varphi(\rho_1^{-1}(L_1) \cap \rho_2^{-1}(L_2) \cap \tau^{-1}(T) \cap R).$$

Recall that a *cone* (or *full trio*) is a class of languages closed under morphism, inverse morphism and intersection with regular languages. We have the following corollary:

Corollary 4.1. *Let $\mathcal{L}$ be a cone. Then for all L_1, L_2, T such that two are regular languages and the third is from $\mathcal{L}$, $L_1 \sqcap_T L_2 \in \mathcal{L}$.*

In particular, since the regular languages are a cone, we have the following result (a direct construction using NFAs is also possible):

Corollary 4.2. *Let $L_1, L_2 \subseteq \Sigma^*, T \subseteq \Gamma_\Sigma^*$ be regular languages. Then $L_1 \sqcap_T L_2$ is a regular language.*

Nonclosure properties of other classes of languages are inherited from the syntactic case. In particular, the context-free and linear context-free languages are not closed under $\sqcap_T$.

Theorem 4.2. *Let Σ be an alphabet. There exist weak codings $\rho_1, \rho_2, \tau, \varphi$ and a regular language R such that for all $L_1, L_2 \subseteq \Sigma^*$ and $T \subseteq \Delta_\Sigma^*$,*

$$L_1 \leadsto_T L_2 = \varphi(\rho_1^{-1}(L_1) \cap \rho_2^{-1}(L_2) \cap \tau^{-1}(T) \cap R).$$

4.1 I-Regularity

Let $B = \{i, d\}$, $B_\Sigma = B \cup B \times \Sigma$, $C_\Sigma = \{i\} \cup \{i\} \times \Sigma$ and $D_\Sigma = \{d\} \cup \{d\} \times \Sigma$. We now present a class of trajectories $T \subseteq B_\Sigma^*$ which have the property that for all regular languages R and all languages L (regardless of their complexity), the language $R \leadsto_T L$ is regular.

Let Λ_m be the alphabet $\Lambda_m = \{\#_1, \#_2, \ldots, \#_m\}$ for any $m \geq 1$. We define a class of regular substitutions from $(D_\Sigma + \Lambda_m)^*$ to $2^{B_\Sigma^*}$, denoted $\mathfrak{S}_m$, as follows: a regular substitution $\varphi : (D_\Sigma + \Lambda_m)^* \to 2^{B_\Sigma^*}$ is in $\mathfrak{S}_m$ if both

(a) $\varphi(y) = \{y\}$ for all $y \in D_\Sigma$; and
(b) for all $1 \leq j \leq m$, there exist $u, v, w \in C_\Sigma^*$ such that $\varphi(\#_j) = uv^*w$.

For all $m \geq 1$, we define a class of languages $T \subseteq (D_\Sigma + \Lambda_m)^*$, denoted by $\mathfrak{T}_m$, as the set of all languages $T \subseteq \#_1 D_\Sigma^* \#_2 D_\Sigma^* \cdots \#_{m-1} D_\Sigma^* \#_m$. We then define the desired class of sets of trajectories, $\mathfrak{I}$, as follows:

$$\mathfrak{I} = \{T \subseteq B_\Sigma^* \; : \; \exists m \geq 1, T_m \in \mathfrak{T}_m, \varphi \in \mathfrak{S}_m \text{ such that } T = \varphi(T_m)\}.$$

If $T \in \mathfrak{I}$, we say that T is *i-regular*.

Theorem 4.3. *Let $T \in \mathfrak{I}$. Then for all regular languages R and all languages L, $R \leadsto_T L$ is a regular language.*

5 Examples

We now demonstrate the power of SST and SDT. We first note that SST consists of a valid extension of the shuffle on trajectories: if $T \subseteq \{0, 1\}^*$, then $L_1 \sqcap_T L_2 = L_1 \sqcup_T L_2$, the syntactic shuffle on trajectories operation [24]. We also note that if $T = \sigma^*$, then $\sqcap_T = \cap$.

Infiltration Product. Given the very natural set of trajectories $T = (0 + 1 + \sigma)^*$, $\sqcap_T$ denotes the *infiltration product*, $\uparrow$, see, e.g., Pin and Sakarovitch [28]. The infiltration product is defined as follows: if $x = x_1 \ldots x_n$ is a word of length n and $I = (i_1, i_2, \ldots, i_r)$ is a subsequence of $(1, 2, \ldots, n)$, let $x_I = x_{i_1} x_{i_2} \cdots x_{i_r}$. Then given $x, y \in \Sigma^*$,

$$x \uparrow y = \{z \in \Sigma^* \ : \ \exists I, J \subseteq [|z|] \text{ such that } I \cup J = [|z|], z_I = x \text{ and } z_J = y\}.$$

For example, $ab \uparrow ba = \{aba, bab, baab, baba, abba, abab\}$.

Ciliate Bio-operations. We now show how to use SST and SDT to simulate ciliate bio-operations which have been the subject of recent research in the literature. A model of ciliate bio-operations without circular variants were introduced by Daley and Kari [3] to mimic the manner in which DNA is unscrambled in the DNA of certain uni-cellular ciliates in the process of asexual reproduction. Ciliate bio-operations are also investigated by Ehrenfeucht *et al.* [11] and Daley and McQuillan [4] using different formal language approaches.

Daley and Kari [3] (see also Daley *et al.* [2]) define several language operations which simulate ciliate bio-operations, including synchronized insertion, deletion and bi-polar deletion. Synchronized insertion can be given as follows:

$$\alpha \oplus \beta = \{uavaw \ : \ a \in \Sigma, \alpha = uaw, \beta = va\}.$$

The original definition of all operations in this section allowed for "contexts" – i.e. the positions synchronization occur – longer than $a \in \Sigma$, but it is shown by Daley and Kari that single-letter contexts suffice.) The operation is extended to languages as usual. Let $T = \bigcup_{a \in \Sigma} 0^* \overset{a}{0} 1^* \overset{a}{1} 0^*$. Then for all $L_1, L_2 \subseteq \Sigma^*$, $L_1 \oplus L_2 = L_1 \sqcap_T L_2$. Synchronized deletion can be defined as

$$\alpha \ominus \beta = \{uaw \ : \ a \in \Sigma, \alpha = uavaw, \beta = va\}.$$

Let Σ be an alphabet and $T = \bigcup_{a \in \Sigma} i^* \overset{a}{i} d^* \overset{a}{d} i^*$. Then for all $L_1, L_2 \subseteq \Sigma^*$, $L_1 \ominus L_2 = L_1 \leadsto_T L_2$. Synchronized bi-polar deletion can be defined as

$$\alpha \boxminus \beta = \{va \ : \ a \in \Sigma, \alpha = uavaw, \beta = uaw\}.$$

Let Σ be an alphabet, and $T = \bigcup_{a \in \Sigma} d^* \overset{a}{d} i^* \overset{a}{i} d^*$. Then for all $L_1, L_2 \subseteq \Sigma^*$, $L_1 \boxminus L_2 = L_1 \leadsto_T L_2$.

We note the similarity between the three sets of trajectories in the three previous examples. We expect this, as similar relationships exists between operations and their inverses which are modelled by shuffle and deletion along trajectories [6, 18]. We will formalize this in Section 6 and thus demonstrate the corresponding decidability results for language equations involving the above bio-operations.

We also note that the closure of the regular languages under each of $\oplus, \ominus, \boxminus$ is a direct consequence of the closure properties in Section 4. However, note also that $\ominus, \boxminus$ are the finite union of *i*-regular sets of trajectories. We will use this fact in Section 6.2.

Contextual Insertions and Deletions. Contextual insertion and deletion were introduced by Kari and Thierrin as a simple set of operations which are capable for modelling DNA computing [19].

Let Σ be an alphabet and $[x, y] \in (\Sigma^*)^2$. We call $[x, y]$ a *context*. Then given $v, u \in \Sigma^*$, the $[x, y]$-*contextual insertion* of v into u is given by $u \overleftarrow{[x,y]} v = \{u_1 x v y u_2 : u = u_1 x y u_2, u_1, u_2 \in \Sigma^*\}$. Let $C \subseteq (\Sigma^*)^2$. Then

$$u \overleftarrow{c} v = \bigcup_{[x,y] \in C} u \overleftarrow{[x,y]} v.$$

The operation $\overleftarrow{c}$ is extended to languages monotonically as expected. Let $x = x_1 \cdots x_n$ and $y = y_1 \cdots y_m$ be arbitrary words over Σ. Then define

$$T_{[x,y]} = 0^* \prod_{i=1}^{n} \overset{x_i}{0} 1^* \prod_{i=1}^{m} \overset{y_i}{0} 0^*.$$

We naturally extend this to $T_C = \cup_{[x,y] \in C} T_{[x,y]}$ for all $C \subseteq (\Sigma^*)^2$. Under this definition it is clear that $\sqcap_{T_C} = \overleftarrow{c}$ for all $C \subseteq (\Sigma^*)^2$.

Kari and Thierrin note that if $C \subseteq (\Sigma^*)^2$ is finite, then the regular and context-free languages are closed under $\overleftarrow{c}$. As T_C is regular for all finite C, we note that the closure of the regular languages under $\overleftarrow{c}$ is a consequence of Corollary 4.2. Further investigation of the closure properties of SST are necessary to show, as a corollary, the known closure of the CFLs under $\overleftarrow{c}$ [19].

For $[x, y] \in (\Sigma^*)^2$, the $[x, y]$-contextual deletion of a word v from u is defined as $u \overrightarrow{[x,y]} v = \{u_1 x y u_2 : u = u_1 x v y u_2, u_1, u_2 \in \Sigma^*\}$. This operation is extended to sets of contexts $C \subseteq (\Sigma^*)^2$ as in the case of contextual insertion. Note that in this case, if $x = x_1 \cdots x_n$ and $y = y_1 \cdots y_m$, we define $T_{[x,y]} \subseteq \Delta_\Sigma^*$ as

$$T_{[x,y]} = i^* \prod_{j=1}^{n} \overset{x_j}{i} d^* \prod_{j=1}^{m} \overset{y_j}{i} i^*.$$

In this case, if C is finite, we may immediately conclude that the regular and context-free languages are closed under C-contextual deletion with a regular language. This was established by Kari and Thierrin [19, Cor. 2.1].

Finally, we note that $[x, y]$-contextual dipolar deletion, defined by $u \overrightarrow{[x,y]} v = \{w \in \Sigma^* : u = u_1 x w y u_2, v = u_1 x y u_2\}$, is simulated by

$$T_{[x,y]} = d^* \prod_{j=1}^{n} \overset{x_j}{d} i^* \prod_{j=1}^{m} \overset{y_j}{d} d^*,$$

if $x = x_1 \cdots x_n$ and $y = y_1 \cdots y_m$. The closure of the regular languages under C-contextual dipolar deletion for finite $C \subseteq (\Sigma^*)^2$ is an instance of Corollary 4.2; this was first established by Kari and Thierrin [19, Prop. 4.1].

Synchronized Shuffle. Recall that *synchronized shuffle* (see, e.g., Latteux and Roos [22]) is defined as follows. Let Σ_1, Σ_2 be alphabets, not necessarily disjoint.

Let $\rho_i : (\Sigma_1 \cup \Sigma_2) \to \Sigma_i$ be the projection onto Σ_i given by $\rho_i(a) = a$ for all $a \in \Sigma_i$ and $\rho_i(a) = \epsilon$ for all $a \in (\Sigma_1 \cup \Sigma_2) - \Sigma_i$.

Let $L_i \subseteq \Sigma_i$ for $i = 1, 2$. Then the synchronized shuffle of L_1 and L_2, denoted $L_1 \parallel L_2$, is given by

$$L_1 \parallel L_2 = \rho_1^{-1}(L_1) \cap \rho_2^{-1}(L_2).$$

Lemma 5.1. *Let* Σ_1, Σ_2 *be alphabets. Let* $T \subseteq \Gamma^*_{\Sigma_1 \cup \Sigma_2}$ *be given by*

$$T = ((\bigcup_{a \in \Sigma_1 \cap \Sigma_2} \overset{a}{\sigma}) + (\bigcup_{a \in \Sigma_1 - \Sigma_2} \overset{a}{0}) + (\bigcup_{a \in \Sigma_2 - \Sigma_1} \overset{a}{1}))^*.$$

Then for all L_1, L_2 *such that* $L_i \subseteq \Sigma_i$ *for* $i = 1, 2$,

$$L_1 \parallel L_2 = L_1 \sqcap_T L_2.$$

We can generalize the notion of synchronized shuffle to develop a notion of synchronized shuffle along trajectories, which we denote by $\parallel_T$ for arbitrary $T \subseteq \{0, 1\}^*$. Specifically, let Σ_1, Σ_2 be alphabets and let $T \subseteq \{0, 1\}^*$ be a chosen set of trajectories (which will give the desired set of trajectories for non-synchronized events). Let $\zeta(T)$ be the set of trajectories which we obtain from T under the finite substitution ζ which maps 0 to $\cup_{a \in \Sigma_1 - \Sigma_2} \overset{a}{0}$ and 1 to $\cup_{a \in \Sigma_2 - \Sigma_1} \overset{a}{1}$. Finally, let

$$T_s = \zeta(T) \sqcup (\bigcup_{a \in \Sigma_1 \cap \Sigma_2} \overset{a}{\sigma})^*.$$

Then let $\parallel_T$ be given by $\sqcap_{T_s}$. We note that $\parallel_T$ is a natural generalization of $\parallel$ in the same way that $\sqcap_T$ is a generalization of $\sqcup$: in both cases, with $T = \{0, 1\}^*$, we are left with the operation we have generalized.

The operation $\parallel_T$ is a useful generalization of $\parallel$, especially for modelling discrete event systems (DES) in the Ramadge-Wonham framework [1]: by altering T, the operation $\parallel_T$ allows us to specify restrictions to the way in which plants (which are modelled by regular languages) are allowed to interleave their processes, without restricting the way in which their synchronized actions are required to behave. This has applications in modelling where constraints such as, e.g., fairness, are necessary. Fairness has previously been discussed by Mateescu et al. [24, Sect. 6.1].

Latin Product. Mateescu and Salomaa [25] define the operation of Latin product as follows: let $u = u_1 u_2 \cdots u_n$ and $v = v_1 \cdots v_m$ be words over Σ. Then the Latin product $\star$ is given by

$$u \star v = \begin{cases} u_1 u_2 \cdots u_n v_2 v_3 \cdots v_m & \text{if } u_n = v_1; \\ u_1 u_2 \cdots u_n v_1 v_2 \cdots v_m & \text{otherwise.} \end{cases}$$

The Latin product is used in modelling parallel processes with re-entrant routines [25]. We note that with $T_\star = 0^* \sigma 1^* + \cup_{a,b \in \Sigma, a \neq b} 0^* \overset{a\,b}{01} 1^*$, the operation $\sqcap_{T_\star}$ is exactly the Latin product.

6   Decidability of Equations

We now turn to language equations involving SST and SDT. Equations of this form have been previously studied by, e.g., Daley *et al.* [2, 3] and Kari and Thierrin [19], as well as by the author [6], Kari and Sosík [18] and the author and Salomaa [9, 10] for syntactic shuffle on trajectories. As in these previous works, we focus on the decidability of the existence of solutions to certain language equations.

Before beginning with language equations, we establish the left- and right-inverse properties between SST and SDT, in the sense established by Kari [14].

Theorem 6.1. *Let Σ be an alphabet. There exists a morphism $\tau : \Gamma_\Sigma^* \to \Delta_\Sigma^*$ such that for all $x, y, z \in \Sigma^*$ and for all $t \in \Gamma_\Sigma^*$, $x \in y \sqcap_t z \iff y \in x \leadsto_{\tau(t)} z$.*

Theorem 6.2. *Let Σ be an alphabet. There exists a morphism $\pi : \Gamma_\Sigma^* \to \Delta_\Sigma^*$ such that for all $x, y, z \in \Sigma^*$ and all $t \in \Gamma_\Sigma^*$, $x \in y \sqcap_t z \iff z \in x \leadsto_{\pi(t)} y$.*

Theorem 6.3. *Let Σ be an alphabet. There exists a morphism $sym_\Delta : \Delta_\Sigma^* \to \Delta_\Sigma^*$ such that for all $x, y, z \in \Sigma^*$ and all $t \in \Delta_\Sigma^*$, $x \in y \leadsto_t z \iff z \in y \leadsto_{sym_\Delta(t)} x$.*

6.1   Solving Linear Equations

Given two binary word operations $\diamond, \star : (\Sigma^*)^2 \to 2^{\Sigma^*}$, we say that $\diamond$ is a *left-inverse* of $\star$ [14, Defn. 4.5] (resp., a *right-inverse* of $\star$ [14, Defn. 4.1]) if, for all $u, v, w \in \Sigma^*$, $w \in u \star v \iff u \in w \diamond v$ (resp., for all $u, v, w \in \Sigma^*$, $w \in u \star v \iff v \in u \diamond w$). The following results of Kari [14, Thms. 4.2 and 4.6] allow us to find solutions to equations involving shuffle on trajectories.

Theorem 6.4. *Let L, R be languages over Σ and $\diamond, \star$ be two binary word operations, which are left-inverses (resp., right-inverses) to each other. If the equation $X \diamond L = R$ (resp., $L \diamond X = R$) has a solution $X \subseteq \Sigma^*$, then the language $R' = \overline{\overline{R} \star L}$ (resp., $R' = \overline{L \star \overline{R}}$) is also a solution of the equation. Moreover, R' is a superset of all other solutions of the equation.*

Thus, the following theorem is immediate:

Theorem 6.5. *Let Σ be an alphabet, Let $L, R \subseteq \Sigma^*$ be regular languages and $T \subseteq \Gamma_\Sigma^*$ be a set of trajectories. Then for each of the following equations, it is decidable whether there exists $X \subseteq \Sigma^*$ such that equality holds: (a) $L \sqcap_T X = R$; (b) $X \sqcap_T L = R$; (c) $L \leadsto_T X = R$; (d) $X \leadsto_T L = R$.*

The particular cases of $\oplus$ and $\ominus$ of Theorem 6.5 were given by Daley and Kari [3, Prop. 11, Prop. 18]. For a finite set of contexts $C \subseteq (\Sigma^*)^2$, the particular cases of Theorem 6.5 were given by Kari and Thierrin [19, Prop. 5.3].

6.2   Solving Decompositions

If $T \subseteq B_\Sigma^*$ is a regular set of trajectories which is a finite union of regular, i-regular sets of trajectories, we say that T is a *pseudo-letter-bounded* (regular) set of trajectories. Let $V = \{0,1\}$ and $V_\Sigma = V \cup V \times \Sigma$. If $T \subseteq V_\Sigma^*$ is a set of trajectories such that $\tau(T)$ and $\pi(T)$ are pseudo-letter-bounded (where π and τ are from Theorem 6.1 and 6.2), we also say that T is pseudo-letter-bounded.

For instance, let Σ be arbitrary and $T = \cup_{a \in \Sigma} i^* \overset{a}{i} d^* \overset{a}{d} i^*$, as given in Section 5. Then T is pseudo-letter-bounded as $T_a = 0^* \overset{a}{0} 1^* \overset{a}{1} 0^*$ is i-regular for all $a \in \Sigma$ (by choosing $T_a' = \#_1 d^* \overset{a}{d} \#_2$ and φ with $\varphi(\#_1) = i^* \overset{a}{i}$ and $\varphi(\#_2) = i^*$). Our main result is:

Theorem 6.6. *Let $T = \cup_{i=1}^n T_i \subseteq V_\Sigma^*$ be a pseudo-letter-bounded regular set of trajectories. Then given a regular language R, it is decidable whether there exist X_1, X_2 such that $X_1 \sqcap_T X_2 = R$.*

The following is an interesting particular case, which has not been examined with respect to synchronized insertion:

Corollary 6.1. *Let R be a regular language. Then it is decidable whether there exist languages X_1, X_2 such that $R = X_1 \oplus X_2$.*

Further, the following is also a corollary of Theorem 6.6 not previously noted:

Corollary 6.2. *Let $C \subseteq (\Sigma^*)^2$ be a finite set of contexts. Let R be a regular language. Then it is decidable whether there exist languages X_1, X_2 such that $R = X_1 \overset{\leftarrow}{c} X_2$.*

References

1. CASSANDRAS, C., AND LAFORTUNE, S. *Introduction to Discrete Event Systems.* Kluwer Academic Publishers, 1999.
2. DALEY, M., IBARRA, O., AND KARI, L. Closure properties and decision questions of some language classes under ciliate bio-operations. *Theor. Comp. Sci. 306*, 1 (2003), 19–38.
3. DALEY, M., AND KARI, L. Some properties of ciliate bio-operations. In *Developments in Language Theory* (2003), vol. 2450 of *LNCS*, Springer, pp. 116–127.
4. DALEY, M., AND MCQUILLAN, I. Template-Guided DNA Recombination. In Proc. 5th Workshop DCFS 2003 (2003), pp. 234–244.
5. DIEKERT, V., AND MÉTIVIER, Y. *Partial Commutation and Traces.* In [29], vol. 3, pp. 457–534.
6. DOMARATZKI, M. Deletion along trajectories. *Theor. Comp. Sci. 320* (2004), 293–313.
7. DOMARATZKI, M. Trajectory-based codes. *Acta Inf. 40*, 6–7 (2004), 491–527.
8. DOMARATZKI, M. Trajectory-based embedding relations. *Fund. Inf. 59*, 4 (2004), 349–363.
9. DOMARATZKI, M., AND SALOMAA, K. Decidability of trajectory-based equations. In *Mathematical Foundations of Computer Science 2004*, J. Fiala, V. Koubek, and J. Kratochvil, Eds., vol. 3153, in *LNCS*. Springer-Verlag, 2004, pp. 723–734.

10. DOMARATZKI, M., AND SALOMAA, K. Restricted sets of trajectories and decidability of shuffle decompositions. In *Descriptional Complexity of Formal Systems 2004*, L. Ilie and D. Wotschke, Eds., pp. 37–51.

11. EHRENFEUCHT, A., HARJU, T., PETRE, I., PRESCOTT, D., AND ROZENBERG, G. Computation in Living Cells: Gene assembly in ciliates. Springer-Verlag (2004).

12. GINSBURG, S., AND SPANIER, E. H. Bounded regular sets. *Proc. Amer. Math. Soc. 17* (1966), 1043–1049.

13. HARJU, T., MATEESCU, A., AND SALOMAA, A. Shuffle on trajectories: The Schützenberger product and related operations. In *Mathematical Foundations of Computer Science 1998*, L. Brim, J. Gruska, and J. Zlatuska, Eds., vol. 1450 in *LNCS*. Springer-Verlag, 1998, pp. 503–511.

14. KARI, L. On language equations with invertible operations. *Theor. Comp. Sci. 132* (1994), 129–150.

15. KARI, L., KONSTANTINIDIS, S., AND SOSÍK, P. On properties of bond-free DNA languages. Tech. Rep. 609, Comp. Sci. Dept., Univ. Western Ontario, 2003. Submitted for publication.

16. KARI, L., KONSTANTINIDIS, S., AND SOSÍK, P. Bond-free languages Tech. Rep. 2004–001, St. Mary's University Dept. of Math. and Comp. Sci., 2004.

17. KARI, L., KONSTANTINIDIS, S., AND SOSÍK, P. Substitutions, trajectories and noisy channels. In *Pre-proceedings of CIAA 2004*, M. Domaratzki, A. Okhotin, K. Salomaa and S. Yu, Eds., pp. 154–162.

18. KARI, L., AND SOSÍK, P. Language deletions on trajectories. Tech. Rep. 606, Comp. Sci. Dept., Univ. of Western Ontario, 2003. Submitted for publication.

19. KARI, L., AND THIERRIN, G. Contextual insertions/deletions and computability. *Inf. and Comp. 131* (1996), 47–61.

20. KUDLEK, M., AND MATEESCU, A. On distributed catenation. *Theor. Comp. Sci. 180* (1997), 341–352.

21. KUDLEK, M., AND MATEESCU, A. On mix operation. In *New Trends in Formal Languages* (1997), G. Păun and A. Salomaa, Eds., vol. 1218 of *LNCS*, Springer, pp. 430–439.

22. LATTEUX, M., AND ROOS, Y. Synchronized shuffle and regular languages. In *Jewels are Forever: Contributions on Theoretical Computer Science in Honor of Arto Salomaa* (1999), J. Karhumäki, H. Maurer, G. Păun, and G. Rozenberg, Eds., Springer, pp. 35–44.

23. MATEESCU, A. Splicing on routes: a framework of DNA computation. In *Unconventional Models of Computation* (1998), C. Calude, J. Casti, and M. Dinneen, Eds., Springer, pp. 273–285.

24. MATEESCU, A., ROZENBERG, G., AND SALOMAA, A. Shuffle on trajectories: Syntactic constraints. *Theor. Comp. Sci. 197* (1998), 1–56.

25. MATEESCU, A., AND SALOMAA, A. Parallel composition of words with re-entrant symbols. *An. Univ. Bucureşti Mat. Inform. 45*, 1 (1996), 71–80.

26. MATEESCU, A., AND SALOMAA, A. Nondeterministic trajectories. In *Formal and Natural Computing*, vol. 2300 of *LNCS*. Springer-Verlag, 2002, pp. 96–106.

27. MATEESCU, A., SALOMAA, A., AND YU, S. Factorizations of languages and commutativity conditions. *Acta Cyb. 15*, 3 (2002), 339–351.

28. PIN, J.-E., AND SAKAROVITCH, J. Some operations and transductions that preserve rationality. In *6th GI Conference* (1983), A. Cremers and H.-P. Kriegel, Eds., vol. 145 of *LNCS*, Springer-Verlag, pp. 277–288.

29. ROZENBERG, G., AND SALOMAA, A., Eds. *Handbook of Formal Languages*. Springer-Verlag, 1997.

30. YU, S. Regular languages. In [29], vol. 1, pp. 41–110.

Sturmian Graphs and a Conjecture of Moser[*]

Chiara Epifanio[1], Filippo Mignosi[1], Jeffrey Shallit[2], and Ilaria Venturini[3]

[1] Dip. Mat. e Appl., Univ. di Palermo, Italy
{epifanio, mignosi}@math.unipa.it
[2] School of Computer Science, University of Waterloo, Ontario, Canada
shallit@graceland.math.uwaterloo.ca
[3] TSI, ENST, Paris, France
venturi@tsi.enst.fr

Abstract. In this paper we define Sturmian graphs and we prove that all of them have a "counting" property. We show deep connections between this counting property and two conjectures, by Moser and by Zaremba, on the continued fraction expansion of real numbers. These graphs turn out to be the underlying graphs of CDAWGs of central Sturmian words. We show also that, analogously to the case of Sturmian words, these graphs converge to infinite ones.

1 Introduction

Sturmian words are aperiodic infinite words over a binary alphabet of minimal complexity, i.e., with exactly $n+1$ factors of length n. They have been extensively studied for their properties and equivalent definitions. Moreover, the well-known Fibonacci word is Sturmian.

Among the different definitions, one is that obtained by considering the intersections with a square-lattice of a ray having an irrational slope $\alpha > 0$. The word obtained by coding each vertical intersection with an a, each horizontal intersection by a b and each corner with ab or ba is Sturmian. If the ray starts from the origin, the word obtained is called *characteristic*. Another way for constructing characteristic Sturmian words is by applying the *standard method*. Define inductively the two sequences of words $\{A_n\}$ and $\{B_n\}$ by

$$\begin{cases} A_0 = a \\ B_0 = b \end{cases}$$

and by the *Rauzy's two rules* [20]

$$R_1 : \begin{cases} A_{n+1} = A_n \\ B_{n+1} = A_n B_n \end{cases} \qquad R_2 : \begin{cases} A_{n+1} = B_n A_n \\ B_{n+1} = B_n \end{cases}$$

These two sequences converge, when each of the two rules is applied infinitely often, to the same infinite word which is characteristic and conversely each characteristic word is obtained in this way.

---

[*] Partially supported by MIUR National Project PRIN "Linguaggi Formali e Automi: teoria ed applicazioni."

C.S. Calude, E. Calude and M.J. Dinneen (Eds.): DLT 2004, LNCS 3340, pp. 175–187, 2004.

Given a pair (A_n, B_n), we can associate with it its *directive sequence* (cf. [8]), that is, the sequence of integers $[a_0, a_1, \ldots, a_s]$ such that $\sum_{i=0}^{s} a_i = n$, representing the fact that the final sequences A_n and B_n are obtained by applying R_1 to A_0 and B_0 a_0 consecutive times, after that R_2 a_1 consecutive times, etc. Words obtained by removing last two characters from A_n or B_n are called *central Sturmian words*. Given a pair (A_n, B_n) having directive sequence $[a_0, a_1, \ldots, a_s]$, it is possible to define recursively $\max(|A_n|, |B_n|)$ as the $(s+1)$-th element of the following sequence (l_j):

$$\begin{cases} l_0 = 1 \\ l_1 = a_0 + 1 \qquad\qquad j = 1 \ldots s \\ l_{j+1} = a_j \cdot l_j + l_{j-1} \end{cases}$$

For references on Sturmian words and their geometric representation see [17, Chap. 2] and [14].

If the directive sequence $[a_0, a_1, \cdots]$ is infinite, the infinite word to which A_n and B_n converge represents a ray having slope α, where α has $[a_0, a_1, \ldots]$ as its simple continued fraction expansion.

Let us recall some basic notations and results on continued fractions.

If α is a real number, we can expand α as a *simple continued fraction*

$$\alpha = a_0 + \cfrac{1}{a_1 + \cfrac{1}{a_2 + \cfrac{1}{a_3 + \cdots}}}$$

which is usually abbreviated as $\alpha = [a_0, a_1, a_2, a_3, \ldots]$.

In this paper, we only discuss the case where a_0 is a non-negative integer and a_i is a positive integer for $i \geq 1$; the expansion may or may not terminate. For references to continued fractions, see [10–Chap. 10], [18], [5] and [21].

If α is irrational, this representation is infinite and unique. If α is rational, there are two possible finite representations of it. Indeed, it is well known that $[a_0, a_1, \ldots, a_{s-1}, a_s, 1] = [a_0, a_1, \ldots, a_{s-1}, a_s + 1]$.

The integers in the continued fraction expansion of a real number are called partial quotients.

Given the continued fraction expansion of α, it is possible to construct a sequence of rationals $\frac{P_s}{Q_s}$, called "convergents", that converges to α, by the following rules

$$\begin{cases} P_0 = a_0 \qquad\qquad Q_0 = 1 \\ P_1 = a_1 \cdot a_0 + 1 \qquad Q_1 = a_1 \\ P_{s+1} = a_n \cdot P_s + P_{s-1} \qquad Q_{s+1} = a_s \cdot Q_s + Q_{s-1} \end{cases}$$

It is easy to see that $l_{j+1} = P_j + Q_j$.

The *directed acyclic word graph* of a word w, $DAWG(w)$, is the smallest finite state automaton that recognizes all the suffixes of the word. DAWGs have linear size and can be built in linear time with respect to the size of the word. They are involved in several combinatorial algorithms on strings and have many applications, such as full-text indexing. If the last letter in w is a letter $ that does not appear elsewhere in w, DAWG(w) coincides, apart from the set of final states, with the factor automaton of w, i.e., with the minimal deterministic automaton that recognizes the factors of w. Blumer et al. (cf. [1, 2, 3]) first introduced the *compact directed acyclic word graph* of a word w, *CDAWG(w)*, a space efficient variant of *DAWG(w)*, obtained by compacting it. Arcs in the obtained structure are labeled by representations of the factors of the word. More precisely, each arc is labeled by the initial position and the length of the factor represented by the arc. For a reference on CDAWGs, see also [6, 7, 12, 13].

In this paper we will define a new data structure, the Sturmian Graph of a directive sequence $[a_0, \ldots, a_s]$, $G([a_0, \ldots, a_s])$, and we will show how it coincides with the CDAWG of the word w obtained by the longest word in the pair (A_n, B_n) of directive sequence $[a_0, \ldots, a_s]$ replacing last two letters with a \$ symbol, where the label of each arc is replaced by the length of the factor it represents. More exactly we will show that $G([a_0, \ldots, a_s])$ coincides with the CDAWG of the word obtained in such a way, where arcs are labeled only by the lengths of the factors they represent.

The proofs of all results in this paper can be found in [9].

2   Special (or Finite) Sturmian Graphs

A *weighted DAG* is a directed acyclic graph, where each arc is weighted by a real number.

For any rational $\frac{P}{Q} = [a_0, \ldots, a_s]$ with $\sum_{i=0}^{s} a_i \geq 2$ we inductively define a graph $G(\frac{P}{Q}) = G([a_0, \ldots, a_s])$ that we call the *Sturmian graph of* $\frac{P}{Q} = [a_0, \ldots, a_s]$. This graph is a weighted DAG where weights are positive integers.

If $a_0 = 0$ we set $G([a_0, a_1, \ldots, a_s]) = G([a_1, \ldots, a_s])$. Therefore in what follows we will suppose that $a_0 \geq 1$.

The first Sturmian graph, the base case, is the graph $G([1, 1]) = G([2])$. It consists of only two states and two arcs, both going from state 1 to the final state F and having weights respectively 1 and 2. It can be seen in Figure 1. To give the inductive step, let us recall the definition of the sequence (l_j):

$$\begin{cases} l_0 = 1 \\ l_1 = a_0 + 1 \\ l_{j+1} = a_j \cdot l_j + l_{j-1} \end{cases}$$

Fig. 1. Base case

Given the Sturmian graph of $[a_0, \ldots, a_s], s \geq 0, \sum_{i=0}^{s} a_i \geq 2, G([a_0, \ldots, a_s])$, we define the Sturmian graph $G([a_0, \ldots, a_s, 1])$ in the following way. Each arc of maximal length in $G([a_0, \ldots, a_s])$ (all of them end at the final state) is split in one arc of that length minus 1 from the same outgoing state to a new state (the same for each arc) and two arcs from this new state towards the final one, one labeled 1 and the other labeled $l_s + 1$.

Moreover, if $a_s = 1$, then for each state of out-degree 2, except the new one, one must add a new outgoing arc labeled $l_s + 1$ towards the final state, with the exception of the new state that has already one such arc.

As $[a_0, a_1, \ldots, a_s, 1] = [a_0, a_1, \ldots, a_s + 1]$, the previously defined inductive step lets us construct every Sturmian graph $G([a_0, \ldots, a_k]), k \geq 0$.

Let us give some examples. Figure 2 shows graphs $G([3, 1])$ and $G([1, 1, 1, 1])$. The first one is obtained starting from $G([3])$, inductively built from the base case $G([2])$, being $G([3]) = G([2, 1])$. The second one is derived starting from $G([1, 1, 1])$ that, in turns, comes out from the base case $G([1, 1])$.

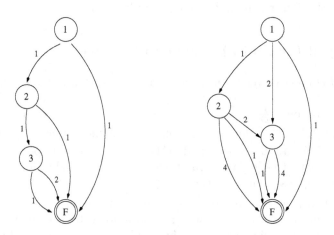

Fig. 2. Graphs $G([3, 1])$ and $G([1, 1, 1, 1])$

Proposition 1. *The Sturmian graph* $G([a_0, \ldots, a_s, 1]), s \geq 0, n = \sum_{i=0}^{s} a_i + 1 \geq 2$, *contains exactly* n *states, among them* a_s *of out-degree 2, and* $3(n-1) - a_s$ *arcs.*

Definition 1. *A DAG having a unique smallest state with respect to the order induced by the arcs is called semi-normalized. If it has also a unique greatest state it is called normalized. The smallest state is called the initial state and the greatest is called the final one.*

Note that any normalized DAG is also semi-normalized. Note also that any DAG can always be semi-normalized by adding at most one new state and can be normalized by adding at most two new states.

Definition 2. *A normalized weighted DAG G has the $(1,n)$-counting property, or, in short, it counts from 1 to n if any path from the initial state to the final one has weight in the range $1 \ldots n$ and for any i, $1 \leq i \leq n$ there exists just a unique path from the initial state to the final one having weight i. A semi-normalized weighted DAG G' has the $(1,n)$-counting property, or, in short, it counts from 1 to n if any nonempty path from the initial state has weight w in the range $1 \ldots n$ and for any i, $1 \leq i \leq n$ there exists just a unique path that starts from the initial state and have weight i.*

Remark 1. Note that a normalized graph is also semi-normalized and that it can have the counting property as semi-normalized but not as normalized. Indeed, if G' is semi-normalized and it counts from 1 to n, then we can build a normalized DAG G that counts from 1 to $n + 1$ in the following way. Add a final state F to G', and, for any state $q \in G'$ add also an arc (q, F) labeled by 1. If G' has out-degree at most h then G has out-degree at most $h + 1$.

Suppose, conversely, that G is normalized with final state F with positive integer weights, that it counts from 1 to $n + 1$ and that from any state q there is an arc (q, F), then we can build a DAG G' semi-normalized that counts from 1 to n in the following way. For any arc (q, F) decrease its label by 1, and, if this label is now 0, erase the arc. If G has out-degree at most $h + 1$ then G' has out-degree at most h.

It is easy to prove that Sturmian graphs are normalized weighted DAG with positive integer weights and out-degree at most 3.

Indeed, Sturmian graphs turn out to have also the $(1,n)$-counting property for some n, as pointed out by the following theorem, whose proof is given in next section. The proof is based on the fact that Sturmian graphs are CDAWGs of Sturmian words and any CDAWG has the counting property, i.e., it is a direct consequence of Proposition 6 and Theorem 2 together with its remark.

Theorem 1. $G(\frac{P}{Q} = [a_0, \ldots, a_s])$ *can count from 1 up to $P + Q - 1$.*

The reader can check that $G(\frac{5}{3} = [1, 1, 1, 1])$ can count from 1 up to 7.

Remark 2. Notice that in Sturmian graphs having final state F, from any state q there is an arc (q, F). Therefore we can apply the procedure described in Remark 1 and obtain a semi-normalized DAG $G'(\frac{P}{Q})$ with positive integer weights, of out-degree at most 2, that can count from 1 up to $P + Q - 2$. By extension, these graphs are also called Sturmian graphs.

We are now interested in the "inverse problem".

Problem 1. Given a positive integer m, find a normalized DAG with positive integer weights, where each state has out-degree at most 3, having minimal number of states and that can count from 1 up to m.

The reader can check that for any positive integer m there are just two (up to isomorphisms) normalized DAGs with positive integer weights, where each state has out-degree at most 2, that can count from 1 up to m. They have respectively m and $m + 1$ states. Therefore the hypothesis on the out-degree 3 makes sense. The same problem can be analogously stated for semi-normalized DAGs with out-degree at most 2. Notice also that if we do not impose a bound on the out-degree above problem has the trivial solution given by a graph having just the initial and the final states and m arcs labeled from 1 to m going from the initial to the final.

We do not know whether above problem can be settled in polynomial time in the size of $\log(m)$ (recall that the number of bits needed to describe m is $O(\log(m))$. We do not even know whether the minimal number of states is $O(\log(m))$, and, concerning this fact, we make the following conjecture.

Conjecture 1. Given a number m, the minimal number of states of a normalized DAG with positive integer weights, where each state has out-degree at most 3, that can count from 1 up to m is $O(\log(m))$.

For some special classes of numbers above conjecture is a consequence of Theorem 1 and Proposition 1. For instance, if $m = f_s - 1$, where f_s is the s-th Fibonacci number, then $G(\frac{f_{s-1}}{f_{s-2}})$ has $s - 1$ states, because $\frac{f_{s-1}}{f_{s-2}} = [a_0, a_1, \ldots, a_s]$ with, for any i, $0 \leq i \leq s$, $a_i = 1$. Since it is well known that $f_s = O(\varphi^s)$, where φ is the golden ratio, the Conjecture 1 holds true.

By using Theorem 1 and Proposition 1, with the same ideas used to prove above conjecture for $m = f_s - 1$ we can prove the following proposition.

Proposition 2. *If there exists an integer K such that for every integer $m \geq 1$ there exist integers $1 \leq p < q$ with $\gcd(p, q) = 1$ and $p + q = m$ such that every partial quotient in the continued fraction expansion of p/q is $\leq K$ then Conjecture 1 is true.*

We conjecture further that the hypothesis of previous proposition always holds.

Conjecture 2. There exists an integer K such that for every integer $m \geq 1$ there exist integers $1 \leq p < q$ with $\gcd(p, q) = 1$ and $p + q = m$ such that every partial quotient in the continued fraction expansion of p/q is $\leq K$.

We do not know if this conjecture is true, but it turns out to be equivalent to the following celebrated conjecture of Zaremba.

Conjecture 3 (Zaremba). There exists an integer K such that for every integer $m \geq 1$ there exists an integer i, $1 \leq i \leq m$, $\gcd(i, m) = 1$, such that every partial quotient in the continued fraction expansion of i/m is $\leq K$.

In [4] it is reported that Zaremba's conjecture has been verified with constant $K = 5$ up to 3200000 by D. Knuth.

Proposition 3. *Conjecture 2 and Zaremba's conjecture are logically equivalent. The same K can be used in both cases.*

Alternatively we can consider the sum of the partial quotients. Moser made the following conjecture that is weaker than Zaremba's one, in the sense that if Zaremba conjecture is true then also next conjecture is true.

Conjecture 4 (Moser). There exists a constant c such that for all integers $m \geq 2$ there exists an integer i, $0 \leq i \leq m$, $\gcd(i, m) = 1$, such that $i/m \leq c \log m$.

As above, Moser's conjecture is logically equivalent to a similar conjecture about the sum of p and q.

Indeed, analogously to Proposition 2 we have the following proposition.

Proposition 4. *If the Moser's conjecture is true then Conjecture 1 is also true.*

Larcher [15–Corollary 2] proved that this conjecture holds if $\log n$ is replaced by $(\log n)(\log \log n)^2$. Hence we get

Proposition 5. *There exists a constant c such that for all integers $n \geq 2$ there exist integers p, q with $\gcd(p, q) = 1$ and $p + q = n$ such that $p/q = [0, a_1, \ldots, a_n]$ and $\sum_i a_i < c(\log n)(\log \log n)^2$.*

This result implies a weak form of our conjecture.

Corollary 1. *Given a number m, there exists a constant c such that the minimal number of states of a normalized DAG with positive integer weights, where each state has out-degree at most 3 and that can count from 1 up to m, is smaller than $c(\log m)(\log \log m)^2$.*

3   Indexing, DAWGS and Sturmian Graphs

The *directed acyclic word graph* of a word w, $DAWG(w)$, is the smallest finite state automaton that recognizes all the suffixes of the word. DAWGS have several applications, such as indexing. Blumer et al. (cf. [1, 2, 3]) introduced the *compact directed acyclic word graph* of a word w, $CDAWG(w)$, that is obtained by compacting $DAWG(w)$, i.e., by deleting all states of out-degree 1 and their corresponding edges, joining all consecutive arcs in a path including such states in an unique arc. Thus arcs are labeled by representations of the factors of the word. More precisely, each arc is labeled by the initial position and the length of the factor represented by the arc. For a reference on CDAWGs, see also [6, 7, 12, 13]. We just recall that the underlining DAG of the CDAWG of w, that is $CDAWG(w)$ without labels, is a semi-normalized one. If the last character of w is a symbol never encountered before in w, then the underlining DAG of the CDAWG of w is a normalized one, i.e., it has also a unique final state.

In this section we will show how the Sturmian DAG $G([a_0, \ldots, a_s])$ defined in last section coincides with the CDAWG of the word w obtained by the longest word in the pair (A_n, B_n) of directive sequence $[a_0, \ldots, a_s]$ replacing last two letters with a $ symbol, where the label of each arc is replaced by the length of the factor it represents.

CDAWGs can be used in indexing. Indeed a CDAWG of a word w can give the list of all occurrences of a factor of w in time proportional to the size of this list. Indeed, by reading the factor in the CDAWG we reach a position t in it. This position either can be a state or can correspond to a proper prefix of the word representing the label of an arc. Each occurrence of the required factor is the length of w minus the length of any path from position t to any final state. The reason of this relies on the fact that all possible paths represent a branch of the suffix-tree of w corresponding to the required factor.

Since the empty word is a factor of any word w, the list of its occurrences in w is the set $\{1, 2, \ldots, |w|\}$. Hence we have proved the following.

Proposition 6. *Suppose that the last character of w is a symbol never encountered before in w. If we label each arc of CDAWG(w) just with the length of the factor it represents, the obtained weighted DAG can count from 1 up to $|w|$.*

Let us go in details. Consider the s-uple $[a_0, \ldots, a_s]$ and apply the Rauzy rule R_1 a_0 times to the pair $(A_0, B_0) = (a, b)$. We will obtain a pair (A_{a_0}, B_{a_0}). Let us now apply a_1 times R_2 rule to (A_{a_0}, B_{a_0}), in such a way we obtain $(A_{a_0+a_1}, B_{a_0+a_1})$. Let us continue by alternating the two rules and at the end we will have the pair $(A_{a_0+\cdots+a_s}, B_{a_0+\cdots+a_s})$ of directive sequence $[a_0, \ldots, a_s]$. Pick the longest between these two words and replace last two letters with a $ symbol. The word obtained is the one whose CDAWG we are interested in.

Let us give an example. Consider the directive sequence $[1, 1, 1, 1]$, the word we obtain is $abaaba\$$. In fact

$$(a, b) \to_{R_1} (a, ab) \to_{R_2} (aba, ab) \to_{R_1} (aba, abaab) \to_{R_2} (abaababa, abaab)$$

Figure 3 shows *CDAWG(abaaba\$)* and, next to it, the DAG obtained by it labelling each arc only with the length of the factor it represents. In order to give a better idea of which factor each arc represents we have labelled each arc not with the initial position and the length of the factor, but by the factor itself. This kind of representation will be also used in Figure 5. Remember that it is not the right representation, because this last representation requires, in the worst case, quadratic space, while the right one requires only linear space. As we can see, the DAG obtained coincides with Sturmian graph $G([1, 1, 1, 1])$, i.e., with the Sturmian graph of the same sequence from which we have obtained word $abaaba\$$.

What is surprising in CDAWGs of Sturmian words is that they have a relatively "small" number of nodes, compared to the length of the word itself, as shown by Proposition 1 and next theorem.

Theorem 2. *Let w_n be the word obtained by replacing in the longest word of the pair (A_n, B_n) of directive sequence $[a_0, \ldots, a_s, 1]$ last two letters with a $ symbol and CDAWG(w_n) be its CDAWG. Let now code each arc with the length of the factor it represents. The obtained DAG always coincides with the Sturmian graph $G([a_0, \ldots, a_s, 1])$.*

Remark 3. Notice that if $\frac{P}{Q} = [a_0, \ldots, a_s, 1]$ then the length of w_n defined in Theorem 2 is $P + Q$.

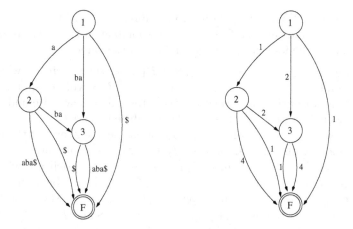

Fig. 3. The CDAWG of word *abaaba$* and the one obtained by coding each arc with the length of the factor it represents

The proof of the theorem will not be reported here. It requires some combinatorial results described in [8], the main result of [19] and it requires an inductive characterization of $CDAWG(w_n)$ that we describe in what follows and that, by itself, represents a significative contribution of this paper.

Given the directive sequence $[1, 1]$, the pair (A_{1+1}, B_{1+1}) obtained by applying to pair (a, b) once Rauzy rule R_1 and once R_2 rule is $(A_{1+1}, B_{1+1}) = (aba, ab)$. Let us pick the longest between the two words and replace last two letters with a $ symbol. We obtain word a. The CDAWG of a is represented in Figure 4. It is the same of the CDAWG corresponding to the directive sequence $[2]$ and represents the base case. Before introducing the inductive step, let us give some

Fig. 4. $CDAWG(a\$)$

new notation. Let w be a word, we will denote $w^=$ the word obtained by w deleting its two last letters and by $w\$$, the catenation of w and $ symbol. Moreover given two words $u = \alpha_1 \ldots \alpha_i$ and $v = \alpha_1 \ldots \alpha_j$ the first one being prefix of the second one, we will denote with $u^{-1}v$ the factor $\alpha_{i+1} \ldots \alpha_j$ of v. Finally, given a pair (A_n, B_n), we will denote with M_n the longest between the two words A_n and B_n.

Given the CDAWG corresponding to the directive sequence $[a_0, \ldots, a_s], s \geq 0, \sum_{i=0}^{s} a_i \geq 2$, we define the CDAWG corresponding to the directive sequence $[a_0, \ldots, a_s, 1]$ in the following way, depending on the value of s.

1. If s is even, i.e., we have just applied R_1 rule, then each arc which label corresponds to the factor $S_n\$ = (M_{n-1}{}^=)^{-1}B_n^=\$$ will be split in an arc labeled S_n from the same outgoing state towards a new state and two arcs from this new state towards the final one, labeled $\$$ and $T_n = (B_n^=)^{-1}A_{n+1}^=\$$. Moreover, if $a_s = 1$, then for each state of out-degree 2, except the new one, there is a new outgoing arc labeled T_n towards the final state.

2. If s is odd, i.e., we have just applied R_2 rule, then each arc which label corresponds to the factor $S'_n\$ = (M_{n-1}{}^=)^{-1}A_n^=\$$ will be split in an arc labeled S'_n from the same outgoing state towards a new state and two arcs from this new state towards the final one, labeled $\$$ and $T'_n = (A_n^=)^{-1}B_{n+1}^=\$$. Moreover, if $a_s = 1$, then for each state of out-degree 2, except the new one, there is a new outgoing arc labeled T'_n towards the final state.

In such a way we can construct a CDAWG corresponding to each directive sequence $[a_0, \ldots, a_k], k \geq 0$. Indeed, it is a folklore result that the central Sturmian word of directive sequence $[a_0, \ldots, a_s, 1]$ coincides with the one of directive sequence $[a_0, \ldots, a_{s+1}]$.

Let us give some examples. Figure 5 shows CDAWGs corresponding to directive sequences $[3]$ and $[3, 1]$, the first one obtained directly from the base case, and the second one obtained starting from the first one.

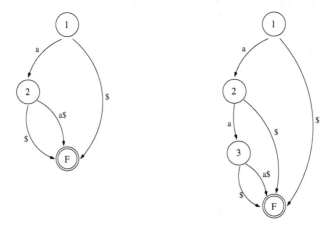

Fig. 5. CDAWGs corresponding to directive sequences $[3]$ and $[3, 1]$

4   Infinite Graphs

Analogously as done with finite and infinite words, we can define a convergence of semi-normalized weighted DAWGs. More precisely, we say that a sequence

G_n, $n = 0 \ldots \infty$, of semi-normalized weighted DAWGs with positive weights, converges to the infinite weighted DAG G if for any constant $c \geq 0$ there exists a number $\hat{n}$ such that for any $n \geq \hat{n}$ the restriction of G_n and G to states having distance from the initial state smaller than K are isomorphic. Moreover the number of outgoing arcs from such states together their weights must be the same.

Notice that the second condition implies that the number of outgoing arcs from any state in G is finite, while it is still possible for one state to have infinitely many ingoing arcs, as it will be shown in what follows.

Definition 3. *For any irrational number $\alpha > 0$ we define the Sturmian graph $G(\alpha)$ as the (unique) limit of the sequence of graphs $G(\frac{P_n}{Q_n})$, $n = 0 \ldots \infty$, where $\frac{P_n}{Q_n}$, $n = 0 \ldots \infty$, is the sequence of convergents to α.*

The reader can check that above definition is a good one, i.e., that $G(\alpha)$ exists and it is unique. It is worth noticing the behavior of the final state F. Its distance from the initial state is 1, and, consequently it belongs to $G(\alpha)$. Moreover, eventually any state will have an arc toward F of weight 1. Therefore in $G(\alpha)$ it has no outgoing arcs but infinitely many ingoing arcs, each weighted by 1. As Sturmian words represents geometrical ray, this state F can be thought as the analogous of the vanishing point in projective geometry. Analogously as in the finite case, we call F the final state of the DAWG.

The extension of Definition 2 is left to the reader, as well the proof of next two propositions.

Theorem 3. *For any positive irrational α, $G(\alpha)$ can count from 1 up to infinity.*

If $\alpha = \frac{\sqrt{5}+1}{2}$, that is the golden ratio, we call $G(\alpha)$ the Golden graph.

Proposition 7. *For every $m \geq 1$ the Golden graph, uses $O(\log_\varphi(m))$ states to count from one to m.*

Since any state reaches the vanishing state with an arc of length 1, we can eliminate the vanishing state and the arc going to it and the new graph $G'(\alpha)$ can count from 0 to infinity, supposing each state "terminal". Indeed $G'(\alpha)$ is the limit graph of the sequence $G'(\frac{P_n}{Q_n})$, $n = 0 \ldots \infty$, where $\frac{P_n}{Q_n}$, $n = 0 \ldots \infty$, is the sequence of convergents to α and $G'(\frac{P_n}{Q_n})$ is defined in Remark 2.

Definition 4. *An infinite graph having a denumerable number of states has the "local" property, or, in short, it is local, with constant k if there exists a way of numbering states such that for any state i all outgoing arcs (i, j) are such that $i - k \leq j \leq i + k$.*

Next proposition makes a link between the structure of a Sturmian graph and the continued fraction expansion of α.

Proposition 8. *$G'(\alpha)$ is local if and only if α has bounded partial quotients in its continued fraction expansion.*

References

1. A. Blumer, J. Blumer, D. Haussler, A. Ehrenfeucht, M. T. Chen, and J. Seiferas. *The Smallest Automaton Recognizing the Subwords of a Text.* Theoretical Computer Science, **40**, 1, 1985, 31–55.

2. A. Blumer, J. Blumer, D. Haussler, R. McConnell, A. Ehrenfeucht. *Complete inverted files for efficient text retrieval and analysis.* Journal of the ACM, **34**, 3, 1987, 578–595.

3. A. Blumer, D. Haussler, A. Ehrenfeucht. *Average sizes of suffix trees and dawgs.* Discrete Applied Mathematics, **24**, 1989, 37–45.

4. I. Borosh, H. Niederreiter. *Optimal multipliers for pseudo-random number generation by the linear congruential method.* BIT **23** (1983), 65–74.

5. C. Brezinski. *History of continued fractions and Padé approximants.* Springer Series in Computational Mathematics, **12**, Springer-Verlag (1991).

6. M. Crochemore, *Reducing space for index implementation.* Theoretical Computer Science, **292**, 1, 2003, 185–197.

7. M. Crochemore, R. Vérin, *Direct Construction of Compact Directed Acyclic Word Graphs.* CPM97, A. Apostolico and J. Hein, eds., LNCS **1264**, Springer-Verlag, 1997, 116–129.

8. A. de Luca, F. Mignosi, *Some combinatorial properties of Sturmian words.* TCS **136**, 1994, 361–385.

9. C. Epifanio, F. Mignosi, J. Shallit, I. Venturini, *Sturmian Graphs and a Conjecture of Moser.* Technical Report **262**, 2004, Dip. Mat. ed Appl., Università di Palermo. Available at URL http:\\math.unipa.it\~mignosi

10. G. H. Hardy and E. M. Wright. *An Introduction to the Theory of Numbers.* Oxford University Press, 5th edition (1989).

11. J. Holub, M. Crochemore. *On the Implementation of Compact DAWG's.* Proceedings of the *7th Conference on Implementation and Application of Automata*, University of Tours, France, 2002, LNCS **2608**, Springer-Verlag, 2003, 289–294.

12. S. Inenaga, H. Hoshino, A. Shinohara, M. Takeda, S. Arikawa, G. Mauri, G. Pavesi. *On-Line Construction of Compact Directed Acyclic Word Graphs.* To appear in Discrete Applied Mathematics (special issue for CPM'01).

13. S. Inenaga, H. Hoshino, A. Shinohara, M. Takeda, S. Arikawa, G. Mauri, G. Pavesi. *On-Line Construction of Compact Directed Acyclic Word Graphs.* Proceedings of *CPM 2001*, LNCS **2089**, Springer-Verlag, 2001, 169–180.

14. R. Klette, A. Rosenfeld *Digital straightness – a review.* Discrete Applied Mathematics **139 (1-3)**, 2004, 197–230.

15. G. Larcher. *On the distribution of sequences connected with good lattice points.* Monatshefte Math. **101** (1986), 135–150.

16. M. Lothaire. *Combinatorics on Words.* Encyclopedia of Mathematics and its Applications, **17**, Addison-Wesley, (1983). Reprinted in the Cambridge Mathematical Library, Cambridge University Press (1997).

17. M. Lothaire. *Algebraic Combinatorics on Words.* Encyclopedia of Mathematics and its Applications, **90**, Cambridge University Press (2002).

18. O. Perron. *Die Lehre von den Kettenbrüchen.* B. G. Teubner, Stuttgart (1954).

19. M. Raffinot. *On maximal repeats in strings.* Inf. Proc. Letters, **83** (2001), 165–169.

20. G. Rauzy. *Mots infinis en arithmétique*. In M. Nivat, D. Perrin (Eds.) *Automata on Infinite Words*, LNCS **192**, Springer-Verlag, 1985, 165–171.
21. J. Shallit. *Real numbers with bounded partial quotients*. Enseignement Math. **38** (1992), 151–187.
22. S. K. Zaremba. *La méthode de "bons treillis" pour le calcul des intégrales multiples*. In S. K. Zaremba, ed., *Applications of Number Theory to Numerical Analysis*, Academic Press, 1972, pp. 39–119.

P Systems Working in the Sequential Mode on Arrays and Strings

Rudolf Freund

Faculty of Informatics, Vienna University of Technology, Wien, Austria
rudi@emcc.at

Abstract. Based on a quite general definition of P systems where the rules are applied in a sequential way (and not in the maximally parallel way as it usually happens in most models of P systems considered so far in the literature), we investigate the generative power of various models of such P systems working in the sequential mode on arrays and strings, respectively. P systems working in the sequential mode on arrays/strings without priority relations for the rules reveal the same computational power as the corresponding matrix grammars without appearance checking working on arrays/strings. For obtaining the computational power of matrix grammars with appearance checking, priority relations for the rules (as one of many other possible additional features) are needed.

1 Introduction

In the area of P systems, applying the rules in a maximally parallel way is one of the most common features of many models introduced so far, i.e., when in 1998 Gheorghe Păun in [6] introduced membrane systems (which soon afterwards were called P systems), the way of applying the evolution rules in a maximally parallel way was one of the intrinsic features of this new model. Yet although biological processes in living organisms happen in parallel, they are not synchronized by a universal clock as assumed in the original model of membrane system, instead many processes involve several objects in parallel, but the processes themselves are carried out in an asynchronous way, which feature formally can be captured by letting these processes happen in a sequential/unsynchronized way.

Many variants of P systems have been investigated so far (see [7] for a comprehensive overview as well as [8] for the actual state of research). For a first overview of several examples of sequential P systems we refer to [4]. Most recently, some new results on P systems working in the sequential mode were elaborated in [5] (*asynchronous P systems*) as well as in [1] (*P systems operating in sequential mode*). We assume the reader to be somehow familiar with the original definitions and explanations given for the models considered in [7].

The rest of the paper is organized as follows: In the next section, we start with introducing a general model for grammars, recall some notions for string grammars and array grammars in the general setting used in this paper, and finally define matrix grammars and graph-controlled grammars. In the third section,

C.S. Calude, E. Calude and M.J. Dinneen (Eds.): DLT 2004, LNCS 3340, pp. 188–199, 2004.
© Springer-Verlag Berlin Heidelberg 2004

we define a general model of *P systems working in the sequential mode* (*sequential P systems* for short) which allows us to consider both array productions as well as string productions; moreover, we illustrate the definition by giving an example with context-free string productions. For matrix grammars with and without appearance checking working on arrays and strings, respectively, a special normal form (we call it activator normal form) is elaborated in the fourth section. In the fifth and sixth section, we show that sequential P systems without/with priorities as introduced in this paper allow for the characterization of matrix grammars without/with appearance checking for different variants of context-free array productions as well as different variants of context-free string productions. A short summary of the results obtained in this paper as well as an outlook to future research conclude the paper.

2   Definitions

The set of integers is denoted by $\mathbf{Z}$, the set of positive integers by $\mathbf{N}$. An *alphabet* V is a finite non-empty set of abstract *symbols*. Given V, the free monoid generated by V under the operation of concatenation is denoted by V^*; the elements of V^* are called strings, and the *empty string* is denoted by λ; $V^* \setminus \{\lambda\}$ is denoted by V^+. For more details on formal language theory we refer to [2] and [10].

2.1   Grammars

As we deal with various types of objects and grammars in the following, we first introduce a general model of a grammar:

A *grammar* G is a construct

$$(O, O_T, P, \Longrightarrow_G, w) \text{ where}$$

- O is the set of *objects*;
- $O_T \subseteq O$ is the set of *terminal* objects;
- P is a finite set of *productions*;
- $\Longrightarrow_G \subseteq O \times O$ is the *derivation relation* of G induced by the productions in P;
- $w \in O$ is the *axiom*.

The derivation relation $\Longrightarrow_G$ is obtained as the union of all $\Longrightarrow_p \subseteq O \times O$, i.e., $\Longrightarrow_G := \bigcup_{p \in P} \Longrightarrow_p$, where each $\Longrightarrow_p$ is a relation that we assume at least to be recursive. The reflexive and transitive closure of $\Longrightarrow_G$ is denoted by $\Longrightarrow_G^*$. The *language generated by* G is the set of all terminal objects (we also assume $v \in O_T$ to be decidable for every $v \in O$) derivable from the axiom, i.e.,

$$L(G) = \{v \in O_T \mid w \Longrightarrow_G^* v\}.$$

Depending on the components of G, especially with respect to different types of productions, we consider different types of grammars. The family of languages generated by grammars of type X is denoted by $\mathcal{L}(X)$.

2.2   String Grammars

Usually, a string grammar is defined as a construct (N, T, P, S) where

- N is the alphabet of *non-terminal symbols*;
- T is the alphabet of *terminal* symbols, $N \cap T = \emptyset$;
- P is a finite set of *productions* of the form $u \to v$ with $u \in V^+$ and $v \in V^*$, where $V := N \cup T$;
- $S \in N$ is the *start symbol*.

In the general notion of the preceding section, a string grammar now is represented as $(V^*, T^*, P, \Longrightarrow_G, S)$ where the derivation relation for $u \to v \in P$ is defined as usual by $xuy \Longrightarrow_{u \to v} xvy$ for all $x, y \in V^*$, thus yielding the well-known derivation relation $\Longrightarrow_G$ for the string grammar G.

As special types of string grammars we consider string grammars with arbitrary productions, context-free productions of the form $A \to v$ with $A \in N$ and $v \in V^*$, and λ-free context-free productions of the form $A \to v$ with $A \in N$ and $v \in V^+$, the corresponding types of grammars denoted by $ENUM$, CF, and $CF_{-\lambda}$, thus yielding the families of languages $\mathcal{L}(ENUM)$, i.e, the family of recursively enumerable languages, as well as $\mathcal{L}(CF)$ and $\mathcal{L}(CF_{-\lambda})$, i.e., the families of context-free and λ-free context-free languages, respectively.

2.3   Arrays and Array Grammars

In this subsection we introduce the basic notions for n-dimensional arrays and array grammars (e.g., see [3], [9], [11]).

Let $d \in \mathbf{N}$. Then a d-*dimensional array* $\mathcal{A}$ over an alphabet V is a function $\mathcal{A} : \mathbf{Z}^d \to V \cup \{\#\}$, where $shape(\mathcal{A}) = \{v \in \mathbf{Z}^d \mid \mathcal{A}(v) \neq \#\}$ is finite and $\# \notin V$ is called the *background* or *blank-symbol*. We usually write $\mathcal{A} = \{(v, \mathcal{A}(v)) \mid v \in shape(\mathcal{A})\}$.

The set of all d-dimensional arrays over V is denoted by V^{*d}. The *empty array* in V^{*d} with empty shape is denoted by Λ_d. Moreover, we define $V^{+d} = V^{*n} \setminus \{\Lambda_d\}$.

Let $v \in \mathbf{Z}^d$, $v = (v_1, \dots, v_d)$. The *translation* $\tau_v : \mathbf{Z}^d \to \mathbf{Z}^d$ is defined by $\tau_v(w) = w + v$ for all $w \in \mathbf{Z}^d$, and for any array $\mathcal{A} \in V^{*d}$ we define $\tau_v(\mathcal{A})$, the corresponding d-dimensional array translated by v, by $(\tau_v(\mathcal{A}))(w) = \mathcal{A}(w - v)$ for all $w \in \mathbf{Z}^d$. The vector $(0, \dots, 0) \in \mathbf{Z}^d$ is denoted by Ω_d.

A d-*dimensional array production* p over V is a triple $(W, \mathcal{A}_1, \mathcal{A}_2)$, where $W \subseteq \mathbf{Z}^d$ is a finite set and $\mathcal{A}_1$ and $\mathcal{A}_2$ are mappings from W to $V \cup \{\#\}$ such that $shape(\mathcal{A}_1) \neq \emptyset$, $shape(\mathcal{A}_1) = \{v \in \mathbf{W} \mid \mathcal{A}(v) \neq \#\}$. We say that the array $\mathcal{B}_2 \in V^{*d}$ is *directly derivable* from the array $\mathcal{B}_1 \in V^{*d}$ by the d-dimensional array production $(W, \mathcal{A}_1, \mathcal{A}_2)$, i.e., $\mathcal{B}_1 \Longrightarrow_p \mathcal{B}_2$, if and only if there exists a vector $v \in \mathbf{Z}^d$ such that $\mathcal{B}_1(w) = \mathcal{B}_2(w)$ for all $w \in \mathbf{Z}^d \setminus \tau_v(W)$ as well as $\mathcal{B}_1(w) = \mathcal{A}_1(\tau_{-v}(w))$ and $\mathcal{B}_2(w) = \mathcal{A}_2(\tau_{-v}(w))$ for all $w \in \tau_v(W)$, i.e., the sub-array of $\mathcal{B}_1$ corresponding to $\mathcal{A}_1$ is replaced by $\mathcal{A}_2$, thus yielding $\mathcal{B}_2$.

A d-*dimensional array grammar* is a grammar

$$\left((N \cup T)^{*d}, T^{*d}, P, \Longrightarrow_G, w \right) \text{ where}$$

- N is the alphabet of *non-terminal symbols*;
- T is the alphabet of *terminal* symbols, $N \cap T = \emptyset$;
- P is a finite set of *array productions* over V, $V := N \cup T$;
- $\Longrightarrow_G$ is the derivation relation induced by the array productions in P according to the explanations given above, i.e., for arbitrary $\mathcal{B}_1, \mathcal{B}_2 \in V^{*d}$, $\mathcal{B}_1 \Longrightarrow_G \mathcal{B}_2$ if and only if there exists a d-dimensional array production $p = (W, \mathcal{A}_1, \mathcal{A}_2)$ in P such that $\mathcal{B}_1 \Longrightarrow_p \mathcal{B}_2$;
- $\{(v_0, S)\}$ with $S \in N$ and $v_0 \in \mathbf{Z}^d$ is the *start array (axiom)*.

A d-dimensional array production $p = (W, \mathcal{A}_1, \mathcal{A}_2)$ in P is called

- *#-context-free*, if $shape\,(\mathcal{A}_1) = \{\Omega_d\}$;
- *context-free*, if it is #-context-free and $shape\,(\mathcal{A}_1) \subseteq shape\,(\mathcal{A}_2)$;
- *strictly context-free*, if it is context-free and $shape\,(\mathcal{A}_2) = W$.

A d-dimensional #-context-free array production $p = (W, \mathcal{A}_1, \mathcal{A}_2)$ in the following will be represented in the form $\mathcal{A}_1\,(\Omega_d) \to \mathcal{A}_2\,(\Omega_d)\,\{(v, \mathcal{A}_2\,(v)) \mid v \in U\}$ with $U = W - \{\Omega_d\}$.

An array grammar is said to be of type $d\text{-}ENUMA$, $d\text{-}\#\text{-}CFA$, $d\text{-}CFA$, $d\text{-}SCFA$, respectively, if every array production in P is of the corresponding type, i.e., a d-dimensional arbitrary, #-context-free, context-free, or strictly context-free array production, respectively. The corresponding families of d-dimensional array languages of type X are denoted by $\mathcal{L}\,(X)$. $\mathcal{L}\,(d\text{-}ENUMA)$ is the family of recursively enumerable d-dimensional array languages.

2.4   Matrix Grammars

A *matrix grammar with appearance checking (with* ac *for short)* G_M of type X is a construct

$$(O, O_T, P, \Longrightarrow_G, w, M, F)$$

where $G = (O, O_T, P, \Longrightarrow_G, w)$ is a grammar of type X, M is a finite set of finite sequences of productions (an element of M is called a *matrix*), and $F \subseteq P$. For a matrix $m_i = [m_{i,1}, \ldots, m_{i,n_i}]$ in M and $v, u \in O$ we define $v \Longrightarrow_{m_i} u$ if and only if there are $w_0, w_1, \ldots, w_{n_i} \in O$ such that $w_0 = v$, $w_{n_i} = u$, and for each $j, 1 \leq j \leq n_i$,

- **either** $w_{j-1} \Longrightarrow_{m_{i,j}} w_j$ according to $\Longrightarrow_G$,
- **or** $m_{i,j}$ is not applicable to w_{j-1} according to $\Longrightarrow_G$, $w_j = w_{j-1}$, $m_{i,j} \in F$.

The language generated by G_M is

$$L\,(G_M) = \{v \in O_T \mid w \Longrightarrow_{m_{i_1}} w_1 \ldots \Longrightarrow_{m_{i_k}} w_k,\ w_k = v,$$
$$w_j \in O,\ m_{i_j} \in M\ \text{ for } 1 \leq j \leq k, k \geq 1\}.$$

The matrix grammar G_M is said to be of type MAT_{ac}; it is said to be of type MAT - to be *without appearance checking (without* ac*)* - if $F = \emptyset$. The corresponding families of languages are denoted by $\mathcal{L}\,(X\text{-}MAT_{\mathsf{ac}})$ and $\mathcal{L}\,(X\text{-}MAT)$, respectively.

2.5   Graph-Controlled Grammars

A *graph-controlled grammar* G_C of type X is a construct

$$(O, O_T, P, \Longrightarrow_G, w, R, L_{in}, L_{fin})$$

where $G = (O, O_T, P, \Longrightarrow_G, w)$ is a grammar of type X, R is a finite set of rules r of the form $(l(r) : p(l(r)), \sigma(l(r)), \varphi(l(r)))$, where $l(r) \in Lab(G_C)$, $Lab(G_C)$ being a set of labels associated (in a one-to-one manner) with the rules r in R, $p(l(r)) \in P$, $\sigma(l(r)) \subseteq Lab(G_C)$ is the *success field* of the rule r, and $\varphi(l(r)) \subseteq Lab(G_C)$ is the *failure field* of the rule r; $L_{in} \subseteq Lab(G_C)$ is the set of initial labels, and $L_{fin} \subseteq Lab(G_C)$ is the set of final labels. For $r = (l(r) : p(l(r)), \sigma(l(r)), \varphi(l(r)))$ and $v, u \in O$ we define $(v, l(r)) \Longrightarrow_{G_C} (u, k)$ if and only if

- **either** $p(l(r))$ is applicable to v, $v \Longrightarrow_G u$, and $k \in \sigma(l(r))$,
- **or** $p(l(r))$ is not applicable to v, $u = v$, and $k \in \varphi(l(r))$.

The language generated by G_C is

$$L(G_C) = \{v \in O_T \mid (w_0, l_0) \Longrightarrow_{G_C} (w_1, l_1) \ldots \Longrightarrow_{G_C} (w_k, l_k), \ k \geq 1,$$
$$w_j \in O \ \text{and} \ l_j \in Lab(G_C) \ \text{for} \ 0 \leq j \leq k,$$
$$w_0 = w, \ w_k = v, \ l_0 \in L_{in}, \ l_k \in L_{fin}\}.$$

The graph-controlled grammar G_C is said to be of type GC_{ac}; it is said to be of type GC - to be *without appearance checking (without* ac*)* - if $\varphi(l) = \emptyset$ for all $l \in Lab(G_C)$. The corresponding families of languages are denoted by $\mathcal{L}(X\text{-}GC_{ac})$ and $\mathcal{L}(X\text{-}GC)$, respectively.

Theorem 2.5.1. For any arbitrary type X,

$$\mathcal{L}(X\text{-}MAT) \subseteq \mathcal{L}(X\text{-}GC) \ \text{and} \ \mathcal{L}(X\text{-}MAT_{ac}) \subseteq \mathcal{L}(X\text{-}GC_{ac}).$$

3   A General Model of Sequential P Systems

The reader is assumed to be familiar with the main features and variants of the basic models of P systems, we especially refer to [7] and the original papers cited there. The most important feature is the membrane structure, which usually is represented by pairs of matching brackets and forms a tree structure.

In this section we define the general model of P systems working in the sequential mode that we will consider with array and string productions in this paper; moreover, we illustrate the definition by giving an example of a sequential P system with priorities generating the string language $\{a^{2^n} \mid n \geq 1\}$ using context-free productions in three membranes.

A *sequential P system with priorities of type X* (an *X-sP(pri)* system for short) is a construct

$$\Pi = (O, O_T, P, \Longrightarrow, w, \mu, R_\mu, \rho_\mu) \text{ where}$$

- $G = (O, O_T, P, \Longrightarrow, w)$ is a *grammar* of type X,
- O is the set of *objects*;
- $O_T \subseteq O$ is the set of *terminal* objects;
- P is a finite set of *productions*;
- $\Longrightarrow$ is the *derivation relation* of the grammar G;
- $w \in O$ is the *axiom* of the underlying grammar G and is put as initial object into the skin membrane;
- μ is the *membrane structure* of Π; usually, we shall label the membranes with $1, ..., n$ in a bijective way; the outermost membrane is always labelled by 1 and is called the *skin membrane*;
- R_μ ($= (R_1, ..., R_n)$) is a function assigning a finite set of *rules* to each region i of the membrane structure μ, where each rule is of the form $p\,(tar)$ for some production p from P and $tar \in \{here, in, out\}$;
- ρ_μ ($= (\rho_1, ..., \rho_n)$) is a function assigning a *priority relation* for the rules in R_i to each region i of the membrane structure μ.

A derivation in Π works as follows: we start with the axiom w in the skin membrane, whereas all other membranes are empty. At any stage of the derivation, there will be exactly one object v in some region i; to this object, a rule $p(tar)$ from R_i can only be applied if no other rule of higher priority according to the priority relaion ρ_i could be applied, too, where applying $p\,(tar)$ to v means applying p to v according to the derivation relation $\Longrightarrow$ and moving the result according to the target tar, where $tar = here$ keeps the result in the current membrane i (and we shall usually omit this target $here$), $tar = in$ moves the result into a membrane j inside membrane i, and $tar = out$ moves the result into the membrane j outside membrane i. In this paper, we shall not allow an object to move outside the skin membrane, hence for all rules in R_1 the target out is not allowed.

All terminal objects from O_T ever appearing at any step in any membrane contribute to the language $L(\Pi)$ generated by Π. The family of languages generated by X-sP(pri) systems (with membrane structure μ) is denoted by $\mathcal{L}(X\text{-}sP(pri))$ ($\mathcal{L}(X\text{-}sP(\mu, pri))$), respectively). If all priority relations in an X-sP(pri) system are empty, we call it an X-sP system and denote the corresponding families of languages generated by X-sP systems (with membrane structure μ) by $\mathcal{L}(X\text{-}sP)$ ($\mathcal{L}(X\text{-}sP(\mu))$), respectively).

Example 3.1 Consider the sequential P system working on strings

$$\Pi = (\{a, B, C\}^*, \{a\}^*, P, \Longrightarrow, B, [_1[_2[_3\ ]_3]_2]_1, R_1, R_2, R_3, \rho_1, \rho_2, \rho_3)$$

where $(\{a, B, C\}, \{a\}, P, \Longrightarrow, B)$ is a context-free string grammar with

$P = \{B \to CC, C \to B, C \to a\}$ and
$R_1 = \{C \to B\,(here), B \to CC\,(in)\}$,
$R_2 = \{B \to CC\,(here), C \to B\,(out), C \to a\,(in)\}$,
$R_3 = \{C \to a\,(here)\}$,
$\rho_1 = \{C \to B\,(here) > B \to CC\,(in)\}$,
$\rho_2 = \{B \to CC\,(here) > C \to B\,(out), B \to CC\,(here) > C \to a\,(in)\}$,
$\rho_3 = \emptyset$.

By repeatedly applying the rule $B \to CC$ (*here*) in membrane 2 the number of symbols is doubled, the first doubling being already made when moving the string from membrane 1 to membrane 2 by applying the rule $B \to CC$ (*in*), which according to the priority relation ρ_1 is only possile when the current string consists of symbols B only. After doubling all symbols B to CC, the string consisting of symbols C only now allows the application of the rule $C \to B$ (*out*) or of the rule $C \to a$ (*in*). $C \to B$ (*out*) sends the string out into the skin membrane, where every symbol C is replaced by a symbol B by repeatedly applying the rule $C \to B$ (*here*). On the other hand, $C \to a$ (*in*) sends the string into membrane 3, where after applying $C \to a$ (*here*) as long as possible we finally obtain a string consisting of terminal symbols a only. According to these explanations it becomes obvious that $L(\Pi) = \left\{ a^{2^n} \mid n \geq 0 \right\}$, i.e.,

$$\left\{ a^{2^n} \mid n \geq 0 \right\} \in \mathcal{L}\left(CF_{-\lambda}\text{-}sP\left([_1[_2[_3 \ ]_3]_2]_1, pri \right) \right).$$

4   Matrix Grammars in Activator Normal Form

For matrix grammars of array types $n\text{-}\#\text{-}CFA$, $n\text{-}CFA$, and $n\text{-}SCFA$ as well as of string types CF and $CF_{-\lambda}$ we prove a special normal form, where after each application of a matrix exactly one position in the array/string is marked. Due to lack of space, we cannot give detailed proofs of the theorems stated in this section.

4.1   Working on Arrays

A matrix grammar $\left((N'' \cup T)^{*d}, T^{*d}, P, \Longrightarrow_G, w, M, F \right)$ of array types $d\text{-}\#\text{-}CFA$, $d\text{-}CFA$, and $d\text{-}SCFA$, respectively, is said to be in *activator normal form* (*anf* for short) if $N'' = N' \times L \cup N$ for two disjoint sets N' and L with $N' = N \cup T (\cup \{\#\})$, $w \in N \times L$, and every matrix in M is of one of the following forms:

1. $[(X, r) \to (Y, s) \{(v, Y_v) \mid v \in U\}]$, $X \in N$, $Y \in N'$, $Y_v \in N \cup T \cup \{\#\}$ for $v \in U$, $r, s \in L$;
2. $[(X \to (X, r), (Y, r) \to Y)]$, $X \in N$, $Y \in N'$, $r \in L$;
3. $[(a, r) \to a]$, $a \in T (\cup \{\#\})$, $r \in L$;
4. $[X \to H \{(v, Y_v) \mid v \in U\}, (Y, r) \to (Y, s)]$, $Y \in N'$, $Y_v \in N \cup T \cup \{\#\}$ for $v \in U$, $r, s \in L$, $X, H \in N$, where H is a trap symbol, i.e., H cannot evolve any more.

The rules of type 4 only appear in the case of a matrix grammar with ac, i.e., if $F \neq \emptyset$. Moreover, the notation $(\cup \{\#\})$ indicates that we have to take $\#$ into account only for the case of type $d\text{-}\#\text{-}CFA$.

Theorem 4.1.1. For any matrix grammar without/with ac of type X, we can effectively construct an equivalent matrix grammar without/with ac in anf of the same type, for any $X \in \{d\text{-}\#\text{-}CFA, d\text{-}CFA, d\text{-}SCFA\}$.

Proof. Given any matrix grammar G' without/with ac of type X, the main idea is to construct the corresponding graph-controlled grammar G_M according to Theorem 2.5.1 and then to simulate G_M by a matrix grammar G without/with ac in anf of the same type X. The set L in G then corresponds to the set of labels of the control graph of G_M, i.e., we keep track of the current label r in G_M by the second component in the activated symbol (X, r). Matrices of type 1 simulate the corresponding successful applications of productions in G_M, whereas matrices of type 4 simulate the failure case. Matrices of type 3 are to be used in the last step of a derivation in G, and matrices of type 2 allow the activator, i.e., the current label, to move around within the underlying array. Following these explanations, G can be constructed from G' via G_M.   □

4.2   Working on Strings

A matrix grammar $((N'' \cup T)^*, T^*, P, \Longrightarrow_G, w, M, F)$ of string types CF and $CF_{-\lambda}$, respectively, is said to be in *activator normal form (*anf for short) if $N'' = N' \times L \cup N$ for two disjoint sets N' and L with $N' = N \cup T (\cup \{\lambda\})$, $w \in N \times L$, and every matrix in M is of one of the following forms:

1. $[(X, r) \rightarrow (Y, s) Y_1...Y_k]$, $X \in N$, $Y \in N'$, $Y_1, ..., Y_k \in N \cup T$ for $1 \leq i \leq k$, $k \geq 0$, $r, s \in L$;
2. $[(X \rightarrow (X, r), (Y, r) \rightarrow Y)]$, $X \in N$, $Y \in N'$, $r \in L$;
3. $[(a, r) \rightarrow a]$, $a \in T (\cup \{\lambda\})$, $r \in L$;
4. $[X \rightarrow HY_1...Y_k, (Y, r) \rightarrow (Y, s)]$, $Y \in N'$, $Y_1, ..., Y_k \in N \cup T$ for $1 \leq i \leq k$, $k \geq 0$, $r, s \in L$, $X, H \in N$, where H is a trap symbol, i.e., H cannot evolve any more.

The rules of type 4 only appear in the case of a matrix grammar with ac, i.e., if $F \neq \emptyset$. Moreover, the notation $(\cup \{\lambda\})$ indicates that we have to take the empty string λ into account only for the case of type CFA.

Theorem 4.2.1. For any matrix grammar without/with ac of type X, we can effectively construct an equivalent matrix grammar without/with ac in anf of the same type, for any $X \in \{CF, CF_{-\lambda}\}$.

Proof. As in the proof of Theorem 4.1.1, the main idea is to simulate the corresponding graph-controlled grammar by the matrix grammar without/with ac in anf. Instead of considering #-context-free array productions of the form $X \rightarrow Y \{(v, Y_v) \mid v \in U\}$ we now have to deal with context-free string productions of the form $X \rightarrow YY_1...Y_k$, yet this makes no difference for the main parts of the construction.   □

5   The Generative Power of Sequential P Systems Without Priorities

Using the activator normal form for matrix grammars elaborated in the preceding section we now are able to show that sequential P systems without priorities generate the same families of languages as matrix grammars without ac in the case of array grammars of types d-#-CFA, d-CFA, and d-$SCFA$ as well as in the case of string grammars of types CF and $CF_{-\lambda}$.

5.1   Working on Arrays

Theorem 5.1.1. For any $X \in \{d\text{-}\#\text{-}CFA, d\text{-}CFA, d\text{-}SCFA\}$,

$$\mathcal{L}(X\text{-}MAT) = \mathcal{L}(X\text{-}sP([_1[_2[_3]_3]_2]_1)).$$

Proof. Let the array language $L \in \mathcal{L}(X\text{-}MAT)$ be given by a matrix grammar in anf $\left((N'' \cup T)^{*d}, T^{*d}, P, \Longrightarrow_G, w, M, F\right)$ according to Theorem 4.1.1. Then we construct the corresponding $X\text{-}sP([_1[_2[_3]_3]_2]_1)$ system Π as follows:

$$\Pi = \left(O', T^{*d}, P', \Longrightarrow, w, [_1[_2[_3]_3]_2]_1, R_1, R_2, R_3\right);$$
$$O' = \left(N' \times L \cup N' \times L \times L \cup \bar{N}' \times L \times L \cup N \cup T\right)^{*d};$$

P' contains all productions from P as well as all the new productions introduced in the sets of rules R_1, R_2, and R_3, which allow for directly simulating the matrices $[(X, r) \to (Y, s) \{(v, Y_v) \mid v \in U\}]$ and $[(a, r) \to a]$ in the first membrane, whereas for moving the activator to another position, i.e., for simulating a matrix $[(X \to (X, r), (Y, r) \to Y)]$, we also need the other two membranes to synchronize the change:

$$
\begin{aligned}
R_1 = & \{(X, r) \to (Y, s) \{(v, Y_v) \mid v \in U\} \, (\textit{here}) \mid \\
& \quad [(X, r) \to (Y, s) \{(v, Y_v) \mid v \in U\}] \in M\} \\
& \cup \{(a, r) \to a \, (\textit{here}) \mid [(a, r) \to a] \in M\} \\
& \cup \{X \to (X, r, r) \, (\textit{in}) \mid [(X \to (X, r), (Y, r) \to Y)] \in M\}, \\
R_2 = & \{(Y, r) \to (\bar{Y}, r, r) \, (\textit{in}), (X, r, 0) \to (X, r) \, (\textit{out}), \\
& \quad (\bar{Y}, r, i) \to (\bar{Y}, r, i - 1) \, (\textit{in}) \mid \\
& \quad [(X \to (X, r), (Y, r) \to Y)] \in M, 1 \le i \le r\}, \\
R_3 = & (X, r, i) \to (X, r, i - 1) \, (\textit{out}), (\bar{Y}, r, 0) \to Y \, (\textit{out}) \mid \\
& \quad [(X \to (X, r), (Y, r) \to Y)] \in M, 1 \le i \le r\}.
\end{aligned}
$$

For the inverse inclusion $\supseteq$, we just mention the main idea of the proof: according to Theorem 4.1.1, we may simulate the given $X\text{-}sP$ system Π by a graph-controlled grammar of the corresponding type, where the paths in the control graph reflect the movements of the underlying array according to the targets in the rules of Π. $\qquad\square$

In the case of one- and two-dimensional arrays the theorem proved above even allows for a characterization of recursively enumerable array languages by sequential P systems without using priorities:

Theorem 5.1.2. For any $d \in \{1, 2\}$,

$$\mathcal{L}(d\text{-}\#\text{-}ENUMA) = \mathcal{L}(d\text{-}\#\text{-}CFA\text{-}sP([_1[_2[_3]_3]_2]_1)).$$

Proof. The equality $\mathcal{L}(d\text{-}\#\text{-}ENUMA) = \mathcal{L}(d\text{-}\#\text{-}CFA\text{-}MAT)$ is well-known, e.g., see [3]. The claim now immediately follows from Theorem 5.1.1. $\qquad\square$

5.2 Working on Strings

Theorem 5.2.1. For any $X \in \{CF, CF_{-\lambda}\}$,

$$\mathcal{L}(X\text{-}MAT) = \mathcal{L}(X\text{-}sP([_1[_2[_3 \ ]_3]_2]_1)).$$

Proof. We can use a similar construction as the one given in the proof of Theorem 5.1.1, i.e., the proof follows from the proof of Theorem 5.1.1 in the same way as the proof of Theorem 4.1.2 followed from the proof of Theorem 4.1.1. □

6 The Generative Power of Sequential P Systems with Priorities

Sequential P systems with priorities generate the same families of languages as matrix grammars with ac in the case of array grammars of types d-#-CFA, d-CFA, and d-$SCFA$ as well as in the case of string grammars of types CF and $CF_{-\lambda}$.

6.1 Working on Arrays

We first show that when using priorities on the rules only two membranes are needed for simulating matrix grammars without ac by the corresponding sequential P systems:

Theorem 6.1.1. For any $X \in \{d\text{-}\#\text{-}CFA, d\text{-}CFA, d\text{-}SCFA\}$,

$$\mathcal{L}(X\text{-}MAT) \subseteq \mathcal{L}(X\text{-}sP([_1[_2 \ ]_2]_1, pri)).$$

Proof. Consider the X-sP system without priorities constructed in the proof of Theorem 5.1.1. We can construct an equivalent X-sP system with priorities and only two membranes: first we move all rules from membrane 3 into membrane 1 thereby changing each target indication from *out* to *in*; moreover, for every rule $X \to (X, r, r)\,(in)$ we add $\alpha \to H\,(in)$ in R_1 as a rule with higher priority (with repect to ρ_1) for every non-terminal symbol α of the form (Y, s, j) and $(\bar{Y}, s, j)$, where H is a new non-terminal symbol (a trap symbol, which cannot evolve any more). □

When allowing priorities, in sum we even get a characterization of matrix grammars with ac by the corresponding sequential P systems:

Theorem 6.1.2. For any $X \in \{d\text{-}\#\text{-}CFA, d\text{-}CFA, d\text{-}SCFA\}$,

$$\mathcal{L}(X\text{-}sP([_1[_2 \ ]_2]_1, pri)) = \mathcal{L}(X\text{-}MAT_{ac}).$$

Proof. For proving the inclusion $\supseteq$, consider the X-sP system with priorities constructed in the proof of Theorem 6.1.1 and, for every matrix of the form

$$[X \rightarrow H\{(v, \#) \mid v \in U\}, (Y, r) \rightarrow (Y, s)]$$

(where H is the trap symbol), add these two rules in R_1 as well as the priority

$$X \rightarrow H\{(v, \#) \mid v \in U\} > (Y, r) \rightarrow (Y, s)$$

to ρ_1. The details of this construction as well as the proof of the inverse inclusion are rather obvious and therefore omitted.   □

Hence, even for arbitrary $d \in \mathbf{N}$ we now obtain the following characterization of $\mathcal{L}(d\text{-}ENUMA)$:

Corollary 6.1.3. For any $d \in \mathbf{N}$,

$$\mathcal{L}(d\text{-}ENUMA) = \mathcal{L}(d\text{-}\#\text{-}CFA\text{-}sP([_1[_2\ ]_2]_1, pri)).$$

6.2   Working on Strings

The following results are immediate consequences of the constructions given for the results obtained in subsection 6.1 and the corresponding results from subsection 5.2.

Theorem 6.2.1. $\mathcal{L}(CF_{-\lambda}\text{-}sP([_1[_2\ ]_2]_1, pri)) = \mathcal{L}(CF_{-\lambda}\text{-}MAT_{ac})$

As $\mathcal{L}(CF\text{-}MAT_{ac}^{\lambda}) = \mathcal{L}(ENUM)$ (see [2]), we even obtain a characterization of recursively enumerable string languages by sequential P systems with priorities using context-free string productions in only two membranes:

Theorem 6.2.2. $\mathcal{L}(CF\text{-}sP([_1[_2\ ]_2]_1, pri)) = \mathcal{L}(CF\text{-}MAT_{ac}^{\lambda}) = \mathcal{L}(ENUM)$

Observe that this result is already optimal with respect to the number of membranes in $CF\text{-}sP(pri)$ systems being able to generate any arbitrary string language, as $\mathcal{L}(CF\text{-}sP([_1\ ]_1, pri))$ only corresponds with the family of string languages generated by ordered grammars, which family is strictly included in $\mathcal{L}(ENUM)$, e.g., see [2].

7   Summary and Future Research

We have investigated a specific variant of P systems where the rules are not applied in the maximally parallel way, but instead in a sequential/asynchronous way. For related results of sequential/asynchronous P systems working on string objects and on symbol objects, we refer to [5] and [1].

The model of sequential P systems with/without priorities introduced in this paper allowed for the characterization of matrix grammars with/without appearance checking for different variants of context-free array productions as well as different variants of context-free string productions.

Many other models of P systems, e.g., one may simply consider some variants of the big variety of models of membrane systems described in the book of Gheorghe Păun, [7], deserve to be investigated for the case of the sequential application of rules, too, as the sequential/asynchronous way for the application of rules promises interesting new results and applications for the future.

Acknowledgements

The author acknowledges inspiring discussions with Gheorghe Păun and Oscar Ibarra on P systems working in the sequential mode.

References

1. Z. Dang, O. H. Ibarra: On P systems operating in sequential mode. In: L. Ilie, D. Wotschke (eds.): *Pre-proceedings of the Workshop Descriptional Complexity of Formal Systems (DCFS) 2004*. Report No. **619**, Univ. of Western Ontario, London, Canada (2004) 164–177
2. J. Dassow, Gh. Păun: *Regulated Rewriting in Formal Language Theory*. Springer-Verlag, Berlin (1989)
3. R. Freund: Control mechanisms on #-context-free array grammars. In: Gh. Păun (ed.): *Mathematical Aspects of Natural and Formal Languages*. World Scientific Publ., Singapore (1994) 97–137
4. R. Freund: Sequential P-systems. *Romanian Journal of Information Science and Technology*, Vol. **4**, Numbers 1-2 (2001) 77–88
5. R. Freund: Asynchronous P systems. *Proceedings of WMC 5*, Milano, Italy, June 2004 (2004)
6. Gh. Păun: Computing with membranes. *Journal of Computer and System Sciences* **61**, 1 (2000) 108–143, and
   TUCS Research Report **208** (1998) (`http://www.tucs.fi`)
7. Gh. Păun: *Membrane Computing: an Introduction*. Springer-Verlag, Berlin (2002)
8. The P Systems Web Page: `http://psystems.disco.unimib.it/`
9. A. Rosenfeld: *Picture Languages*. Academic Press, Reading, MA (1979)
10. A. Salomaa, G. Rozenberg (eds.): *Handbook of Formal Languages*. Springer-Verlag, Berlin (1997)
11. P. S.-P. Wang (ed.): *Array Grammars, Patterns and Recognizers*. World Scientific Series in Computer Science **18**. World Scientific Publ., Singapore (1989)

Optimal Time and Communication Solutions of Firing Squad Synchronization Problems on Square Arrays, Toruses and Rings[*]

Jozef Gruska[1], Salvatore La Torre[2], and Mimmo Parente[2]

[1] Faculty of Informatics, Masaryk University, Brno, Slovakia
gruska@informatics.muni.cz
[2] Dipartimento di Informatica ed Applicazioni,
Università degli Studi di Salerno, Italy
{slatorre, parente}@unisa.it

Abstract. A new solution for the Firing Squad Synchronization Problem (FSSP) on two-dimensional square arrays is presented and its correctness is demonstrated in detail. Our new solution is time as well as communication optimal (the so-called minimal time 1-bit solution). In addition, it is shown that the technique developed and the results obtained allow also to solve in optimal time & communication FSSP for several other variants of this problem on networks shaped as square grids (with four Generals), square toruses and rings.

1 Introduction

Cellular automata, as networks of identical finite automata (cells), are perhaps the most fascinating model of computation. In any case, no other model of computation has received so far so concentrated attention and no other model of computation has been used to deal with such an enormous variety of problems from all areas of science, technology and applications. Recent books by Wolfram [20, 21] and Ilachinski [4] demonstrate in an impressive way the richness of this model and its broad use in general. Surprising capability of cellular automata, even with very small cells, to synchronize themselves very fast is one of the most peculiar and at the same time most important properties of this model – introduced by von Neumann about 50 years ago.

Firing Squad Synchronization Problem, originally formulated by Myhill (1957), is one of the most famous synchronization problems for cellular automata. In the original formulation, we are given a line of n identical cells (finite state machines) that work synchronously at discrete time steps, initially a distinguished cell (called the *General*) starts computing while all others are in a

---

[*] This research has been completed while the first author was visiting the Dipartimento di Informatica ed Applicazioni, Università degli Studi di Salerno. Work partially supported by MIUR grant ex-60% 2003 Università di Salerno. The first author is also supported by the grant GAČR, 201/04/1153.

C.S. Calude, E. Calude and M.J. Dinneen (Eds.): DLT 2004, LNCS 3340, pp. 200–211, 2004.

quiescent state; at each time step any cell sends/receives to/from its neighbours some information about their state at the preceding time: the problem is to let all cells in the line enter the same state for the first time and at the very same instant. This problem has been investigated for various underlying communication graphs: for one-dimensional arrays, called Lines here, and rings (Culik [2]), for one-dimensional paths in two-dimensional arrays (Kobayashi [8]), for two-dimensional and more dimensional arrays, especially squares, and toruses (Szwerinski [17], Kobayashi [7]), for arbitrary graphs with all nodes of the same degree and for classes of graphs (see, for example, Roka [15] for Cayley graphs), for various number and positions of Generals, for various restrictions on the types of underlying cellular automata, for example reversible, in a special way, (see Imai and Morita [5]), or with a number-conserving property (Imai et al. [6]), for the case of minimal capacities of communication links (Mazoyer [12], Umeo et al. [18]) and also for the case that some of the underlying finite automata are faulty to some extent (Umeo et al. [14]) and so on. For several of these variants ingenious and beautiful solutions have been found.

Of a special interest is always to find solutions that are minimal with respect to one or, better, simultaneously with respect to several important characteristics (resources) – especially with respect to time, size (number of states or number of transitions) or the capacity of the communication links between neighbouring cells.

Concerning the minimal time, Minsky [13] has shown that $2n - 1$ is a lower bound and Goto [3], Waksman [19] and Balzer [1] gave the minimal time solution for the FSSP on one-dimensional arrays of n automata. The argument in this case for the lower bound on synchronization time is very simple: time $2n - 1$ is needed for a signal from the General at one corner of the array (the cell activated by an external intervention) to reach the last soldier and to come back to the General.[1] Using this argument it would seem that, for the case of two dimensional $n \times n$ arrays with the General in a corner, the minimal synchronization time is $4n - 3$ because this is the time needed for a signal from the General to reach the most distant soldier and come back. Surprisingly, this intuition turned out to be wrong and, as shown by Shinar [16], the minimum synchronization time in this case is just $2n - 1$. The basic idea how to achieve synchronization in such a short time is very simple and beautiful: at time $2i - 1$ the signal from the General can reach the i-th diagonal cell and initiates separate synchronization of one-dimensional sub-row and sub-column arrays of $n - i + 1$ cells in the topmost row and leftmost column of the sub-square with diagonal cell (i,i) in the left-top corner as a new General (see Figure 1). The synchronization on such lines can be done in time $2(n - i) + 1$ using the optimal time solution for the one-dimensional

---

[1] It is important to make clear that when discussing synchronization we express time by the overall number of configurations used. In other words, the initial configuration corresponds to time 1. This is, of course, not the only way to count time that is used in the literature. Sometimes the number of steps is counted which is one less than the number of configurations. The reader should realize the existence of these unfortunate differences.

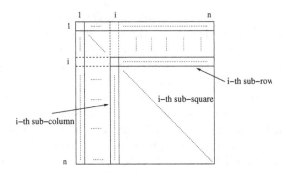

Fig. 1. The splitting of the $n \times n$ array processors in Shinar's solution

basic variant of the FSSP. This is the first key idea used also in our solution of FSSP on Squares.

Concerning the minimal communication (1-bit) solution, a clever and elaborated simultaneously minimal time and minimal communication solution, for one-dimensional array, is due to Mazoyer [12]. Details of Mazoyer's reasoning about the correctness of his solution are not always easy to follow and handle formally, but Mazoyer provides also less formal, but quite clear illustrations and demonstrations, also through space-time diagrams, of his solution. Main properties of his solution are clearly summarized in [11, 12]. In designing our solution, we have exploited some of these properties that therefore become crucial in the correctness proof of our solution.

Minimal (that is 1-bit inter-cell) communication cellular automata are of a special interest because they represent one of the less powerful and simplest models of cellular automata, especially for synchronization tasks. Such cellular automata are then handy primitives for more complex synchronization tasks.

It is quite natural to try to combine the ideas of Shinar's and Mazoyer's solutions to get simultaneously a minimal time and minimal communication links solution for the FSSP on two-dimensional square arrays. However, it does not seem to be a simple task to embed Mazoyer's solution in the Shinar's algorithm maintaining both minimal time and minimal communication in the solution. In fact, in addition to the problems coming from both separate tasks, a new non-trivial problem is that of achieving at the right time and without time delays an isolation of the sub-rows and sub-columns that need to be synchronized separately, using only one-bit communication links between neighbouring cells. (Such an isolation is not hard, in principle, if no restriction on the capacity of inter-cell communications is given). It is even far from obvious that this can be done, indeed.

In the literature, solutions to the FSSP are either presented in full detail, by giving transition tables, or solutions are just sketched, and very rarely proofs of correctness are carried over. In this paper, we try to find a compromise between too informal and too formal approaches to the description of solutions and of their correctness. This has turned out, we think, not only possible, to the large extent due to a very detailed presentation and illustration of Mazoyer's solution in his paper, but also stimulating and rewarding. Indeed, we could use the insights obtained to solve a variety of other FSSPs for toruses and rings as they are dealt with in Section 5.

The idea to adopt Shinar's idea also to the minimal communication FSSP on Squares has appeared first in La Torre et al. [9], but not sufficiently worked out. An important step has been done by Umeo at al. [18]. They reported a (different) way to construct an almost optimal solution (with time $2n$) for the FSSP on Squares with one-bit inter-cell communications, for both standard left-corner position of the General and also for an arbitrary position of the General. However, instead of providing a detailed description and a proof of correctness of their solution, they reported successful testing of their solution for $2 \leq n \leq 1000$. Our solution therefore not only closes the gap between known upper and lower bounds, but we present the basic idea in a way that its application to other underlying interconnection structures is then quite straightforward, as demonstrated later.

2   Models and Preliminaries

We deal with models of cellular automata consisting of identical finite automata interconnected into a line, a ring, a two-dimensional array and a torus. All particular finite automata of a cellular automata are usually considered to be identical (and those that are on a border of the interconnection structure are assumed to get on the sides with no neighbour from the environment always a special signal(s) that allow(s) them to recognize their special position). Another way to deal with such an inhomogeneity, that will be followed in this paper, is to consider several types of the underlying finite automata distinguished by their position in the underlying interconnection structure - that is by the number and position of the inter-cell interconnections.

Any cellular automaton works in discrete steps and at each unit of time a state of any of its finite automaton is determined by its former state and signals received from its neighbours and from the environment (if any). A *configuration* of a cellular automaton is an assignment of states to its particular finite automata.

In the case of the FSSP, we assume that the set of states of the finite automata contains always special states G (General), λ (the sleeping state) and F (the firing state). We assume the sleeping state to have two basic properties.

1. If an automaton is in the sleeping state and receives only inputs 0, then it stays in the sleeping state and outputs 0 on all its output links;
2. no automaton ever goes from a non-sleeping state into the sleeping state.

If an automaton goes from the sleeping state to a non-sleeping state it is considered to be *awakened*. From the second property above it follows that if an automaton gets awakened it never starts to sleep again.

To consider FSSP, it is assumed that in the initial configuration some pre-determined cells are in the state G, all the others are in the sleeping state λ and the goal of the synchronization is that at some time step all cells come simultaneously and for the first time, to the firing state F.

In the following we deal with FSSP for cellular automata with the following underlying interconnection networks (graphs).

- One-dimensional networks:
  - *Line*: a one-dimensional array as underlying graph with n cells (numbered $1, \ldots, n$), and a General at the left end (at the cell 1).
  - *Two-End-Line*: the underlying graph is again a Line but with a General at each end (i.e., at cells 1 and n).
  - *Ring*: with a one-dimensional circular array of n nodes as the underlying graph (where nodes 1 and n are directly connected) and with a General at the position 1.
- Two-dimensional networks:
  - *Square*: a two-dimensional array of $n \times n$ nodes with a General in the upper-leftmost corner (i.e., in the position $(1,1)$).
  - *Four-End-Square*: as the Square but with Generals at the four corners (i.e., positions $(1,1)$, $(1,n)$, $(n,1)$ and (n,n)).
  - *Torus*: as a Square but with cells connected to form a torus (for example, cell $(1,1)$ is directly connected also to the cell $(1,n)$ and $(n,1)$ besides $(1,2)$ and $(2,1)$ as in the Square).

The key result from [12] that will be used in the following is now formally stated.

Theorem 1. *There is a 1-bit solution for the FSSP on any Line of n cells at time $2n - 1$.*

3   Lower Bounds on Synchronization Time for FSSP

Synchronization of a Line requires at least time $2n - 1$. Intuitively, this is the minimal time for the first cell to wake up all other cells and to get back the message that all cells have been awakened. Recall that in any starting configuration each cell, except the first one, is in the sleeping state and that the i-th cell can not leave the sleeping state before time i. Thus all the cells are awakened at time n, and the first cell gets this information back at time $2n - 1$.

Regarding two-dimensional cellular automata, Shinar [16] has shown that the minimum time for synchronizing a rectangular array of $m \times n$ cells is $n + m + max\{n, m\} - 2$, but this time reduces to $2n - 1$ in the case of a Square. The following lemma summarizes this result.

Lemma 1. *Every synchronization of a Line or a Square for FSSP requires at least time $2n - 1$.*

The minimum time to synchronize a Ring is at least the time required by a cell to send a message to all the other cells and to get back information from all cells. The lower bound for Rings then obviously holds for Toruses as well.

Lemma 2. *Every synchronization of a Ring or of a Torus for FSSP requires time at least $n + 1$.*

4   A New Method to Solve FSSP on Squares

In this section we present a new minimal time and minimal communication synchronization of $n \times n$ square arrays with one General at the left-top corner. As already mentioned, the basic idea is to apply Mazoyer's minimal communication synchronization method for all sub-rows and sub-columns [2]. The main new problems are: to make each cell to learn, using only one-bit communication and without any delay, whether it is participating in the synchronization process of a sub-row or of a sub-column or of both; to achieve that synchronization of sub-rows and of sub-columns is done in minimal time; to arrange that synchronization of one sub-line does not interfere with synchronization of other sub-lines. We show that all this can be done in such a way that once a cell learns that it participates in a synchronization process of a sub-row, it ignores inputs from the upper and lower adjacent cells and, moreover, it does not send (non-zero) outputs upwards or downwards. The same holds for cells participating in synchronization of the sub-columns.

To show that the above way of synchronization can be carried out with 1-bit inter-cell communication only and without any delay (with respect to a minimal time synchronization), we use several simple properties of Mazoyer's synchronization, that capture situations during the first two steps of the synchronization process. The properties, listed below, are implicitly contained in [12] and explicitly formulated in [11] (Section 2.10.1). However, in order to facilitate the understanding of our synchronization algorithm, let us observe that in Mazoyer's solution the first cell, the General, transmits the information about its name to the second cell via two consecutive bits 1. The second cell, originally in the sleeping state, learns, at time 3, after receiving two consecutive bits 1, that it is the cell number 2. Also, any cell in the sleeping state that receives from the left a bit 1 learns that it is the last awakened cell. Moreover, if it receives immediately after that bit 0, it learns that its number in the line is greater than 2.

We summarize now the properties of Mazoyer's synchronization that are used in our solution.

P0 Each just awakened cell sends bit 1 to the not yet activated neighbouring cell.

P1 The first cell knows that it is the General, and sends at time 2 bit 1 to its right neighbour;

P2 Each cell number $1 < i \leq n$ knows at time $t = i$ that it is the last cell awakened so far (since it receives bit 1 in the sleeping state);

---

[2] In the following sub-row(i) is the line of cells from (i, i) to (i, n) and sub-column(j) is the line of cells from (j, j) to (n, j). The term sub-line is used to refer either to a sub-row or to a sub-column.

P3 The second cell knows after 2 steps (that is at time 3) that it is the cell number 2 (since it receives two consecutive bits 1 – the first while it is in the sleeping state and the second immediately after that) and sends to the right bit 0;

P4 Each cell number $2 < i < n$ knows at time $t = i+1$ that it is not the second cell (since it receives first a bit 1 while in the sleeping state and immediately after that bit 0);

P5 The last cell when awakened realizes that it is the cell number n (the last cell).

Now we are ready to show:

Theorem 2. *There is a 1-bit communication synchronization for FSSP on any $n \times n$ Square at time $2n - 1$.*

Proof. Let the signal from the General (in the cell $(1,1)$), start at time 1 and move with maximal speed right and down through the Square, thus reaching each cell (i,j) at time $i + j - 1$. This is done by letting each awakened cell to send bit 1 right and down – in accordance with the behaviour of any awakened cell in Mazoyer's synchronization of the Line (see property P0).

Now we want that each awakened cell (i,j), for $i < n$ and $j < n$, can play the role either of a new General, if $i = j$, or of the second cell of a sub-row(i) (resp. sub-column(j)) to be synchronized, if $j = i+1$ (resp. $i = j+1$), or of a different cell of the sub-line. Since all such cells in Mazoyer's solution behave in the same way (it follows from the property P0 that at the next time step they all send bit 1 to their right neighbour), it is not needed at this stage to establish which of the above three cases holds. To capture all that we say that each awakened cell enters a "three options state" $3OP$. According to the properties (P2-P4) listed above, in the next step each cell in the state $3OP$ can determine exactly which of the three above types of cells it is and it can do this behaving always as a cell in Mazoyer's synchronization would behave within the sub-line to which it belongs. The rules are as follows.

1. If a cell in the state $3OP$ receives bits 0 both from above and from the left, then it learns that it is a new General (that is a cell (i,i)). These bits need to be sent by the second cells of the sub-row(i-1) and sub-column(i-1). From that moment new General starts to ignore signals coming from the left and from above. As an immediate output it sends bits 1 to the right and below (see the property P1, this is in accordance with Mazoyer's synchronization algorithm for sub-row(i) and sub-column(i) starting from the cell (i,i)).

2. If a cell in the state $3OP$ receives bit 1 (resp. 0) from the left and bit 0 (resp. 1) from above, then it knows that it is neither a new General nor the cell next to a new General in a sub-column (resp. sub-row) to be synchronized (see properties P0 and P4). After that it will ignore inputs from the left (resp. from above) and as output sends the bit 1 to the right (resp. below) and bit 0 below (resp. to the right). That is, it starts to behave as a cell $i > 2$ in Mazoyer's solution along a sub-column (resp. a sub-row), and bit 1

sent to the right (resp. below) is used by the neighbouring cell to determine that it is not a new General of its sub-line.

3. If a cell in the state 3OP receives bits 1 both from above and from the left, then it knows it is the second cell in its sub-line – for our considerations it does not matter whether in a sub-row or a sub-column. As outputs it sends 0 in both directions, to the right and below (see the property P3). This way it behaves again as a cell in Mazoyer's solution, and communicates to its neighbour in the next inner sub-line that it is the General of such a sub-line.

Let us add that those outputs of each cell that are not explicitly described are similar to those of the corresponding sub-lines of Mazoyer's solution.

Each cell belonging to the last row and column knows, due to the property P5, that it is the last cell in a sub-line and behaves immediately (without entering the 3OP state) as the last cell in Mazoyer's solution. In particular, the cell (n, n) will enter the firing state F when it is awakened.

To finish the proof let us note that each sub-line(i), whose length is $n - i + 1$, synchronizes in $2(n-i)$ steps after the cell (i, i) has been awakened at time $2i-1$. The overall synchronization is therefore obtained at time $2n - 1$ (note that if a synchronization requires time $2n - 1$ then it makes $2n - 2$ steps). $\qquad\square$

5 FSSP Solutions for Rings and Toruses

In this section we deal with minimal time and minimal communication synchronizations for FSSP on circularly shaped networks. We start with the synchronization of Rings of n cells. Later we discuss synchronization for FSSP on Toruses of $n \times n$ cells.

5.1 One-Dimensional Networks

We obtain a minimal time synchronization for FSSP on a Ring by simulating a minimal time and communication synchronization for FSSP on Two-End-Lines. Such a synchronization can be designed by adapting Mazoyer's minimal time 1-bit synchronization [12], as shown in the proof of the following lemma.

Lemma 3. *There is a 1-bit communication synchronization of FSSP on Two-End-Lines of n cells at time n.*

Proof. A 1-bit synchronization for FSSP on Two-End-Lines of n cells can be obtained by starting Mazoyer's minimal time 1-bit synchronization for FSSP on two Lines, with Generals at both ends. Thus imagine the Line as being split into two halves: the first half is a Line from cell 1 to cell $\lfloor (n + 1)/2 \rfloor$ (with the General at position 1) and the second half is a Line from cell n to cell $\lceil (n+1)/2 \rceil$ (with the General at position n). Note that in case n is even the two Lines are disjoint, otherwise they share the last cell. Moreover, each cell knows the sub-line to which it belongs starting from the time it is awakened.

If n is odd, each sub-line is composed of exactly $(n + 1)/2$ cells and the central cell of the original Line needs to behave as the last cell for both halves.

Clearly, this can be easily done since the last cell in a solution sends/receives only bits to/from its single neighbour. Thus, from Theorem 1 the total time to synchronize each half is $2((n+1)/2) - 1$, that is n.

If n is even, cells $n/2$ and $n/2 + 1$ are aware of being the central cells only with one step delay since they are awakened. This has as a consequence that the synchronization on each half gets delayed by 1 and thus by Theorem 1 the total time to synchronize each half is $(2(n/2) - 1) + 1$. The overall synchronization time is therefore again n. □

We are ready now to deal with the FSSP on Rings. Note that the outcome will be an improvement of the result presented in [2].

Theorem 3. *There is a 1-bit communication synchronization of the FSSP on any Ring of n cells in time $n + 1$.*

Proof. A FSSP on a Ring of n cells can be seen as the FSSP on Two-End-Line of $n + 1$ cells where cells 1 and $n + 1$ both correspond to the cell 1 of the Ring. From Lemma 3, there is a 1-bit communication synchronization of the FSSP for Two-End-Line of $n + 1$ cells at time $n + 1$. Thus, we have the theorem. □

5.2   Two-Dimensional Networks

In this section, we prove as the main result a tight upper bound on the 1-bit synchronization for FSSP on networks of cells arranged as toruses. To show that, we use the following result of interest by itself.

Lemma 4. *There is a 1-bit communication synchronization for the FSSP on Four-End-Squares of $n \times n$ cells in time n.*

Proof. Consider a Square as consisting of n concentric "frames", where the i-th inner frame is formed by the four sub-lines $(i, i) \ldots (i, n - i + 1)$, $(i, n - i + 1) \ldots (n-i+1, n-i+1)$, $(i, i) \ldots (n-i+1, i)$ and $(n-i+1, i) \ldots (n-i+1, n-i+1)$, see Figure 2.

Since the cells $(1, 1)$, $(1, n)$, $(n, 1)$ and (n, n) are all Generals, the four lines of the first frame can all synchronize in time n using Lemma 3. If we start with the Generals in the four corners of the i-th frame, $i \leq \lceil n/2 \rceil$, again by Lemma 3, we can synchronize this frame at time $n - 2(i - 1)$ (each sub-line of the frame has exactly $n - 2(i - 1)$ cells). Thus, to achieve the synchronization of the Four-End-Square at time n we just need to have the corners of the i-th frame entering the General at time $2i + 1$ (note that $2i + 1$ corresponds to time 1 of the synchronization on the frame, thus the total time is $2i + n - 2i$, that is n). To obtain this, we can design a synchronization process that behaves similarly to that we have constructed for the minimal time and minimal communication synchronization of the FSSP on Squares in Section 4. All that can be clearly done with one-bit communications and therefore we omit further details here.

□

Now it is easy to show the main result of this section.

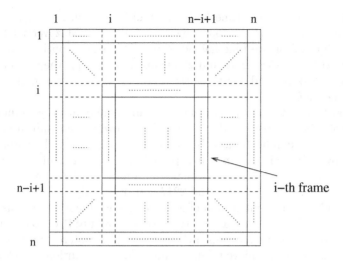

Fig. 2. The frames in a Square of $n \times n$ processors

Theorem 4. *There is a 1-bit communication synchronization for FSSP on any Torus of $n \times n$ cells at time $n + 1$.*

Proof. A synchronization for FSSP on a Torus of $n \times n$ cells at time $n + 1$ can be obtained by looking at the underlying interconnection graph as being split into three parts: the first row, the first column and the remaining part of the array – a sub-square of the size$(n - 1) \times (n - 1)$. By Theorem 3, the first row and the first column can be synchronized in time $n + 1$. During these synchronizations (in the first two steps) cells $(2, 2)$, $(2, n)$, $(n, 2)$, (n, n) can enter new states that can be seen as Generals of a Four-End-Square with $(n - 1) \times (n - 1)$ cells. Using Lemma 4 and considering that this last synchronization starts with two steps of delay, the overall synchronization time is $n + 1$. The technical problems to cope with in designing this algorithm are all related to the limitation that each cell can only send/receive 1-bit to/from each neighbouring cell at a time. One can easily verify that it is possible to implement our algorithm using techniques similar to those used in Section 4 for the synchronization of the Square, and thus we omit further details here.   □

6   Conclusions and Open Problems

FSSP got so famous also for the fact that it is still an open problem to determine even for the simplest case of the Line a solution with minimal time and minimal size (in terms of states).[3] Related problems are to minimize simultaneously time and the number of transitions. These size-minimization problems are of interest

---

[3] A beautiful result by Mazoyer [10] shows that minimal size for minimal time solution is at most 6. By Balzer [1], it is at least 5 and the gap has not been resolved yet.

also for the case of minimal time and communication solutions. They may be quite difficult problems as the following facts indicate.[4] Mazoyer's 1-bit communication solution for the Line has 56 states. The not-yet optimal solution in [18] is reported to have 78 states and 208 transition rules for the General in one end-line and 127 states and 405 transition rules for the General in an arbitrary position of the line.

A natural task is to find an optimal time and communication solution also for the case of arbitrary rectangular arrays. The following facts indicate that this may be a much more difficult task. In case of the $n \times m$ array the lower bound, established by Shinar, is $m + n + \max\{m, n\} - 3$ steps and his upper bound for the size is 28 states. Almost optimal solution in [18] has, for an arbitrary position of the General 862 states and 2217 transition rules.

As we could see our combination of Mazoyer's and Shinar's solutions for two quite different FSSP provides a way to solve several FSSP on one- and two-dimensional arrays and their modifications. It would be interesting to see if our approach can be used to achieve minimal time and 1-bit solutions also for arrays with more than two dimensions and their variants.

Acknowledgments. We thank Margherita Napoli for many helpful discussions.

References

1. R. Balzer, *An 8-states minimal time solution to the firing squad synchronization problem*, Information and Control, 10 (1967), 22–42.
2. K. Culik, *Variations of the firing squad problem and applications*, Information Processing Letters, 30 (1989), 153–157.
3. E. Goto, *A Minimal Time Solution of the Firing Squad Problem*, Lecture Notes for Applied Mathematics 298 (1962), Harvard University, 52–59.
4. A. Ilachinski, *Cellular Automata: A Discrete Universe*, World Scientific, 2001.
5. K. Imai and K. Morita, *Firing squad synchronization problem in reversible cellular automata*, Theoretical Computer Science, 165 (1996), 475–482.
6. K. Imai, K. Morita, K. Sako, *Firing squad synchronization problem in number-conserving cellular automata*, Proc. of the IFIP Workshop on Cellular Automata, Santiago (Chile), 1998.
7. K. Kobayashy, *The Firing Squad Synchronization Problem for Two Dimensional Arrays*, Information and Control 34 (1977), 153–157.
8. K. Kobayashy, *On Time Optimal Solutions of the Firing Squad Synchronization Problem for Two-Dimensional Paths*, Theoretical Computer Science 259 (2001), 129–143.
9. S. La Torre and M. Parente, *Different Time Solutions for the Firing Squad Synchronization Problem on Various Networks*, Tech. Rep. DIA - Università degli Studi di Salerno, (2001).
10. J. Mazoyer, *A six states minimal time solution to the firing squad synchronization problem*, Theoretical Computer Science 50 (1987), 183–238.

---

[4] However, one should keep in mind that the first Goto's solution was believed to have about 1 million states (that came then to about 100 000 as Umeo et al. [14] showed).

11. J. Mazoyer, *A minimal time solution to the firing squad synchronization problem with only one bit of information exchanged*, Laboratoire de l'Information du Parallelisme, Ecole Normale Superieure de Lyon, Rapport n.89-03, April 1989.

12. J. Mazoyer, *On optimal solutions to the firing squad synchronization problem*, Theoretical Computer Science 168(2) (1996) 367-404.

13. M. Minsky, *Computation: Finite and Infinite Machines*, Prentice-Hall, 1967.

14. J. Nishimura, T. Sogabe, H. Umeo, *A Design of Optimum-Time Firing Squad Synchronization Algorithm on 1-Bit Cellular Automaton* Tech. Rep. of IPSJ, (2000).

15. Z. Roka, *The Firing Squad Synchronization Problem on Cayley Graphs*, in Proc. of the 20-th International Symposium on Mathematical Foundations of Computer Science MFCS'95, Prague, Czech Republic, 1995. *Lecture Notes in Computer Science*, 969 (1995) 402–411.

16. I. Shinar, *Two and Three-Dimensional Firing Squad Synchronization Problems*, Information and Control 24 (1974), 163–180.

17. H. Szwerinski, *Time-Optimal Solution of the Firing-Squad-Synchronization-Problem for n-Dimensional Rectangles with the General at an Arbitrary Position*, Theoretical Computer Science, 19 (1982), 305-320.

18. H. Umeo, K. Michisaka, N. Kamikawa, *A Synchronization Problem on 1-Bit Communication Cellular Automata*, in Proc. of ICCS'03, *Lecture Notes in Computer Science*, 2657 (2003) 492–500.

19. A. Waksman, *An optimum solution to the firing squad synchronization problem*, Information and Control 9 (1966), 66–78.

20. S. Wolfram, *Cellular Automata and Complexity: Collected Papers*, Wolfram Research Inc., 1994.

21. S. Wolfram, *A New Kind of Science*, Wolfram Media, 2002.

The Power of Maximal Parallelism in P Systems[*]

Oscar H. Ibarra[1], Hsu-Chun Yen[2], and Zhe Dang[3]

[1] Department of Computer Science,
University of California, Santa Barbara, CA, USA
ibarra@cs.ucsb.edu
[2] Department of Electrical Engineering,
National Taiwan University, Taipei, Taiwan, R.O.C.
[3] School of Electrical Engineering and Computer Science,
Washington State University, Pullman, WA, USA

Abstract. We consider the following definition (different from the standard definition in the literature) of "maximal parallelism" in the application of evolution rules in a P system G: Let $R = \{r_1, ...r_k\}$ be the set of (distinct) rules in the system. G operates in maximal parallel mode if at each step of the computation, a maximal subset of R is applied, and at most one instance of any rule is used at every step (thus at most k rules are applicable at any step). We refer to this system as a maximally parallel system. We look at the computing power of P systems under three semantics of parallelism. For a positive integer $n \le k$, define:

n-**Max-Parallel:** At each step, nondeterministically select a maximal subset of at most n rules in R to apply (this implies that no larger subset is applicable).

$\le n$-**Parallel:** At each step, nondeterministically select any subset of at most n rules in R to apply.

n-**Parallel:** At each step, nondeterministically select any subset of exactly n rules in R to apply.

In all three cases, if any rule in the subset selected is not applicable, then the whole subset is not applicable. When $n = 1$, the three semantics reduce to the **Sequential** mode.

We focus on two popular models of P systems: multi-membrane catalytic systems and communicating P systems. We show that for these systems, n-**Max-Parallel** mode is strictly more powerful than any of the following three modes: **Sequential**, $\le n$-**Parallel**, or n-**Parallel**. For example, it follows from the result in [7] that a maximally parallel communicating P system is universal for $n = 2$. However, under the three limited modes of parallelism, the system is equivalent to a vector addition system, which is known to only define a recursive set. These generalize and refine the results for the case of 1-membrane systems recently reported in [3]. Some of the present results are rather surprising. For example, we show that a **Sequential** 1-membrane communicating P

[*] The research of Oscar H. Ibarra was supported in part by NSF Grants IIS-0101134, CCR-0208595, and CCF-0430945. The research of Zhe Dang was supported in part by NSF Grant CCF-0430531.

C.S. Calude, E. Calude and M.J. Dinneen (Eds.): DLT 2004, LNCS 3340, pp. 212–224, 2004.

system can only generate a semilinear set, whereas with k membranes, it is equivalent to a vector addition system for any $k \geq 2$ (thus the hierarchy collapses at 2 membranes - a rare collapsing result for nonuniversal P systems). We also give another proof (using vector addition systems) of the known result [6] that a 1-membrane catalytic system with only 3 catalysts and (non-prioritized) catalytic rules operating under 3-**Max-Parallel** mode can simulate any 2-counter machine M. Unlike in [6], our catalytic system needs only a *fixed* number of noncatalysts, independent of M.

A simple cooperative system (SCO) is a P system where the only rules allowed are of the form $a \rightarrow v$ or of the form $aa \rightarrow v$, where a is a symbol and v is a (possibly null) string of symbols not containing a. We show that a 9-**Max-Parallel** 1-membrane SCO is universal.

1   Introduction

There has been a flurry of research activities in the area of membrane computing (a branch of molecular computing) initiated five years ago by Gheorghe Paun [13]. Membrane computing identifies an unconventional computing model, namely a P system, from natural phenomena of cell evolutions and chemical reactions. Due to the built-in nature of maximal parallelism inherent in the model, P systems have a great potential for implementing massively concurrent systems in an efficient way that would allow us to solve currently intractable problems (in much the same way as the promise of quantum and DNA computing) once future bio-technology (or silicon-technology) gives way to a practical bio-realization (or chip-realization).

The Institute for Scientific Information (ISI) has recently selected membrane computing as a fast "Emerging Research Front" in Computer Science (*http://esi-topics.com/ erf/october2003.html*). A P system is a computing model, which abstracts from the way the living cells process chemical compounds in their compartmental structure. Thus, regions defined by a membrane structure contain objects that evolve according to given rules. The objects can be described by symbols or by strings of symbols, in such a way that multisets of objects are placed in regions of the membrane structure. The membranes themselves are organized as a Venn diagram or a tree structure where one membrane may contain other membranes. By using the rules in a nondeterministic, maximally parallel manner, transitions between the system configurations can be obtained. A sequence of transitions shows how the system is evolving. Various ways of controlling the transfer of objects from a region to another and applying the rules, as well as possibilities to dissolve, divide or create membranes have been studied. P systems were introduced with the goal to abstract a new computing model from the structure and the functioning of the living cell (as a branch of the general effort of Natural Computing – to explore new models, ideas, paradigms from the way nature computes). Membrane computing has been quite successful: many models have been introduced, most of them Turing complete and/or able to solve computationally intractable problems (NP-complete, PSPACE-complete) in a feasible time (polynomial), by trading space for time. (See the P system website at *http://psystems.disco.unimib.it* for a large collection of papers in the area, and in particular the monograph [14].)

As already mentioned above, in the standard semantics of P systems [13–15], each evolution step of a system G is a result of applying all the rules in G in a maximally parallel manner. More precisely, starting from the initial configuration, w, the system goes through a sequence of configurations, where each configuration is derived from the directly preceding configuration in one step by the application of a multi-set of rules, which are chosen nondeterministically. For example, a catalytic rule $Ca \rightarrow Cv$ in membrane q is applicable if there is a catalyst C and an object (symbol) a in the preceding configuration in membrane q. The result of applying this rule is the evolution of v from a. If there is another occurrence of C and another occurrence of a, then the same rule or another rule with Ca on the left hand side can be applied. Thus, in general, the number of times a particular rule is applied at anyone step can be unbounded. We require that the application of the rules is maximal: all objects, from all membranes, which *can be* the subject of local evolution rules *have to* evolve simultaneously. Configuration z is reachable (from the starting configuration) if it appears in some execution sequence; z is halting if no rule is applicable on z.

In this paper, we study a different definition of maximal parallelism. Let G be a P system and $R = \{r_1, ..., r_k\}$ be the set of (distinct) rules in all the membranes. (Note that r_i uniquely specifies the membrane the rule belongs to.) We say that G operates in maximal parallel mode if at each step of the computation, a maximal subset of R is applied, and at most one instance of any rule is used at every step (thus at most k rules are applicable at any step). For example, if r_i is a catalytic rule $Ca \rightarrow Cv$ in membrane q and the current configuration has two C's and three a's in membrane q, then only one a can evolve into v. Of course, if there is another rule r_j, $Ca \rightarrow Cv'$, in membrane q, then the other a also evolves into v'. Throughout the paper, we will use this definition of maximal parallelism. Here, we look at the computing power of P systems under three semantics of parallelism. For a positive integer $n \leq k$, define:

n-**Max-Parallel:** At each step, nondeterministically select a maximal subset of at most n rules in R to
apply (this implies that no larger subset is applicable).

$\leq n$-**Parallel:** At each step, nondeterministically select any subset of at most n rules in R to apply.

n-**Parallel:** At each step, nondeterministically select any subset of exactly n rules in R to apply.

In all three cases, if any rule in the subset selected is not applicable, then the whole subset is not applicable. When $n = 1$, the three semantics reduce to the **Sequential** mode.

In the next four sections, we investigate the computing power of two popular models of P systems with respect to the above semantics of parallelism – the catalytic P systems and the communicating P systems.

We should mention some related work on P systems operating in sequential and limited parallel modes. Sequential variants of P systems have been studied, in a different framework, in [5]. There, generalized P systems (GP-systems) were considered and were shown to be able to simulate graph controlled grammars. Our notion of limited parallelism seems to correspond to "cooperation modes" in cooperating distributed grammar systems, investigated in [2].

Because of space limitation, many of the proofs are omitted here. They will appear in the journal version of this paper.

2   Multi-membrane Catalytic Systems

2.1   Maximally Parallel CS

First we recall the definition of a catalytic system (CS). The membranes (regions) are organized in a hierarchical (tree) structure and are labeled 1, 2, .., m for some m, with the outermost membrane (the skin membrane) labeled 1. At the start of the computation, there is a distribution of *catalysts* and *noncatalysts* in the membranes (the distribution represents the initial configuration of the system). Each membrane may contain a finite set of catalytic rules of the form $Ca \rightarrow Cv$, where C is a catalyst, a is a noncatalyst, and v is a (possibly null) string of noncatalysts. When this rule is applied, the catalyst remains in the membrane the rule is in, symbol a is deleted from the membrane, and the symbols comprising v (if nonnull) are transported to other membranes in the following manner. Each symbol b in v has a designation or target, i.e., it is written b_x, where x can be *here*, *out*, or *in$_j$*. The designation *here* means that the object b remains in the membrane containing it (we usually omit this target, when it is understood). The designation *out* means that the object is transported to the membrane directly enclosing the membrane that contains the object; however, we do not allow any object to be transported out of the skin membrane. The designation *in$_j$* means that the object is moved into a membrane, labeled j, that is directly enclosed by the membrane that contains the object.

It is important to note that our definition of catalytic system is different from what is usually called catalytic system in the literature. Here, we do not allow rules without catalysts, i.e., rules of the form $a \rightarrow v$. Thus our systems use only purely catalytic rules.

Suppose that S is a CS with m membranes. Let $\{a_1, ..., a_n\}$ be the set of noncatalyst symbols (objects) that can occur in the configurations of S. Let $w = (w_1, ..., w_m)$ be the initial configuration, where w_i represents the catalysts and noncatlysts in membrane i. (Note that w_i can be null.) Each reachable configuration of S is an nm-tuple $(v_1, ..., v_m)$, where v_i is an n-tuple representing the multiplicities of the symbols $a_1, ..., a_n$ in membrane i. Note that we do not include the catalysts in considering the configuration as they are not changed (i.e., they remain in the membranes containing them, and their numbers remain the same during the computation). Hence the set of all reachable configurations of S, denoted by $R(S)$ is a subset of $\mathbf{N}^{mn}$. The set of all halting reachable configurations is denoted by $R_h(S)$.

2.2   Sequential CS

In a sequential multi-membrane CS, each step of the computation consists of an application of a single nondeterministically chosen rule, i.e., the membrane and rule within the membrane to apply are chosen nondeterministically. We show below that sequential multi-membrane CS's define exactly the semilinear sets.

We need the definition of a vector addition system. An n-dimensional *vector addition system* (VAS) is a pair $G = \langle x, W \rangle$, where $x \in \mathbf{N}^n$ is called the *start point* (or *start vector*) and W is a finite set of transition vectors in $\mathbf{Z}^n$, where $\mathbf{Z}$ is the set of all integers (positive, negative, zero). Throughout this paper, for a $w \in \mathbf{Z}^n$ we write $w \geq 0$ to mean that w has only nonnegative components (i.e., $w \in \mathbf{N}^n$). The *reachability set* of the VAS $\langle x, W \rangle$ is the set $R(G) = \{z \mid$ for some j, $z = x + v_1 + \ldots + v_j$, where, for all $1 \leq i \leq j$, each $v_i \in W$ and $x + v_1 + \ldots + v_i \geq 0\}$. Note that $R(G)$ is the smallest set satisfying the following two properties: (1) $x \in R(G)$, and (2) whenever $z \in R(G)$, $v \in W$, and $z + v \in \mathbf{N}^n$, then $z + v \in R(G)$. The *halting reachability set* $R_h(G) = \{z \mid z \in R(G), z + v \not\geq 0$ for every v in $W\}$.

An n-dimensional *vector addition system with states* (VASS) is a VAS $\langle x, W \rangle$ together with a finite set T of transitions of the form $p \to (q, v)$, where q and p are states and v is in W. The meaning is that such a transition can be applied at point y in state p and yields the point $y + v$ in state q, provided that $y + v \geq 0$. The VASS is specified by $G = \langle x, W, T, p_0 \rangle$, where p_0 is the starting state. Assuming that the set of states of G is $\{p_0, \ldots, p_k\}$ (for some $k \geq 0$), the *reachability set* of VASS G is $R(G) = \{(i, w) \in \mathbf{N}^{n+1} \mid 1 \leq i \leq k, (p_i, w)$ is reachable from $(p_0, x)\}$. The *halting reachability set* $R_h(G) = \{(i, w) \mid (i, w) \in R(G)$ and no transition is applicable in $(p_i, w)\}$.

The *reachability problem* for a VASS (respectively, VAS) G is to determine, given a vector y, whether y is in $R(G)$. The *equivalence problem* is to determine given two VASS (respectively, VAS) G and G', whether $R(G) = R(G')$. Similarly, one can define the reachability problem and equivalence problem for halting configurations.

The following summarizes the known results concerning VAS and VASS [17, 8, 1, 9, 12]:

Theorem 1.

1. Let G be an n-dimensional VASS. We can effectively construct an $(n + 3)$-dimensional VAS G' that simulates G.
2. If G is a 2-dimensional VASS, then $R(G)$ is an effectively computable semilinear set.
3. There is a 3-dimensional VASS G such that $R(G)$ is not semilinear.
4. If G is a 5-dimensional VAS, then $R(G)$ is an effectively computable semilinear set.
5. There is a 6-dimensional VAS G such that $R(G)$ is not semilinear.
6. The reachability problem for VASS (and hence also for VAS) is decidable.
7. The equivalence problem for VAS (and hence also for VASS) is undecidable.

Clearly, it follows from part 6 of the theorem above that the halting reachability problem for VASS (respectively, VAS) is decidable.

A *communication-free VAS* is a VAS where in every transition, at most one component is negative, and if negative, its value is -1. Communication-free VAS's are equivalent to communication-free Petri nets, which are also equivalent to commutative context-free grammars [4, 10]. It is known that they have effectively computable semilinear reachability sets [4].

Our first result shows that a sequential CS is weaker than a maximally parallel CS.

Theorem 2. *Every sequential multi-membrane CS S can be simulated by a communication-free VAS G, and vice versa.*

Proof. Let S be an m-membrane CS with noncatalysts $a_1, ..., a_n$. Suppose that the start configuration of $w = (w_1, ..., w_m)$ has k catalysts $C_1, ..., C_k$. We may assume, without loss of generality by adding new catalysts and rules if necessary, that each C_i occurs at most once in w_i $(1 \leq i \leq m)$. Number all the rules in S by $1, ..., s$. Note that the rule number uniquely determines the membrane where the rule is applicable.

We first transform S to a new system S' by modifying the rules and the initial configuration w. S' will now have catalysts $C_1, ..., C_k, Q_1, ..., Q_s$ and noncatalysts $a_1, ..., a_n, d_1, ..., d_s$. The component w_q of the initial configuration in membrane q will now be w_q plus each Q_h for which rule number h is in membrane q. The rules of S' are defined as follows:

Case 1: Suppose that $C_j a_i \rightarrow C_j v$ is a rule in membrane q of S, and a_i does not appear in v with designation (target) *here*. Then this rule is in membrane q of S'.

Case 2: Suppose that $C_j a_i \rightarrow C_j a_i^t v$ is rule number r and $t \geq 1$. Suppose that this rule is in membrane q, with the target of each a_i in a_i^t being *here*, and v does not contain any a_i with target *here*. Then the following rules are in membrane q of S': $C_j a_i \rightarrow C_j d_r^t v$ and $Q_r d_r \rightarrow Q_r a_i$. In the above rules, the target for a_i and each d_r in the right-hand side of the rules is *here*.

Clearly, S' simulates S, and S' has the property that in each rule $Xb \rightarrow Xv$ (where X is a catalyst, b is a noncatalyst, and v a string of noncatalysts), v does not contain a b with target *here*. It is now obvious that each rule $Xb \rightarrow Xv$ in S' can be transformed to a VAS transition rule of $mn + s$ components, where the component of the transition corresponding to noncatalyst b is -1, and the other components (corresponding to the target designations in v) are nonnegative. Thus, the VAS is communication free.

Conversely, let G be a communication-free VAS. We construct a sequential 1-membrane CS S which has one catalyst C, noncatalysts $\#, a_1, ..., a_k$, and starting configuration $C\#w$, where w corresponds to the starting vector of G. Suppose that $(j_1, ..., j_{m-1}, j_m, j_{m+1}, ..., j_k)$ is a transition in G.

Case 1: $j_m = -1$ and all other j_i's are nonnegative. Then the following rule is in S: $Ca_m \to Ca_1^{j_1}...a_{m-1}^{j_{m-1}} a_{m+1}^{j_{m+1}}...a_k^{j_k}$.

Case 2: All the j_i's are nonnegative. Then the following rule is in S: $C\# \to C\#a_1^{j_1}...a_k^{j_k}$.

Clearly, S simulates G. In fact, $R(G) = R(S) \times \{1\}$. $\qquad\square$

Corollary 1.

1. If S is a sequential multi-membrane CS, then $R(S)$ and $R_h(S)$ are effectively computable semilinear sets.
2. The reachability problem (whether a given configuration is reachable) for sequential multi-membrane CS's is NP-complete.

Since a communication-free VAS can be simulated by a sequential 1-membrane CS (from conversely in the proof of Theorem 2), we have:

Corollary 2. *The following are equivalent: communication-free VAS, sequential multi-membrane CS, sequential 1-membrane CS.*

2.3   CS Under Limited Parallelism

Here we look at the computing power of the multi-membrane CS under three semantics of parallelism (namely, n-**Max-Parallelism**, $\le n$-**Parallelism**, and n-**Parallelism**) defined in Section 1. We can show the following:

Theorem 3. *For $n = 3$, a 1-membrane CS operating under the n-**Max-Parallel** mode can define any recursively enumerable set. For any n, a multi-membrane CS operating under $\le n$-**Parallel** mode or n-**Parallel** mode can be simulated by a VASS.*

2.4   3-Max-Parallel 1-Membrane CS

As noted above, it is known that a 3-**Max-Parallel** 1-membrane CS is universal [6] in that it can simulate any 2-counter machine M. We show in this section that the universality result of [6] can be obtained in terms of communication-free VAS. Later we improve this result by showing that, in fact, the 1-membrane CS need no more than k noncatalysts for some *fixed* k, independent of M.

Consider an n-dimensional communication-free VAS $G = \langle x, W \rangle$ with its set of addition vectors W partitioned into three disjoint sets W_1, W_2 and W_3. Under the 3-**Max-Parallel** mode, at each step G nondeterministically applies a maximal set of at most 3 addition vectors simultaneously to yield the next vector; however, from each set $W_i, 1 \le i \le 3$, at most one addition vector can be chosen.

Acting as either *acceptors* or *generators*, the following result shows the equivalence of 2-counter machines and communication-free VAS operating under the 3-**Max-Parallel** mode.

Theorem 4. *Let M be a 2-counter machine with two counters C_1 and C_2. There exist an n-dimensional VAS $G = \langle x, (W_1, W_2, W_3) \rangle$ under the* **3-Max-Parallel** *mode and a designated coordinate l such that M accepts on initial counter values $C_1 = m$ and $C_2 = 0$ iff*

1. *(generator:) $m \in \{v(l) \mid v \in R_h(G)\}$;*
2. *(acceptor:) from start vector x with $x(l) = m$, $R_h(G) \neq \emptyset$, i.e., G has a halting computation.*

   Here n is bounded by a function of the number of states of M.

By assigning for each $1 \leq i \leq 3$, a catalyst C_i for the set of addition vectors W_i, defining a distinct noncatalyst symbol for each position in the addition vector, and converting each vector in W_i to a rule of the form $C_i a \rightarrow C_i v$, where a is a noncatalyst and v is a (possibly null) string of noncatalysts, the following result can easily be obtained from Theorem 4.

Corollary 3. *Let M be a 2-counter machine with two counters. There exists a 1-membrane 3-**Max-Parallel** CS S with catalysts C_1, C_2, C_3 and n noncatalysts with a designated noncatalyst symbol a_l such that M accepts on initial counter values m and 0, respectively, iff*

1. *(generator:) $m \in \{\#_{a_l}(y) \mid y \in R_h(S)\}$;*
2. *(acceptor) if S starts with initial configuration $(a_l)^m y$, for some y not containing a_l, then $R_h(S) \neq \emptyset$.*

We note that in the corollary above, the CS operates in 3-**Max-Parallel** mode. Now the catalyst C_1 (resp. C_2) is needed to make sure that at most one addition vector in W_1 (resp. W_2) is simulated by the CS at each step. However, catalyst C_3 is not really needed in that we can convert each addition vector in W_3 to a rule of the form $a \rightarrow v$, i.e., a *noncooperative rule* (without a catalyst). Thus, the system can be constructed to have only *two* catalysts with catalytic rules and noncooperative rules. This was also shown in [6]. However, the degree of maximal parallelism in the system is no longer 3 (because now more than one noncooperative rule may be applicable at each step). It can be shown that at any point, no more than 3 noncooperative rules are applicable. This in turn implies that the degree of maximal parallelism now becomes 5 (two catalysts plus 3 noncooperative rules). Note also that n, the dimension of the communication-free VAS, which translates to the number of noncatalysts for the system, is also a function of the number of states, hence is unbounded.

We can improve the above results. We need the following lemma.

Lemma 1. *There exists a 2-counter machine U with counters C_1 and C_2 that is universal in the following sense. When U is given a description of an arbitrary 2-counter machine M as a positive integer in C_2 and an input m in C_1, U accepts iff M with input m on its first counter and 0 on its other counter accepts.*

Therefore, we have:

Corollary 4. *There exists a* fixed positive integer n *such that if* $L \subseteq \mathbf{N}$ *is any recursively enumerable set of nonnegative integers, then:*

1. *L can be generated (accepted) by a 1-membrane* 3-**Max-Parallel** *CS with 3 catalysts and n noncatalysts.*
2. *L can be generated (accepted) by a 1-membrane* 5-**Max-Parallel** *P system with 2 catalysts and n noncatalysts with catalytic and noncooperative rules.*

2.5   9-Max-Parallel 1-Membrane CS with One Catalyst

We now look at a model of a 1-membrane CS with only *one* catalyst C with initial configuration $C^k x$ for some string x of noncatalysts (thus, there are k copies of C). The rules allowed are of the form $Ca \to Cv$ or of the form $Caa \to Cv$, i.e., C catalyzes two copies of an object. Clearly the system operates in maximally parallel mode, but uses no more than k rules in any step. We call this system 1GCS. This system is equivalent to a restricted form of cooperative P system [13, 14]. A simple cooperative system (SCO) is a P system where the rules allowed are of the form $a \to v$ or of the form $aa \to v$. Moreover, there is some fixed integer k such that the system operates in maximally parallel mode, but uses no more than k rule instances in any step. We can show the following:

Theorem 5. *1GCS (hence, also SCO) operating under the* 9-**Max-Parallel** *mode is universal.*

3   Sequential 1-Membrane Communicating P Systems

Consider the model of a communicating P system (CPS) with only *one* membrane, called the skin membrane [16]. The rules are of the form: (1) $a \to a_x$, (2) $ab \to a_x b_y$, (3) $ab \to a_x b_y c_{come}$, where a, b, c are objects, x, y (which indicate the directions of movements of a and b) can only be *here* (i.e., the object remains in the membrane) or *out* (i.e., the object is expelled into the environment). The third rule brings in an object c from the environment into the skin membrane. In the sequel, we omit the designation *here*, so that objects that remain in the membrane will not have this subscript. There is a fixed finite set of rules in the membrane. At the beginning, there is a fixed configuration of objects in the membrane.

Assume that the computation is *sequential*; i.e., at each step there is only one application of a rule (to one instance). So, e.g., if nondeterministically a rule like $ab \to a_{here} b_{out} c_{come}$ is chosen, then there must be at least one a and one b in the membrane. After the step, a remains in the membrane, b is thrown out of the membrane, and c comes into the membrane. There may be several a's and b's, but only one application of the rule is applied. Thus, there is *no* parallelism involved. The computation halts when there is no applicable rule. We are interested in the multiplicities of the objects when the system halts.

One can show that a 1-membrane CPS can be simulated by a vector addition system (VAS) (this is a special case of a theorem in the next section). However, the converse is not true – it was shown in [3] that a sequential 1-membrane CPS can only define a semilinear set.

4   Sequential 1-Membrane Extended CPS (ECPS)

Recall that 1-membrane CPS's operating sequentially define only semilinear sets. In contrast, we shall see in the next section that sequential 2-membrane CPS's are equivalent to VASS.

There is an interesting generalization of a 1-membrane CPS, we call extended CPS (or ECPS) – we add a fourth type of rule of the form: $ab \rightarrow a_x b_y c_{come} d_{come}$. That is, two symbols can be imported from the environment. We shall see below that ECPS's are equivalent to VASS's.

Let G be an n-dimensional VASS. Clearly, by adding new states, we may assume that all transitions in G have the form: $p_i \rightarrow (p_j, +1_h)$, $\quad p_i \rightarrow (p_j, -1_h)$. The above is a short-hand notation. The $+1_h$ is addition of 1 to the h-th coordinate, and -1_h is subtraction of 1 from the h-th coordinate. All other coordinates are unchanged. Note at each step, the state uniquely determines whether it is a '+1 transition' or a '-1 transition'.

For constructing the ECPS S equivalent to G, we associate symbol p_i for every state of the VASS, a_h for every coordinate (i.e., position) h in the transition. We also define a new special symbol c. So the ECPS has symbols $p_1, ... p_s$ (s is the number of states), $a_1, ..., a_n$ (n is dimension of the VASS), and c.

Then a transition of the form $p_i \rightarrow (p_j, +1_h)$ in G is simulated by the following rule in S: $p_i c \rightarrow p_{i(out)} c_{here} p_{j(come)} a_{h(come)}$

A transition of the form $p_i \rightarrow (p_j, -1_h)$ in G is simulated by the following rule in S: $p_i a_h \rightarrow p_{i(out)} a_{h(out)} p_{j(come)}$.

If the VASS G has starting point $\langle p_1, v \rangle$, where $v = (i_1, ..., i_n)$ and p_1 is the start state, then ECPS S starts with the word $p_1 a_1^{i_1} ... a_n^{i_n} c$. Clearly, S simulates G.

Conversely, suppose we are given an ECPS S over symbols $a_1, ..., a_n$ with initial configuration w and rules $R_1, ..., R_k$. The VASS G has states $R_0, R_1, R_1', ... R_k, R_k'$ and starting point $\langle R_0, v_0 \rangle$, where v_0 is the n-dimensional vector in $\mathbf{N}^n$ representing the multiplicities of the symbols in the initial configuration w. The transitions of G are defined as follows:

1. $R_0 \rightarrow (R_i, zero)$ for every $1 \leq i \leq k$ is a transition, where $zero$ represents the zero vector.
2. If R_i is a rule of the form $a_h \rightarrow a_{hx}$, then the following are transitions:
   $R_i \rightarrow (R_i', -1_h)$
   $R_i' \rightarrow (R_j, d_{hx})$ for every $1 \leq j \leq k$, where $d_{hx} = 0_h$ if $x = (out)$ and $d_{hx} = +1_h$ if $x = (here)$.
   (As before, $-1_h, 0_h, +1_h$ mean subtract 1, add 0, add 1 to h, respectively; all other coordinates are unchanged.)
3. If R_i is a rule of the form $a_h a_r \rightarrow a_{hx} a_{ry}$, then the following are transitions:
   $R_i \rightarrow (R_i', -1_h, -1_r)$
   $R_i' \rightarrow (R_j, d_{hx}, d_{ry})$ for every $1 \leq j \leq k$, where d_{hx} and d_{ry} are as defined above.
   (Note that if $h = r$, then $(R_i', -1_h, -1_r)$ means $(R_i', -2_h)$, i.e., subtract 2 from coordinate h.)

4. If R_i is a rule of the form $a_h a_r \rightarrow a_{hx} a_{ry} a_{s(come)}$, then the following are transitions:

$R_i \rightarrow (R_i', -1_h, -1_r)$

$R_i' \rightarrow (R_j, d_{hx}, d_{ry}, +1_s)$ for every $1 \le j \le k$, where d_{hx} and d_{ry} are as defined above.

5. If R_i is a rule of the form $a_h a_r \rightarrow a_{hx} a_{ry} a_{s(come)} a_{t(come)}$, then the following are transitions:

$R_i \rightarrow (R_i', -1_h, -1_r)$

$R_i' \rightarrow (R_j, d_{hx}, d_{ry}, +1_s, +1_t)$ for every $1 \le j \le k$, where d_{hx} and d_{ry} are as defined above.

It follows from the construction above that G simulates S. Thus, we have:

Theorem 6. *Sequential 1-membrane ECPS and VASS are equivalent.*

We can generalize rules of an ECPS further as follows:

1. $a_{i_1} ... a_{i_h} \rightarrow a_{i_1 x_1} ... a_{i_1 x_h}$
2. $a_{i_1} ... a_{i_h} \rightarrow a_{i_1 x_1} ... a_{i_1 x_h} c_{j_1 come} ... c_{j_l come}$

where $h, l \ge 1$, and $x_m \in \{here, out\}$ for $1 \le m \le h$, and the a's and c's are symbols. Call this system ECPS+. Generalizing the constructions in the proof of Theorem 6, we can show ECPS+ is still equivalent to a VASS. Thus, we have:

Corollary 5. *The following systems are equivalent: Sequential 1-membrane ECPS, sequential 1-membrane ECPS+, and VASS.*

Using rules of types of 1 and 2 above, we can define the three versions of parallelism as in Section 2.3, and we can prove the following result. Note that the first part was shown in [7].

Theorem 7. *For $n = 2$, a 1-membrane CPS (and, hence, also 1-membrane ECPS+) operating under the n-**Max-Parallel** mode can define a recursively enumerable set. For any n, a 1-membrane ECPS+ operating under $\le n$-**Parallel** mode or n-**Parallel** mode is equivalent to a VASS.*

5   Multi-membrane CPS and ECPS

In this section, we look at CPS and ECPS with multiple membranes. Now the subscripts x, y in the CPS rules $a \rightarrow a_x$, $ab \rightarrow a_x b_y$, $ab \rightarrow a_x b_y c_{come}$ (and $ab \rightarrow a_x b_y c_{come} d_{come}$ in ECPS) can be *here*, *out*, or *in$_j$*. As before, *here* means that the object remains in the membrane containing it, *out* means that the object is transported to the membrane directly enclosing the membrane that contains the object (or to the environment if the object is in the skin membrane), and *come* can only occur within the outermost region (i.e., skin membrane). The designation *in$_j$* means that the object is moved into a membrane, labeled j, that is directly enclosed by the membrane that contains the object.

In Section 3, we saw that a sequential 1-membrane CPS can only define a semilinear set. We can show that if the system has two membranes, it can simulate a vector addition system. Thus, we have:

Theorem 8. *A sequential 2-membrane CPS S can simulate a VASS G.*

In Theorem 6, we saw that a sequential 1-membrane ECPS can be simulated by a VASS. The construction can be extended to multi-membrane ECPS. Recall that we now allow rules of the form: $ab \rightarrow a_x b_y c_{come} d_{com}$.

Suppose that S has m membranes. Let $\{a_1, ..., a_n\}$ be the set of symbols (objects) that can occur in the configurations of S. Then each reachable configuration of S is an mn-tuple $(v_1, ..., v_m)$, where v_q is an n-tuple representing the multiplicities of the symbols $a_1, ..., a_n$ in membrane q. Then the set of all reachable configurations of S is a subset of $\mathbf{N}^{mn}$. Let $R_1, ..., R_k$ be the rules in S. Note that R_i not only gives the rule but also the membrane where it appears. The construction of the VASS G simulating S is similar to the construction described in the second part of the proof of Theorem 6. In fact the construction also works for ECPS+. Since a sequential 2-membrane CPS can simulate a VASS, we have:

Theorem 9. *The following are equivalent: VASS, sequential 2-membrane CPS, sequential 1-membrane ECPS, sequential multi-membrane ECPS, and sequential multi-membrane ECPS+.*

Finally, we observe that Theorem 7 extends to multi-membrane CPS:

Theorem 10. *For any n, a multi-membrane ECPS+ operating under $\leq n$-**Parallel** mode or n-**Parallel** mode is equivalent to a VASS.*

6 Conclusion

We showed in this paper that P systems that compute in sequential or limited parallel mode are strictly weaker than systems that operate with maximal parallelism for two classes of systems: multi-membrane catalytic systems and multi-membrane communicating P systems. Our proof techniques can be used to show that many of the P systems that have been studied in the literature (including ones with membrane dissolving rules) operating under sequential or limited parallelism with unprioritized rules can be simulated by vector addition systems.

References

1. H. G. Baker. Rabin's proof of the undecidability of the reachability set inclusion problem for vector addition systems. In *C.S.C. Memo 79, Project MAC, MIT*, 1973.
2. E. Csuhaj-Varju, J. Dassow, J. Kelemen, and Gh. Paun. *Grammar Systems: A Grammatical Approach to Distribution and Cooperation.* Gordon and Breach Science Publishers, Inc., 1994.
3. Z. Dang and O. H. Ibarra. On P systems operating in sequential and limited parallel modes. In *Pre-Proc. 6th Workshop on Descriptional Complexity of Formal Systems*, 2004.

4. J. Esparza. Petri nets, commutative context-free grammars, and basic parallel processes. In *Proc. Fundamentals of Computer Theory*, volume 965 of *Lecture Notes in Computer Science*, pages 221–232. Springer, 1995.

5. R. Freund. Sequential P-systems. Available at *http://psystems.disco.unimib.it*, 2000.

6. R. Freund, L. Kari, M. Oswald, and P. Sosik. Computationally universal P systems without priorities: two catalysts are sufficient. Available at *http://psystems.disco.unimib.it*, 2003.

7. R. Freund and A. Paun. Membrane systems with symport/antiport rules: universality results. In *Proc. WMC-CdeA2002*, volume 2597 of *Lecture Notes in Computer Science*, pages 270–287. Springer, 2003.

8. M. H. Hack. The equality problem for vector addition systems is undecidable. In *C.S.C. Memo 121, Project MAC, MIT*, 1975.

9. J. Hopcroft and J.-J. Pansiot. On the reachability problem for 5-dimensional vector addition systems. *Theoretical Computer Science*, 8(2):135–159, 1979.

10. D.T. Huynh. Commutative grammars: The complexity of uniform word problems. *Information and Control*, 57:21–39, 1983.

11. O. Ibarra, Z. Dang, and O. Egecioglu. Catalytic P systems, semilinear sets, and vector addition systems. *Theoretical Computer Science*, 312(2-3): 379–399, 2004.

12. E. Mayr. Persistence of vector replacement systems is decidable. *Acta Informat.*, 15:309–318, 1981.

13. Gh. Paun. Computing with membranes. *Journal of Computer and System Sciences*, 61(1):108–143, 2000.

14. Gh. Paun. *Membrane Computing: An Introduction*. Springer-Verlag, 2002.

15. Gh. Paun and G. Rozenberg. A guide to membrane computing. *Theoretical Computer Science*, 287(1):73–100, 2002.

16. P. Sosik. P systems versus register machines: two universality proofs. In *Pre-Proceedings of Workshop on Membrane Computing (WMC-CdeA2002), Curtea de Arges, Romania*, pages 371–382, 2002.

17. J. van Leeuwen. A partial solution to the reachability problem for vector addition systems. In *Proceedings of STOC'74*, pages 303–309.

An Efficient Pattern Matching Algorithm on a Subclass of Context Free Grammars

Shunsuke Inenaga[1], Ayumi Shinohara[2,3], and Masayuki Takeda[2,4]

[1] Department of Computer Science, P.O. Box 68 (Gustaf Hällströmin katu 2b)
FIN-00014, University of Helsinki, Finland
inenaga@cs.helsinki.fi
[2] Department of Informatics, Kyushu University 33, Fukuoka 812-8581, Japan
{ayumi, takeda}@i.kyushu-u.ac.jp
[3] PRESTO, Japan Science and Technology Agency (JST)
[4] SORST, Japan Science and Technology Agency (JST)

Abstract. There is a close relationship between formal language theory and data compression. Since 1990's various types of *grammar-based text compression* algorithms have been introduced. Given an input string, a grammar-based text compression algorithm constructs a context-free grammar that only generates the string. An interesting and challenging problem is pattern matching on context-free grammars $\mathcal{P}$ of size m and $\mathcal{T}$ of size n, which are the descriptions of pattern string P of length M and text string T of length N, respectively. The goal is to solve the problem in time proportional *only* to m and n, *not* to M nor N. Kieffer et al. introduced a very practical grammar-based compression method called *multilevel pattern matching code* (*MPM code*). In this paper, we propose an efficient pattern matching algorithm which, given two MPM grammars $\mathcal{P}$ and $\mathcal{T}$, performs in $O(mn^2)$ time with $O(mn)$ space. Our algorithm outperforms the previous best one by Miyazaki et al. which requires $O(m^2n^2)$ time and $O(mn)$ space.

1 Introduction

In 1990's formal language theory found text data compression to be a very promising application area; data compression is the discipline which aims to reduce space consumption of the data by removing its redundancy, and this is achievable by constructing a *context-free grammar* $\mathcal{G}$ which only generates the input text string w. Namely, the grammar $\mathcal{G}$ is such that its language $L(\mathcal{G})$ is $\{w\}$. Such a context-free grammar adroitly extracts, and succinctly represents, repeated segments of the input string, and thus gives a superbly compact representation of the string. According to this observation, many types of ingenious *grammar-based text compression* algorithms have been introduced so far. Examples of grammar-based text compressions are SEQUITUR [15, 17], Re-Pair [12], byte pair encoding (BPE) [5], grammar transform [9, 10], and straight-line programs (SLPs) [8].

C.S. Calude, E. Calude and M.J. Dinneen (Eds.): DLT 2004, LNCS 3340, pp. 225–236, 2004.

As strings are the most basic type for data storage, *pattern matching* has been an omnipresent problem in Computer Science [3]. Due to rapid spread and increase of compressed data, we naturally face the pattern matching problem with compressed strings. Namely, we are here required to do pattern matching on two compressed strings (text and pattern) that are described in the form of a context-free grammar. This problem is also called the *fully compressed pattern matching problem* [18]. The problem is formalized as follows:

Input: context-free grammars $\mathcal{P}$ and $\mathcal{T}$ generating only pattern P and text T, respectively.

Output: all occurrences of P in T.

Let m and n be the sizes of the grammars $\mathcal{P}$ and $\mathcal{T}$ respectively, and M and N be the lengths of the strings P and T, respectively. What should be emphasized here is that the goal is to solve this problem in time proportional *only* to m and n, *not* to M nor N. Although there exist a number of $O(M + N)$-time algorithms that solve the pattern matching problem for uncompressed strings P and T [6, 4], none of them supplies us with a polynomial time solution to the compressed version of the problem since M (resp. N) can be *exponentially large* with respect to m (resp. n). Therefore, in order for us to develop a polynomial time solution, quite a limited amount of computational space is avaliable, and this makes the problem by far harder to solve.

The first polynomial-time solution to the problem was given by Karpinski et al. for straight-line programs (SLPs) [8]. SLPs are a grammar-based compression method which constructs a context-free grammar in the Chomsky normal form. They proposed an algorithm which runs in $O((m + n)^4 \log(m + n))$ time using $O((m + n)^3)$ space. Later on, Miyazaki et al. [13] gave an improved algorithm running in $O(m^2 n^2)$ time using $O(mn)$ space.

Since computing a minimal SLP that generates a given string is known to be NP-complete, it is of great significance to develop approximative algorithms for generating small grammars [19, 2]. One of those algorithms is the *multilevel pattern matching code (MPM code)* introduced by Kieffer et al. [11]. MPM code is attractive in that it performs in linear time with respect to the input string size, and is capable of exponential compression - the generated grammar size can be exponentially small with respect to the input string size. It is also noteworthy that MPM grammars have a hierarchical structure, which suggests that MPM code has a potential for recognizing lexical and grammatical structures in strings similarly to SEQUITUR [14, 16].

In this paper, we consider the pattern matching problem on MPM grammars. Although the algorithm by Miyazaki et al. [13] for general SLPs requires $O(m^2 n^2)$ time and $O(mn)$ space, our algorithm specialized for MPM grammars performs in $O(mn^2)$ time within $O(mn)$ space.

2   Preliminaries

Let $\mathcal{N}$ be the set of natural numbers, and $\mathcal{N}^+$ be positive integers. Let Σ be a finite *alphabet*. An element of Σ^* is called a *string*. The length of a string T is

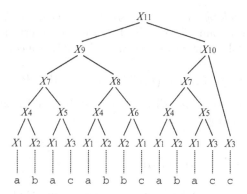

Fig. 1. Derivation tree of the MPM for string `abacabbcabacc`

denoted by $|T|$. The i-th character of a string T is denoted by $T[i]$ for $1 \le i \le |T|$, and the substring of a string T that begins at position i and ends at position j is denoted by $T[i:j]$ for $1 \le i \le j \le |T|$.

A *period* of a string T is an integer p ($1 \le p \le |T|$) such that $T[i] = T[i+p]$ for any $i = 1, 2, \ldots, |T| - p$.

Let X be any variable of a context-free grammar. We define the *length* of X to be the length of the string X produces, and denote it by $|X|$.

A *multilevel pattern matching grammar* (*MPM grammar*) T is a sequence of assignments such that

$$X_1 = expr_1, X_2 = expr_2, \ldots, X_n = expr_n,$$

where X_i are variables and $expr_i$ are expressions of the form either:

- $expr_i = a$ ($a \in \Sigma$), or
- $expr_i = X_\ell X_r$ ($\ell, r < i$) where $|X_\ell| \ge |X_r|$ and $|X_\ell|$ is a power of 2,

and $T = X_n$. MPM grammar T is a context-free grammar in the Chomsky normal form such that its language $L(T)$ is $\{T\}$. The size of T is n and is denoted by $\|T\|$. For example, MPM grammar T for $T = $ `abacabbcabacc` is:

$$X_1 = \mathsf{a}, X_2 = \mathsf{b}, X_3 = \mathsf{c}, X_4 = X_1 X_2, X_5 = X_1 X_3, X_6 = X_2 X_3, X_7 = X_4 X_5,$$
$$X_8 = X_4 X_6, X_9 = X_7 X_8, X_{10} = X_7 X_3, X_{11} = X_9 X_{10},$$

and $T = X_{11}$. Note $\|T\| = 11$. Fig. 1 illustrates the derivation tree of T.

The *height* of variable X, denoted by $height(X)$, is defined as follows:

$$height(X) = \begin{cases} 1 & \text{if } X = a \ (a \in \Sigma), \\ \max(height(X_\ell), height(X_r)) + 1 & \text{if } X = X_\ell X_r. \end{cases}$$

That is, $height(X)$ is the length of the longest path from X to a leaf. In the running example, $height(X_{10}) = 4$, $height(X_{11}) = height(T) = 5$, and so on (see Fig. 1). It is easy to see $height(T) \le n$.

The *pattern matching problem for strings in terms of MPM grammars* is, given two MPM grammars $\mathcal{T}$ and $\mathcal{P}$ that are the descriptions of text T and pattern P, to find all occurrences of P in T. Namely, we compute the following set:

$$Occ(T, P) = \{i \mid T[i : i + |P| - 1] = P\}.$$

In the sequel, we use X and X_i for variables of $\mathcal{T}$, and Y and Y_j for variables of $\mathcal{P}$. Let $\|\mathcal{T}\| = n$ and $\|\mathcal{P}\| = m$.

3   Overview of Algorithm

In this section, we show an overview of our algorithm that outputs a compact representation of $Occ(T, P)$ for given MPM grammars $\mathcal{T}$ and $\mathcal{P}$.

For strings $X, Y \in \Sigma^*$ and integer $k \in \mathcal{N}$, we define the set of all occurrences of Y that cover or touch the position k in X by

$$Occ^\uparrow(X, Y, k) = \{i \in Occ(X, Y) \mid k - |Y| \leq i \leq k\}.$$

In the following, $[i, j]$ denotes the set $\{i, i+1, \ldots, j\}$ of consecutive integers.

Observation 1 ([7]). *For any strings $X, Y \in \Sigma^*$ and integer $k \in \mathcal{N}$,*

$$Occ^\uparrow(X, Y, k) = Occ(X, Y) \cap [k - |Y|, k].$$

Lemma 1 ([7]). *For any strings $X, Y \in \Sigma^*$ and integer $k \in \mathcal{N}$, $Occ^\uparrow(X, Y, k)$ forms a single arithmetic progression.*

For positive integers $a, d, t \in \mathcal{N}^+$, we define $\langle a, d, t \rangle = \{a + (i-1)d \mid i \in [1, t]\}$. Assume that for $t = 0$, $\langle a, d, t \rangle = \emptyset$. Note that t denotes the cardinality of the set $\langle a, d, t \rangle$. By Lemma 1, $Occ^\uparrow(X, Y, k)$ can be represented as the triple $\langle a, d, t \rangle$ with the minimum element a, the common difference d, and the length t of the progression. By '*computing $Occ^\uparrow(X, Y, k)$*', we mean to calculate the triple $\langle a, d, t \rangle$ such that $\langle a, d, t \rangle = Occ^\uparrow(X, Y, k)$.

For a set U of integers and an integer k, we denote $U \oplus k = \{i + k \mid i \in U\}$ and $U \ominus k = \{i - k \mid i \in U\}$. For MPM variables $X = X_\ell X_r$ and Y, we denote $Occ^\triangle(X, Y) = Occ^\uparrow(X, Y, |X_\ell| + 1)$.

Lemma 2 ([13]). *For any MPM variables $X = X_\ell X_r$ and Y,*

$$Occ(X, Y) = Occ(X_\ell, Y) \cup Occ^\triangle(X, Y) \cup (Occ(X_r, Y) \oplus |X_\ell|).$$

(See Fig. 2.)

Lemma 2 implies that $Occ(X_n, Y)$ can be represented by a combination of

$$\{Occ^\triangle(X_i, Y)\}_{i=1}^n = Occ^\triangle(X_1, Y), Occ^\triangle(X_2, Y), \ldots, Occ^\triangle(X_n, Y).$$

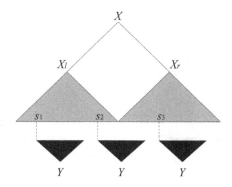

Fig. 2. $s_1, s_2, s_3 \in Occ(X, Y)$, where $s_1 \in Occ(X_\ell, Y)$, $s_2 \in Occ^\triangle(X, Y)$ and $s_3 \in Occ(X_r, Y)$

Thus, the desired output $Occ(T, P) = Occ(X_n, Y_m)$ can be expressed as a combination of $\{Occ^\triangle(X_i, Y_m)\}_{i=1}^n$ that requires $O(n)$ space. Hereby, computing $Occ(T, P)$ is reduced to computing $Occ^\triangle(X_i, Y_m)$ for every $i = 1, 2, \ldots, n$. In computing each $Occ^\triangle(X_i, Y_j)$ recursively, the same set $Occ^\triangle(X_{i'}, Y_{j'})$ might repeatedly be referred to, for $i' < i$ and $j' < j$. Therefore we take the dynamic programming strategy. We use an $m \times n$ table App where each entry $App[i, j]$ at row i and column j stores the triple for $Occ^\triangle(X_i, Y_i)$. We compute each $App[i, j]$ in a bottom-up manner, for $i = 1, \ldots, n$ and $j = 1, \ldots, m$. In Section 4, we will show each $App[i, j]$ is computable in $O(height(X_i))$ time. Since $height(X_i) \leq n$, we can construct the whole table App in $O(mn^2)$ time. The size of the whole table is $O(mn)$, since each triple occupies $O(1)$ space. We therefore have the main result of the paper, as follows:

Theorem 1. *Given two MPM grammars T and P, $Occ(T, P)$ can be computed in $O(mn^2)$ time with $O(mn)$ space.*

4 Details of Algorithm

In this section, we show that $Occ^\triangle(X_i, Y_j)$ is computable in $O(height(X_i))$ time for each variable X_i in T and Y_j in P.

The following two lemmas and one observation are necessary to prove Lemma 5 which is one of the key lemmas for our algorithm.

Lemma 3 ([7]). *For strings $X, Y \in \Sigma^*$ and integer $k \in \mathcal{N}$, let $\langle a, d, t \rangle = Occ^\uparrow(X, Y, k)$. If $t \geq 1$, then d is the shortest period of $X[s : b + |Y| - 1]$ for any $s \in \langle a, d, t - 1 \rangle$ and $b = a + (t - 1)d$.*

Proof. First we see that d is a period of $X[a : b + |Y| - 1]$ as follows. Since $\langle a, d, t \rangle = Occ^\uparrow(X, Y, k)$, we know

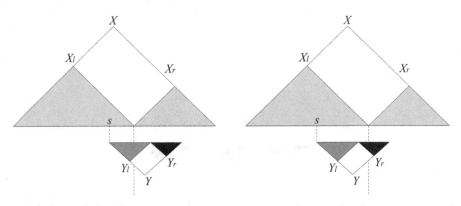

Fig. 3. $s \in Occ^\triangle(X, Y)$ if and only if either $s \in Occ^\triangle(X, Y_\ell)$ and $s + |Y_\ell| \in Occ(X, Y_r)$ (left case), or $s \in Occ(X, Y_\ell)$ and $s + |Y_\ell| \in Occ^\triangle(X, Y_r)$ (right case)

$$Y = X[a : a + |Y| - 1],$$
$$Y = X[a + d : a + d + |Y| - 1],$$
$$\vdots$$
$$Y = X[b : b + |Y| - 1].$$

By these equations, we have

$$X[i] = X[i + d] \text{ for all } i \in [a, b + |Y| - 1 - d],$$

which shows that d is a period of $X[s : b + |Y| - 1]$ for any $s \in \langle a, d, t - 1 \rangle$.

We now suppose that $X[s : b + |Y| - 1]$ has a smaller period $d' < d$ for the contrary. That is, $X[i] = X[i + d']$ for all $i \in [s, b + |Y| - 1 - d']$. Then we have $Y[i] = X[s + i - 1] = X[s + d' + i - 1]$ for all $i \in [1, |Y|]$. Since $b - s \geq b - (a + (t - 2) \cdot d) = b - (b - d) = d > d'$, we have $s + d' \in Occ^\uparrow(X, Y, k)$. However, this contradicts with $\langle a, d, t \rangle = Occ^\uparrow(X, Y, k)$, since $s + d' \notin \langle a, d, t \rangle$. Thus d is the shortest period of $X[s : b + |Y| - 1]$ for any $s \in \langle a, d, t - 1 \rangle$.   □

Observation 2 ([13]). *For any MPM variables X, $Y = Y_\ell Y_r$, and integer $k \in \mathcal{N}$,*

$$Occ^\triangle(X, Y) = \left(Occ^\triangle(X, Y_\ell) \cap (Occ(X, Y_r) \ominus |Y_\ell|) \right)$$
$$\cup \left(Occ(X, Y_\ell) \cap (Occ^\triangle(X, Y_r) \ominus |Y_\ell|) \right).$$

(See Fig. 3.)

Lemma 4 ([7]). *For any strings $X, Y_1, Y_2 \in \Sigma^*$ and integers $k_1, k_2 \in \mathcal{N}$, $Occ^\uparrow(X, Y_1, k_1) \cap (Occ^\uparrow(X, Y_2, k_2) \ominus |Y_1|)$ can be computed in $O(1)$ time, provided that $Occ^\uparrow(X, Y_1, k_1)$ and $Occ^\uparrow(X, Y_2, k_2)$ are already computed.*

For strings $X, Y \in \Sigma^*$ we consider the two following queries:

Single-Match Query: Given integer $s \in \mathcal{N}$, return if $s \in Occ(X,Y)$ or not.

Covering-Match Query: Given integer $k \in \mathcal{N}$, return triple $\langle a, d, t \rangle$ which represents $Occ^{\uparrow}(X, Y, k)$.

Lemma 5. *For any MPM variables X and $Y = Y_\ell Y_r$ and integer $k \in \mathcal{N}$, computing $Occ^{\triangle}(X,Y)$ is reducible in constant time to the following queries:*

(1) covering-match query $Occ^{\uparrow}(X, Y_\ell, |X_\ell| + 1) = Occ^{\triangle}(X, Y_\ell)$,

(2) covering-match query $Occ^{\uparrow}(X, Y_r, |X_\ell| + 1) = Occ^{\triangle}(X, Y_r)$,

(3) at most two covering-match queries $Occ^{\uparrow}(X, Y', k_1)$ and $Occ^{\uparrow}(X, Y', k_2)$ for some integers k_1, k_2, where Y' is either Y_ℓ or Y_r, and

(4) at most two single-match queries $s_1, s_2 \in Occ(X, Y')$ for some integers s_1, s_2, where Y' is either Y_ℓ or Y_r.

Proof. We perform two covering-match queries $Occ^{\triangle}(X, Y_\ell)$ and $Occ^{\triangle}(X, Y_r)$, and let $\langle a_1, d_1, t_1 \rangle$ and $\langle a_2, d_2, t_2 \rangle$ be answers of them, respectively. Depending on the cardinalities of triples, we have the four following cases:

(a) when $t_1 \leq 1$ and $t_2 \leq 1$.

At most two single-match queries are necessary for the following reasons. If $t_1 = 0$, we know $Occ^{\triangle}(X, Y_\ell) = \emptyset$. If $t_1 = 1$, we perform a single-match query $a_1 + |Y_\ell| \in Occ(X, Y_r)$, and we have

$$Occ^{\triangle}(X, Y_\ell) \cap (Occ(X, Y_r) \ominus |Y_\ell|) = \{a_1\} \cap (Occ(X, Y_r) \ominus |Y_\ell|)$$
$$= \begin{cases} \{a_1\} & \text{if } a_1 + |Y_\ell| \in Occ(X, Y_r), \\ \emptyset & \text{otherwise.} \end{cases}$$

Similarly, if $t_2 = 0$ we know $Occ^{\triangle}(X, Y_r) = \emptyset$. If $t_2 = 1$, we have

$$Occ(X, Y_\ell) \cap (Occ^{\triangle}(X, Y_r) \ominus |Y_\ell|) = Occ(X, Y_\ell) \cap (\{a_2\} \ominus |Y_\ell|)$$
$$= \begin{cases} \{a_2 - |Y_\ell|\} & \text{if } a_2 - |Y_\ell| \in Occ(X, Y_\ell), \\ \emptyset & \text{otherwise.} \end{cases}$$

By Observation 2, $Occ^{\triangle}(X, Y)$ is a union of these two sets. Trivially, the union operation can be done in constant time since each of these two sets is either singleton or empty.

(b) when $t_1 \geq 2$ and $t_2 \leq 1$.

First we compute $A = Occ^{\triangle}(X, Y_\ell) \cap (Occ(X, Y_r) \ominus |Y_\ell|) = \langle a_1, d_1, t_1 \rangle \cap (Occ(X, Y_r) \ominus |Y_\ell|)$, by using one covering-match query and at most one single-match query. Let $b_1 = a_1 + (t_1 - 1)d_1$. We consider two sub-cases depending on the length of Y_r with respect to $b_1 - a_1 = (t_1 - 1)d_1 \geq d_1$, as follows.

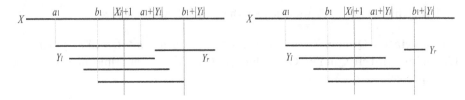

Fig. 4. Long case (left) and short case (right)

- the case $|Y_r| \geq b_1 - a_1$ (see the left of Fig. 4). By this assumption, we have $b_1 - |Y_r| \leq a_1$, which implies $[a_1, b_1] \subseteq [b_1 - |Y_r|, b_1]$. Thus

$$
\begin{aligned}
A &= \langle a_1, d_1, t_1 \rangle \cap (Occ(X, Y_r) \ominus |Y_\ell|) \\
&= (\langle a_1, d_1, t_1 \rangle \cap [a_1, b_1]) \cap (Occ(X, Y_r) \ominus |Y_\ell|) \\
&= (\langle a_1, d_1, t_1 \rangle \cap [b_1 - |Y_r|, b_1]) \cap (Occ(X, Y_r) \ominus |Y_\ell|) \\
&= \langle a_1, d_1, t_1 \rangle \cap ([b_1 - |Y_r|, b_1] \cap (Occ(X, Y_r) \ominus |Y_\ell|)) \\
&= \langle a_1, d_1, t_1 \rangle \cap (([b_1 - |Y_r| + |Y_\ell|, b_1 + |Y_\ell|] \cap Occ(X, Y_r)) \ominus |Y_\ell|) \\
&= \langle a_1, d_1, t_1 \rangle \cap (Occ^\uparrow(X, Y_r, b_1 + |Y_\ell|) \ominus |Y_\ell|),
\end{aligned}
$$

where the last equality is due to Observation 1. Here, we perform covering-match query $Occ^\uparrow(X, Y_r, b_1 + |Y_\ell|)$. According to Lemma 4, $\langle a_1, d_1, t_1 \rangle \cap (Occ^\uparrow(X, Y_r, b_1 + |Y_\ell|) \ominus |Y_\ell|)$ can be computed in constant time.
- the case $|Y_r| < b_1 - a_1$ (see the right of Fig. 4). The basic idea is the same as in the previous case, but covering-match query $Occ^\uparrow(X, Y_r, b_1 + |Y_\ell|)$ is not enough, since $|Y_r|$ is 'too short'. However, additional single-match query $a_1 + |Y_\ell| \in Occ(X, Y_r)$ fills up the gap, as follows.

$$
\begin{aligned}
A &= \langle a_1, d_1, t_1 \rangle \cap (Occ(X, Y_r) \ominus |Y_\ell|) \\
&= (\langle a_1, d_1, t_1 \rangle \cap [a_1, b_1]) \cap (Occ(X, Y_r) \ominus |Y_\ell|) \\
&= (\langle a_1, d_1, t_1 \rangle \cap ([a_1, b_1 - |Y_r| - 1] \cup [b_1 - |Y_r|, b_1])) \cap (Occ(X, Y_r) \ominus |Y_\ell|) \\
&= \langle a_1, d_1, t_1 \rangle \cap (S \cup Occ^\uparrow(X, Y_r, b_1 + |Y_\ell|)) \ominus |Y_\ell|), \\
&\qquad \text{where } S = [a_1 + |Y_\ell|, b_1 + |Y_\ell| - |Y_r| - 1] \cap Occ(X, Y_r).
\end{aligned}
$$

By Lemma 3, d_1 is the shortest period of $X[a_1 : b_1 + |Y| - 1]$. Therefore, we have $X[a_1 + |Y_\ell| : b_1 + |Y_\ell| - 1] = u^{t_1}$ where u is the suffix of Y_ℓ of length d_1. Thus, if $a_1 + |Y_\ell| \in Occ(X, Y_r)$, $S = \langle a_1 + |Y_\ell|, d_1, t' \rangle$, where t' is the maximum integer satisfying $a_1 + |Y_\ell| + (t' - 1)d_1 \leq b_1 + |Y_\ell| - |Y_r| - 1$. Since $Occ^\uparrow(X, Y_r, b_1 + |Y_\ell|)$ forms a single arithmetic progression by Lemma 1, the union operation can be done in constant time. Otherwise (if $a_1 + |Y_\ell| \notin Occ(X, Y_r)$), we have $S = \emptyset$ for the same reason, and thus the union operation can be done in constant time.

We now consider set $B = Occ(X, Y_\ell) \cap (Occ^\triangle(X, Y_r) \ominus |Y_\ell|)$. Since $t_2 \leq 1$, $Occ^\triangle(X, Y_r)$ is either singleton or empty. If it is empty, $B = \emptyset$. If it is singleton $\{a_2\}$, we just perform single-match query $a_2 - |Y_\ell| \in Occ(X, Y_\ell)$. If the answer is 'yes', then $B = \{a_2 - |Y_\ell|\}$, and otherwise $B = \emptyset$.

The union operation for $Occ^\triangle(X,Y) = A \cup B$ can be done in constant time since B is at most singleton.

In total, a covering-match query and at most two single-match queries are enough to compute $Occ^\triangle(X,Y)$ in this case.

(c) when $t_1 \leq 1$ and $t_2 \geq 2$.

Symmetric to Case (b).

(d) when $t_1 \geq 2$ and $t_2 \geq 2$.

We can compute $A = Occ^\triangle(X,Y_\ell) \cap (Occ(X,Y_r) \ominus |Y_\ell|)$ in the same way as Case (b), since the proof for Case (b) does not depend on the cardinality of $Occ(X,Y_r)$. Also, computing $B = Occ(X,Y_\ell) \cap (Occ^\triangle(X,Y_r) \ominus |Y_\ell|)$ is symmetric to computing A. Recall that each of A and B is an intersection of two sets both form a single arithmetic progression. This implies that A and B also form a single arithmetic progression (it can be proven in a similar manner to Lemma 4). Hence the union operation for $Occ^\triangle(X,Y) = A \cup B$ can be done in constant time. Thus, two covering-match queries and at most two single-match queries are enough in this case.

$\square$

The time complexity of a single-match query is the following:

Lemma 6 ([13]). *For any MPM variables X,Y and integer $s \in \mathcal{N}$, single-match query $s \in Occ(X,Y)$ can be done in $O(height(X))$ time.*

Now the only remaining thing is how to efficiently perform covering-match query $Occ^\uparrow(X,Y,k)$. We will show it in Lemma 7.

For any MPM variable $X = X_\ell X_r$, we recursively define the *leftmost descendant* $lmd(X,h)$ and the *rightmost descendant* $rmd(X,h)$ of X with respect to height h ($\leq height(X)$), as follows:

$$lmd(X,h) = \begin{cases} lmd(X_\ell,h) & \text{if } height(X) > h, \\ X & \text{if } height(X) = h, \end{cases}$$

$$rmd(X,h) = \begin{cases} rmd(X_r,h) & \text{if } height(X) > h, \\ X & \text{if } height(X) = h. \end{cases}$$

In the example of Fig. 1, $lmd(X_{10},3) = X_7$, $rmd(X_9,2) = X_6$, $rmd(X_7,1) = X_3$, and so on. For variable X_i ($1 \leq i \leq n$) and height h ($< height(Y)$), we precompute two tables storing $lmd(X_i,h)$ and $rmd(X_i,h)$ respectively. By using these tables, we can refer to any $lmd(X_i,h)$ and $rmd(X_i,h)$ in constant time. These tables can be constructed in $O(mn)$ time in a bottom-up manner.

Lemma 7. *For any MPM variables X,Y and integer $k \in \mathcal{N}$, covering-match query $Occ^\uparrow(X,Y,k)$ is reducible in $O(height(X))$ time to at most three covering-match queries $Occ^\triangle(L,Y)$, $Occ^\triangle(C,Y)$, and $Occ^\triangle(R,Y)$ where L,C,R are a descendant of X or X itself.*

Proof. Let $X = X_\ell X_r$ and $Y = Y_\ell Y_r$. If $k = |X_\ell| + 1$, then only one covering-match query $Occ^\triangle(X,Y)$ is enough. Now we assume $k \neq |X_\ell| + 1$.

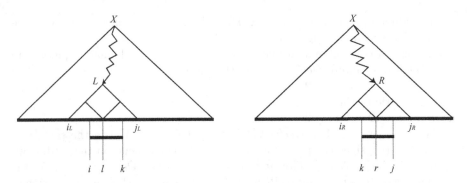

Fig. 5. Given integer k, the left (right, resp.) illustrates how to find L (R, resp.)

Let $i = \max(k-|Y|, 1)$ and $j = \min(k+|Y|-1, |X|)$. We consider the possibly shortest descendant L of X which covers the range $[i, k]$. (see the left of Fig. 5.) Let i_L, j_L be the integers such that $X[i_L : j_L] = L$. Let $l = i_L + |L_\ell|$. Similarly, we consider the possibly shortest descendant R of X which covers the range $[k, j]$. (see the right of Fig. 5.) Let i_R, j_R be the integers such that $X[i_R : j_R] = R$. Let $r = i_R + |R_\ell|$.

Assume $l = r$, that is, $L = R$. In this case only one covering-match query $Occ^\triangle(L, Y)$ is enough, since $k = l = i_L + |L_\ell|$ and thus

$$Occ^\uparrow(X, Y, k) = Occ^\uparrow(L, Y, |L_\ell| + 1) \oplus (i_L - 1)$$
$$= Occ^\triangle(L, Y) \oplus (i_L - 1).$$

In case $l < r$, we have the following sub-cases.

(1) when L is a descendant of R.
Depending on the shapes of $R = R_\ell R_r$ and $Y = Y_\ell Y_r$, we have the four following sub-cases:
  (a) when $|R_\ell| = |R_r|$ and $|Y_\ell| = |Y_r|$. In this case, $L = rmd(R_\ell, height(Y) + 1)$. Then,

$$Occ^\uparrow(X, Y, k) = ((Occ^\triangle(L, Y) \cap [k-|Y|-i_L+1 : k-i_L+1]) \oplus (i_L-1))$$
$$\cup ((Occ^\triangle(R, Y) \cap [k-|Y|-i_R+1 : k-i_R+1]) \oplus (i_R-1)).$$

  Since $Occ^\triangle(L, Y)$ and $Occ^\triangle(R, Y)$ form a single arithmetic progression by Lemma 1, the intersection and union operations take $O(1)$ time.
  (b) when $|R_\ell| > |R_r|$ and $|Y_\ell| = |Y_r|$. Since $|R_\ell|$ and $|Y|$ are a power of 2, we have $L = rmd(R_\ell, height(Y) + 1)$. Thus we have the same equation as in Case (1)-(a).
  (c) when $|R_\ell| = |R_r|$ and $|Y_\ell| > |Y_r|$. We have the two following sub-cases:
    (i) when $r-k+|Y| \leq 2 \times |Y_\ell|$. In this case, $L = rmd(R_\ell, height(Y_\ell)+1)$. Thus we have the same equation as in Case (1)-(a).

(ii) when $r - k + |Y| > 2 \times |Y_\ell|$. In this case, $L = rmd(R_\ell, height(Y_\ell) + 2)$. Let $C = L_r$. Then,

$$
\begin{aligned}
&Occ^\uparrow(X, Y, k) \\
&= ((Occ^\triangle(L, Y) \cap [k - |Y| - i_L + 1 : k - i_L + 1]) \oplus (i_L - 1)) \\
&\quad \cup ((Occ^\triangle(C, Y) \cap [k - |Y| - p + 1 : k - p + 1]) \oplus (p - 1)) \\
&\quad \cup ((Occ^\triangle(R, Y) \cap [k - |Y| - i_R + 1 : k - i_R + 1]) \oplus (i_R - 1)),
\end{aligned}
$$

where $p = i_L + |L_\ell|$. By Lemma 1, the intersection and union operations can be done in $O(1)$ time.

(d) when $|R_\ell| > |R_r|$ and $|Y_\ell| > |Y_r|$. Since $|R_\ell|$ is a power of 2, we can use the same equations as in Case (1)-(c).

(2) when L is an ancestor of R.

Depending on the shapes of $L = L_\ell L_r$ and $Y = Y_\ell Y_r$, we have the four following sub-cases:

(a) when $|L_\ell| = |L_r|$ and $|Y_\ell| = |Y_r|$. This is symmetric to Case (1)-(a).
(b) when $|L_\ell| > |L_r|$ and $|Y_\ell| = |Y_r|$. Let $L_r = L_{\ell(r)} L_{r(r)}$. Since $|L_{\ell(r)}|$ is a power of 2, we can use the same strategy as in Case (2)-(a).
(c) when $|L_\ell| = |L_r|$ and $|Y_\ell| > |Y_r|$. This is a symmetric to Case (1)-(c).
(d) when $|L_\ell| > |X_r|$ and $|Y_\ell| > |Y_r|$. Let $L_r = L_{\ell(r)} L_{r(r)}$. Since $|L_{\ell(r)}|$ is a power of 2, we can use the same strategy as in Case (2)-(c).

Since each of R, L is a descendant of X or X itself, we can find them in $O(height(X))$ time by a top-down traversal on X. Moreover, C can be found in constant time from L or R.

$\square$

By Lemmas 5, 6 and 7, we conclude that each entry $App[i, j]$ representing $Occ^\triangle(X_i, Y_j)$ can be computed in $O(height(X_i))$ time. Since $height(X_i) \leq n$, given two MPM grammars T and P, we can compute $Occ(T, P)$ in $O(mn^2)$ time.

5   Conclusions and Further Discussions

This paper considered the pattern matching problem on a subclass of context-free grammars called *multilevel pattern matching grammars (MPM grammars)*. MPM code was developed by Kieffer et al. [11] for efficient grammar-based text compression. Since MPM grammar sizes can be exponentially small with respect to the original string sizes, it is a rather hard task to solve the pattern matching problem in time proportional only to the grammar sizes. In this paper, we developed an efficient pattern matching algorithm which, given two MPM grammars P and T, runs in $O(mn^2)$ time with $O(mn)$ space, where $m = \|P\|$ and $n = \|T\|$. Our algorithm outperforms the previous best algorithm of [13] running in $O(m^2n^2)$ time using $O(mn)$ space. An interesting open problem is whether an $O(mn)$-time solution is achievable or not.

As a final remark we mention that MPM grammars can be seen as text compression by *ordered binary decision diagrams* (*OBDDs*) [1]. OBDDs were originally developed to represent a Boolean function as a directed acyclic graph. OBDDs are also used for *symbolic* or *implicit* graph algorithms [20]. MPM code turns out to reveal yet another application of OBDDs to text compression.

References

1. R. E. Bryant. Symbolic boolean manipulation with ordered binary decision diagrams. *ACM Computing Surveys*, 24:293–318, 1992.
2. M. Charikar, E. Lehman, D. Liu, R. Panigrahy, M. Prabhakaran, A. Rasala, A. Sahai, and A. Shelat. Approximating the smallest grammar: Kolmogorov complexity in natural models. In *Proc. STOC'02*, pages 792–801, 2002.
3. M. Crochemore and W. Rytter. *Text Algorithms*. Oxford University Press, New York, 1994.
4. M. Crochemore and W. Rytter. *Jewels of Stringology*. World Scientific, 2002.
5. P. Gage. A new algorithm for data compression. *The C Users Journal*, 12(2), 1994.
6. D. Gusfield. *Algorithms on Strings, Trees, and Sequences*. Cambridge University Press, New York, 1997.
7. S. Inenaga, A. Shinohara, and M. Takeda. A fully compressed pattern matching algorithm for simple collage systems. In *Proc. PSC'04*, pages 98–113. Czech Technical University, 2004.
8. M. Karpinski, W. Rytter, and A. Shinohara. An efficient pattern-matching algorithm for strings with short descriptions. *Nordic J. Comput.*, 4(2):172–186, 1997.
9. J. Kieffer and E. Yang. Grammar-based codes: a new class of universal lossless source codes. *IEEE Transactions on Information Theory*, 46(3):737–754, 2000.
10. J. Kieffer and E. Yang. Grammar-based codes for universal lossless data compression. *Communications in Information and Systems*, 2(2):29–52, 2002.
11. J. Kieffer, E. Yang, G. Nelson, and P. Cosman. Universal lossless compression via multilevel pattern matching. *IEEE Transactions on Information Theory*, 46(4):1227–1245, 2000.
12. J. Larsson and A. Moffat. Offline dictionary-based compression. In *Proc. DCC'99*, pages 296–305. IEEE Computer Society, 1999.
13. M. Miyazaki, A. Shinohara, and M. Takeda. An improved pattern matching algorithm for strings in terms of straight line programs. *Journal of Discrete Algorithms*, 1(1):187–204, 2000.
14. C. Nevill-Manning and I. Witten. Compression and explanation using hierarchical grammars. *Computer Journal*, 40(2/3):103–116, 1997.
15. C. Nevill-Manning and I. Witten. Identifying hierarchical structure in sequences: a linear-time algorithm. *J. Artificial Intelligence Research*, 7:67–82, 1997.
16. C. Nevill-Manning and I. Witten. Inferring lexical and grammatical structure from sequences. In *Proc. DCC'97*, pages 265–274. IEEE Computer Society, 1997.
17. C. Nevill-Manning and I. Witten. Phrase hierarchy inference and compression in bounded space. In *Proc. DCC'98*, pages 179–188. IEEE Computer Society, 1998.
18. W. Rytter. Algorithms on compressed strings and arrays. In *Proc. SOFSEM'99*, volume 1725 of *LNCS*, pages 48–65. Springer-Verlag, 1999.
19. W. Rytter. Application of Lempel-Ziv factorization to the approximation of grammar-based compression. *Theoretical Comput. Sci.*, 302(1–3):211–222, 2003.
20. P. Woelfel. Symbolic topological sorting with OBDDs. In *Proc. MFCS'03*, volume 2747 of *LNCS*, pages 671–680. Springer-Verlag, 2003.

On the Complexity of 2-Monotone Restarting Automata[*]

T. Jurdziński[1,**], F. Otto[1], F. Mráz[2], and M. Plátek[2]

[1] Fachbereich Mathematik/Informatik, Universität Kassel, Kassel, Germany
{tju, otto}@theory.informatik.uni-kassel.de
[2] Department of Computer Science, Charles University, Prague, Czech Republic
mraz@ksvi.ms.mff.cuni.cz, platek@ksi.ms.mff.cuni.cz

Abstract. The R-automaton is the weakest form of the *restarting automaton*. It is shown that the class of languages accepted by these automata is incomparable under set inclusion to the class of *growing context-sensitive languages*. In fact, this already holds for the class of languages that are accepted by *2-monotone* R-automata. Further it is shown that already this class contains NP-*complete languages*. Thus, already the 2-monotone R-automaton has a surprisingly large expressive power.

1   Introduction

The restarting automaton was introduced by Jancar et. al. as a formal tool to model the analysis by reduction, which is a technique used in linguistics to analyse sentences of natural languages [4]. This *analysis by reduction* consists of a stepwise simplification of a given sentence so that the (in-) correctness of the sentence is not affected.

A restarting automaton, RRWW-automaton for short, is a device M that consists of a finite state control, a tape containing a word delimited by sentinels, and a read/write window of a fixed size. This window moves from left to right along the tape until the control decides (nondeterministically) that the content of the window should be rewritten by some shorter string. After a rewrite, M can continue to move its window to the right until it either halts and accepts, or halts and rejects, or restarts, that is, it moves its window to the leftmost position, re-enters the initial state, and continues with the computation. Thus, each computation of M can be described through a sequence of cycles.

In addition to the general model outlined above, also various restricted versions of the restarting automaton have been considered. First of all there is the

---

[*] The first two authors were partially supported by a grant from the Deutsche Forschungsgemeinschaft. The third and the fourth authors were supported by the Grant Agency of the Czech Republic, Grant No. 201/02/1456, and Grant No. 201/04/2102.
[**] On leave from the Institute of Computer Science, University of Wrocław, Poland.

C.S. Calude, E. Calude and M.J. Dinneen (Eds.): DLT 2004, LNCS 3340, pp. 237–248, 2004.

RWW-automaton, which is required to make a restart immediately after each rewrite operation. Thus, in each cycle of a computation an RWW-automaton only sees the part of the tape between the left sentinel and the position where the rewrite operation is performed. Then there is the RRW-automaton, which only uses the letters of the input alphabet in its rewrite operations. A further restriction leads to the RR-automaton, where each rewrite operation simply deletes some letters from the content of the read/write window. Obviously the restriction on the restart operation and the restrictions on the rewrite operation can be combined leading to the RW-automaton and the R-automaton.

In addition, a *monotonicity* property was introduced for the various types of restarting automata which is based on the idea that from one cycle to the next in a computation, the actual place where a rewrite operation is performed must not increase its distance from the *right* end of the tape. The monotone restarting automata essentially model bottom-up one-pass parsers. The monotone RWW- and RRWW-automaton characterize the class CFL of context-free languages, and the monotone and deterministic versions of all the above mentioned types of restarting automata characterize the class DCFL of deterministic context-free languages [5]. On the other hand it is immediate from the definition that the class $\mathcal{L}(\mathsf{RRWW})$ of languages that are accepted by RRWW-automata is contained in the class CSL of context-sensitive languages, and that it is contained in the complexity class NP.

The monotone grammars generate the *context-sensitive languages*, and the strictly monotone grammars generate the *growing context-sensitive languages* (GCSL). In [2] the class GCSL is characterized by a nondeterministic machine model, the *shrinking two-pushdown automaton*, sTPDA for short. The class GCSL strictly contains the context-free languages, but the membership problem for each growing context-sensitive language can still be solved in polynomial time [3]. Therefore GCSL can be seen as an interesting generalization of the context-free languages.

In [10] it is shown that the deterministic variant of the sTPDA characterizes the class CRL of *Church Rosser languages* [9], which also coincides with the family of languages that are accepted by the deterministic RWW- and RRWW-automata [11, 12]. On the other hand, the class GCSL is properly included in the class $\mathcal{L}(\mathsf{RWW})$ of languages that are accepted by the (nondeterministic) RWW-automata. In fact GCSL coincides with the class of languages that are accepted by the so-called *weakly monotone* (nondeterministic) RWW- and RRWW-automata [6].

The context-free languages do not have sufficient expressive power to capture all issues of the analysis by reduction of natural languages. Thus, neither do the monotone restarting automata. On the other hand, the general RRWW- and RWW-automata even accept NP-complete languages [6, 12], which means that they cannot be implemented efficiently.

Therefore, in [14, 15] the notion of *j-monotonicity* $(j \geq 1)$ was introduced as a generalization of the notion of monotonicity. It models the generalization from bottom-up one-pass parsers to bottom-up multi-pass parsers, and it allows to measure the level of non-monotonicity of a language. Further, this new notion seems to be much better suited to the task of modelling the analysis by reduction. A restarting automaton is called j-*monotone* for an integer $j \geq 1$ if, for each of its computations, the corresponding sequence of cycles can be partitioned into at most j subsequences that are each monotone. It is shown in [15] that the expressive power of the j-monotone RRW-automaton increases with the value of the parameter j.

Naturally this raises the question of the computational power of the various types of j-monotone restarting automata and their relationships to other language classes. From a practical point of view one would like to obtain a device with an expressive power that is as large as possible, but which can still be implemented efficiently, that is, at least in polynomial time.

Here we study the computational power of the j-monotone restarting automata and their relationship to the growing context-sensitive languages. We will proceed as follows. After giving the necessary definitions in Section 2, we present a general reduction from RWW-automata first to RW- and then to R-automata in Section 3. Using these reductions as a tool we can translate many results concerning the language class $\mathcal{L}(\mathsf{RWW})$ to the class $\mathcal{L}(\mathsf{R})$. In this way we will see that already R-automata accept some NP-complete languages. Further, we will show that the class $\mathcal{L}(2\text{-mon-R})$ of languages that are accepted by 2-monotone R-automata is incomparable under set inclusion to the language class GCSL (Section 4), thus improving on the previously known result that $\mathcal{L}(\mathsf{R})$ and CFL are incomparable.

Finally, we will present a generic construction that shows that already 2-monotone RWW-automata accept NP-complete languages (Section 5). By the reductions above this will then imply that already the 2-monotone R-automata accept some NP-complete languages. In the concluding section we will shortly discuss the consequences of this (unexpected) result.

Because of the page limit we cannot possibly present the constructions and proofs in detail. The interested reader can find them in the corresponding technical report [7].

2   Definitions

For any class A of automata, $\mathcal{L}(\mathsf{A})$ will denote the class of languages that can be accepted by the automata from A, and for a particular automaton M, $L(M)$ denotes the language that is accepted by M. Further, we will sometimes use regular expressions instead of the corresponding regular languages. Finally, ε denotes the empty word, and $\mathbb{N}_+$ denotes the set of all positive integers.

A *restarting automaton*, RRWW-automaton for short, is a nondeterministic one-tape machine M with a finite-state control Q and a read/write window of a fixed size $k \geq 1$. The work space is limited by the left sentinel $\math345$ and the right sentinel \$, which cannot be removed from the tape. In addition to the input alphabet Σ, the tape alphabet Γ of M may contain a finite number of auxiliary symbols. The behaviour of M is described by a transition relation δ

that associates to a pair (q, u) consisting of a state q and a possible content u of the read/write window a finite set of possible transition steps. There are four types of transition steps (see, e.g., [5] or [13] for details):

1. A *move-right step* (MVR) causes M to shift the read/write window one position to the right and to change the state. However, the read/write window cannot move across the right sentinel $.
2. A *rewrite step* causes M to replace the content u of the read/write window by a shorter string v, and to change the state. Further, the read/write window is placed immediately to the right of the string v.
3. A *restart step* causes M to move its read/write window to the left end of the tape, so that the first symbol it sees is the left sentinel $\cent$, and to re-enter the initial state q_0.
4. An *accept step* causes M to halt and accept.

If $\delta(q, u) = \emptyset$ for some pair (q, u), then M necessarily halts, and we say that M *rejects* in this situation.

A *configuration* of M is a string $\alpha q \beta$ where $q \in Q$, and either $\alpha = \varepsilon$ and $\beta \in \{\cent\} \cdot \Gamma^* \cdot \{\$\}$ or $\alpha \in \{\cent\} \cdot \Gamma^*$ and $\beta \in \Gamma^* \cdot \{\$\}$; here q represents the current state, $\alpha\beta$ is the current content of the tape, and it is understood that the head scans the first k symbols of β or all of β when $|\beta| \leq k$. A *restarting configuration* is of the form $q_0 \cent w \$$, where $w \in \Gamma^*$; if $w \in \Sigma^*$, then $q_0 \cent w \$$ is an *initial configuration*. Thus, initial configurations are special restarting configurations.

We observe that any computation of a restarting automaton M consists of certain phases. A phase, called a *cycle*, starts in a restarting configuration, the head moves along the tape performing MVR and Rewrite operations until a Restart operation is performed and thus a new restarting configuration is reached. If no further Restart operation is performed, then the computation necessarily halts after finitely many steps – such a phase is called a *tail*. We require that M performs *exactly one* Rewrite operation during each cycle – thus each new phase starts on a shorter word than the previous one. During a tail at most one Rewrite operation may be executed.

An input word $w \in \Sigma^*$ is *accepted by M*, if there is a computation which, starting with the initial configuration $q_0 \cent w \$$, finishes by executing an Accept instruction. Then $L(M)$ is the language consisting of all words accepted by M.

An RWW-*automaton* is an RRWW-automaton which restarts immediately after rewriting. An R(R)W-*automaton* is an R(R)WW-automaton whose working alphabet coincides with its input alphabet. Note that each restarting configuration is initial in this case. Finally, an R(R)-*automaton* is an R(R)W-automaton whose rewriting instructions can be viewed as deletions, that is, if $(q', v) \in \delta(q, u)$, then v is obtained by deleting some symbols from u.

Each cycle of a computation of an RWW-automaton consists of two phases. In the first phase the automaton scans the tape from left to right, behaving like a finite-state acceptor. In the second phase it performs a Rewrite transition followed by a Restart step. Accordingly the transition relation of an RWW-automaton M can be described through a finite sequence of so-called *meta-instructions* of the form $(R, u \rightarrow v)$, where R is a regular expression, called the *regular constraint* of this instruction, and u and v are strings such that $u \rightarrow v$ stands for a Rewrite step [13]. On trying to execute this meta-instruction M will get stuck (and so reject) starting from the restarting configuration $q_0 \mathcal{c} w\$$, if w does not admit a factorization of the form $w = w_1 u w_2$ such that $\mathcal{c} w_1 \in R$. On the other hand, if w does admit a factorization of this form, then one such factorization is chosen nondeterministically, and $q_0 \mathcal{c} w\$$ is transformed into the restarting configuration $q_0 \mathcal{c} w_1 v w_2\$$. In order to be able to also describe the tails of accepting configurations we use meta-instructions of the form (R, Accept), accepting the strings from the regular language defined by R in tail computations.

Finally we come to the notion of *monotonicity*. Each cycle C of a computation of a restarting automaton contains a unique configuration $\mathcal{c} \alpha q \beta\$$ in which a Rewrite instruction is applied. Then $|\beta\$|$ is the *right distance* of C, denoted by $D_r(C)$.

We say that a *sequence of cycles* $Sq = (C_1, C_2, \cdots, C_n)$ is *monotone* if $D_r(C_1) \geq D_r(C_2) \geq \cdots \geq D_r(C_n)$. A *computation* is *monotone* if the corresponding sequence of cycles is monotone, and an RRWW-automaton is called *monotone* if all its computations that start with initial configurations are monotone. Observe that the tails of the computations do not play any role here. We use the prefix mon- to denote the classes of monotone restarting automata.

Let j be a positive integer. A sequence of cycles $Sq = (C_1, C_2, \cdots, C_n)$ is called *j-monotone* if there is a partition of Sq into j (scattered) subsequences that are monotone [15]. A *computation* is *j-monotone* if the corresponding sequence of cycles is *j*-monotone. Finally, an RRWW-automaton is called *j-monotone* if all its computations that start with initial configurations are *j*-monotone. The prefix *j*-mon- is used to denote the corresponding classes of restarting automata.

3   A Reduction from RWW- to R-Automata

Here we describe a sequence of two complexity preserving reductions that first replace an RWW-automaton by an RW-automaton and then replace the latter by an R-automaton.

Theorem 1. [12] *A language L is accepted by a (deterministic) RWW-automaton if and only if there exist a (deterministic) RW-automaton M_1 and a regular language R such that $L = L(M_1) \cap R$ holds.*

The RW-automaton M_1 and the regular language R are obtained as follows. Let M be an RWW-automaton with input alphabet Σ and tape alphabet Γ for a language $L \subseteq \Sigma^*$. The corresponding RW-automaton M_1 is obtained from M by simply taking Γ as input alphabet, and the regular language R is chosen as $R := \Sigma^*$. The identity mapping on Σ^* is obviously a reduction from the language L to the language $L(M_1)$. Thus, we have the following consequence.

Corollary 1. *The language class $\mathcal{L}(\mathsf{RWW})$ is reducible in linear time and constant space to the language class $\mathcal{L}(\mathsf{RW})$.*

Continuing our discussion of M and M_1, assume that the RWW-automaton M is j-monotone for some integer $j \geq 1$. Then each computation of the RW-automaton M_1 that begins with an initial configuration of the form $q_0 \mathcal{c} w\$$ with $w \in \Sigma^*$ is j-monotone. However, this does not necessarily mean that M_1 is also j-monotone, as there could exist computations of M_1 that start with an initial configuration of the form $q_0 \mathcal{c} w\$$ with $w \in \Gamma^* \setminus \Sigma^*$ and that are *not* j-monotone. Hence, Corollary 1 does not immediately carry over to j-monotone RWW-automata. If we want to establish a corresponding reduction for a particular j-monotone RWW-automaton, then we must inspect the resulting RW-automaton in detail to verify j-monotonicity.

Next we turn to a reduction that replaces arbitrary **Rewrite** operations by delete operations. Let $\Gamma_1 = \{a_1, \ldots, a_m\}$ be a finite alphabet, k a positive integer, and $\Gamma_2 := \{0, 1, c, d\}$. We define the encoding $\varphi_{k,m} : \Gamma_1^* \to \Gamma_2^*$ as the morphism that is induced by the mapping

$$a_i \mapsto c1^{m+1-i}0^i(cd1^{m+1}0^{m+1})^k \quad (1 \leq i \leq m).$$

Then, for all $1 \leq i \leq m$, $|\varphi_{k,m}(a_i)| = (m+2) \cdot (2k+1)$. Observe that $\varphi_{k,m}$ is indeed an encoding. It has the following important property.

Lemma 1. *For all $u \in \Gamma_1^k$ and $v \in \Gamma_1^*$, if $|v| < k$, then $\varphi_{k,m}(v)$ is a scattered subword of $\varphi_{k,m}(u)$, that is, $\varphi_{k,m}(v)$ is obtained from $\varphi_{k,m}(u)$ by simply deleting some factors from the latter.*

Based on encodings of this form we obtain our second reduction.

Theorem 2. *If a language L is accepted by an RW-automaton M with tape alphabet Γ_1 of size m and read/write window of size k, then there exists an R-automaton M' which accepts the language $\varphi_{k,m}(L) \subseteq \Gamma_2^*$. In addition, if M is j-monotone for some $j \geq 1$, then so is M'.*

Proof. Let $L := L(M)$, and let $p := (m+2) \cdot (2k+1)$. We construct an R-automaton M' with the tape alphabet Γ_2 and the read/write window of size $k \cdot p$. Whenever the distance of the read/write window of M' from the left end of its tape is a multiple of p, then M' checks whether the read/write window contains a string that is the image $\varphi_{k,m}(u)$ of some $u \in \Gamma_1^*$. In the negative it rejects immediately, in the affirmative it performs an action that simulates the actual transition of M. It follows that M' accepts the language $\varphi_{k,m}(L)$.

In order to show that the above construction preserves j-monotonicity, it is enough to notice that, for each computation of M' on each word $\varphi_{k,m}(w)$ ($w \in \Gamma_1^*$), there exists a computation of M on w such that the sequence of right distances of M' is obtained by multiplying each element in the appropriate sequence of right distances of M by the constant p. For an input x such that $x \notin \varphi_{k,m}(\Gamma_1^*)$, M' works in a j-monotone fashion on the longest prefix of x that belongs to the set $\varphi_{k,m}(\Gamma_1^*)$. As soon as M' proceeds beyond this prefix, it realizes that its tape content is not of the correct form, and it rejects at this point. Thus, the overall computation of M' on x is j-monotone.   □

Thus, we have the following result.

Corollary 2.

(a) *The language class* $\mathcal{L}(\mathsf{RW})$ *is reducible in linear time and constant space to the language class* $\mathcal{L}(\mathsf{R})$.

(b) *For each* $j \geq 1$, *the language class* $\mathcal{L}(j\text{-}\mathrm{mon\text{-}RW})$ *is reducible in linear time and constant space to the language class* $\mathcal{L}(j\text{-}\mathrm{mon\text{-}R})$.

As the class $\mathcal{L}(\mathsf{RWW})$ contains NP-complete languages [6], Corollaries 1 and 2 yield the following result.

Corollary 3. *The class* $\mathcal{L}(\mathsf{R})$ *contains* NP-*complete languages.*

4 $\mathcal{L}(2\text{-}\mathbf{mon\text{-}R})$ Versus GCSL

The language $L_a := \{\, a^n b^n \mid n \geq 0 \,\} \cup \{\, a^n b^m \mid m > 2n \geq 0 \,\}$ is not accepted by any RRW-automaton [5]. However, the following result can be shown.

Lemma 2. $L_a \in \mathsf{CRL}$.

This example yields the following non-inclusion result.

Theorem 3. CRL *is not contained in the class* $\mathcal{L}(\mathsf{RRW})$.

As a consequence we see that neither CRL nor GCSL is contained in any of the classes $\mathcal{L}(\mathsf{R})$, $\mathcal{L}(\mathsf{RR})$, or $\mathcal{L}(\mathsf{RW})$, either.

Each growing context-sensitive language is accepted by a one-way auxiliary pushdown automaton that is simultaneously logarithmically space- and polynomially time-bounded [2], while the language $L_{\mathrm{copy}} := \{\, ww \mid w \in \{a,b\}^* \,\}$ is not accepted by any one-way auxiliary pushdown automaton within logarithmic space [8], and so $L_{\mathrm{copy}} \notin \mathsf{GCSL}$. As GCSL is closed under bounded morphisms [1], it follows that $L'_{\mathrm{copy}} := \{\, w\#w \mid w \in \{a,b\}^* \,\}$ is not in GCSL, either. However, the following can be shown by using a technique from [6].

Lemma 3. $L'_{\mathrm{copy}} \in \mathcal{L}(\mathsf{RWW})$.

Proof. The language L'_{copy} is accepted by an RWW-automaton M that works as follows, given an input of the form $u\#v$ with $u, v \in \{a,b\}^*$:

Phase 1: The prefix u and the factor v are compressed into words of the form u_1 and v_1, respectively. This compression is performed from left to right, encoding two symbols into one auxiliary symbol, leaving the $\#$ symbol unchanged. Here a problem arises from the fact that M restarts immediately after each Rewrite step. Hence, we need a special technique that directs M to try to alternate the compression steps on u and on v. Actually we can only force M to perform one or more compression steps on u between any two compression steps on v. Accordingly, this process succeeds if and only if u and v are both of even length or both of uneven length, and if v is a subsequence of u (see [6] for details).

Phase 2: Once the compression has succeeded, M checks whether u_1 and v_1 are of the same length. This is done by simply erasing the two symbols surrounding the symbol # in each cycle.   □

Hence, $\mathcal{L}(\mathsf{RWW})$ is not contained in GCSL. By applying the reductions from the previous section this observation yields the following stronger result.

Theorem 4. $\mathcal{L}(\mathsf{R})$ *is not contained in* GCSL.

Proof. As $L'_{\text{copy}} \in \mathcal{L}(\mathsf{RWW})$, there exists, by Theorem 1, an RW-automaton M_1 and a regular language R such that $L'_{\text{copy}} = L(M_1) \cap R$. As GCSL is closed under the operation of intersection with regular languages [1], it follows that $\hat{L} := L(M_1) \notin$ GCSL.

Then, by Theorem 2, the language $L' := \varphi_{k,m}(\hat{L})$ is in $\mathcal{L}(\mathsf{R})$, where k is the size of the read/write window of M_1 and m is the size of its tape alphabet. As the class GCSL is closed under inverse morphisms [1], it follows that $L' \notin$ GCSL, thus completing the proof.   □

Actually the following stronger version of Theorem 4 holds.

Theorem 5. $\mathcal{L}(\text{2-mon-R})$ *is not contained in* GCSL.

Proof. The RWW-automaton M for L'_{copy} described in the proof of Lemma 3 can be shown to actually be 2-monotone. Then we use the reductions described in Theorems 1 and 2. As the latter reduction preserves j-monotonicity in general, it remains to verify that the RW-automaton M_1 that is obtained from the RWW-automaton M is indeed 2-monotone.   □

As a consequence we obtain the following.

Corollary 4. $\mathcal{L}(\text{2-mon-R(R)})$, $\mathcal{L}(\text{2-mon-R(R)W})$, $\mathcal{L}(\mathsf{R(R)})$, *and* $\mathcal{L}(\mathsf{R(R)W})$ *are incomparable under set inclusion to the language classes* CRL *and* GCSL.

5 $\mathcal{L}$(2-mon-R) Contains NP-Complete Languages

As seen in Section 3 the class $\mathcal{L}(\mathsf{R})$ contains NP-complete languages. Here we will improve upon this result considerably by showing that already 2-monotone R-automata accept NP-complete languages. This will be done by presenting a generic proof that an encoded version of each language from NP is contained in $\mathcal{L}(\text{2-mon-RWW})$, and by then applying the reductions from Section 3.

5.1   An Encoding of Languages from NP

Let L be a language that belongs to the complexity class NP, and let M be a nondeterministic one-tape Turing machine which accepts L in polynomial time, that is, $L(M) = L$, and there exists a fixed polynomial $p(n)$ such that, for each input $x \in L$ of length n, M has an accepting computation of length at most $p(n)$.

Without loss of generality we may assume that the tape of M is infinite only to the right, and that the input is on the initial section of the tape. We will prove that an encoded version of the language L is accepted by a 2-monotone RWW-automaton M_2.

Let $L \subseteq \Sigma_0^*$, let $\&, \# \notin \Sigma_0$ be two additional symbols, and let $\Sigma := \Sigma_0 \cup \{\&, \#\}$. We define the language $L_1 \subset \Sigma^*$ as follows:

$$L_1 := \{\, w\&^{m-n}\#(\&^m\#)^p \mid w \in L, |w| = n, m \geq n, \text{ and there exists}$$
$$\text{a computation of } M \text{ on } w \text{ that reaches an accepting}$$
$$\text{state in at most } p \text{ steps, using less than } m \text{ tape cells}\}.$$

The function $f : \Sigma_0^* \rightarrow \Sigma^*$ that maps a word $w \in \Sigma_0^n$ onto the string $w\&^{p(n)-n+1}\#(\&^{p(n)+1}\#)^{p(n)}$ is clearly a log-space reduction from L to L_1. In particular, this means that L_1 is NP-complete, if L is. Below we will outline our arguments that show that the language L_1 is accepted by an RWW-automaton M_2 that is 2-monotone.

Let Λ be the tape alphabet of the Turing machine M. In order to describe configurations of M, we define the additional alphabet $\Lambda_{\text{state}} := Q \times \Lambda$, where Q denotes the set of states of M. Then a *possible configuration* of M is presented in the obvious way by a word of the form $y_1 H y_2$, where $y_1, y_2 \in \Lambda^*$ and $H = \langle q, s \rangle \in \Lambda_{\text{state}}$. The *initial configuration* of M for an input word $w := w_1 w_2 \ldots w_n$, where $w_1, \ldots, w_n \in \Sigma_0$, is written as $\langle q_0, w_1 \rangle w_2 \cdots w_n$, where $q_0 \in Q$ denotes the initial state of M.

5.2   A High-Level Description of the RWW-Automaton M_2

Let $x = x_0 \# \ldots \# x_p \#$, where $x_i \in (\Sigma_0 \cup \{\&\})^*$, $0 \leq i \leq p$. For checking that $x \in L_1$, we will use the following overall strategy:

(1) We check that x_0 is of the form $x_0 = y_0 \&^{m_0}$ for some word $y_0 \in \Sigma_0^n$, $n \geq 0$, and some integer $m_0 \geq 0$. In the affirmative y_0 is transformed into the initial configuration of M on input y_0. This yields the syllable $x_0' := \langle q_0, y_{0,1} \rangle y_{0,2} \cdots y_{0,n} \&^{m_0}$.

(2) Then the following steps are performed iteratively for $i = 1, 2 \ldots, p$:

    (a) Check that $|x_{i-1}'| \geq |x_i|$ and that $x_i \in \&^+$.

    (b) In the previous round the syllable x_{i-1} was transformed into a word of the form $x_{i-1}' = y_{i-1}\&^{m_{i-1}}$ for some configuration y_{i-1} of the Turing machine M and some integer $m_{i-1} \geq 0$. If y_{i-1} is already a final configuration of M, then y_i is taken to be y_{i-1}, otherwise a transition of M is chosen nondeterministically that is to be performed in the configuration y_{i-1}, and y_i is chosen as the configuration that M reaches from y_{i-1} by performing this step. If $|y_i| \geq |x_i|$, then the input is rejected, otherwise m_i is taken to be the integer $m_i := |x_i| - |y_i|$, and the syllable x_i is transformed into the word $x_i' := y_i \&^{m_i}$.

(3) If y_p is an accepting configuration, and if $|x_0'| = |x_p'|$, then the input is accepted, otherwise it is rejected.

Unfortunately this strategy cannot be implemented on an RWW-automaton in the form it is described above, because of the restrictions of the RWW-model. Here the facts that an RWW-automaton performs a Restart immediately after each Rewrite transition, and that it forgets its actual state whenever it makes a Restart pose some problems. In addition we want to ensure that the RWW-automaton is 2-monotone.

In particular, a problem is caused by step 2(b), which replaces the syllable x_i by (a slightly revised form of) the factor $y_{i-1}\&^{m_{i-1}}$. This task is similar to the task of verifying that the syllables u and v of a given string $u\#v$ coincide. The simplest way known (to us) for solving the latter task by a 2-monotone RWW-automaton is described in the proof of Lemma 3. Fortunately, the task described in step 2(b) above can be solved similarly.

However, between these two tasks there are two major differences:

- We must realize a sequence of 'copying' moves that work on an unbounded sequence of consecutive syllables.
- In each round we do not simply make a copy of a factor, but we also make some changes to the copy, as we have to take into consideration the effect of the next step of the Turing machine M.

Concerning the first issue, we observe the following, returning to the similar but simpler task of accepting the language L'_{copy}. For each integer $j \geq 3$, the language $L^{(j)}_{\text{copy}} := \{ (w\#)^{j-1}w \mid w \in \{a,b\}^* \}$ is accepted by a 2-monotone RWW-automaton that employs the following strategy:

For an input w of the form $w_1\# \ldots \#w_j$, where $w_1, \ldots, w_j \in \{a,b\}^*$, it is checked sequentially whether w_{i+1} is a subsequence of w_i for all $i = 1, \ldots, j-1$, by using only two monotone sequences. For each value of i, the first sequence performs Rewrite steps within the syllable w_i, and the second sequence performs Rewrite steps within the syllable w_{i+1}. Finally, the final versions of the syllables $w_2, \ldots, w_{j-1}$ are removed, and it is checked whether $|w_1| = |w_j|$ holds. The corresponding Rewrite transitions can be combined with the first of the above sequences into a single monotone sequence. Together with the condition that w_{i+1} is a subsequence of w_i for all values of i, $|w_1| = |w_j|$ implies that all the factors w_i are identical, that is, the word w belongs to the language $L^{(j)}_{\text{copy}}$.

The same technique is used in steps (2) and (3) above to check that all syllables $x_0, x_1, \ldots, x_p$ have the same length.

Concerning the second problem, we observe that the syllables $y_{i-1}\&^{m_{i-1}}$ and $y_i\&^{m_i}$ differ at most on a factor of length two. Moreover, there are only three possible positions for this factor, depending on whether the head of M moves left, right or stays at the same position during the step from y_{i-1} to y_i.

Assume that the envisioned RWW-automaton M_2 has discovered two consecutive syllables x'_{i-1} and x_i (bordered by #-symbols) of the current tape content such that $x'_{i-1} = y_{i-1}\&^{m_{i-1}}$, where y_{i-1} is a possibly correct configuration of the Turing machine M. Then M_2 needs to verify that $x_i \in \&^*$ satisfying $|x_i| \leq |x'_{i-1}|$, to guess a step of M, and to replace x_i by $y_i\&^{m_i}$.

For solving this task, M_2 will realize the following algorithm, where we take $x'_{i-1}[0]$ to denote the left endmarker $\mathopen{\mbox{\textcent}}$ (for $i = 1$) or the #-symbol to the left of x_{i-1} (for $i > 1$), and $x_i[0]$ to denote the #-symbol to the left of x_i for all $i \geq 1$:

algorithm *copy_and_modify*;
begin $j := 0$;
  while $j < |x'_{i-1}|$ **do**
  begin
    if $x'_{i-1}[j+1] \notin \Lambda_{\text{state}}$ **then**
      $x_i[j] := x'_{i-1}[j]$; $j := j + 1$
    else
      guess a transition step Δ of M;
      if Δ is not applicable to y_{i-1} **then halt** and **reject**;
      $x_i[j, j+2] := \Delta(x'_{i-1}[j, j+2])$;
      $j := j + 3$
  end
end

An implementation of the above algorithm on an RWW-automaton requires some specific techniques. In order to remember the actual value of j, the RWW-automaton needs to apply Rewrite steps alternatingly to the syllables x'_{i-1} and x_i. In fact, we cannot ensure that these Rewrite steps alternate between the two syllables, but we can at least make sure that between any two Rewrite steps on x_i, there is at least one Rewrite step on x'_{i-1} (compare the proof of Lemma 3). A further complication arises from the requirement that each Rewrite step must be length-reducing. However, this can be overcome by using an appropriate encoding compressing several symbols into a single new symbol. In summary we obtain the following technical result (see [7] for details).

Theorem 6. *If L belongs to the complexity class* NP*, then the language L_1 that is obtained from L as described above is accepted by an* RWW*-automaton that is 2-monotone.*

By taking the tape alphabet as input alphabet, we obtain an RW-automaton M_3 from the RWW-automaton M_2 above. We claim that with M_2 also M_3 is 2-monotone. To verify this claim we need to show that also those computations of M_3 that start from a restarting configuration $q_0 \mathopen{\mbox{\textcent}} w\$$ that is not an initial configuration of M_2 are 2-monotone. By Theorem 2 this yields the following.

Theorem 7. *The class* $\mathcal{L}(2\text{-mon-R})$ *contains* NP*-complete languages.*

6   Conclusions

As the class $\mathcal{L}(\text{RRWW})$ is contained in the complexity class NP, the result above gives the following complexity theoretical characterization, where $\text{LOG}(\mathcal{L})$ denotes the closure of a language class $\mathcal{L}$ under log-space reductions.

Corollary 5. $\text{LOG}(\mathcal{L}(2\text{-mon-R})) = \text{NP}$.

Thus, modulo log-space reductions the 2-monotone R-automata have the same expressive power as the RRWW-automata. Therefore a different generalization of the notion of monotonicity is needed to capture the phenomena of natural languages, a notion that is stronger than that of monotonicity (as we need to express non-context-free phenomena), but that is weaker than that of 2-monotonicity (as we want an efficient implementation).

References

1. G. Buntrock. *Wachsende kontext-sensitive Sprachen.* Habilitationsschrift, Fakultät für Mathematik und Informatik, Universität Würzburg, 1996.
2. G. Buntrock and F. Otto. Growing context-sensitive languages and Church-Rosser languages. *Inform. and Comput.*, 141:1–36, 1998.
3. E. Dahlhaus and M. Warmuth. Membership for growing context-sensitive grammars is polynomial. *J. Comput. System Sci.*, 33:456–472, 1986.
4. P. Jančar, F. Mráz, M. Plátek, J. Vogel. Restarting automata. In: H. Reichel (ed.), *FCT'95, Proc., LNCS* 965, 283–292. Springer, Berlin, 1995.
5. P. Jančar, F. Mráz, M. Plátek, J. Vogel. On monotonic automata with a restart operation. *J. Autom. Lang. Comb.*, 4:287-311, 1999.
6. T. Jurdziński, K. Loryś, G. Niemann, F. Otto. Some Results on RRW- and RRWW-Automata and Their Relationship to the Class of Growing Context-Sensitive Languages. *Mathematische Schriften Kassel*, no. 14/01, 2001. Also: *J. Autom. Lang. Comb.*, to appear.
7. T. Jurdziński, F. Otto, F. Mráz, M. Plátek. On the Complexity of 2-Monotone Restarting Automata. *Mathematische Schriften Kassel*, no. 4/04, 2004. Available at http://www.theory.informatik.uni-kassel.de/techreps/TR-2004-4.ps.
8. C. Lautemann. One pushdown and a small tape. In: K.W. Wagner (ed.), *Dirk Siefkes zum 50. Geburtstag*, 42–47. Technische Universität Berlin and Universität Augsburg, 1988.
9. R. McNaughton, P. Narendran, F. Otto. Church-Rosser Thue systems and formal languages. *J. Assoc. Comput. Mach.*, 35:324–344, 1988.
10. G. Niemann and F. Otto. The Church-Rosser languages are the deterministic variants of the growing context-sensitive languages. In: M. Nivat (ed.), *FoSSaCS 1998, Proc., LNCS* 1378, 243–257. Springer, Berlin, 1998.
11. G. Niemann and F. Otto. Restarting automata, Church-Rosser languages, and representations of r.e. languages. In: G. Rozenberg and W. Thomas (eds.), *DLT 1999, Proc.*, 103–114. World Scientific, Singapore, 2000.
12. G. Niemann and F. Otto. Further results on restarting automata. In: M. Ito and T. Imaoka (eds.), *Words, Languages and Combinatorics III, Proc.*, 352–369. World Scientific, Singapore, 2003.
13. F. Otto. Restarting automata and their relations to the Chomsky hierarchy. In: Z. Ésik and Z. Fülöp (eds.), *DLT 2003, Proc., LNCS* 2710, 55-74. Springer, Berlin, 2003.
14. M. Plátek. Two-way restarting automata and j-monotonicity. In: L. Pacholski and P. Ružička (eds.), *SOFSEM'01, Proc., LNCS* 2234, 316–325. Springer, Berlin, 2001.
15. M. Plátek and F. Mráz. Degrees of (non)monotonicity of RRW-automata. In: J. Dassow and D. Wotschke (eds.), *Preproceedings of the 3rd Workshop on Descriptional Complexity of Automata, Grammars and Related Structures*, Report No. 16, 159–165. Fakultät für Informatik, Universität Magdeburg, 2001.

On Left-Monotone Deterministic Restarting Automata*

T. Jurdziński[1,**], F. Otto[1], F. Mráz[2], and M. Plátek[2]

[1] Fachbereich Mathematik/Informatik, Universität Kassel, Kassel, Germany
{tju, otto}@theory.informatik.uni-kassel.de
[2] Department of Computer Science, Charles University, Prague, Czech Republic
mraz@ksvi.ms.mff.cuni.cz, platek@ksi.ms.mff.cuni.cz

Abstract. The notion of *left-monotonicity* is introduced for the restarting automaton, and the expressive power of the various types of left-monotone restarting automata is studied. We concentrate on the deterministic classes, as here the results differ greatly from those for the corresponding classes of (right-) monotone restarting automata.

1 Introduction

The original motivation for introducing the restarting automata in [1] was the desire to model the so-called *analysis by reduction* of natural languages. In fact, many aspects of the work on restarting automata are motivated by the basic tasks of computational linguistics. The notions developed in the study of restarting automata give a rich taxonomy of contraints for various models of analysers and parsers. Already several programs based on the idea of restarting automata are being used in Czech and German (corpus) linguistics (cf., e.g., [6, 10]).

A (two-way) restarting automaton, RLWW-automaton for short, is a device M with a finite-state control and a read/write window of a fixed size. This window moves along a tape containing a word delimited by sentinels until the control decides (nondeterministically) that the content of the window should be rewritten by some shorter string. After a rewrite, M continues to move its window until it either halts and accepts, halts and rejects, or restarts, that is, it moves its window to the leftmost position, enters the initial state, and continues with the computation. Thus, each computation of M can be described through a sequence of cycles. In fact, M cannot only be considered as a device for accepting a language, but it can also be interpreted as a 'rewriting system,' as each cycle replaces a factor of the tape content by a shorter factor, in this way performing a rewrite of the tape content.

---

* The work of the first two authors was supported by a grant from the Deutsche Forschungsgemeinschaft. The third and the forth authors were supported by the Grant Agency of the Czech Republic, Grant-No. 201/02/1456, and Grant No. 201/04/2102.
** On leave from the Institute of Computer Science, University of Wrocław, Poland.

C.S. Calude, E. Calude and M.J. Dinneen (Eds.): DLT 2004, LNCS 3340, pp. 249–260, 2004.

Also various restricted versions of the restarting automaton have been considered. Here the RRWW-automata, which can only move their window from left to right along the tape, and the RWW-automata, which are in addition required to perform a restart step immediately after executing a rewrite operation are of particular interest. Further a *monotonicity* property was introduced for RLWW-automata which is based on the idea that from one cycle to the next in a computation, the actual place where a rewrite is performed does not increase its distance from the *right* end of the tape. The monotone restarting automata essentially model bottom-up one-pass parsers. As it turned out the monotone RRWW- and RWW-automata characterize the class CFL of context-free languages, while the monotone deterministic RRWW- and RWW-automata as well as several restricted versions thereof all characterize the class DCFL of deterministic context-free languages [2]. Also a generalization of the notion of monotonicity was introduced, which models the generalization from bottom-up one-pass parsers to bottom-up multi-pass parsers [11]. For an integer $j \geq 1$, a restarting automaton is called j-monotone if, for each of its computations, the corresponding sequence of cycles can be partitioned into at most j subsequences such that each of these subsequences is monotone. It is shown in [11] that by increasing the value of the parameter j, the expressive power of the (nondeterministic) restarting automata without auxiliary symbols is increased. However, for the various types of deterministic RRWW-automata, the parameter j does not influence the class of accepted languages [12, 13].

Here we consider the notion of *left-monotonicity* for restarting automata, which is another constraint motivated by linguistic considerations. This notion is based on the idea that from one cycle to the next in a computation, the actual place where a rewrite takes place does not increase its distance from the *left* end of the tape. Although the notions of (j-) monotonicity and (j-) left-monotonicity seem to be symmetric to each other, it turns out that for deterministic restarting automata, these notions lead to completely different forms of behaviour. The combination of various types of j-monotonicity and j'-left-monotonicity for these types of restarting automata has been studied in [3].

After restating the basic definitions and results in Section 2, we will show in Section 3 that various types of left-monotone deterministic restarting automata with auxiliary symbols are equally powerful. In particular, we introduce restarting automata that are *shrinking*, that is, they are *weight-reducing* with respect to an arbitrary weight function, and we will see that even this generalization does not increase the power of left-monotone deterministic restarting automata. In Section 4 we will establish hierarchies with respect to the degree of left-monotonicity for the various types of deterministic restarting automata without auxiliary symbols, and in Section 5 we will compare these hierarchies to each other. Finally we will separate the classes of languages that are accepted by the 2-left-monotone deterministic RLWW-automata and R(R)WW-automata from the classes of languages that are accepted by the left-monotone deterministic RLWW-automata and R(R)WW-automata, respectively (Section 6). In the concluding section several problems concerning left-monotone restarting automata are outlined that still remain open.

2   Definitions and Notation

For an alphabet Δ, we denote by Δ^+ the set of non-empty words over Δ, while Δ^* denotes the set of all words over Δ including the empty word ε. Further, $\mathbb{N}_+$ will denote the set of positive integers. Finally, for a class A of automata, $\mathcal{L}(\mathsf{A})$ will denote the class of languages that can be accepted by the automata from A, and for a particular automaton M, $L(M)$ denotes the language that is accepted by M.

We start by restating in short the definitions of the various models of the restarting automaton that will be considered in this paper.

A *two-way restarting automaton*, RLWW-automaton for short, is a nondeterministic one-tape machine M with a finite-state control and a read/write window of a fixed size $k \geq 1$. The work space is delimited by the left sentinel ¢ and the right sentinel \$, which cannot be removed from the tape. The tape alphabet Γ of M contains the input alphabet Σ and possibly a finite number of auxiliary symbols. The behaviour of M is described by a transition relation δ that associates to a pair (q, u) consisting of a state q and a possible content u of the read/write window a finite set of possible transition steps. There are five types of transition steps (see, e.g., [7, 9] for details):

1. A *move-right step* (MVR) causes M to shift the read/write window one position to the right and to change the state. However, the window cannot move across the right sentinel \$.

2. A *move-left step* (MVL) causes M to shift the read/write window one position to the left and to change the state. Of course, the window cannot move across the left sentinel ¢.

3. A *rewrite step* (Rewrite) causes M to replace the content u of the read/write window by a shorter string v, and to change the state. Further, the read/write window is placed immediately to the right of the string v.

4. A *restart step* (Restart) causes M to move its read/write window to the left end of the tape, so that the first symbol it sees is the left sentinel ¢, and to re-enter the initial state q_0.

5. An *accept step* (Accept) causes M to halt and accept.

If $\delta(q, u) = \emptyset$ for some pair (q, u), then M necessarily halts, and we say that M *rejects* in this situation.

A *configuration* of M is a string $\alpha q \beta$, where q represents the current state, $\alpha\beta$ is the current content of the tape, and it is understood that the head scans the first k symbols of β or all of β when $|\beta| \leq k$. Here either $\alpha = \varepsilon$ and $\beta \in \{¢\} \cdot \Gamma^* \cdot \{\$\}$ or $\alpha \in \{¢\} \cdot \Gamma^*$ and $\beta \in \Gamma^* \cdot \{\$\}$. A *restarting configuration* is of the form $q_0 ¢ w \$$, where q_0 is the initial state of M and $w \in \Gamma^*$; if $w \in \Sigma^*$, then $q_0 ¢ w \$$ is an *initial configuration*. Thus, initial configurations are special restarting configurations.

In general, the automaton M is *nondeterministic*, that is, there can be two or more instructions with the same left-hand side (q, u). If that is not the case, the automaton is *deterministic*. For brevity, the prefix det- will be used to denote classes of deterministic restarting automata.

We observe that any computation of a two-way restarting automaton M consists of certain phases. A phase, called a *cycle*, starts in a restarting configuration, the head moves along the tape performing MVR, MVL, and Rewrite operations until a Restart operation is performed and thus a new restarting configuration is reached. If no further Restart operation is performed, any finite computation necessarily finishes in a halting configuration – such a phase is called a *tail*. We require that M performs *exactly one* Rewrite operation during any cycle – thus each new phase starts on a shorter word than the previous one. During a tail at most one Rewrite operation may be executed.

An input word $w \in \Sigma^*$ is *accepted by* M, if there is a computation which, starting with the initial configuration $q_0 \text{\textcent} w\$$, finishes by executing an Accept instruction.

Now we define those subclasses of RLWW-automata that are relevant for our investigation.

An RLW-*automaton* is an RLWW-automaton whose working alphabet coincides with its input alphabet. Note that each restarting configuration is initial in this case.

An RL-*automaton* is an RLW-automaton whose Rewrite instructions can be viewed as deletions, that is, if $(q', v) \in \delta(q, u)$, then v is obtained by deleting some symbols from u.

An RRWW-*automaton* is an RLWW-automaton which does not use any MVL instructions. Analogously, we obtain RRW-*automata* and RR-*automata*.

Finally, an RWW-*automaton* is an RRWW-automaton which restarts immediately after rewriting, that is, for these automata each Rewrite transition is immediately followed by a Restart transition. Analogously, we obtain RW-*automata* and R-*automata*.

The transition relation of an RWW-automaton M can be described by a sequence of *meta-instructions* of the form $(R, u \to v)$, where R is a regular expression. On trying to execute this meta-instruction M will get stuck (and so reject) starting from the configuration $q_0 \text{\textcent} w\$$, if w does not admit a factorization of the form $w = w_1 u w_2$ such that $\text{\textcent} w_1 \in L(R)$. On the other hand, if w does admit such a factorization, then one such factorization is chosen nondeterministically, and $q_0 \text{\textcent} w\$$ is transformed into $q_0 \text{\textcent} w_1 v w_2\$$.

Now we come to the various notions of *monotonicity*. Each cycle C contains a unique configuration $\text{\textcent} \alpha q \beta\$$ in which a Rewrite instruction is applied. Then $|\beta\$|$ is the *right distance* of C, denoted by $D_r(C)$, and $|\text{\textcent}\alpha|$ is the *left distance* of C, denoted by $D_l(C)$.

We say that a *sequence of cycles* $Sq = (C_1, C_2, \cdots, C_n)$ is *monotone* (or *right-monotone*) if $D_r(C_1) \geq D_r(C_2) \geq \ldots \geq D_r(C_n)$, and we say that this sequence is *left-monotone* if $D_l(C_1) \geq D_l(C_2) \geq \ldots \geq D_l(C_n)$.

For each prefix X $\in$ {right, left}, a *computation* is X-*monotone* if the corresponding sequence of cycles is X-monotone. Observe that the tail of the computation does not play any role here. An RLWW-automaton M is called X-*monotone* if all its computations are X-monotone. The prefix X-mon- will be used to denote the corresponding classes of restarting automata. Observe that right-monotonicity is the concept called monotonicity in [2].

3   Left-Monotone Deterministic Restarting Automata

Here we concentrate on the expressive power of the left-monotone deterministic restarting automaton. In order to simplify the discussion of the technical details we introduce a slight generalization of the restarting automaton. A restarting automaton M with working alphabet Γ is called *shrinking* if there exists a *weight function* $\varphi : \Gamma \to \mathbb{N}_+$ such that, for each Rewrite step $u \to v$ of M, $\varphi(u) > \varphi(v)$ holds. Here φ is extended to a morphism $\varphi : \Gamma^* \to \mathbb{N}$ by taking $\varphi(\varepsilon) := 0$ and $\varphi(wa) := \varphi(w) + \varphi(a)$ for all $w \in \Gamma^*$ and $a \in \Gamma$. The prefix s- will be used to denote the classes of shrinking restarting automata.

Our first result states that for shrinking deterministic restarting automata that are left-monotone, the RWW-model is as powerful as the RLWW-model.

Theorem 1. $\mathcal{L}(\text{det-left-mon-sRLWW}) = \mathcal{L}(\text{det-left-mon-sRWW})$.

Proof. Let M be a deterministic RLWW-automaton that is left-monotone and shrinking with respect to the weight function φ. We will construct a left-monotone deterministic sRWW-automaton M' such that M' accepts the same language as M. In fact, given an input w, M' will simulate the computation of M on input w. For defining M' we need to analyze the behaviour of M in detail.

Each cycle of a computation of M consists of three phases:

1. M scans its tape by repeatedly performing MVR-steps and MVL-steps.
2. M executes a Rewrite step, replacing a factor u of the current tape content by a string v satisfying $\varphi(v) < \varphi(u)$.
3. M rescans its tape by repeatedly performing MVR-steps and MVL-steps until it eventually accepts, rejects, or restarts.

To simplify the following discussion we may assume without loss of generality that in phase 1, M first scans its tape completely from left to right by performing a sequence of MVR-steps.

In contrast to the behaviour of M described above, M' simply scans its tape from left to right, performing a number of MVR-steps, until it decides to execute a Rewrite transition, thus ending the current cycle. Hence, if M' is to simulate a cycle of M, then it needs to determine the information that M collects during phases 1 and 3 *before* it can execute the simulation of the actual Rewrite step. Thus, M' will have to perform some preparatory cycles before it can actually execute the simulation of the Rewrite step of the current cycle of M.

Assume that the actual configuration of M at the start of the current cycle is $q_0 \mathfrak{c} xuy\$$, where $x, u, y \in \Gamma^*$, and u is the factor that M is about to replace by the word v in this cycle. In order to simulate this cycle M' will first encode information of the behaviour of M on the suffix y by performing a number of preparatory cycles that replace this suffix letter by letter from right to left by an encoding of y. This encoding replaces each letter c of y by a symbol that together with the letter c encodes a crossing sequence describing the possible behaviour of M at the tape square containing the letter c.

The actual simulation of the Rewrite step of M, which replaces the syllable u by the string v, is performed by M' on the border between the prefix of the tape content that is still unencoded, and the suffix of the tape content that has already been encoded.

As M is left-monotone, the Rewrite step in the next cycle of M is performed at the same position or to the left of the position of the current Rewrite step. Hence, M' may first have to execute some further encoding cycles before it can simulate the next Rewrite step of M. Thus, we see that with M, also M' is left-monotone. Details can be found in [8].   □

Our second technical result shows that, for deterministic RWW-automata that are left-monotone, the standard (length-reducing) variant is as powerful as the shrinking variant.

Theorem 2. $\mathcal{L}(\text{det-left-mon-sRWW}) = \mathcal{L}(\text{det-left-mon-RWW})$.

Proof. We will essentially follow a simulation technique presented in [4]. This method was initially used for two-pushdown automata, but because of the correspondence between the class of Church-Rosser languages and $\mathcal{L}(\text{det-RWW})$ [5], it can be generalized to restarting automata. However, if one adjusts this simulation directly to RWW-automata, then the resulting automaton is not left-monotone, even if the automaton being simulated is. Thus, we must follow through the steps of the simulation from [4] and discuss the changes that are required in order to guarantee that the property of being left-monotone is preserved by the simulation.

Here we just give a high level description of the simulation of a left-monotone deterministic RWW-automaton M that is shrinking with respect to a weight function φ by a left-monotone (length-reducing) deterministic RWW-automaton. This simulation consists of three major steps, which are outlined below. Again details can be found in [8].

1. For the shrinking deterministic RWW-automaton M, we first construct a shrinking deterministic RWW-automaton M' such that $L(M') = L(M)$ and each Rewrite transition of M' reduces the weight of the actual tape content exactly by one (using a method from [4] Lemma 4). Thus, we can assume in the following that each Rewrite transition of M reduces the weight exactly by one.

2. Let $\#$ be a new symbol, and let $h : \Gamma^* \to (\Gamma \cup \{\#\})^*$ be the morphism that is induced by the mapping $h(a) := a\#^{\varphi(a)-1}$ ($a \in \Gamma$). Thus, for each string $w \in \Gamma^*$, the string $h(w) \in (\Gamma \cup \{\#\})^*$ satisfies the condition $|h(w)| = \varphi(w)$. We construct a standard (length-reducing) RWW-automaton M_1 that simulates the computation of M on the tape content $\math0/c w\$ ($w \in \Gamma^*$) step by step on the tape content $\math0/c h(w)\$. Hence, $L(M_1) = h(L(M))$. Further, if M is left-monotone, then so is M_1. Moreover, M_1 is length-reducing, as each Rewrite step of M reduces the weight by one, and so each Rewrite step of M_1 reduces the length of the actual tape content by one.

3. To complete the construction we would now like to simulate the automaton M_1 by an RWW-automaton that, instead of processing an input of the form

$h(x)$ $(x \in \Sigma^*)$, works directly with the original input x. However, it might be impossible to simulate the computation of M_1 on $h(x)$ in a length-reducing manner on the input x itself, as already the length of $h(x)$ will in general be larger than the length of x. In order to overcome this problem, we follow a strategy from [4] (Proof of Theorem 5):

(a) First an automaton M_2 is used to replace the input string x by a 'compressed' version of $h(x)$. As compression ratio we take the number 2μ, where $\mu := \max_{a \in \Gamma} \varphi(a)$.

(b) Then an automaton M_3 is used that, in each cycle, simulates 2μ cycles of M_1.

The compression of ratio 2μ guarantees that the compressed version x_c of $h(x)$ satisfies $|x_c| \leq |x|$, and by simulating 2μ cycles of M_1 in a single cycle of M_3 we guarantee that the length of the actual tape content of M_3 is reduced by exactly one per cycle. Thus, we see that the composition of M_2 and M_3 is length-reducing. Further, M_2 can clearly be realized in left-monotone manner, and also M_3 is left-monotone, as M_1 is.

However, the composition of M_2 and M_3 is clearly not left-monotone, as the 'compression phase' realized by M_2, in which the tape content is completely rewritten, precedes the 'real' simulation. In order to make this part of the simulation left-monotone, we skip the compression phase. Instead we adopt the strategy of 'lazy' compression, that is, we start on the uncompressed input and use the compression in combination with the simulation of Rewrite steps (that is, we combine M_2 and M_3 into one automaton). In fact, we can make sure that the tape content always consists of an uncompressed prefix that is followed by a compressed suffix, and that each Rewrite step just replaces a prefix of the compressed part. □

From Theorem 1 and Theorem 2 we immediately obtain the following result.

Corollary 1. $\mathcal{L}(\text{det-left-mon-sRLWW}) = \mathcal{L}(\text{det-left-mon-RLWW}) = \mathcal{L}(\text{det-left-mon-RRWW}) = \mathcal{L}(\text{det-left-mon-RWW})$.

It remains to derive a characterization of the class $\mathcal{L}(\text{det-left-mon-RWW})$ in terms of other language classes. As $\mathcal{L}(\text{det-mon-RLWW})$ properly contains the class of deterministic context-free languages [9], it follows that the class $\mathcal{L}(\text{det-left-mon-RWW})$ properly contains the reversals of all deterministic context-free languages, but still remains to classify the additional expressive power of the left-monotone deterministic RWW-automata.

4   Degrees of Non-monotonicity for Restarting Automata

A sequence of cycles $Sq = (C_1, C_2, \cdots, C_n)$ is called *j-right-monotone* for some $j \in \mathbb{N}$, if there is a partition of Sq into j (scattered) subsequences that are right-monotone. Analogously, the notion of *j-left-monotonicity* is defined. Obviously a sequence of cycles $(C_1, C_2, \cdots, C_n)$ is not j-right-monotone if and only if there exist $1 \leq i_1 < i_2 < \ldots < i_{j+1} \leq n$ such that $D_r(C_{i_1}) < D_r(C_{i_2}) < \cdots < D_r(C_{i_{j+1}})$. A corresponding observation holds for j-left-monotonicity.

Let $j \geq 1$, and let $X \in \{\mathsf{right}, \mathsf{left}\}$. A *computation* is j-X-*monotone* if the corresponding sequence of cycles is j-X-monotone. Again the tail of the computation does not play any role here. An RLWW-automaton is called j-X-*monotone* if all its computations are j-X-monotone. The prefixes j-X-mon- are used to denote the corresponding classes of restarting automata. Observe that 1-X-monotonicity coincides with X-monotonicity.

It is known that the degree of right-monotonicity does not influence the expressive power of deterministic restarting automata.

Theorem 3. [12, 13] DCFL $= \mathcal{L}(\mathsf{det\text{-}right\text{-}mon\text{-}X}) = \mathcal{L}(\mathsf{det\text{-}}j\text{-}\mathsf{right\text{-}mon\text{-}X})$ *for each* $j \in \mathbb{N}_+$ *and for each* $X \in \{\mathsf{R}, \mathsf{RR}, \mathsf{RW}, \mathsf{RRW}, \mathsf{RWW}, \mathsf{RRWW}\}$.

On the other hand, based on the degree of left-monotonicity, infinite hierarchies for the various types of deterministic restarting automata without auxiliary symbols have been obtained.

Theorem 4. [3] $\mathcal{L}(\mathsf{det\text{-}}j\text{-}\mathsf{left\text{-}mon\text{-}X}) \subsetneq \mathcal{L}(\mathsf{det\text{-}}(j{+}1)\text{-}\mathsf{left\text{-}mon\text{-}X})$ *for each* $j \in \mathbb{N}_+$ *and for each* $X \in \{\mathsf{R}, \mathsf{RR}, \mathsf{RW}, \mathsf{RRW}\}$.

To derive a corresponding separation result for deterministic RL(W)-automata, we consider the languages

$$\bar{L}_j := \{\, a^{n_1} b^{n_1} a^{n_2} b^{n_2} \ldots a^{n_j} b^{n_j} \mid n_1 \geq n_2 \geq \ldots \geq n_j \geq 1 \,\} \quad (j \geq 1).$$

Lemma 1. *For* $j \geq 2$, $\bar{L}_j \in \mathcal{L}(\mathsf{det\text{-}}j\text{-}\mathsf{left\text{-}mon\text{-}RL}) \smallsetminus \mathcal{L}((j-1)\text{-}\mathsf{left\text{-}mon\text{-}RLW})$.

Proof. For a word of the form $(a^+ b^+)^j$, we denote the factors from $a^+ b^+$ as 'blocks.' We construct a deterministic RL-automaton M for the language $\bar{L}_j$ as follows. In each cycle M removes a factor ab from the r-th block if r is the largest index such that the parity of the number of a's (and b's) in all previous blocks is equal to the parity of the number of a's in the block r. In this way we ensure that it is not possible to make two rewrites in the r-th block without any rewrite in the previous block. M accepts if and only if the tape content is $(ab)^j$. It is easily seen that M does indeed accept the language $\bar{L}_j$, and that it is j-left-monotone.

On the other hand, by analysing the possible computations of an RLW-automaton M' for the language $\bar{L}_j$ on inputs of the form $(a^n b^n)^j$ for large values of n, it can be shown [13] that M' is not $(j-1)$-left-monotone. □

This gives the following separation results.

Theorem 5. $\mathcal{L}(\mathsf{det\text{-}}j\text{-}\mathsf{left\text{-}mon\text{-}X}) \subsetneq \mathcal{L}(\mathsf{det\text{-}}(j+1)\text{-}\mathsf{left\text{-}mon\text{-}X})$ *for each* $j \in \mathbb{N}_+$ *and for each* $X \in \{\mathsf{RL}, \mathsf{RLW}\}$.

5   Comparing the Classes of the Same Degree of Left-Monotonicity to Each Other

It is shown in [3] that, for any $j \geq 2$, the language

$$L^{(j)} := \{\, a^{m_1} b^{m_1} a^{m_2} b^{m_2} \ldots a^{m_j} b^{m_j} \mid m_1, m_2, \ldots, m_j > 0 \,\}$$

cannot be accepted by any $(j-1)$-left-monotone deterministic RRW-automaton. Contrasting this result, we see below that auxiliary symbols do help in accepting this language.

Proposition 1. *For each $j \geq 2$, $L^{(j)} \in \mathcal{L}(\text{det-left-mon-RWW})$.*

Proof. Let $j \geq 2$. We describe an RWW-automaton M for the language $L^{(j)}$.

(1.) First M moves its read/write window all the way to the right, verifying that the given input is of the form $w := a^{m_1} b^{n_1} a^{m_2} b^{n_2} \ldots a^{m_j} b^{n_j}$ for some positive integers $m_i, n_i, 1 \leq i \leq j$. In the negative it rejects immediately, in the affirmative it goes to (2.).

(2.) Using the auxiliary symbol B_j, M rewrites the suffix b^{n_j} into $B_j^{n_j/2}$ within $n_j/2$ cycles. If n_j is not an even number, then the factor ab of $a^{m_j} b^{n_j}$ is deleted in this process.

(3.) Then within the next $n_j/2$ cycles it is checked whether $m_j = n_j$ holds by deleting factors of the form $a^2 B_j$. This phase ends by generating an occurrence of the auxiliary symbol A_j in the affirmative.

(4.) Now steps (2.) and (3.) are repeated for the factors $a^{m_{j-1}} b^{n_{j-1}}$, $a^{m_{j-2}} b^{n_{j-2}}$ down to $a^{m_1} b^{n_1}$.

Obviously M is a deterministic RWW-automaton, and it is easily seen that M accepts the language $L^{(j)}$ and that it is left-monotone. Hence, we see that $L^{(j)} \in \mathcal{L}(\text{det-left-mon-RWW})$. □

Together with the fact that $L^{(j)} \notin \mathcal{L}(\text{det-}(j-1)\text{-left-mon-RRW})$, this yields the following separation results.

Theorem 6.
For each $j \in \mathbb{N}_+$, $\mathcal{L}(\text{det-}j\text{-left-mon-R(R)W}) \subsetneq \mathcal{L}(\text{det-}j\text{-left-mon-R(R)WW})$.

Actually the above example languages will give us still further separation results. Let $j > 1$, and let M be a det-left-mon-RWW-automaton for the language $L^{(j)}$. By including all the auxiliary symbols of M in the input alphabet, we obtain a det-left-mon-RW-automaton M' for some language $\hat{L}^{(j)}$. Observe that $\hat{L}^{(j)} \cap \{a, b\}^* = L^{(j)}$ holds, which allows to derive the following fact.

Lemma 2. *For each $j > 1$, $\hat{L}^{(j)} \notin \mathcal{L}(\text{det-}(j-1)\text{-left-mon-RR})$.*

Thus, we obtain the following separation results.

Theorem 7.
For each $j \in \mathbb{N}_+$, $\mathcal{L}(\text{det-}j\text{-left-mon-R(R)}) \subsetneq \mathcal{L}(\text{det-}j\text{-left-mon-R(R)W})$.

For deriving a corresponding result separating the R(W)-classes from the RR(W)-classes, we consider the example language $L_t := L_{t1} \cup L_{t2} \cup L_{t3} \cup L_{t4}$, where

$$L_{t1} := \{\, a^m a^n (bc)^n f a^m \mid m, n > 0 \,\},$$
$$L_{t2} := \{\, a^m a^n (bc)^i b^j a^m \mid m, n, j > 0, \, i \geq 0, \, n = i + j \,\},$$
$$L_{t3} := \{\, a^n (bc)^i c^j \quad\;\; \mid n, j > 0, \, i \geq 0, \, n = 2(i + j) \,\},$$
$$L_{t4} := \{\, a^m (bc)^n f a^k f \mid m, n, k > 0 \,\}.$$

For this language we have the following technical result.

Lemma 3. $L_t \in \mathcal{L}(\text{det-left-mon-RR}) \smallsetminus \mathcal{L}(\text{det-RW})$.

This technical result yields the following proper inclusion results.

Theorem 8.
For each $j \geq 1$, $\mathcal{L}(\text{det-}j\text{-left-mon-R(W)}) \subsetneqq \mathcal{L}(\text{det-}j\text{-left-mon-RR(W)})$.

For the automata without auxiliary symbols, we have the following separation results in contrast to Corollary 1.

Theorem 9.
For each $j \in \mathbb{N}_+$, $\mathcal{L}(\text{det-}j\text{-left-mon-RR(W)}) \subsetneqq \mathcal{L}(\text{det-}j\text{-left-mon-RL(W)})$.

Proof. The language $L_a := \{ a^n b^n c \mid n \geq 0 \} \cup \{ a^n b^{2n} d \mid n \geq 0 \}$ is easily seen to be accepted by a deterministic left-mon-RL-automaton. On the other hand, this language is not accepted by any deterministic RRW-automaton, because in each cycle such an automaton must rewrite a string of L_a into another string belonging to L_a. This means that on a long input word, it would have to delete a factor of the form $a^i b^i$ or $a^i b^{2i}$ for some $i > 0$ without knowing which of the two cases applies. $\square$

Let $L_c := \{ cwcw \mid w \in \{a, b\}^* \text{ and } |w| = 2n \text{ for some } n \geq 0 \}$. Concerning this language we have the following result.

Proposition 2. $L_c \in \mathcal{L}(\text{det-2-left-mon-RLWW}) \smallsetminus \mathcal{L}(\text{RLW})$.

Further, let M_c be a 2-left-monotone deterministic RLWW-automaton for L_c, let M_c' be the RLW-automaton that is obtained from M_c by including the auxiliary symbols of M_c into the input alphabet, and let $L_c' := L(M_c')$.

Proposition 3. $L_c' \in \mathcal{L}(\text{det-2-left-mon-RLW}) \smallsetminus \mathcal{L}(\text{RL})$.

Together, these technical results yield the following consequences.

Theorem 10.
For each $j \geq 2$, $\mathcal{L}(\text{det-}j\text{-left-mon-RL(W)}) \subsetneqq \mathcal{L}(\text{det-}j\text{-left-mon-RL(W)W})$.

Here it remains open whether a corresponding result also holds for left-monotone RL(W)-automata.

6   Separating the Second from the First Level of Left-Monotonicity for Restarting Automata with Auxiliary Symbols

As the language L_c is not context-free, while $\mathcal{L}(\text{left-mon-RLWW}) = \text{CFL}$ [13], Proposition 2 yields the following additional result.

Corollary 2. $\mathcal{L}(\text{det-left-mon-RLWW}) \subsetneq \mathcal{L}(\text{det-2-left-mon-RLWW}).$

In the following we will extend this result to RWW- and RRWW-automata. For doing so we consider the language

$$L := \{\, a_5^n a_4^n a_3^m a_2^m a_1^p \mid n, m, p > 0 \,\} \cup \{\, a_5^n a_4^l a_3^m a_2^p a_1^p \mid n \neq l,\; n, l, m, p > 0 \,\}.$$

Proposition 4. $L \in \mathcal{L}(\text{det-2-left-mon-R}).$

Proof. We describe a deterministic R-automaton M for L by a sequence of meta-instructions:

(1) $(\text{¢} \cdot a_5^*,\ a_5^2 a_4^2 \rightarrow a_5 a_4)$,

(2) $(\text{¢} a_5 a_4 \cdot a_3^*,\ a_3^2 a_2^2 \rightarrow a_3 a_2)$,

(3) $(\text{¢} a_5 a_4 a_3 a_2 \cdot a_1^+ \cdot \$,\ \text{Accept})$,

(4) $(\text{¢} a_5 a_4 \cdot a_4^+ \cdot a_3^+ \cdot a_2^*,\ a_2^2 a_1^2 \rightarrow a_2 a_1)$,

(5) $(\text{¢} a_5 \cdot a_5^+ \cdot a_4 \cdot a_3^+ \cdot a_2^*,\ a_2^2 a_1^2 \rightarrow a_2 a_1)$,

(6) $(\text{¢} a_5 a_4 \cdot a_4^+ \cdot a_3^+ \cdot a_2 a_1 \$,\ \text{Accept})$,

(7) $(\text{¢} a_5 \cdot a_5^+ \cdot a_4 \cdot a_3^+ \cdot a_2 a_1 \$,\ \text{Accept})$.

It is easily seen that M accepts the language L. Notice further that M makes two sequences of Rewrite steps: first it rewrites on the border between a_5^* and a_4^* and then it rewrites on the border between a_3^* and a_2^* or on the border between a_2^* and a_1^*. Thus, M is 2-left-monotone. $\qquad\square$

The announced separation results are a consequence of the following result.

Theorem 11. [8] $L \notin \mathcal{L}(\text{det-left-mon-RRWW}).$

Thus, we obtain the following proper inclusions.

Corollary 3. $\mathcal{L}(\text{det-left-mon-R(R)WW}) \subsetneq \mathcal{L}(\text{det-2-left-mon-R(R)WW}).$

Further, as the language L is obviously deterministic context-free, we also obtain the following non-inclusion result.

Corollary 4. $\mathcal{L}(\text{det-mon-RRWW}) = \text{DCFL} \not\subseteq \mathcal{L}(\text{det-left-mon-RRWW}).$

7 Concluding Remarks

We have seen that between the language classes defined by the various types of left-monotone deterministic restarting automata very different relations hold than between the classes that are defined by the corresponding types of right-monotone deterministic restarting automata. However, it is still open whether the degree of left-monotonicity gives infinite hierarchies also for the deterministic automata with auxiliary symbols. Further, we conjecture that the equality expressed by Corollary 1 even extends to the language classes $\mathcal{L}(\text{det-left-mon-RLW})$ and $\mathcal{L}(\text{det-left-mon-RL})$ in contrast to the situation for j-left-monotone deterministic RL- and RLW-automata ($j \geq 2$) as expressed by Theorem 10.

References

1. P. Jančar, F. Mráz, M. Plátek and J. Vogel. Restarting automata. In: H. Reichel (ed.), *FCT'95, Proc.*, *LNCS* 965, pages 283–292. Springer, Berlin, 1995.
2. P. Jančar, F. Mráz, M. Plátek and J. Vogel. On monotonic automata with a restart operation. *J. Autom., Lang. and Comb.* 4, pages 287–311, 1999.
3. F. Mráz, F. Otto and M. Plátek. Restarting automata and combinations of constraints. In: P. Vojtáš (ed.), *ITAT 2003: Information Technologies - Applications and Theory, Proc.*, pages 83–91. Department of Computer Science, Faculty of Science, Pavol Josef Šafárik University, Košice, 2004.
4. G. Niemann and F. Otto. The Church-Rosser languages are the deterministic variants of the growing context-sensitive languages. In: M. Nivat (ed.), *Foundations of Software Science and Computation Structures, Proc. FoSSaCS'98, LNCS* 1378, pages 243–257. Springer Verlag, Berlin, 1998.
5. G. Niemann and F. Otto. Restarting automata, Church-Rosser languages, and representations of r.e. languages. In: G. Rozenberg and W. Thomas (eds.), *Developments in Language Theory - Foundations, Applications, and Perspectives, Proc. DLT 1999*, pages 103–114. World Scientific, Singapore, 2000.
6. K. Oliva, P. Květoň and R. Ondruška. The computational complexity of rule-based part-of-speech tagging. In: V. Matoušek and P. Mautner (eds.), *TSD 2003, Proc.*, *LNCS* 2807, pages 82–89. Springer, Berlin, 2003.
7. F. Otto. Restarting automata and their relations to the Chomsky hierarchy. In: Z. Esik and Z. Fülöp (eds.), *Developments in Language Theory, Proc. DLT'2003, LNCS* 2710, pages 55–74. Springer, Berlin, 2003.
8. F. Otto and T. Jurdziński. On left-monotone restarting automata. *Mathematische Schriften Kassel* 17/03, Universität Kassel, 2003.
9. M. Plátek. Two-way restarting automata and j-monotonicity. In: L. Pacholski and P. Ružička (eds.), *Theory and Practice of Informatics, Proc. SOFSEM 2001, LNCS* 2234, pages 316–325. Springer-Verlag, Berlin, 2001.
10. M. Plátek, M. Lopatková and K. Oliva. Restarting automata: motivations and applications. In: M. Holzer (ed.), *Workshop 'Petrinetze' and 13. Theorietag 'Formale Sprachen und Automaten', Proc.*, pages 90–96. Institut für Informatik, Technische Universität München, 2003.
11. M. Plátek and F. Mráz. Degrees of (non)monotonicity of RRW-automata. In: J. Dassow and D. Wotschke (eds.), *Preproceedings of the 3rd Workshop on Descriptional Complexity of Automata, Grammars and Related Structures*, Report No. 16, pages 159–165. Fakultät für Informatik, Universität Magdeburg, 2001.
12. M. Plátek, F. Otto and F. Mráz. Restarting automata and variants of j-monotonicity. In: E. Csuhaj-Varjú, C. Kintala and D. Wotschke (eds.), *Descriptional Complexity of Formal Systems, Proceedings DCFS 2003*, pages 303–312. MTA SZTAKI, Budapest, 2003.
13. M. Plátek, F. Otto, F. Mráz and T. Jurdziński. Restarting automata and variants of j-monotonicity. *Mathematische Schriften Kassel* 9/03, Universität Kassel, 2003.

On the Computation Power of Finite Automata in Two-Dimensional Environments

Oleksiy Kurganskyy[1,*] and Igor Potapov[2,**]

[1] Institute of Applied Mathematics and Mechanics,
Ukrainian National Academy of Sciences, Donetsk, Ukraine
kurgansk@gmx.de
[2] Department of Computer Science,
University of Liverpool, Liverpool, U.K.
igor@csc.liv.ac.uk

Abstract. In this paper we study the model of a finite state automaton interacting with infinite two-dimensional geometric environments. We show that the reachability problem for a finite state automaton interacting with a quadrant of the plane extended by a power function, a polynomial function or a linear function is algorithmically undecidable, by simulating a Minsky machine. We also consider the environment defined by a parabola which impedes the direct simulation of multiplication. However we show that the model of a finite automaton interacting inside a parabola is also universal.

1   Introduction

Finite state automata arose as models of transducers of discrete information, i.e. models interacting with their environments [9]. Automata on picture languages [3, 17], automata in labyrinths [2, 10, 11], communicating automata [14], multi-counter automata [15, 7], a model of a computer in the form of interaction of a control automaton [4] are examples of such an interaction.

The fundamental problem for systems where an automaton interacts with (possibly infinite) environment is the reachability problem: "Does a global state S (state of the automaton and state of the environment) belong to the set of states reachable from an initial global state". The reachability problem has connections to many classical problems in automata theory such as diagnostic problems, distinguishability problems, searching in labyrinths, etc. One of the standard methods to show the undecidablity of the reachability problem for some model is to prove that this computational model is universal.

In this paper we consider the computational power of the reactive system, where an input/output automaton interacts with a two-dimensional geometric environment. In particular we consider the reactive system that includes a finite state automaton (FSA) A and an infinite environment E where

---

* Work partially supported by RDF grant RDF/4417.
** Work partially supported by The Nuffield Foundation grant NAL/00684/G.

C.S. Calude, E. Calude and M.J. Dinneen (Eds.): DLT 2004, LNCS 3340, pp. 261–271, 2004.

- an automaton A is a kind of well-known 4-way finite state automaton introduced by Blum and Hewitt [1] that reads the symbol from the environment and reacts by changing its internal state and then moving up, down, left or right to a neighbouring cell of the environment E;
- an environment E is a two-dimensional language over one-letter alphabet [13, 3] which is defined as a subspace of two-dimensional integer grid bounded by a number of integer value functions.

It was shown in [12, 13] that the problem of checking indistinguishable states for two finite automata in two-dimensional environment E over a one-letter alphabet (geometric environment) is decidable if and only if the reachability problem for a finite state automaton in E is decidable. The geometric environment is called *efficient* if the reachability problem is decidable and *non-efficient* otherwise. It is known from [5, 6, 13] that the environment defined by a set of rectangles of fixed height and the environment represented by a regular or context-free expression are efficient. There are many other classes of efficient environments with or without holes that can be constructed by substitution of one cell in efficient environment by another efficient environment [13].

It was proved in [2] that the sets of input-output words generated by automata interacting with geometric environments without holes, namely, rectangles of unlimited height are, in general, context-sensitive languages. Thus the reachability problem for finite automata interacting with geometric environments is fundamentally difficult and algorithmically unsolvable, in the general case. In spite of known undecidability results it is interesting to identify new classes of non-efficient environments, i.e. where a model of finite automaton interacting with these environments is universal.

We start from the environment defined by a quadrant of the plane that corresponds to the Minsky machine model. First we consider an extension of quadrant of the plane, changing the vertical border by power, polynomial or linear functions. In particular we show that the finite automaton interacting with these environments can simulate a Minsky machine. Then we consider an environment which impedes the direct simulation of multiplication. However we show on example of an environment defined by parabola:

$$D_{n^2,n^2} = \{(x,y) \in \mathbb{Z} \times \mathbb{Z} | y \geq x^2\}$$

that FSA interacting with D_{n^2,n^2} is again a universal model of computation.

The additional motivation for the investigation of automaton dynamics in geometric environments is that the different constraints on the environments can be also seen as constraints on the values of counters in multicounter automata. So we hope that the new results about n-dimensional geometric environments could lead to the new results for n-counter automata.

The long-term goal of this work is to characterize the whole class of geometric environments to obtain a better understanding of the border between decidability and undecidability for the reachability problems. While such an ambitious goal is not feasible at the moment, we instead investigate several special cases of geometric environments, that we believe is a sound and reasonable first step towards our ultimate goal.

2   Automata Interacting with Two-Dimensional Environments

In what follows we use traditional denotations $\mathbb{N}, \mathbb{Z}, \mathbb{Z}_n, \mathbb{Q}^+$ for the sets of naturals (non-negative integers), integers, bounded integers such that $\{i | i \in \mathbb{Z}, |i| \leq n\}$ and the set of positive rational numbers respectively.

Definition 1. *Let $A = (S, I, O, \delta_A, \lambda_A, s_0)$ be a finite deterministic everywhere defined Mealy automaton, where S, I and O are the sets of states, input symbols, and output symbols, respectively, and $\delta_A : S \times I \to S$ and $\lambda_A : S \times I \to O$ are transition function and function of outputs respectively and $s_0 \in S$ is an initial state.*

Definition 2. *The geometric environment is defined by possibly infinite (countable) Moore automaton $E = (D, O, I, \delta_E, \lambda_E)$, where $D \subseteq \mathbb{Z} \times \mathbb{Z}$ is a set of states, $O = \mathbb{Z}_1 \times \mathbb{Z}_1$ is a set of input symbols, $I = 2^{\mathbb{Z}_1 \times \mathbb{Z}_1}$ is a set of output symbols, $\delta_E : D \times O \to D$ such that*

$$\delta_E((x, y), (d_1, d_2)) = \begin{cases} (x + d_1, y + d_2), & (x + d_1, y + d_2) \in D \\ undefined, & (x + d_1, y + d_2) \notin D \end{cases}$$

is the partial transition function and $\lambda_E : D \to I$ such that

$$\lambda_E(x, y) = \{(d_1, d_2) \in \mathbb{Z}_1 \times \mathbb{Z}_1 | (x + d_1, y + d_2) \in D\}$$

is the function of outputs.

We call the set of states D - the nodes of the environment, the symbols of the output alphabet I - the labels of nodes, and the function of outputs λ_E - the function of labels of nodes. In such case we can say that two-dimensional geometric environment is defined by an automaton E.

Two nodes (x_1, y_1) and (x_2, y_2) of the same environment are neighbours iff $(x_2 - x_1, y_2 - y_1) \in \mathbb{Z}_1 \times \mathbb{Z}_1$. All neighbours of the node (x, y) form the Moore neighbourhood of range 1.

We can also consider finite automaton in two-dimensional environment as a kind of 4-way finite state automaton introduced by Blum and Hewitt [1] that reads the symbol (Moore neighbourhood of range 1) from the environment and reacts by changing its internal state and then moving up, down, left or right to a neighbouring cell of an environment E.

Let an automaton A and environment E interact with each other then an output signal of each of them coincides with an input signal of the other at each instant of time. So in case of 4-way finite state automaton the set of output symbols of an automaton A (set of input symbols of an environment E) is $\{Left, Right, Up, Down\}$ and the input symbols of A (the output symbols of E) form a matrix $O^{3 \times 3}$ that represents the Moore neighbourhood of range 1 for a position (x, y):

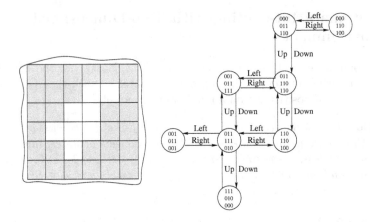

Fig. 1. An example of a geometric environment represented by Moore automaton

$$\begin{pmatrix} o_{-1,1} & o_{0,1} & o_{1,1} \\ o_{-1,0} & o_{0,0} & o_{1,0} \\ o_{-1,-1} & o_{0,-1} & o_{1,-1} \end{pmatrix},$$

where $o_{i,j} = 1$ if $(x+i, y+j) \in D$, otherwise $o_{i,j} = 0$ and $i, j \in \mathbb{Z}_1$ (see Figure 1).

Let us describe the process of interaction between automaton and environment. An automaton A initiates the interaction with an environment E starting from an initial state s_0 and a node $r \in D$. Let A be in a state s and a node r, then automaton moves to the state $\delta_A(s, \lambda_E(r))$ and the node $\delta_E(r, \lambda_A(s, \lambda_E(r)))$.

The *configuration* of an automaton in an environment we denote by pair (s, r), where s is state of an automaton, and r is a node of the environment We also say that configuration (s', r') is directly reachable from (s, r) if $s' = \delta_A(s, \lambda_E(r))$ and $r' = \delta_E(r, \lambda_A(s, \lambda_E(r)))$ and we denote it by $(s, r) \rightarrow (s', r')$. We define $\rightarrow$ as a binary relation over the set configurations.

The *trajectory* of an automaton A in an environment E is a sequence of reachable configurations. By $(s, r)..(s', r')$ we denote a trajectory of an automaton that reach configuration (s', r') from (s, r).

3   From Minsky Machine to More Exotic Models of Computation

In this section we start from geometric interpretation of a well-known model of two-counter Minsky machine that can increment and decrement counters by one and test them for zero. It is known that Minsky machine is a universal model of computations and it is equivalent to Turing machine [16].

It is easy to see that the behaviour of Minsky machine can be interpreted as a 4-way finite state automaton that interacts with (or moves in) the quadrant of the plane $D_Q = \{(x,y)|x \in \mathbb{N}, y \in \mathbb{N}\}$, where the border of a quadrant is the following set of nodes: $G_Q = \{(x,y)|x \geq 0, y = 0, x \in \mathbb{N}, y \in \mathbb{N}\} \cup \{(x,y)|x = 0, y \geq 0, x \in \mathbb{N}, y \in \mathbb{N}\}$. The node (x,y) of the environment $E_Q = (D_Q, O, I, \delta_E, \lambda_E)$ represents the values x and y of two counters and the empty counters of Minsky machine corresponds to the situation when the automaton is on the borders of the geometric environment E_Q (i.e. in a cell of the environment that does not allow to move at least in one direction).

Now we can define some exotic models of computation by changing the shape of the geometric environment. Let us consider an extension of the quadrant of the plane, changing the vertical border by a power function, a polynomial function, a linear function or a sublinear function. Let $a \in \mathbb{Q}^+$ then by $D_{a \cdot n}$, D_{n^a} and D_{a^n} we denote the following environments:

$$D_{a \cdot n} = \{(x,y) \in \mathbb{Z} \times \mathbb{Z}|y \geq a \cdot |x|, x < 0; y \geq 0, x \geq 0\}$$

$$D_{n^a} = \{(x,y) \in \mathbb{Z} \times \mathbb{Z}|y \geq |x|^a, x < 0; y \geq 0, x \geq 0\}$$

$$D_{a^n} = \{(x,y) \in \mathbb{Z} \times \mathbb{Z}|y \geq a^{|x|}, x < 0; y \geq 0, x \geq 0\}$$

In particular in the next section we show that a 4-way finite state automaton in the following three types of geometric environments $D_{a \cdot n}$, D_{n^a} and D_{a^n} where $a = 2$ can simulate a Minsky machine (see Figure 2).

It is not difficult to see that if the left border of the quadrant is extended by a sublinear function (in other words it cannot be covered by a sector in the half plane), for example:

$$D_{sub} = \{(x,y) \in \mathbb{Z} \times \mathbb{Z}|y \geq log|x|, x < 0; y \geq 0, x \geq 0\}$$

then a 4-way automaton in this environment cannot simulate a universal computational model. This environment actually is not *essentially two-dimensional* environment, i.e. the dynamics in this model can be simulated by a finite state automaton interacting with one-dimensional environment.

Another interesting case is an environment which impedes the direct simulation of multiplication as it can be done in the environment D_Q, for example the environment bounded by a parabola. However we show that a finite state automaton interacting with the environment defined by parabola:

$$D_{n^2,n^2} = \{(x,y) \in \mathbb{Z} \times \mathbb{Z}|y \geq x^2\}$$

is a universal model of computation.

The main aim of these results is to support our thesis that the reachability problem is undecidable for any *essentially two-dimensional* environments.

3.1   Undecidability Results

In this subsection we consider several types of geometric environments with borders defined by integral functions $\lceil a \cdot x \rceil$, $\lceil x^a \rceil$ and $\lceil a^x \rceil$.

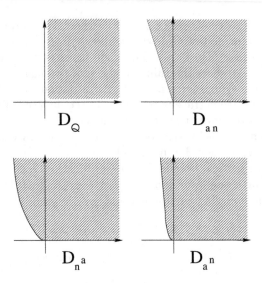

Fig. 2. Universal models of computation represented by different geometric environments

Theorem 1. *A finite state machine in the geometric environments D_{2n}, D_{n^2}, D_{2^n} can simulate a Minsky machine.*

Proof: The proof of this fact is based on a simulation of Minsky machine by a 4-way finite state automaton in the above geometric environments.

First of all we use a well-known trick [19] to get an equivalent model of two counter machine where one of the counters is used as a scratchpad. Another, counter holds an integer whose prime factorization is $2^a \cdot 3^b$. The exponents a, b can be thought of as two virtual counters that are being simulated. If the real counter is set to zero then incremented once, that is equivalent to setting all the virtual counters to zero. If the real counter is doubled, that is equivalent to incrementing a, and if it's halved, that's equivalent to decrementing a. By a similar procedure, it can be multiplied or divided by 3, which is equivalent to incrementing or decrementing b.

Let the finite state machine A interacting with D_Q can reach the configuration (s, a', b') from configuration (s, a, b) by one step. Now we can construct another FSA A' interacting with D_Q, that can reach the configuration $(s, 2^{a'} \cdot 3^{b'}, 0)$ from configuration $(s, 2^a \cdot 2^b, 0)$ by a finite number of states.

To check if a virtual counter such as a (b) is equal to zero, just divide the real counter by 2 (3), see what the remainder is, then multiply by 2 (3) and add back the remainder. That leaves the real counter unchanged. The remainder will have been nonzero if and only if a (b) was zero.

The Geometric Environment D_{2n}. The straightforward modification of the FSA A' gives the result for any sector environment formed by two lines in the half plane. Since the integral line defines the regular shifts of the border, we only need to amortise these shifts by adding a fixed number of right or left moves according to the chosen direction.

The Geometric Environment D_{n^2}. The case where one of the borders does not have periodic shifts is less trivial. However we use again the same scheme and prove that interaction of finite state automaton A' with the quadrant environment D_Q can be reduced in some sense to the interaction of finite state automaton B with environment D_{n^2}. In case of nonperiodic border we need to choose another method to code the counter in the new environment, and more sophisticated method of amortisation during the operations of multiplication and division.

Let the boundary point $(-x, x^2) \in G_{n^2}$ represents a number x. Now let us show that we can convert any point $(-x, x^2)$ to the point $(-2x, (2x)^2)$, that stands for multiplication of x by 2.

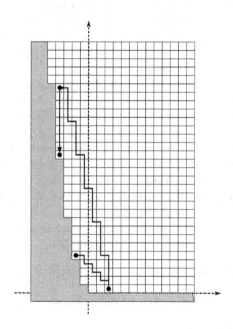

Fig. 3. Simulation of the multiplication by 2 in the geometric environment D_{n^2}

First, FSA converts point $(-x, x^2)$ to point $(x^2 - x, 0)$ by moving right and down until it reaches the border. Then it converts the point $(x^2 - x, 0)$ to the $(-2x, 4x^2 + 4x)$ by repeating the following pattern "moving four times up and one time left" until it reaches the border. Since the inequality $(2x)^2 \leq 4x^2 + 4x \leq (2x + 1)^2$ holds for any natural number we can state that the point $(-2x, 4x^2 + 4x)$ belongs to the border G_{n^2}. Finally the point $(-2x, 4x^2 + 4x)$ can be converted to the point $(-2x, 4x^2)$ by a finite number of moves down along the border, since all points $\{(-2x, w) | (2x)^2 < w \le (2x+1)^2\}$ are the boundary points. Let us consider how to perform multiplication on 3 in this environment. In other words we need to show that from any point $(-x, x^2)$ we can reach the point $(-3x, (3x)^2)$. We start just as in previous case. We convert point $(-x, x^2)$ into $(x^2 - x, 0)$ and then to $(-3x - 3, 9x^2 + 18x + 27)$, which is the boundary point according to the inequality $(3x+3)^2 < 9x^2 + 18x + 27 < (3x+4)^2$, which holds for all $x \in \mathbb{N}$. After that FSA moves from point $(-3x - 3, 9x^2 + 18x + 27)$ to point $(-3x, (3x)^2)$ along the border via 3 corners $(-3x - 3, (3x+3)^2)$, $(-3x - 2, (3x+2)^2)$ and $(-3x - 1, (3x+1)^2)$. In a similar way we construct automaton that can divide on 2 and 3 in D_{n^2}.

The Geometric Environment D_{2^n}. Let us prove that we can perform the operations of multiplication and division by a finite state automaton in the geometric environment D_{2^n}.

Let the boundary point $p = (-\lfloor log_2 x \rfloor, x)$ stands for a positive number x in the geometric environment D_{2^n}. The finite automaton can move from the point $(-\lfloor log_2 x \rfloor, x)$ to the new point $(0, x - \lfloor log_2 x \rfloor)$ by repeating a pair of operations move left and down until it reaches the border. Then the finite automaton can reach the border point r by repeating two moves up and one move left. Thus, r is either $(-\lfloor log_2 x \rfloor, 2x)$ or $(-\lfloor log_2 x \rfloor - 1, 2x + 2)$.

Now let us show how to check by a finite automaton in which part of E it reaches the boundary point r from the initial point p. We use the simple property that if n is an even number then $(2n \bmod 4) = 0$ and $(2n+2 \bmod 4) \ne 0$, if n is an odd number then $(2n \bmod 4) \ne 0$ and $(2n+2 \bmod 4) = 0$. We first check by finite automaton if the ordinate of the point p is odd or even and then we check the ordinate of the point r on divisibility by 4. So if r's ordinate is divisible on 4 and p's ordinate is odd then the automaton is in the point $(-\lfloor log_2 x \rfloor - 1, 2x + 2)$ and it can move down to the point $(-\lfloor log_2 x \rfloor - 1, 2x) = (-\lfloor log_2 2x \rfloor, 2x)$, otherwise it is in the point $(-\lfloor log_x \rfloor, 2x) = (-\lfloor log_2 2x \rfloor, 2x)$.

In a similar way we can show that finite automaton can multiply on 3 and divide on 2 and 3 in the geometric environment D_{2^n}.

Theorem 2. *A finite state automaton in the geometric environment*

$$D_{n^2, n^2} = \{(x, y) \in \mathbb{Z} \times \mathbb{Z} | y \ge x^2\}$$

can simulate a Minsky machine.

Proof: Let the boundary point (x, x^2) stands for a positive number x in the geometric environment D_{n^2, n^2}. A finite automaton cannot multiply in D_{n^2, n^2} if it touches the border only by a constant number of times as we have in case of D_{n^2} or D_{2^n}. So in case of D_{n^2, n^2} we introduce some kind of cycle that will be used for multiplication and division.

Let us prove that a finite automaton will always reach a point $(2x, (2x)^2)$ from a point (x, x^2) using th following procedure:

Multiplication by 2 in D_{n^2,n^2}

   Input: A point (x, x^2)
   Output: A point $(2x, (2x)^2)$

  repeat
    repeat
        Do 1 move Up and 1 move Left;
    until FSA is on the border
    repeat
        Do move Right;
    until FSA is on the border
    Do 4 moves Up;

    until FSA is on the cell $\begin{pmatrix} 1\ 1\ 1 \\ 1\ 1\ 0 \\ 1\ 1\ 0 \end{pmatrix}$

  repeat
    Do move Down;

    until FSA is on the cell $\begin{pmatrix} 1\ 1\ 0 \\ 1\ 1\ 0 \\ 1\ 0\ 0 \end{pmatrix}$

A finite automaton can reach the boundary point $(-(x + 1), (x + 1)^2) = (-(x + 1), x^2 + 2x + 1)$ in D_{n^2,n^2} from a point (x, x^2) by repeating the pattern "up and left", since it will move left for exactly $2x + 1$ cells. It is easy to see that an automaton will reach the point $(-(x + 1), (x + 1)^2 + c)$ from $(x, x^2 + c)$ for any $0 \le c < 2x + 1$ by repeating the same pattern "up and left". Then it can reach the point $(x + 1, (x + 1)^2 + c)$ by moving "right" until it reaches the border and the point $(x + 1, (x + 1)^2 + c + 4)$ doing additional 4 moves "up".

Let us call the sequence of moves from the point $(x, x^2 + c)$ to the point $(x + 1, (x + 1)^2 + c + 4)$ a *cycle*. In the above procedure of multiplication by 2, starting from the point (x, x^2) the finite automaton will meet the cell

$$\begin{pmatrix} 1\ 1\ 1 \\ 1\ 1\ 0 \\ 1\ 1\ 0 \end{pmatrix}$$

for the first time exactly after x cycles since it corresponds to the point $(2x, 4x^2 + 4x)$. If the automaton will move down from the cell

$$\begin{pmatrix} 1\ 1\ 1 \\ 1\ 1\ 0 \\ 1\ 1\ 0 \end{pmatrix} \text{ to } \begin{pmatrix} 1\ 1\ 0 \\ 1\ 1\ 0 \\ 1\ 0\ 0 \end{pmatrix}$$

it reaches the point $(2x, 4x^2)$ and the multiplication by 2 is completed (see Figure 4). Similarly we can perform multiplication by 3 if we slightly change the previous procedure. If we change our cycle in a such way that the automaton will do 3 additional moves up instead of 4 it reaches the cell

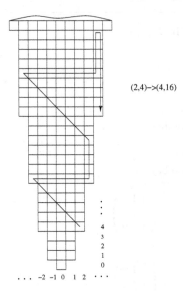

(2,4)->(4,16)

Fig. 4. An example of multiplication of $x = 2$ by 2 in the geometric environment D_{n^2,n^2}. The operation of multiplication is simulated by the movement of a finite automaton from (x, x^2) to $(2x, (2x)^2)$

$$\begin{pmatrix} 1\ 1\ 1 \\ 1\ 1\ 0 \\ 1\ 1\ 0 \end{pmatrix}$$

for the first time after $2x$ cycles and will be in the position $(3x, 9x^2 + 6x)$. After that it moves down in the same way as in previous case to reach the point $(3x, (3x)^2)$. Another pair of operation such that division by 2 and 3 an automaton can perform in the following way. The automaton moves from point (x, x^2) to the point $(-x, x^2)$ and back. In such way it perform $4x$ steps or moves. In order to check the divisibility of x by 2 (or 3) it needs to check the divisibility of $4x$ by 8 (or 12) that can be done using a finite memory.

4   Future Work and Conclusion

As we mentioned before the long-term goal of this work is to characterize the class of non-efficient geometric environments. It follows from the developed methods of simulating a Minsky machine in the considered geometric environments that there are several factors that have an influence on the status of reachability problem: border markers such as corners in the environments, the continued growth expansion of the environment, etc. These factors create essentially two-dimensional environment that cannot be modelled by one-dimensional systems, like push-down or one-counter automata. So it could be interesting to define more formal condition for essentially two-dimensional systems. On the other hand we still have unanswered questions about the reachability problem for a number of specific environments. One of these cases is the environment in $\mathbb{N}^2$ bounded by logarithmic function:

$$D_{log} = \{(x, y) \in \mathbb{N} \times \mathbb{N} | y \leq log(x)\}.$$

References

1. Manuel Blum and Carl Hewitt. Automata on a 2-Dimensional Tape. FOCS (1967) 155-160
2. V.I. Grunskaya. Some properties of trajectories of automata in labyrinths, Thesis [in Russian]. Moscow State University, Moscow (1994)
3. Dora Giammarresi. Finite State Recognizability for Two-Dimensional Languages: A Brief Survey. Developments in Language Theory (1995) 299-308
4. V.M. Glushkov, G.E. Tseitlin and E.L.Yushchenko. Algebra, Languages and Programming [in Russian]. Naukova Dumka, Kiev (1989)
5. I.S. Grunsky and Kurganskyy. Indistinguishability of finite automata interacting with an environment [in Russian]. Dokl. AN Ukraine, v11 (1993), 31-33
6. I.S. Grunsky and A.N. Kurganskyy. Indistinguishability of finite automata with constrained behaviour, Cybernetics and systems analysis, Vol5, (1996) 58-72
7. M. Jantzen and O. Kurganskyy. Refining the hierarchy of blind multicounter languages and twist-closed trios. *Information and Computation*, 185 (2003) 159-181.
8. Katsushi Inoue and Itsuo Takanami. A survey of two-dimensional automata theory. Inf. Sci. 55(1-3) (1991) 99-121
9. V.B. Kudryavtsev, S.V. Alyoshin and A.S. Podkolzin. An introduction to the Theory of Automata [in Russian], Nauka, Moscow, (1985)
10. V.B. Kudryavtsev, Sh. Ushchumlich, and G. Kilibarda. On the behavior of automata in labyrinths. Discrete Math. and Applications, 3 (1993) 1-28
11. G. Kilibarda, V. B. Kudryavtsev and Sh. Ushchumlich. Collectives of automata in labyrinths Discrete Mathematics and Applications 13(5) (2003) 429-466
12. A.N. Kurganskyy. Indistinguishability of the finite automata interacting with an environment. Thesis [in Russian]. Saratov State University, Russia (1997)
13. A.N. Kurganskij. Indistinguishability of finite-state automata with respect to some environments. (Russian, English) [J] 37, No.1, (2001) 33-41 ; translation from Kibern. Sist. Anal. 2001, No.1, (2001) 43-55
14. A. Kurganskyy, I. Potapov. On the bound of algorithmic resolvability of correctness problems of automaton interaction through communication channels. *Cybernetics and System Analysis*, 3 (1999) 49-57
15. Maurice Mergenstern. Frontier between decidability and undecidability: a survey. Theoretical Computer Science, v.231 (2000) 217-251.
16. M. Minsky. Computation: Finite and Infinite Machines. Englewood Cliffs, N.J.: Prentice-Hall (1967)
17. C. Moore and J. Kari. New Results on Alternating and Non-Deterministic Two-Dimensional Finite-State Automata. International Symposium on Theoretical Aspects of Computer Science, STACS (2001) 396-406
18. Tokio Okazaki, Katsushi Inoue, Akira Ito, Yue Wang. Space Hierarchies of Two-Dimensional Alternating Turing Machines, Pushdown Automata and Counter Automata. IJPRAI 13(4) (1999) 503-521
19. Wikipedia, the free encyclopedia. http://en.wikipedia.org/wiki/Counter

The Role of the Complementarity Relation in Watson-Crick Automata and Sticker Systems

Dietrich Kuske[1] and Peter Weigel[2]

[1] Institut für Algebra, Technische Universität Dresden, Dresden, Germany
`kuske@math.tu-dresden.de`
[2] Institut für Informatik, Martin-Luther-Universität Halle-Wittenberg,
Halle, Germany
`mail@peter-weigel.de`

Abstract. In [4, page166], it is asked what influence the complementarity relation plays as far as the expressiveness of sticker systems and Watson-Crick automata are concerned. Here, we give the answer: (almost) none! More precisely, we show that every language L of a sticker system or a Watson-Crick automaton is the language of such a system with a one-to-one complementarity relation. Our second group of results shows that L is the inverse block coding of a language from the same family over any nontrivial fixed complementarity relation. Finally, we prove that any Watson-Crick automaton can be transformed into an equivalent simple and all-final one. This implies the collapse of parts of the hierarchy introduced in [4].

1 Introduction

The advent of bioinformatics has blessed the formal language community with several new mechanisms that generate or accept words. Often, these new mechanisms explore the interplay of standard mechanisms with the Watson-Crick complementarity. Sticker systems [2] use ligation of incomplete DNA-molecules to perform computations. The idea goes back to Adleman's experiment where ligation can take place at both ends of the molecule independently. In order to go beyond regular languages, sticker systems synchronize the extension at the two ends of a molecule. While these systems generate molecules, Watson-Crick automata [1] accept DNA-molecules. They are based on the idea of finite automata running on a complete DNA-molecule. Since such a molecule is not a single word but consists of two strands, a Watson-Crick automaton is equipped with two heads that move with different speeds on the two strands.

As usual in theoretical computer science, the conceptual content of a phenomenon becomes clear only if one abstracts some aspects. In case of sticker systems and Watson-Crick automata, the obvious candidates for abstraction are the concrete alphabet and the form of the complementarity relation.[1] The

---

[1] On the other hand, to the knowledge of the authors, Watson-Crick D0L-systems [3, 6] have only been considered with one-to-one complementarity relations.

C.S. Calude, E. Calude and M.J. Dinneen (Eds.): DLT 2004, LNCS 3340, pp. 272–283, 2004.
© Springer-Verlag Berlin Heidelberg 2004

main topic of this paper is the question (raised in [4–page 166]) what influence these abstractions have. The answer is the same for both, sticker systems and Watson-Crick automata. If one fixes the alphabet and varies the complementarity relation, the language family does not change significantly. More precisely, any sticker system or Watson-Crick automaton over the alphabet V can be replaced by an equivalent such device over V with a one-to-one complementarity relation. This result also holds for subfamilies of sticker systems and Watson-Crick automata as defined in the literature. If the alphabet V is allowed to change as well, we run into the trouble of alphabets of different size and therefore need squeezing mechanisms. Here, we use block codes, i.e., words of a fixed length over V that represent letters of V'. Using such block codes, we obtain that any alphabet with any non-trivial complementarity relation generates all sticker or Watson-Crick languages. Again, this result holds for most of the subfamilies considered in the literature as well.

The final section of this paper deals with the hierarchy of Watson-Crick languages defined by the above mentioned subfamilies of these automata. We show that any Watson-Crick automaton can be transformed effectively into an equivalent one where any state is accepting and where any transition of the automaton involves the movement of at most one of the two heads. This result implies the collapse of parts of the hierarchy presented in [4].

2   Basic Definitions

Let V be an alphabet with $V = \{C, G, T, A\}$ being the most prominent example. This alphabet V comes equipped with a binary relation ρ called *complementarity*.[2] The standard complementarity on the alphabet $\{C, G, T, A\}$ is the relation $\{(A, T), (T, A), (G, C), (C, G)\}$. In our investigations, we will come across the set $V^* \times V^*$ of pairs of words and the set ρ^* of finite sequences of pairs from ρ. Elements from the latter can naturally be considered as elements from $V^* \times V^*$, namely the sequence $\begin{pmatrix} a_1 \\ b_1 \end{pmatrix} \begin{pmatrix} a_2 \\ b_2 \end{pmatrix} \cdots \begin{pmatrix} a_n \\ b_n \end{pmatrix}$ corresponds to the pair $\begin{pmatrix} a_1 a_2 \ldots a_n \\ b_1 b_2 \ldots b_n \end{pmatrix}$. For notational simplicity, we write $\begin{bmatrix} a_1 a_2 \ldots a_n \\ b_1 b_2 \ldots b_n \end{bmatrix}_\rho$ for the word over ρ. Considering the standard alphabet, it makes sense to call these words from ρ^* *complete molecules* over (V, ρ). By $O(V)$, we denote the set $(V^* \times \{\varepsilon\}) \cup (\{\varepsilon\} \times V^*)$ - its elements are *one-stranded molecules* (where we distinguish molecules with an empty second strand - i.e., elements of $V \times \{\varepsilon\}$ - from those with an empty first strand). Finally, $W(V, \rho)$ denotes the set $W(V, \rho) = (O(V) \times \rho^+ \times O(V)) \cup O(V)$ of *molecules*, its elements will usually be denoted by small Greek letters from the beginning of the alphabet. The molecule

$$\alpha = \left(\begin{pmatrix} u_1 \\ u_2 \end{pmatrix}, \begin{bmatrix} v_1 \\ v_2 \end{bmatrix}_\rho, \begin{pmatrix} w_1 \\ w_2 \end{pmatrix} \right) \text{ will also be abbreviated } \begin{pmatrix} u_1 \\ u_2 \end{pmatrix} \begin{bmatrix} v_1 \\ v_2 \end{bmatrix}_\rho \begin{pmatrix} w_1 \\ w_2 \end{pmatrix}.$$

---

[2] Note that, differently from [4], we do not require ρ to be symmetric - this is done just for convenience and does not affect the results.

If $u_1 = u_2 = w_1 = w_2 = \varepsilon$, we consider α as an element of ρ^*, i.e., as a complete molecule. A partial concatenation $\cdot_\rho$ on $W(V, \rho)$ is defined as follows:

- $\begin{pmatrix} u_1 \\ \varepsilon \end{pmatrix} \cdot_\rho \begin{pmatrix} v_1 \\ \varepsilon \end{pmatrix} = \begin{pmatrix} u_1 v_1 \\ \varepsilon \end{pmatrix}$ and $\begin{pmatrix} \varepsilon \\ u_2 \end{pmatrix} \cdot_\rho \begin{pmatrix} \varepsilon \\ v_2 \end{pmatrix} = \begin{pmatrix} \varepsilon \\ u_2 v_2 \end{pmatrix}$.

- $\begin{pmatrix} u_1 \\ u_2 \end{pmatrix} \begin{bmatrix} v_1 \\ v_2 \end{bmatrix}_\rho \begin{pmatrix} w_1 \\ w_2 \end{pmatrix} \cdot_\rho \begin{pmatrix} x_1 \\ x_2 \end{pmatrix} = \begin{pmatrix} u_1 \\ u_2 \end{pmatrix} \begin{bmatrix} v_1 y_1 \\ v_2 y_2 \end{bmatrix}_\rho \begin{pmatrix} z_1 \\ z_2 \end{pmatrix}$ if y_i is the prefix of $w_i x_i$ of

  length $\min\{|w_1 x_1|, |w_2 x_2|\}$, $y_i z_i = w_i x_i$, and $\begin{pmatrix} y_1 \\ y_2 \end{pmatrix} \in \rho^*$.

- $\begin{pmatrix} x_1 \\ x_2 \end{pmatrix} \cdot_\rho \begin{pmatrix} u_1 \\ u_2 \end{pmatrix} \begin{bmatrix} v_1 \\ v_2 \end{bmatrix}_\rho \begin{pmatrix} w_1 \\ w_2 \end{pmatrix}$ is defined analogously.

- $\begin{pmatrix} u_1 \\ u_2 \end{pmatrix} \begin{bmatrix} v_1 \\ v_2 \end{bmatrix}_\rho \begin{pmatrix} w_1 \\ w_2 \end{pmatrix} \cdot_\rho \begin{pmatrix} x_1 \\ x_2 \end{pmatrix} \begin{bmatrix} y_1 \\ y_2 \end{bmatrix}_\rho \begin{pmatrix} z_1 \\ z_2 \end{pmatrix} = \begin{pmatrix} u_1 \\ u_2 \end{pmatrix} \begin{bmatrix} v_1 w_1 x_1 y_1 \\ v_2 w_2 x_2 y_2 \end{bmatrix}_\rho \begin{pmatrix} z_1 \\ z_2 \end{pmatrix}$ if

  $\begin{pmatrix} u_1 \\ u_2 \end{pmatrix} \begin{bmatrix} v_1 \\ v_2 \end{bmatrix}_\rho \begin{pmatrix} w_1 \\ w_2 \end{pmatrix} \cdot_\rho \begin{pmatrix} x_1 \\ x_2 \end{pmatrix} = \begin{pmatrix} u_1 \\ u_2 \end{pmatrix} \begin{bmatrix} v_1 w_1 x_1 \\ v_2 w_2 x_2 \end{bmatrix}_\rho \begin{pmatrix} \varepsilon \\ \varepsilon \end{pmatrix}$.

If none of these cases applies, the concatenation $\alpha \cdot_\rho \beta$ is undefined. One can check that this operation, if defined, is associative and that any molecule is a finite product of elements from $\rho \cup (V \times \{\varepsilon\}) \cup (\{\varepsilon\} \times V)$.

Definition 1. *A* sticker system *(over (V, ρ)) is a tuple $S = (V, \rho, A, D)$ consisting of a finite set $A \subseteq W(V, \rho) \setminus O(V)$ of* axioms *and a finite set $D \subseteq W(V, \rho) \times W(V, \rho)$ of* rules.

Let $C^0(S) = A$ and, for $k \in \mathbb{N}$, define $C^{k+1}(S)$ to be the set of molecules $\alpha \cdot_\rho \beta \cdot_\rho \gamma$ with $\beta \in C^k(S)$ and $(\alpha, \gamma) \in D$, i.e., $C^k(S)$ contains those molecules that can be derived from an axiom in A using k derivation steps.

Definition 2. *Let $S = (V, \rho, A, D)$ be a sticker system. Its* molecule language *$LM(S) \subseteq \rho^*$ is the set of complete molecules $\beta \in \bigcup_{k \in \mathbb{N}} C^k(S) \cap \rho^*$. The* language *of S is the set*

$$L(S) = \{u \in V^+ \mid \exists v \in V^+ : \begin{bmatrix} u \\ v \end{bmatrix}_\rho \in LM(S)\}.$$

Definition 3. *A* Watson-Crick automaton *M is a tuple (V, ρ, K, s_0, T, F) where K is a finite set of* states, *$s_0 \in K$ is the* initial state, *$T \subseteq K \times (V^* \times V^*) \times K$ is a finite transition relation, and $F \subseteq K$ is a set of accepting states.*

Note that a Watson-Crick automaton is just a gsm where input- and output-alphabet equal V and this alphabet is equipped with a complementarity relation. Since Watson-Crick automata are special gsms, we can borrow the definition of runs from there:

A run of the Watson-Crick automaton $(V, \rho, K, s_0, \delta, F)$ is a sequence

$$q_1, x_1, q_2, x_2, \ldots, x_n, q_{n+1}$$

with $q_i \in K$ and $x_i \in V^* \times V^*$ such that $(q_i, x_i, q_{i+1}) \in T$ for $1 \leq i \leq n$. Its *label* is the pair of words $x_1 x_2 \ldots x_n$ where we concatenate the elements $x_i \in V^* \times V^*$ componentwise. The run is *successful* if $q_1 = s_0$ and $q_{n+1} \in F$. The language of molecules $LM(M)$ accepted by M is the set of all complete molecules that label a successful run of M. Then, the language of the Watson-Crick automaton M is defined by

$$L(M) = \left\{ u \in V^+ \,\middle|\, \exists v \in V^+ : \begin{bmatrix} u \\ v \end{bmatrix}_\rho \in LM(M) \right\}.$$

Note that writing $\begin{bmatrix} u \\ v \end{bmatrix}_\rho$, we require implicitly that the ith letter of u and the ith letter of v belong to the complementarity relation ρ (in particular, u and v have the same length). Let $\mathrm{AWK}(V, \rho)$ denote the set of languages accepted by some Watson-Crick automaton over (V, ρ) ("A" stands for "arbitrary") and $\mathrm{AWK} = \bigcup_{(V,\rho)} \mathrm{AWK}(V, \rho)$ be the set of languages accepted by some Watson-Crick automaton whatever the complementarity relation is.

A Watson-Crick automaton (V, ρ, K, s_0, T, F) is

- *stateless* if $K = F = \{s_0\}$,
- *all-final* if $K = F$,
- *simple* if $(p, \left(\begin{smallmatrix} u \\ v \end{smallmatrix}\right), q) \in T$ implies $u = \varepsilon$ or $v = \varepsilon$, and
- *1-limited* if $(p, \left(\begin{smallmatrix} u \\ v \end{smallmatrix}\right), q)$ implies $|uv| = 1$.

We abbreviate "stateless" by "N" (for "no state"), "all-final" by "F", "simple" by "S", and "1-limited" by "1". Then $\mathrm{FSWK}(V, \rho)$ is the set of languages accepted by an all-final and simple Watson-Crick automaton over (V, ρ). The terms N1WK, NSWK, F1WK, SWK, 1WK, and NWK can be understood analogously. The left diagram in Fig. 1 (page 281) depicts the inclusion structure as shown in [4] (REG is the family of regular languages in V^+ and CS that of context-sensitive ones).

3   One-to-One Complementarity Relations Suffice

For a set V, let Δ_V denote the identity relation $\{(a, a) \mid a \in V\}$ over the domain V. In this section, we prove that any Watson-Crick language over (V, ρ) is also a Watson-Crick language over (V, Δ_V) and that the same holds for sticker languages. Actually, the result holds for any one-to-one relation ρ' on V in place of Δ_V - in order to keep notations simple, we give the proof for the identity relation, only.

Theorem 4. Let $M = (V, \rho, K, s_0, T, F)$ be a Watson-Crick automaton over the complementarity relation (V, ρ). Then one can construct a Watson-Crick automaton $M' = (V, \Delta_V, K, s_0, T', F)$ over (V, Δ_V) with $L(M) = L(M')$.

We have to construct a Watson-Crick automaton M' over (V, Δ_V) that accepts the molecule $\begin{bmatrix} u \\ u \end{bmatrix}_{\Delta_V}$ if there exists $v \in V^*$ such that $\begin{bmatrix} u \\ v \end{bmatrix}_\rho$ is accepted by M. Hence the second head of M' will read the same word as the first one. When doing so, it will "guess" the word v and simulate the behavior of M. The formal details of the proof are as follows:

Proof. Let

$$T' = \{(p, u, w, q) \in K \times (V^* \times V^*) \times K \mid \exists v \in V^* : (p, u, v, q) \in T \text{ and } \begin{pmatrix} w \\ v \end{pmatrix} \in \rho^* \}.$$

First, let $u = a_1 a_2 \ldots a_n \in L(M)$. Then there exists a word $v = b_1 b_2 \ldots b_n$ with $(a_i, b_i) \in \rho$ and a successful run of M on the pair (u, v). Hence there exist indices $1 = i_1 \leq i_2 \leq \ldots i_k = n + 1$ and $1 = j_1 \leq j_2 \leq \ldots j_k = n + 1$ and states $q_\ell \in K$ $(1 \leq \ell \leq k)$ such that

$$q_1 = s_0, \ (q_\ell, a_{i_\ell} \ldots a_{i_{\ell+1}-1}, b_{j_\ell} \ldots b_{j_{\ell+1}-1}, q_{\ell+1}) \in T \ (1 \leq \ell < k), \text{ and } q_{k+1} \in F.$$

Since $(a_i, b_i) \in \rho$, this implies $(q_\ell, a_{i_\ell} \ldots a_{i_{\ell+1}-1}, a_{j_\ell} \ldots a_{j_{\ell+1}-1}, q_{\ell+1}) \in T'$. Hence we found a successful run in M' labeled (u, u). Since any word is Δ_V-complementary to itself, this implies $u \in L(M')$ and therefore $L(M) \subseteq L(M')$.

Conversely, let $u = a_1 a_2 \ldots a_n \in L(M')$. Since the only word complementary to u is u itself, there exists a successful run of M' labeled (u, u). Hence there exist indices $1 = i_1 \leq i_2 \leq \ldots i_k = n + 1$ and $1 = j_1 \leq j_2 \leq \ldots j_k = n + 1$ and states $q_\ell \in K$ $(1 \leq \ell \leq k)$ such that

$$q_1 = s_0, \ (q_\ell, a_{i_\ell} \ldots a_{i_{\ell+1}-1}, a_{j_\ell} \ldots a_{j_{\ell+1}-1}, q_{\ell+1}) \in T' \ (1 \leq \ell < k), \text{ and } q_{k+1} \in F.$$

The construction of T' implies the existence of letters $b_i \in V$ with $(a_i, b_i) \in \rho$ and $(q_\ell, a_{i_\ell} \ldots a_{i_{\ell+1}-1}, b_{j_\ell} \ldots b_{j_{\ell+1}-1}, q_{\ell+1}) \in T'$. Hence we find a successful run in M' labeled $(u, b_1 b_2 \ldots b_n)$. Since $(a_i, b_i) \in \rho$, the words u and $b_1 b_2 \ldots b_n$ are ρ-complementary implying $u \in L(M)$. Hence, indeed, $L(M') \subseteq L(M)$. $\square$

A short look at the proof reveals that this theorem holds for all subclasses of Watson-Crick automata as defined above, i.e., if M is an automaton having any combination of these properties, then M' satisfies the same combination of properties. So, e.g., SWK$(V, \rho) =$ SWK(V, Δ_V).

Next, we want to prove a result analogous to Thm. 4 for sticker systems. The idea is again quite similar to the one that worked perfectly well for Watson-Crick automata: the sticker system guesses a matching second strand of a molecule over (V, ρ) while building a molecule over (V, Δ_V). The guessing now takes a slightly different form and is somewhat hidden in the mapping $\bar{} : W(V, \rho) \to 2^{W(V, \Delta_V)}$ defined below. For two sets $A, B \subseteq W(V, \Delta_V)$, let $A \cdot_{\Delta_V} B = \{\alpha \cdot_{\Delta_V} \beta \in W(V, \Delta_V) \mid \alpha \in A, \beta \in B\}$. This operation on the power set of $W(V, \Delta_V)$ is associative and defined everywhere.

Lemma 5. *There is a function* $\bar{} : W(V, \rho) \to 2^{W(V, \Delta_V)}$ *satisfying*

(1) $\overline{\begin{bmatrix} a \\ b \end{bmatrix}_\rho} = \left\{ \begin{bmatrix} a \\ a \end{bmatrix}_{\Delta_V} \right\}$, $\overline{\begin{pmatrix} a \\ \varepsilon \end{pmatrix}} = \left\{ \begin{pmatrix} a \\ \varepsilon \end{pmatrix} \right\}$ *and* $\overline{\begin{pmatrix} \varepsilon \\ b \end{pmatrix}} = \left\{ \begin{pmatrix} \varepsilon \\ a \end{pmatrix} \mid \begin{pmatrix} a \\ b \end{pmatrix} \in \rho \right\}$ *and*
(2) $\overline{\alpha} \cdot_{\Delta_V} \overline{\beta} = \overline{\alpha \cdot_\rho \beta}$ *whenever* $\alpha \cdot_\rho \beta$ *is defined, and* $\overline{\alpha} \cdot_{\Delta_V} \overline{\beta} = \emptyset$ *otherwise.*

Furthermore, γ *is complete iff* $\overline{\gamma}$ *contains only complete molecules (for* $\gamma \in W(V, \rho)$).

Proof. First, we define the function in question:

$$\overline{\binom{u}{\varepsilon}} = \left\{ \binom{u}{\varepsilon} \right\} ,$$

$$\overline{\binom{\varepsilon}{v}} = \left\{ \binom{\varepsilon}{u} \,\middle|\, \binom{u}{v} \in \rho^* \right\} ,$$

$$\overline{\binom{u_1}{u_2} \begin{bmatrix} v_1 \\ v_2 \end{bmatrix}_\rho \binom{w_1}{w_2}} = \overline{\binom{u_1}{u_2}} \times \left\{ \begin{bmatrix} v_1 \\ v_1 \end{bmatrix}_{\Delta_V} \right\} \times \overline{\binom{w_1}{w_2}} .$$

Thus, the set $\overline{\alpha}$ is obtained from α by replacing the second strand by a complementary one at both sticky ends (if this second strand exists), and the second by the first in the complete part of α. If $\alpha, \beta \in O(V)$, statement (2) is immediate. We show in detail how to prove (2) in case $\alpha \in W(V, \rho)$ and $\beta \in O(V)$, the remaining cases can be dealt with similarly. So let $\alpha = \binom{u_1}{u_2} \begin{bmatrix} v_1 \\ v_2 \end{bmatrix}_\rho \binom{w_1}{w_2}$ and $\beta = \binom{x_1}{x_2}$.

First, let γ be an element of $\overline{\alpha} \cdot_{\Delta_V} \overline{\beta}$. Then there are $u_2', w_2', x_2' \in V^*$ such that $\binom{u_2'}{u_2}, \binom{w_2'}{w_2}, \binom{x_2'}{x_2} \in \rho^*$ and $\gamma = \binom{u_1}{u_2'} \begin{bmatrix} v_1 \\ v_1 \end{bmatrix}_{\Delta_V} \binom{w_1}{w_2'} \cdot_{\Delta_V} \binom{x_1}{x_2'}$. Because of the special form of the complementarity relation Δ_V, there are $y_1, z_1, z_2' \in V^*$ such that $\gamma = \binom{u_1}{u_2'} \begin{bmatrix} v_1 y_1 \\ v_1 y_1 \end{bmatrix}_{\Delta_V} \binom{z_1}{z_2'}$. Hence y_1 is the prefix of $w_1 x_1$ and of $w_2' x_2'$ of length $\min(|w_1 x_1|, |w_2' x_2'|)$, $w_1 x_1 = y_1 z_1$, and $w_2' x_2' = y_1 z_2'$. Let y_2 be the prefix of $w_2 x_2$ of length $|y_1| = \min(|w_1 x_1|, |w_2 x_2|)$ and define $z_2 \in V^*$ such that $y_2 z_2 = w_2 x_2$. Note that $\binom{y_1 z_2'}{y_2 z_2'} = \binom{w_2' x_2'}{w_2 x_2} \in \rho^*$ and $|y_1| = |y_2|$ imply $\binom{y_1}{y_2} \in \rho^*$. Thus, $\alpha \cdot_\rho \beta = \binom{u_1}{u_2} \begin{bmatrix} v_1 y_1 \\ v_2 y_2 \end{bmatrix}_\rho \binom{z_1}{z_2}$ and $\gamma \in \overline{\alpha \cdot_\rho \beta}$. Thus, we proved the inclusion $\overline{\alpha} \cdot_{\Delta_V} \overline{\beta} \subseteq \overline{\alpha \cdot_\rho \beta}$.

Conversely, let $\gamma \in \overline{\alpha \cdot_\rho \beta}$. Then there are $y_1, y_2, z_1, z_2 \in V^*$ such that

$$\gamma \in \overline{\binom{u_1}{u_2} \begin{bmatrix} v_1 y_1 \\ v_2 y_2 \end{bmatrix}_\rho \binom{z_1}{z_2}}$$

where y_i is the prefix of $w_i x_i$ of length $\min(|w_1 x_1|, |w_2 x_2|)$, $y_i z_i = w_i x_i$, and $\binom{y_1}{y_2} \in \rho^*$. Hence we find $u_2', z_2' \in V^*$ with

$$\gamma = \binom{u_1}{u_2'} \begin{bmatrix} v_1 y_1 \\ v_1 y_1 \end{bmatrix}_{\Delta_V} \binom{z_1}{z_2'}, \text{ and } \binom{u_2'}{u_2}, \binom{z_2'}{z_2} \in \rho^*.$$

We split $y_1 z_2'$ into $w_2' x_2'$ with $|w_2'| = |w_2|$. Then $\binom{w_2' x_2'}{w_2 x_2} = \binom{y_1 z_2'}{y_2 z_2} \in \rho^*$ and $|w_2'| = |w_2|$ imply $\binom{w_2'}{w_2} \in \rho^*$. Hence

$$\binom{u_1}{u_2'} \begin{bmatrix} v_1 \\ v_1 \end{bmatrix}_{\Delta_V} \binom{w_1}{w_2'} \in \overline{\alpha} \text{ and } \binom{x_1}{x_2'} \in \overline{\beta} .$$

Since γ is the product of these two molecules w.r.t. Δ_V, we get $\gamma \in \overline{\alpha} \cdot \overline{\beta}$. $\quad\square$

Theorem 6. *Let $S = (V, \rho, A, D)$ be a sticker system over the complementarity relation (V, ρ). Then one can construct a sticker system S' over (V, Δ_V) with $L(S) = L(S')$.*

In [4], several subclasses of sticker systems are considered (e.g., simple, one-sided, right-sided, with bounded delay, ...). We only remark that the above theorem holds for any of these subclasses equally well since the proof carries over immediately.

Proof. Let $A' = \overline{A} = \bigcup_{\beta \in A} \overline{\beta}$, $D' = \bigcup_{(\alpha, \gamma) \in D} (\overline{\alpha} \times \overline{\gamma})$, and $S' = (V, \Delta_V, A', D')$. Since $\overline{\alpha}$ is a finite subset of $W(V, \Delta_V)$ for any $\alpha \in W(V, \rho)$, the tuple $S' = (V, \Delta_V, A', D')$ is a sticker system over (V, Δ_V).

We show by induction that $C^k(S') = \overline{C^k(S)}$ holds for any $k \in \mathbb{N}$. For $k = 0$, we have $\overline{C^0(S)} = \overline{A} = C^0(S')$ settling the base case. For the inductive argument, suppose $C^k(S') = \overline{C^k(S)}$. First, let $\delta' \in \overline{C^{k+1}(S)}$. Then there is $\delta \in C^{k+1}(S)$ with $\delta' \in \overline{\delta}$. Hence we find $\beta \in C^k(S)$ and $(\alpha, \gamma) \in D$ with $\delta = \alpha \beta \gamma$. But this implies $\delta' \in \overline{\delta} = \overline{\alpha \beta \gamma} = \overline{\alpha} \overline{\beta} \overline{\gamma}$. By the induction hypothesis, $\overline{\beta} \subseteq C^k(S')$. The construction of S' yields $\overline{\alpha} \times \overline{\gamma} \subseteq D'$. Thus, we have $\delta' \in \overline{\alpha} \overline{\beta} \overline{\gamma} \subseteq C^{k+1}(S')$.

Conversely, let $\delta' \in C^{k+1}(S')$. Then there are $\beta' \in C^k(S')$ and $(\alpha', \gamma') \in D'$ with $\delta' = \alpha' \beta' \gamma'$. By the induction hypothesis, there is $\beta \in C^k(S)$ with $\beta' \in \overline{\beta}$. Furthermore, there is $(\alpha, \gamma) \in D$ satisfying $(\alpha', \gamma') \in \overline{\alpha} \times \overline{\gamma}$. Note that $\delta' \in \overline{\alpha} \overline{\beta} \overline{\gamma}$, i.e., the product of these sets is non-empty. Hence $\alpha \beta \gamma \in C^{k+1}(S)$ is defined and satisfies $\delta' \in \overline{\alpha} \overline{\beta} \overline{\gamma} = \overline{\alpha \beta \gamma} \subseteq \overline{C^{k+1}(S)}$.

Now $L(S) = L(S')$ follows from $C^k(S') \cap \rho^* = \overline{C^k(S)} \cap \Delta_V^*$ since the first strands of α and $\overline{\alpha}$ are the same for any $\alpha \in W(V, \rho)$. $\square$

Based on Theorem 6, the second author describes in [7] the relation between the families of the Chomsky hierarchy and the sticker hierarchy from [4] completely. In [7], the relation to multi-head automata is examined as well.

4   Any Non-trivial Complementarity Relation Suffices

In this section, we want to show that any non-trivial (in a sense to be made precise) complementarity relation gives rise to the full expressive power of Watson-Crick automata and of sticker systems. Since the alphabet of a sticker system or a Watson-Crick automaton can be arbitrarily large, we have to choose a mechanism to encode the language over a large alphabet into a language over a small alphabet. For this, we use inverse block codings.

Definition 7. *Let V and W be two alphabets. A* block coding *is a homomorphism $\eta : V^* \to W^*$ such that all elements of $\eta(V)$ have the same length.*

Note that a coding (i.e., a homomorphism with $\eta(a) \in W$ for $a \in V$) is a special case of a block coding with all the images of letters having length one. If a block coding is injective on V, it is injective on the whole of V^*. Furthermore, in this case, the image $\eta(V) \subseteq W$ is a block code.

Definition 8. *A complementarity relation* (V, ρ) *is non-trivial if there are letters* $a, b, c, d \in V$ *(not necessarily distinct) with* $(b, a), (d, c) \in \rho$ *and* $(d, a) \notin \rho$.

Note that, e.g., the relation $\{(x, y), (y, z)\}$ is non-trivial setting $b = x$, $a = d = y$ and $c = z$. On the other hand, the relation $\{(x, y)\}$ is trivial although it is neither complete nor empty. The book [4] restricts attention to symmetric complementarity relations (and mentions the possibility to extend all results there to the non-symmetric case). Suppose ρ is such a symmetric complementarity relation satisfying

1. for any $a \in V$ there is $b \in V$ with $(a, b) \in \rho$, and
2. there are $a, b \in V$ with $(a, b) \notin \rho$.

Then ρ is clearly non-trivial: there are $(d, a) \notin \rho$ and, by the first requirement and the symmetry of ρ, b and c with $(b, a), (d, c) \in \rho$. The first of these conditions rules out useless letters: if there was no $b \in V$ with $(a, b) \in \rho$, then the letter a could not appear in any complete molecule. Hence we could forget about it from the very beginning. Given this elimination of useless letters, the second condition rules out the complete relation $\rho = V^2$ - we consider it of no interest since it does not allow to transmit any information from the first to the second strand (or vice versa) apart from the length.

Let (V, ρ) be a non-trivial complementarity relation with $a, b, c, d \in V$ such that $(b, a), (d, c) \in \rho$ and $(d, a) \notin \rho$. Furthermore, let $V' = \{1, 2, \ldots, k-1\}$ be an alphabet. We define a homomorphism $\eta : V'^* \to V^*$ by $\eta(n) = a^n c^{k-n} a^{k-n} c^n$. Then η is clearly an injective block coding. We will also consider another block coding θ, namely $\theta(n) = b^n d^{k-n} b^{k-n} d^n$.

Lemma 9. *Let* $m, n \in V'$. *Then* $m = n$ *(i.e.,* $(m, n) \in \Delta_{V'}$*) iff the words* $\eta(m)$ *and* $\theta(n)$ *are complementary with respect to* ρ, *i.e., iff* $\left(\begin{smallmatrix} \eta(m) \\ \theta(n) \end{smallmatrix} \right) \in \rho^*$.

Proof. Note that

$$\begin{pmatrix} \eta(m) \\ \theta(n) \end{pmatrix} = \begin{pmatrix} a^m c^{k-m} a^{k-m} c^m \\ b^n d^{k-n} b^{k-n} d^n \end{pmatrix}$$

If $m = n$, then these two words are complementary since $\left(\begin{smallmatrix} a \\ b \end{smallmatrix} \right), \left(\begin{smallmatrix} c \\ d \end{smallmatrix} \right) \in \rho$. If, conversely, these two words are complementary, then $\left(\begin{smallmatrix} a^m c^{k-m} \\ b^n d^{k-n} \end{smallmatrix} \right)$ is a complete molecule. Since $\left(\begin{smallmatrix} a \\ d \end{smallmatrix} \right) \notin \rho$, this implies $m \leq n$. On the other hand, also $\left(\begin{smallmatrix} a^{k-m} c^m \\ b^{k-n} d^n \end{smallmatrix} \right)$ would be a complete molecule, implying $k - m \leq k - n$ and therefore $m \geq n$ by the same reason.   □

Lemma 10. *There is a mapping* $f : W(V', \Delta_{V'}) \to W(V, \rho)$ *satisfying*

- $f \left(\left(\begin{smallmatrix} m \\ \varepsilon \end{smallmatrix} \right) \right) = \left(\begin{smallmatrix} \eta(m) \\ \varepsilon \end{smallmatrix} \right)$ *for* $m \in V'$,
- $f \left(\left(\begin{smallmatrix} \varepsilon \\ m \end{smallmatrix} \right) \right) = \left(\begin{smallmatrix} \varepsilon \\ \theta(m) \end{smallmatrix} \right)$ *for* $m \in V'$,
- $f \left(\left[\begin{smallmatrix} m \\ m \end{smallmatrix} \right]_{\Delta_{V'}} \right) = \left[\begin{smallmatrix} \eta(m) \\ \theta(m) \end{smallmatrix} \right]_{\rho}$ *for* $m \in V'$, *and*

- $f(\alpha \cdot_{\Delta_{V'}} \beta) = f(\alpha) \cdot_{\rho'} f(\beta)$ for all $\alpha, \beta \in W(V', \rho')$.[3]

Proof. The mapping f applies the coding η to the first and the coding θ to the second strand. Now the requirements on f follow easily using Lemma 9 and the fact that η and θ are block codings. $\qquad\square$

Now we formulate and prove the main theorem of this section.

Theorem 11. *For any non-trivial complementarity relation (V, ρ) and any complementarity relation (V', ρ'), there exists an injective block coding $\eta : V'^* \to V^*$ such that the following hold*

1. *If $M' = (V', \rho', K', s_0', T', F')$ is a Watson-Crick automaton over (V', ρ'), then there exists a Watson-Crick automaton $M = (V, \rho, K, s_0, T, F)$ over (V, ρ) with $L(M') = \eta^{-1}(L(M))$.*
2. *If $\gamma' = (V', \rho', A', D')$ is a sticker system over (V', ρ'), then there exists a sticker system $\gamma = (V, \rho, A, D)$ over (V, ρ) with $L(\gamma') = \eta^{-1}(L(\gamma))$.*

Proof. We give a sketch of the proof for Watson-Crick automata, only. The case of sticker systems is dealt with similarly. The injective block coding is the mapping η defined in the construction preceeding Lemma 9 above. By Theorem 4, we can assume $\rho' = \Delta_{V'}$. Further, f is the mapping from Lemma 10. Then the definition of M is simply

- $K = K'$, $s_0 = s_0'$, and $F = F'$,
- $T = \{(p, \eta(u), \theta(v), q) \mid (p, u, v, q) \in T'\}$.

Now it is easy to show that $f(LM(M')) = LM(M)$ using Lemma 10. Since η is injective, the statement of the theorem follows. $\qquad\square$

Looking at the subclasses of Watson-Crick automata defined above, one observes immediately that the property to be stateless, all-final, or simple transfers from M' to M. So, e.g., SWK is contained in the set of block codings of languages in $SWK(V, \rho)$ for any non-trivial complementarity relation (V, ρ). But this transfer does not hold for the property to by 1-limited since we use block codings instead of codings. The restrictions for sticker systems considered in [4] do not mention the size of molecules, hence the theorem applies to these classes equally well.

5   The Hierarchy of Watson-Crick Languages

Above, we defined families of Watson-Crick languages by properties of the accepting Watson-Crick automaton. The diagram on the left in Fig. 1 visualizes those inclusions that have been shown in [4]. Some of these inclusions were shown to be proper, including AWK⊂CS. Since we will prove AWK⊆FSWK, all the classes above F1WK and below CS collapse, i.e., we will obtain the diagram on the right of Fig. 1.

---

[3] This means that the left hand side is defined if, and only if, the right hand side is defined, in which case both sides are equal.

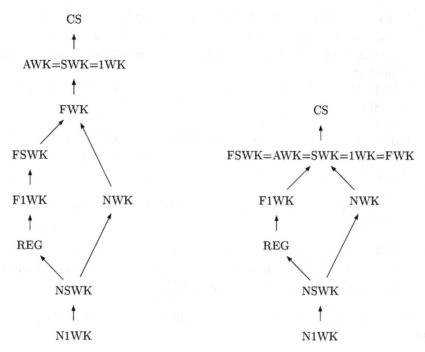

Fig. 1. Inclusions between the Watson-Crick families

Theorem 12. *Let $M = (V, \rho, K, s_0, T, F)$ be a Watson-Crick automaton. Then there exists an all-final and simple Watson-Crick automaton M' over (V, ρ) with $L(M) = L(M')$.*

Proof. The idea of the construction is that the first head of M' will always rest on odd positions while the second head's moves ensure that it only visits even positions (except the very first and very last configuration).

Let $m \in \mathbb{N}$ be such that $(p, \binom{u}{v}, q) \in T$ implies $|u| < m$ and $|v| < m$ for all states $p, q \in K$ and all words u and v over V. Then let the state set of M' be defined by

$$K' = \{(p, u, v) \mid p \in K, u, v \in V^*, |u|, |v| \leq m\} \dot\cup \{\iota', f'\} .$$

Since we want to construct an all-final automaton, we are forced to set $F' = K'$. There are four groups of transitions in T' that we describe next

- initialisation
  For $a \in V$, there is a transition $(\iota', \binom{a}{\varepsilon}, (s_0, a, \varepsilon))$. Since the state ι' will be the initial state of M', this definition forces any run of M' to begin with one step of the first head. After this step, the automaton memorizes the letter a seen on the first string.
- look-ahead
  If $(p, ux, v) \in K'$ and $|x| = 2$, then $((p, u, v), \binom{x}{\varepsilon}, (p, ux, v))$ is a transi-

tion in T'. Symmetrically, if $(p, u, vy) \in K'$ and $|y| = 2$, then the tuple $((p, u, v), \left(\begin{smallmatrix} \varepsilon \\ y \end{smallmatrix}\right), (p, u, vy))$ is a transition in T'. This form of look-ahead ensures that the first head stays on odd and the second one on even positions.

– simulation

If $(p, \left(\begin{smallmatrix} x \\ y \end{smallmatrix}\right), q) \in T$ and $(p, xu, yv) \in K'$, then $((p, xu, yv), \left(\begin{smallmatrix} \varepsilon \\ \varepsilon \end{smallmatrix}\right), (q, u, v))$ is a transition in T'.

– acceptance

If $(p, u, v) \in K'$, $a \in V$, $q \in F$ with $(p, \left(\begin{smallmatrix} ua \\ v \end{smallmatrix}\right), q) \in T$, then $((p, u, v), \left(\begin{smallmatrix} a \\ \varepsilon \end{smallmatrix}\right), f') \in T'$. Symmetrically, if $(p, u, v) \in K'$, $b \in V$, $q \in F$ with $(p, \left(\begin{smallmatrix} u \\ vb \end{smallmatrix}\right), q) \in T$, then $((p, u, v), \left(\begin{smallmatrix} \varepsilon \\ b \end{smallmatrix}\right), f') \in T'$.

Claim 1. Let $x, y \in V^*$ and $(p, u, v) \in K'$. Then the following two statements are equivalent

1. There exists a run from ι' to (p, u, v) in M' labeled $\left(\begin{smallmatrix} x \\ y \end{smallmatrix}\right)$.
2. There are $x', y' \in V^*$ such that $x = x'u$, $y = y'v$, $|x|$ is odd, $|y|$ is even, and there is a run from s_0 to p in M labeled $\left(\begin{smallmatrix} x' \\ y' \end{smallmatrix}\right)$.

This claim can be shown easily by induction on the length of the respective runs.

Claim 2. Let $x, y \in V^+$. Then the following are equivalent

1. There exists a run of M' from ι' to f' labeled $\left(\begin{smallmatrix} x \\ y \end{smallmatrix}\right)$.
2. There is a run of the Watson-Crick automaton M from s_0 to some $q \in F$ with label $\left(\begin{smallmatrix} x \\ y \end{smallmatrix}\right)$, and $|x| \equiv |y| \mod 2$.

Now $L(M) = L(M')$ follows immediately: for $x \in V^+$, we have $x \in L(M)$ iff there exists $y \in V^+$ such that $\left(\begin{smallmatrix} x \\ y \end{smallmatrix}\right) \in \rho^*$ and there exists a run in M from s_0 to some state $q \in F$ labeled $\left(\begin{smallmatrix} x \\ y \end{smallmatrix}\right)$. Since $\left(\begin{smallmatrix} x \\ y \end{smallmatrix}\right) \in \rho^*$ requires in particular that x and y are of the same length, the second claim ensures that $x \in L(M)$ holds iff there is $y \in V^+$ with $\left(\begin{smallmatrix} x \\ y \end{smallmatrix}\right) \in \rho^*$ and there exists a run in M' from ι' to f' labeled $\left(\begin{smallmatrix} x \\ y \end{smallmatrix}\right)$. But this is by definition the case iff $x \in L(M')$. □

6   A Note on ω-Watson-Crick Automata

In [5], Watson-Crick automata are provided with acceptance conditions that allow to accept infinite molecules. In particular, Büchi-, Muller-, Street-, and Rabin-conditions are considered and shown to be equivalent. It is not hard to see that our Thms. 4 and 11 hold for Watson-Crick automata with any of these acceptance conditions. Thus, also in the context of infinite molecules, one-to-one complementarity relations (or non-trivial complementarity relations together with block codings) suffice. This is not clear as far as Theorem 12 is concerned. In the finite case, we used the trick that only at the end of a computation, the two heads rest on the same position. This trick is not available for infinite computations since they do not have an end.

References

1. R. Freund, Gh. Păun, G. Rozenberg, and A. Salomaa. Watson-Crick finite automata. H. Rubin, D. H. Wood (ed.): *DNA Based Computers III*, Discrete Mathematics and Theoretical Computer Science, Vol. 48 (1997), 297-327, American Mathematical Society.

2. L. Kari, Gh. Păun, G. Rozenberg, A. Salomaa, and S. Yu. DNA computing, sticker systems, and universality. *Acta Informatica* 35 (1998), 401-420.

3. V. Mihalache and A. Salomaa. Lindenmayer and DNA: Watson-Crick D0L Systems. EATCS Bulletin 62 (2001), 160-175.

4. Gh. Păun, G. Rozenberg, and A. Salomaa. *DNA computing. New computing paradigms*. EATCS Texts in Theoretical Computer Science. Springer, 1998.

5. I. Petre. Watson-Crick ω-automata. *J. Automata, Languages, and Combinatorics*, 8, (2003), 59–70.

6. A. Salomaa. Watson-Crick walks and roads on D0L graphs. *Acta Cybernetica* 14 (1999), 179-192.

7. P. Weigel. Ausdrucksstärke der Stickersysteme. Diplomarbeit, Martin-Luther-Univesität Halle, 2004.

The Boolean Closure of Linear Context-Free Languages

Martin Kutrib[1], Andreas Malcher[2], and Detlef Wotschke[2]

[1] Institut für Informatik, Universität Giessen, Giessen, Germany
Martin.Kutrib@informatik.uni-giessen.de
[2] Institut für Informatik, Johann Wolfgang Goethe-Universität Frankfurt
Frankfurt am Main, Germany

Abstract. Closures of linear context-free languages under Boolean operations are investigated. The intersection closure and the complementation closure are incomparable. By closing these closures under further Boolean operations we obtain several new language families. The hierarchy obtained by such closures of closures is proper up to level four, where it collapses to the Boolean closure which, in turn, is incomparable with several closures of the family of context-free languages. The Boolean closure of the linear context-free languages is properly contained in the Boolean closure of the context-free languages. A characterization of a class of non-unary languages that cannot be expressed as a Boolean formula over the linear context-free languages is presented.

1 Introduction

Undoubtedly, context-free languages are of great practical importance. They are one of the most important and most developed area of formal language theory. However, on one hand, in many situations appear non-context-free languages in a natural way leading to the observation: "The world is not context-free". A comprehensive discussion of this observation giving "seven circumstances where context-free grammars are not enough" can be found in [4]. So, there is considerable interest in language families that extend the context-free languages, but have similar properties. On the other hand, the known upper bound on the time complexity of context-free language recognition still exceeds $O(n^2)$. So, there is considerable interest in language families that admit efficient recognizers, but decrease the descriptional capacity only slightly.

For example, linear context-free languages have efficient recognition algorithms and are well known to be a proper subfamily of the context-free languages. The family of linear context-free languages (LIN) is closed under union, and intersection with regular sets, but they are not closed under intersection and complementation.

The Boolean closure of LIN offers a significant increase in descriptional capacity compared with the family LIN itself. In addition, it preserves the attractively efficient recognition algorithm taking $O(n^2)$ time and $O(n)$ space.

C.S. Calude, E. Calude and M.J. Dinneen (Eds.): DLT 2004, LNCS 3340, pp. 284–295, 2004.

The latter follows by the family of languages accepted by a certain massively parallel automata model, the so-called deterministic real-time one-way cellular automata. It is known that this family is recognized with $O(n^2)$ time and $O(n)$ space [8, 14], contains the linear context-free languages [13] and is closed under Boolean operations [5, 14].

The significant increase can be seen, e.g., by the languages $L_{wcw} = \{wcw \mid w \in \{a, b\}^*\}$, $\{a^n b^n c^n \mid n \in \mathbb{N}_0\}$ or $\{(ab^n)^n \mid n \in \mathbb{N}_0\}$ that are complements of linear context-free languages, or the sets of valid computations of Turing machines that can be represented by the intersection of two linear languages [1]. From the latter language many undecidability results of the operator problem for LIN follow [2]. E.g., it is not even semidecidable whether the intersection, concatenation, or shuffle of two linear context-free languages, or the complementation, Kleene star, or power of a linear context-free language is linear context free.

The main goal of this paper is to investigate the relationships between the family LIN, their closures under sole Boolean operations, compositions of closures, their Boolean closure and arbitrary context-free languages. The systematic investigation of the Boolean closures of arbitrary and deterministic context-free languages started in [16, 17, 18]. In particular, in [16, 18] deterministic and nondeterministic context-free languages and their Boolean closures are studied motivated by the question "How much more powerful is nondeterminism than determinism?" E.g., it is shown that there are context-free languages that cannot be expressed as a Boolean formula over deterministic context-free languages. Similarly, we may ask "How much more powerful is an unbounded number than just one turn of a pushdown automaton?" motivating the study of the Boolean closures of linear and arbitrary context-free languages.

In [17] machine characterizations for a wide class of Boolean closures of nondeterministic language families are shown. Basically, the characterizations are given by machines for the underlying language families, where the acceptance conditions are modified. In [10] a characterization of deterministic real-time one-way cellular automata by so-called linear conjunctive grammars has been shown. Linear conjunctive grammars are basically linear context-free grammars augmented with an explicit intersection operation, where the number of intersections is, in some sense, not bounded as in a Boolean formula.

The paper is organized as follows. In the next section we present some basic notions and definitions. Section 3 deals with the families of linear languages and their complements as well as with their complementation closure. Section 4 is devoted to the intersection closures of LIN, where a finite number of intersections is distinguished from $k \in \mathbb{N}$ intersections. The results are transfered to the union closures of the complements of linear languages. In Section 5 the Boolean closure of LIN is compared with several other closures. E.g., it is shown that there exists a context-free language not belonging to the Boolean closure of LIN, and, moreover, both families are incomparable. Further, a class of non-unary languages not belonging to the Boolean closure of LIN is characterized. Finally, Section 6 leads us inside the Boolean closure by closing several closures.

It is shown that there are no infinite hierarchies of closures of closures. Instead, the hierarchies collapse at level four. The paper is concluded with a diagram summarizing the inclusions shown.

2   Preliminaries

We denote the positive integers $\{1, 2, ...\}$ by $\mathbb{N}$ and the set $\mathbb{N} \cup \{0\}$ by $\mathbb{N}_0$. The empty word is denoted by λ. For the length of w we write $|w|$. We use $\subseteq$ for inclusions and $\subset$ if the inclusions are strict. Moreover, the reader is assumed to be familiar with basic concepts of formal language theory, in particular with the definition of (linear) context-free grammars as contained, e.g., in [12]. The families of languages that are generated by context-free resp. linear context-free grammars are called *context-free* (CFL) resp. *linear* (LIN) languages. In general, a *family of languages* is a collection of languages containing at least one non-empty language. Let $\mathcal{L}$ be a family of languages and $op_1, op_2, \ldots, op_k$, $k \in \mathbb{N}$, be a finite number of operations defined on $\mathcal{L}$. Then $\Gamma_{op_1, \ldots, op_k}(\mathcal{L})$ denotes the *least family of languages which contains all members of $\mathcal{L}$ and is closed under* $op_1, \ldots, op_k$. In particular, we consider the operations complementation ($\sim$), union ($\cup$), and intersection ($\cap$), which are called *Boolean operations*. Accordingly, we write Γ_{BOOL} for $\Gamma_{\sim, \cup, \cap}$. The intersection (union) with regular languages is denoted by $\cap_R$ ($\cup_R$). It is well known that LIN is closed under intersection and union with regular sets and under union, but is not closed under intersection and complementation, e.g. [12].

3   Complementation Closure

For a family of languages $\mathcal{L}$, the family of complements CO-$\mathcal{L}$ is defined to be $\{\overline{L} \mid L \in \mathcal{L}\}$, where $\overline{L}$ denotes the complement of L. We start by exhibiting briefly the closure properties of CO-LIN under Boolean operations. The next lemma is a simple application of DeMorgan's law but it generalizes the situation.

Lemma 1. *Let $\mathcal{L}$ be a family of languages not closed under complementation. Then (1) CO-$\mathcal{L}$ is not closed under complementation, (2) $\mathcal{L}$ is closed under union (intersection) with regular languages if and only if CO-$\mathcal{L}$ is closed under intersection (union) with regular languages, and (3) $\mathcal{L}$ is closed under union (intersection) if and only if CO-$\mathcal{L}$ is closed under intersection (union).*

By applying Lemma 1, the closure properties of CO-LIN are easily derived from the properties of LIN.

Corollary 2. *The family CO-LIN is not closed under complementation, closed under union and intersection with regular languages, closed under intersection, and not closed under union.*

It follows that LIN and CO-LIN are incomparable. This observation can be strengthened considering the relations with CFL and CO-CFL. To this end, we utilize the language $L_{wcw} = \{wcw \mid w \in \{a,b\}^*\}$ which plays an important role in the sequel. From [7,12] we derive that its complement $\overline{L_{wcw}}$ is a linear context-free language. On the other hand, in [18] it has been shown that L_{wcw} cannot be expressed by any intersection of finitely many context-free languages.

Lemma 3. *The families* LIN *and* CO-CFL *resp.* CFL *and* CO-LIN *are incomparable, respectively.*

Proof. The language $\overline{L_{wcw}}$ is linear context free, but not a member of CO-CFL. Conversely, L_{wcw} is a member of CO-LIN, but not a member of CFL.   □

Now we turn to the complementation closure of LIN, and observe immediately $\Gamma_\sim(\text{LIN}) = \Gamma_\sim(\text{CO-LIN}) = \text{LIN} \cup \text{CO-LIN}$, and the properness of the inclusions LIN $\subset \Gamma_\sim(\text{LIN})$ and CO-LIN $\subset \Gamma_\sim(\text{LIN})$.

In order to disprove the closure of $\Gamma_\sim(\text{LIN})$ under union, we show the following lemma, which generalizes a remark in [18].

Lemma 4. *Let $\mathcal{L}$ be a family of languages not closed under complementation. If $\mathcal{L}$ and CO-$\mathcal{L}$ are closed under intersection with regular languages, then $\Gamma_\sim(\mathcal{L})$ is (1) not closed under union, (2) not closed under intersection, (3) closed under intersection with regular languages, and (4) closed under union with regular languages.*

Proof. (1) Let $L_1 \subseteq A_1^*$ be some language from $\mathcal{L} \setminus$ CO-$\mathcal{L}$. For an alphabet A_2 disjoint from A_1, let $L_2 \subseteq A_2^*$ be some language from CO-$\mathcal{L} \setminus \mathcal{L}$. Assume $\Gamma_\sim(\mathcal{L})$ is closed under union. Then we have $L_1 \cup L_2 \in \Gamma_\sim(\mathcal{L})$. Consider two cases: (i) $L_1 \cup L_2 \in \mathcal{L}$ implies $(L_1 \cup L_2) \cap A_2^* = L_2 \in \mathcal{L}$. (ii) $L_1 \cup L_2 \in$ CO-$\mathcal{L}$ implies $(L_1 \cup L_2) \cap A_1^* = L_1 \in$ CO-$\mathcal{L}$. For both cases a contradiction follows.
(2) Since $\Gamma_\sim(\mathcal{L})$ is trivially closed under complementation, but not closed under union, it cannot be closed under intersection. (3) Since $\mathcal{L}$ and CO-$\mathcal{L}$ are both closed under intersection with regular languages, their union is closed, too. (4) Applying Lemma 1 (2) to $\mathcal{L}$ and CO-$\mathcal{L}$ proves the assertion.   □

An application of Lemma 4 shows the next theorem.

Theorem 5. *The complementation closure $\Gamma_\sim(\text{LIN})$ is not closed under union, not closed under intersection, closed under intersection with regular languages, and closed under union with regular languages.*

From different closure properties we conclude the properness of the inclusion $\Gamma_\sim(\text{LIN}) \subset \Gamma_{\text{BOOL}}(\text{LIN})$.

4   Intersection Closure

Besides the general closure of some family $\mathcal{L}$ under intersection ($\Gamma_\cap(\mathcal{L})$) or union ($\Gamma_\cup(\mathcal{L})$), there is a natural interest in closures under a limited number of intersections or unions. Let $k \in \mathbb{N}$ be some positive integer. Then

$$\Gamma_{\cap_k}(\mathscr{L}) = \{L_1 \cap L_2 \cap \cdots \cap L_k \mid L_i \in \mathscr{L}, 1 \le i \le k\} \text{ and}$$
$$\Gamma_{\cup_k}(\mathscr{L}) = \{L_1 \cup L_2 \cup \cdots \cup L_k \mid L_i \in \mathscr{L}, 1 \le i \le k\}$$

are the *k-intersection and k-union closures* of $\mathscr{L}$, respectively.

Clearly, we have $\Gamma_\cap(\mathscr{L}) = \bigcup_{k \in \mathbb{N}} \Gamma_{\cap_k}(\mathscr{L})$ and $\Gamma_\cup(\mathscr{L}) = \bigcup_{k \in \mathbb{N}} \Gamma_{\cup_k}(\mathscr{L})$.

In [9] an infinite hierarchy of intersections of context-free languages has been shown. For any $k \ge 2$, $\Gamma_{\cap_{k-1}}(\mathrm{CFL}) \subset \Gamma_{\cap_k}(\mathrm{CFL})$ is proved by showing that the witness language $L_k \subseteq \{a_1, \ldots, a_k\}^*$ separates both closures, where

$$L_k = \{a_1^{i_1} a_2^{i_2} \cdots a_k^{i_k} a_1^{i_1} a_2^{i_2} \cdots a_k^{i_k} \mid i_j \in \mathbb{N}, 1 \le j \le k\}.$$

On one hand, since $L_k \notin \Gamma_{\cap_{k-1}}(\mathrm{CFL})$, it does not belong to $\Gamma_{\cap_{k-1}}(\mathrm{LIN})$. On the other hand, in [9] it is observed that $L_k = \bigcap_{1 \le i \le k} L_{k,i}$, where

$$L_{k,i} = \{a_1^* \cdots a_{i-1}^* a_i^n a_{i+1}^* \cdots a_k^* a_1^* \cdots a_{i-1}^* a_i^n a_{i+1}^* \cdots a_k^* \mid n \in \mathbb{N}\}.$$

Since $L_{k,i} \in \mathrm{LIN}$, we conclude $L_k \in \Gamma_{\cap_k}(\mathrm{LIN})$. So, actually in [9] a stronger result has been shown, an infinite hierarchy of intersections of *linear* context-free languages.

Next we derive closure properties of $\Gamma_{\cap_k}(\mathrm{LIN})$ under operations in question.

Theorem 6. *Let $k \in \mathbb{N}$ be some positive integer. Then $\Gamma_{\cap_k}(\mathrm{LIN})$ is (1) closed under intersection and union with regular languages, (2) not closed under intersection, and (3) not closed under complementation.*

Proof. (1) Let $L = L_1 \cap L_2 \cap \cdots \cap L_k$ be an arbitrary language from $\Gamma_{\cap_k}(\mathrm{LIN})$, where L_i are linear languages. For an arbitrary regular language R, the intersection $L \cap R = L_1 \cap R \cap L_2 \cap \cdots \cap L_k$ is represented by $L_{1,R} \cap L_2 \cap \cdots \cap L_k$, where $L_{1,R} = L_1 \cap R$ belongs to LIN, since LIN is closed under intersection with regular languages.

The union $L \cup R = (L_1 \cap L_2 \cap \cdots \cap L_k) \cup R$ is represented by $(L_1 \cup R) \cap \cdots \cap (L_k \cup R)$. Since each $(L_i \cup R)$ is linear, the assertion follows.

(2) The non-closure under intersection is seen by the intersection of $L_{k+1,k+1}$ and $\bigcap_{1 \le i \le k} L_{k+1,i}$.

(3) For the non-closure under complementation we observe that for all $k \in \mathbb{N}$ the language $\overline{L_k}$ is linear. So, $L_{k+1} \notin \Gamma_{\cap_k}(\mathrm{LIN})$ and $\overline{L_{k+1}} \in \Gamma_{\cap_k}(\mathrm{LIN})$. $\square$

We continue with the investigation of the closure properties of the intersection closure $\Gamma_\cap(\mathrm{LIN})$. The properties are similar to the properties of $\Gamma_{\cap_k}(\mathrm{LIN})$ except for the trivial closure under intersection.

Theorem 7. *The intersection closure $\Gamma_\cap(\mathrm{LIN})$ is (1) closed under intersection, (2) closed under union, and (3) not closed under complementation.*

Proof. (2) In order to show the closure under union let $L = L_1 \cap L_2 \cap \cdots \cap L_m$, $m \ge 1$, and $L' = L'_1 \cap L'_2 \cap \cdots \cap L'_n$, $n \ge 1$, be two languages from $\Gamma_\cap(\mathrm{LIN})$, where L_i and L'_j are linear languages. Their union can be written as $L \cup L' = \bigcap_{1 \le i \le m} L_i \cup \bigcap_{1 \le j \le n} L'_j$. By distributive law, this is equivalent to $L \cup L' = \bigcap_{1 \le i \le m, 1 \le j \le n} L_i \cup L'_j$. Since LIN is closed under union, we obtain $L \cup L' = \bigcap_{1 \le i \le mn} L''_i$, where the languages L''_i are linear. Therefore, $L \cup L' \in \Gamma_\cap(\text{LIN})$.

(3) A witness language for the non-closure under complementation is L_{wcw}.

$\square$

Theorem 8. *The intersection closure $\Gamma_\cap(\text{LIN})$ and the complementation closure $\Gamma_\sim(\text{LIN})$ are incomparable.*

Proof. The language L_{wcw} is a candidate for $\Gamma_\sim(\text{LIN}) \setminus \Gamma_\cap(\text{LIN})$. It does not belong to $\Gamma_\cap(\text{LIN})$. Since $\overline{L_{wcw}} \in \text{LIN}$, it follows $L_{wcw} \in \text{CO-LIN}$ and, hence, $L_{wcw} \in \Gamma_\sim(\text{LIN})$. Now consider the language $L_0 = \{c^n d^n c^n \mid n \in \mathbb{N}\}$ whose complement $\overline{L_0}$ is linear. Further set $L_1 = \{a^n b^n a^m \mid m, n \in \mathbb{N}\}$ and $L_2 = \{a^m b^n a^n \mid m, n \in \mathbb{N}\}$. The languages L_1 and L_2 are linear. Therefore, $L = \overline{L_0} \cup (L_1 \cap L_2)$ belongs to $\Gamma_\cap(\text{LIN})$.

Assume $L \in \Gamma_\sim(\text{LIN})$. Then consider two cases: (i) $L \in \text{LIN}$ implies $L \cap \{a^* b^* a^*\} = L_1 \cap L_2$ is linear. (ii) $L \in \text{CO-LIN}$ implies $L \cap \{c^* d^* c^*\} = \overline{L_0}$ belongs to CO-LIN. For both cases we obtained a contradiction, thus, $L \notin \Gamma_\sim(\text{LIN})$. $\square$

So far we have investigated the closures for the Boolean operations under which LIN is not closed. Focusing on CO-LIN we now turn to its union closure. First we observe that the intersection hierarchy nicely induces a union hierarchy by complementation. In fact, we have $\text{CO-}(\Gamma_{\cap_k}(\text{LIN})) = \Gamma_{\cup_k}(\text{CO-LIN})$ and $\text{CO-}(\Gamma_\cap(\text{LIN})) = \Gamma_\cup(\text{CO-LIN})$ by DeMorgan's law. An immediate consequence is the union hierarchy. For any $k \ge 2$: $\overline{L_k} \in \Gamma_{\cup_k}(\text{CO-LIN}) \setminus \Gamma_{\cup_{k-1}}(\text{CO-LIN})$. By Lemma 1 and Theorem 6 resp. Theorem 7 the following properties follow.

Theorem 9. *Let $k \in \mathbb{N}$ be some positive integer. Then $\Gamma_{\cup_k}(\text{CO-LIN})$ is (1) closed under intersection and union with regular languages, (2) not closed under union, and (3) not closed under complementation.*

Theorem 10. *The union closure $\Gamma_\cup(\text{CO-LIN})$ is (1) closed under intersection, (2) closed under union, and (3) not closed under complementation.*

Again we obtained different families as the next theorem shows.

Theorem 11. *The union closure $\Gamma_\cup(\text{CO-LIN})$ and the complementation closure $\Gamma_\sim(\text{LIN})$ are incomparable.*

Proof. By Theorem 8 there exists a language $L \in \Gamma_\sim(\text{LIN}) \setminus \Gamma_\cap(\text{LIN})$. The complement $\overline{L}$ belongs still to $\Gamma_\sim(\text{LIN})$, but does not belong to $\Gamma_\cup(\text{CO-LIN})$.

Similarly, for $L' \in \Gamma_\cap(\text{LIN}) \setminus \Gamma_\sim(\text{LIN})$ we have that $\overline{L'}$ does not belong to $\Gamma_\sim(\text{LIN})$, but does belong to $\Gamma_\cup(\text{CO-LIN})$. $\square$

By means of different closure properties the properness of the inclusions $\Gamma_\cup(\text{CO-LIN}) \subset \Gamma_{\text{BOOL}}(\text{LIN})$ and $\Gamma_\cap(\text{LIN}) \subset \Gamma_{\text{BOOL}}(\text{LIN})$ follows.

	LIN	CO-LIN	$\Gamma_\sim$(LIN)	$\Gamma_{\cap_k}$(LIN)	$\Gamma_{\cup_k}$(CO-LIN)	$\Gamma_\cap$(LIN)	$\Gamma_\cup$(CO-LIN)
$\cup$	+	−	−	?	−	+	+
$\cap$	−	+	−	−	?	+	+
$\sim$	−	−	+	−	−	−	−

Fig. 1. Summary of closure properties I

5   Boolean Closure

In this section we are going to examine the position of the Boolean closure of LIN in the hierarchy of languages. On one end, the deterministic context-sensitive languages (DCSL) are closed under Boolean operations. Therefore, the Boolean closure Γ_{BOOL}(LIN) is contained in DCSL. Moreover, since every unary context-free language is regular, every unary language in Γ_{BOOL}(LIN) is regular, too. So, the inclusion Γ_{BOOL}(LIN) $\subset$ DCSL is proper. On the other end, we compare the Boolean closure of LIN with related closures of the context-free languages and with the family of languages accepted by a certain massively parallel automata model, the so-called deterministic real-time one-way cellular automata. Let us denote these languages by $\mathscr{L}_{rt}$(OCA). It is known that $\mathscr{L}_{rt}$(OCA) contains the linear context-free languages [13] and is closed under Boolean operations [5, 14]. So, it contains Γ_{BOOL}(LIN). Results in [15] show that $\mathscr{L}_{rt}$(OCA) is not closed under concatenation. A point of particular importance for our concern is that the witness language is the concatenation of a linear language. It is shown that the 2-linear language $L_{ab} = LL$ does not belong to $\mathscr{L}_{rt}$(OCA), where $L = \{a^n b^n \mid n \in \mathbb{N}\} \cup \{a^n bxab^n \mid x \in \{a, b\}^*, n \in \mathbb{N}\}$.

Corollary 12. *There are context-free languages not belonging to Γ_{BOOL}(LIN).*

Whenever we compare two closures of the form Γ_{op}(LIN) and Γ_{op}(CFL), where $op \subseteq \{\sim, \cup, \cap\}$, we observe the trivial inclusion Γ_{op}(LIN) $\subseteq \Gamma_{op}$(CFL). The situation for CO-LIN and CO-CFL is similar. In fact, by Corollary 12 the inclusion is proper.

Corollary 13. *The inclusions $\Gamma_\sim$(LIN) $\subset \Gamma_\sim$(CFL), $\Gamma_\cap$(LIN) $\subset \Gamma_\cap$(CFL), $\Gamma_\cup$(CO-LIN) $\subset \Gamma_\cup$(CO-CFL), and Γ_{BOOL}(LIN) $\subset \Gamma_{\text{BOOL}}$(CFL) are proper, respectively.*

On the other hand, we obtain incomparability whenever we relate a CFL-family to a LIN-family.

Lemma 14. *The Boolean closure Γ_{BOOL}(LIN) is incomparable with each of the families (1) CFL, (2) CO-CFL, (3) $\Gamma_\cap$(CFL), (4) $\Gamma_\cup$(CO-CFL), and (5) $\Gamma_\sim$(CFL).*

Proof. (1) $L_{ab} \in \text{CFL} \setminus \Gamma_{\text{BOOL}}$(LIN) and $L = \{a^n b^n c^n \mid n \in \mathbb{N}\} \in \Gamma_{\text{BOOL}}$(LIN) $\setminus$ CFL, since $L \in \text{CO-LIN} \setminus$ CFL.
(2) $\overline{L_{ab}} \in \text{CO-CFL} \setminus \Gamma_{\text{BOOL}}$(LIN) and $\overline{L} \in \Gamma_{\text{BOOL}}$(LIN) $\setminus$ CO-CFL.

(3) $L_{ab} \in \Gamma_\cap(\text{CFL}) \setminus \Gamma_{\text{BOOL}}(\text{LIN})$ and $L_{wcw} \in \Gamma_{\text{BOOL}}(\text{LIN}) \setminus \Gamma_\cap(\text{CFL})$.
(4) $\overline{L_{ab}} \in \Gamma_\cup(\text{CO-CFL}) \setminus \Gamma_{\text{BOOL}}(\text{LIN})$ and $\overline{L_{wcw}} \in \Gamma_{\text{BOOL}}(\text{LIN}) \setminus \Gamma_\cup(\text{CO-CFL})$.
(5) $L_{ab} \in \Gamma_\sim(\text{CFL}) \setminus \Gamma_{\text{BOOL}}(\text{LIN})$. In [18] it has been shown that the language
$L = d_1 L_{wcw} \cup d_2 \{wcx \mid w, x \in \{a, b\}^*, w \neq x\}$ is not a member of $\Gamma_\sim(\text{CFL})$. But
it is easy to show that $L \in \Gamma_{\text{BOOL}}(\text{LIN})$.   □

Similarly we obtain:

Lemma 15. *The closures (1)* $\Gamma_\cap(\text{LIN})$, *(2)* $\Gamma_\cup(\text{CO-LIN})$, *or (3)* $\Gamma_\sim(\text{LIN})$, *and*
CFL or CO-CFL *are incomparable, respectively.*

Proof. For all cases, L_{ab} belongs to CFL, $\overline{L_{ab}}$ does not belong to CO-CFL, but
neither L_{ab} nor $\overline{L_{ab}}$ does belong to one of the closures of LIN.

Conversely, $L = \{a^n b^n c^n \mid n \in \mathbb{N}\}$ as well as $\overline{L}$ do belong to $\Gamma_\cap(\text{LIN})$,
$\Gamma_\sim(\text{LIN})$, or $\Gamma_\cup(\text{CO-LIN})$. But L is not a member of CFL and $\overline{L}$ is not a member
of CO-CFL.   □

The inclusion $\Gamma_{\text{BOOL}}(\text{LIN}) \subseteq \mathscr{L}_{rt}(\text{OCA})$ arises the questions how $\mathscr{L}_{rt}(\text{OCA})$
relates to $\Gamma_{\text{BOOL}}(\text{CFL})$, and whether the inclusion $\Gamma_{\text{BOOL}}(\text{LIN}) \subseteq \mathscr{L}_{rt}(\text{OCA})$ is
proper. Both questions will be answered by showing that there are non-unary
languages not belonging to $\Gamma_{\text{BOOL}}(\text{CFL})$ but belonging to $\mathscr{L}_{rt}(\text{OCA})$. We continue with some preparing remarks and a proposition.

Consider for some fixed positive integer n vectors in $\mathbb{N}^n$. A set of the form
$\{v_0 + x_1 v_1 + \cdots + x_k v_k \mid x_i \geq 0, 1 \leq i \leq k\}$, where $v_0, \ldots, v_k \in \mathbb{N}^n$, is said
to be *linear*. A *semilinear* set is a finite union of linear sets. It is known that
the family of semilinear subsets of $\mathbb{N}^n$ is closed under union, intersection, and
complementation [6]. For an alphabet $A = \{a_1, \ldots, a_n\}$ the *Parikh mapping*
$\Psi : A^* \to \mathbb{N}^n$ is defined by $\Psi(w) = (|w|_{a_1}, \ldots, |w|_{a_n})$, where $|w|_{a_i}$ denotes the
number of occurrences of a_i in the word w.

In [11] a fundamental result concerning the distribution of symbols in the
words of a context-free language has been shown. It says that for any context-
free language L, the Parikh image $\Psi(L) = \{\Psi(w) \mid w \in L\}$ is semilinear. Unfortunately, this is not true for languages in CO-CFL or CO-LIN. For example,
$L = \{(ab^n)^n \mid n \in \mathbb{N}\} \in$ CO-LIN and, clearly, $\Psi(L)$ is not semilinear. Nevertheless, for languages L of a specific form we have the following relation between the complement of the language and the complement of its Parikh image
$\overline{\Psi(L)} = \mathbb{N}^n \setminus \Psi(L)$.

Proposition 16. *Let* $B = \{a_1^{x_1} \cdots a_n^{x_n} \mid x_i \in \mathbb{N}_0, 1 \leq i \leq n\}$ *and* $L \subseteq B$. *Then*
$\overline{\Psi(L)} = \Psi(\overline{L} \cap B)$.

Proof. The complement of L is represented by

$$\overline{L} = \{a_1^{x_1} \cdots a_n^{x_n} \mid a_1^{x_1} \cdots a_n^{x_n} \notin L\} \cup \overline{\{a_1^{x_1} \cdots a_n^{x_n} \mid x_i \in \mathbb{N}_0, 1 \leq i \leq n\}}.$$

Therefore, $\overline{L} \cap B = \{a_1^{x_1} \cdots a_n^{x_n} \mid a_1^{x_1} \cdots a_n^{x_n} \notin L\}$ and

$$\Psi(\overline{L} \cap B) = \{(x_1, \ldots, x_n) \mid a_1^{x_1} \cdots a_n^{x_n} \notin L\}.$$

On the other hand, $\Psi(L) = \{(x_1, \ldots, x_n) \mid a_1^{x_1} \cdots a_n^{x_n} \in L\}$ implies

$$\overline{\Psi(L)} = \mathbb{N}^n \setminus \Psi(L) = \{(x_1, \ldots, x_n) \mid a_1^{x_1} \cdots a_n^{x_n} \notin L\}.$$ □

Now we are prepared to prove that a certain class of languages does not belong to $\Gamma_{\mathrm{BOOL}}(\mathrm{CFL})$.

Theorem 17. *Let* $B = \{a_1^{x_1} \cdots a_n^{x_n} \mid x_i \in \mathbb{N}_0, 1 \le i \le n\}$ *and* $L \subseteq B$. *If* $L \in \Gamma_{\mathrm{BOOL}}(\mathrm{CFL})$ *then* $\Psi(L)$ *is semilinear.*

Proof. Let $L \in \Gamma_{\mathrm{BOOL}}(\mathrm{CFL})$. Then, for some $k, l_1, \ldots, l_k \in \mathbb{N}_0$, language L has a representation $\bigcup_{1 \le i \le k} \bigcap_{1 \le j \le l_i} L_{i,j}$ such that $L_{i,j} \in \mathrm{CFL}$ or $L_{i,j} \in \mathrm{CO\text{-}CFL}$.

Consider the languages $L_i = L_{i,1} \cap \cdots \cap L_{i,l_i}$. Without loss of generality, we may assume $L_{i,1}, \ldots, L_{i,l_i} \subseteq B$, and $L_{i,1} \in \mathrm{CO\text{-}CFL}$ and $L_{i,2}, \ldots, L_{i,l_i} \in \mathrm{CFL}$. So, $\Psi(L_{i,j})$, $2 \le j \le l_i$, is a semilinear set. For $L_{i,1}$ we distinguish two cases.

(1) If $L_{i,1} \in \mathrm{CFL}$, then $\Psi(L_{i,1})$ is a semilinear set.

(2) If $L_{i,1} \in \mathrm{CO\text{-}CFL} \setminus \mathrm{CFL}$, then $\overline{L_{i,1}} \in \mathrm{CFL}$. Since B is a regular language, $\overline{L_{i,1}} \cap B$ belongs to CFL, either. Therefore, $\Psi(\overline{L_{i,1}} \cap B)$ is semilinear. By Proposition 16 it follows that $\overline{\Psi(L_{i,1})}$ is semilinear. Since the family of semilinear subsets of $\mathbb{N}^n$ is closed under Boolean operations, $\Psi(L_{i,1})$ is semilinear. So, due to the closure under intersection and the fact $L_{i,j} \subseteq B$ we obtain $\Psi(L_i) = \Psi(L_{i,1}) \cap \cdots \cap \Psi(L_{i,l_i})$ is semilinear. Finally, due to the closure under union, $\Psi(L) = \Psi(L_1) \cup \cdots \cup \Psi(L_k)$ is a semilinear set.   □

For example, $L_p = \{a^n b^{2^n} \mid n \in \mathbb{N}_0\}$ does not belong to $\Gamma_{\mathrm{BOOL}}(\mathrm{CFL})$. In [3] it has been shown that L_p is a member of $\mathscr{L}_{rt}(\mathrm{OCA})$. This proves the next theorem.

Theorem 18. *The inclusion* $\Gamma_{\mathrm{BOOL}}(\mathrm{LIN}) \subset \mathscr{L}_{rt}(\mathrm{OCA})$ *is proper.*

Theorem 19. *The family* $\mathscr{L}_{rt}(\mathrm{OCA})$ *and the Boolean closure* $\Gamma_{\mathrm{BOOL}}(\mathrm{CFL})$ *are incomparable.*

6   Inside the Boolean Closure

This section is devoted to shed some light on those families inside the Boolean closure of LIN that are obtained by closing some closure under another Boolean operation.

Theorem 20. *The closure* $\Gamma_{\sim}(\Gamma_{\cap}(\mathrm{LIN})) = \Gamma_{\sim}(\Gamma_{\cup}(\mathrm{CO\text{-}LIN}))$ *is properly contained in* $\Gamma_{\mathrm{BOOL}}(\mathrm{LIN})$.

Proof. We may apply Lemma 4. It follows that $\Gamma_{\sim}(\Gamma_{\cap}(\mathrm{LIN}))$ is not closed under union and intersection.   □

Corollary 21. *The closures* $\Gamma_{\cap}(\mathrm{LIN})$, $\Gamma_{\cup}(\mathrm{CO\text{-}LIN})$, *and* $\Gamma_{\sim}(\mathrm{LIN})$ *are properly contained in* $\Gamma_{\sim}(\Gamma_{\cap}(\mathrm{LIN}))$, *respectively.*

Proof. The assertion for $\Gamma_{\cap}(\mathrm{LIN})$ and $\Gamma_{\cup}(\mathrm{CO\text{-}LIN})$ follows from different closure properties. By Theorem 8 there exists a language $L \in \Gamma_{\cap}(\mathrm{LIN}) \setminus \Gamma_{\sim}(\mathrm{LIN})$.   □

Next we close $\Gamma_\sim(\text{LIN}) = \Gamma_\sim(\text{CO-LIN})$ under union and intersection and show that the resulting families $\Gamma_\cap(\Gamma_\sim(\text{LIN}))$ and $\Gamma_\cup(\Gamma_\sim(\text{LIN}))$ are different.

Theorem 22. *The closures $\Gamma_\cap(\Gamma_\sim(\text{LIN}))$ and $\Gamma_\cup(\Gamma_\sim(\text{LIN}))$ are different and properly contained in $\Gamma_{\text{BOOL}}(\text{LIN})$.*

Proof. We show that the closures are different. To this end, we will use the marked concatenation with A^*, where A is some alphabet. LIN is closed under this operation since it is closed under concatenation with regular languages. For $L \in \text{CO-LIN}$, $L \in B^*$, we observe $LcA^* \in \text{CO-LIN}$ if and only if $\overline{LcA^*} \in \text{LIN}$. $\overline{LcA^*}$ can be represented by $L_1 \cup \overline{L}cA^*$, where L_1 is the regular language of words not of the appropriate form, i.e., $L_1 = \overline{B^*cA^*}$. So CO-LIN is closed under this operation, either.

Consider the language $L_d = L_{wcw}d\overline{L'_{wcw}} \subseteq \{a, b, c\}^* d\{a', b', c'\}^*$. The language L_d has the representation $L_{wcw}d\{a', b', c'\}^* \cap \{a, b, c\}^* d\overline{L'_{wcw}}$. So, L_d is the intersection of a linear language and a language from CO-LIN. We conclude $L \in \Gamma_\cap(\Gamma_\sim(\text{LIN}))$.

Now assume $L_d \in \Gamma_\cup(\Gamma_\sim(\text{LIN}))$. Then L_d can be written as $L_1 \cup L_2$, where $L_1 \in \text{LIN}$ and $L_2 \in \Gamma_\cup(\text{CO-LIN})$. Applying the homomorphism $h(a') = h(b') = h(c') = h(d) = \lambda$, $h(a) = a$, $h(b) = b$, and $h(c) = c$ to L_1, we cannot obtain L_{wcw}, since LIN is closed under homomorphisms. Let w_0cw_0 be a fixed word not belonging to $h(L_1)$. Then the intersection $w_0cw_0d\{a', b', c'\}^* \cap L_2 = w_0cw_0d\overline{L_{wcw}}$ must belong to $\Gamma_\cup(\text{CO-LIN})$. Since CO-LIN is closed under left quotient with a singleton, the union closure $\Gamma_\cup(\text{CO-LIN})$ is closed under this operation, either. Therefore, $\overline{L_{wcw}}$ must belong to $\Gamma_\cup(\text{CO-LIN})$. Hence, the complement L_{wcw} would be a member of $\Gamma_\cap(\text{LIN})$, a contradiction. $\qquad\square$

Corollary 23. *The closures $\Gamma_\cap(\Gamma_\sim(\text{LIN}))$ or $\Gamma_\cup(\Gamma_\sim(\text{LIN}))$ are not closed under complementation, and are closed under intersection and union with regular languages. The union closure is not closed under intersection, and the intersection closure is not closed under union.*

The next two theorems are immediate consequences.

Theorem 24. *The closures $\Gamma_\cap(\text{LIN})$ and $\Gamma_\sim(\text{LIN})$ are properly contained in the closure $\Gamma_\cap(\Gamma_\sim(\text{LIN}))$.*

Theorem 25. *The closures $\Gamma_\cup(\text{CO-LIN})$ and $\Gamma_\sim(\text{LIN})$ are properly contained in the closure $\Gamma_\cup(\Gamma_\sim(\text{LIN}))$.*

Continuing the process of composing closures, for $\Gamma_\cap(\Gamma_\cup(\Gamma_\sim(\text{LIN})))$ and $\Gamma_\cup(\Gamma_\cap(\Gamma_\sim(\text{LIN})))$ we reach $\Gamma_{\text{BOOL}}(\text{LIN})$. Let us consider $\Gamma_\sim(\Gamma_\cap(\Gamma_\sim(\text{LIN}))) = \Gamma_\sim(\Gamma_\cup(\Gamma_\sim(\text{LIN})))$.

Theorem 26. *$\Gamma_\sim(\Gamma_\cap(\Gamma_\sim(\text{LIN})))$ is properly contained in $\Gamma_{\text{BOOL}}(\text{LIN})$.*

Proof. By Lemma 4, $\Gamma_\sim(\Gamma_\cap(\Gamma_\sim(\text{LIN})))$ is not closed under union and intersection, but is closed under union and intersection with regular languages. $\qquad\square$

Corollary 27. *The closures $\Gamma_\cup(\Gamma_\sim(\text{LIN}))$ and $\Gamma_\cap(\Gamma_\sim(\text{LIN}))$ are properly contained in the closure $\Gamma_\sim(\Gamma_\cap(\Gamma_\sim(\text{LIN})))$.*

Theorem 28. *$\Gamma_\sim(\Gamma_\cap(\text{LIN}))$ is properly contained in $\Gamma_\sim(\Gamma_\cap(\Gamma_\sim(\text{LIN})))$.*

Proof. Consider $L_{wcw} \subseteq \{a, b, c\}^*$ and let $L'_{wcw} \subseteq \{a', b', c'\}^*$ be the homomorphic image $h(L_{wcw})$, where $h(a) = a'$, $h(b) = b'$, and $h(c) = c'$. Since L_{wcw} does not belong to $\Gamma_\cap(\text{LIN})$, the complement $\overline{L'_{wcw}}$ does not belong to $\Gamma_\cup(\text{CO-LIN})$. Therefore, the union $\overline{L'_{wcw}} \cup L_{wcw}$ cannot belong to $\Gamma_\cap(\text{LIN})$ or $\Gamma_\cup(\text{CO-LIN})$. Thus, it cannot belong to $\Gamma_\sim(\Gamma_\cap(\text{LIN}))$.

On the other hand, it belongs to $\Gamma_\cup(\text{LIN} \cup \text{CO-LIN}) = \Gamma_\cup(\Gamma_\sim(\text{LIN}))$. Therefore, it belongs to $\Gamma_\sim(\Gamma_\cup(\Gamma_\sim(\text{LIN}))) = \Gamma_\sim(\Gamma_\cap(\Gamma_\sim(\text{LIN})))$.   □

Finally, the question whether we obtain an infinite hierarchy is answered negatively. On the next level the closures collapse to $\Gamma_{\text{BOOL}}(\text{LIN})$.

Theorem 29. *$\Gamma_\cap(\Gamma_\sim(\Gamma_\cap(\Gamma_\sim(\text{LIN}))))$ or $\Gamma_\cup(\Gamma_\sim(\Gamma_\cap(\Gamma_\sim(\text{CO-LIN}))))$ are equal to $\Gamma_{\text{BOOL}}(\text{LIN})$.*

	$\Gamma_\sim(\Gamma_\cap(\text{LIN}))$	$\Gamma_\cup(\Gamma_\sim(\text{LIN}))$	$\Gamma_\cap(\Gamma_\sim(\text{LIN}))$	$\Gamma_\sim(\Gamma_\cap(\Gamma_\sim(\text{LIN})))$
$\cup$	−	+	−	−
$\cap$	−	−	+	−
$\sim$	+	−	−	+

Fig. 2. Summary of closure properties II

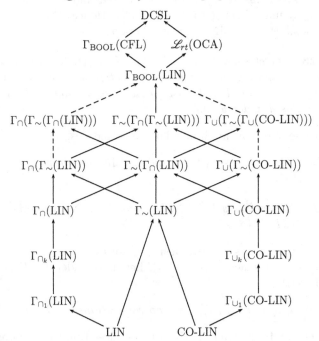

Fig. 3. Solid lines are proper inclusions, dashed lines are inclusions

References

1. Baker, B. S. and Book, R. V. *Reversal-bounded multipushdown machines.* J. Comput. System Sci. 8 (1974), 315–332.

2. Bordihn, H., Holzer, M., and Kutrib, M. *Some non-semi-decidability problems for linear and deterministic context-free languages.* Implementation and Application of Automata (CIAA 2004), LNCS, 2004, to appear.

3. Buchholz, Th. and Kutrib, M. *On time computability of functions in one-way cellular automata.* Acta Inf. 35 (1998), 329–352.

4. Dassow, J. and Păun, G. *Regulated Rewriting in Formal Language Theory.* Springer, Berlin, 1989.

5. Dyer, C. R. *One-way bounded cellular automata.* Inform. Control 44 (1980), 261–281.

6. Ginsburg, S. *The Mathematical Theory of Context-Free Languages.* McGraw Hill, New York, 1966.

7. Ginsburg, S. and Greibach, S. A. *Deterministic context-free languages.* Inform. Control 9 (1966), 620–648.

8. Ibarra, O. H., Jiang, T., and Wang, H. *Parallel parsing on a one-way linear array of finite-state machines.* Theoret. Comput. Sci. 85 (1991), 53–74.

9. Liu, L. Y. and Weiner, P. *An infinite hierarchy of intersections of context-free languages.* Math. Systems Theory 7 (1973), 185–192.

10. Okhotin, A. *Automaton representation of linear conjunctive languages.* Developments in Language Theory (DLT 2002), LNCS 2450, 2003, pp. 393–404.

11. Parikh, R. J. *On context-free languages.* J. Assoc. Comput. Mach. 13 (1966), 570–581.

12. Salomaa, A. *Formal Languages.* Academic Press, New York, 1973.

13. Smith III, A. R. *Cellular automata and formal languages.* IEEE Symposium on Switching and Automata Theory, 1970, pp. 216–224.

14. Smith III, A. R. *Real-time language recognition by one-dimensional cellular automata.* J. Comput. System Sci. 6 (1972), 233–253.

15. Terrier, V. *On real time one-way cellular array.* Theoret. Comput. Sci. 141 (1995), 331–335.

16. Wotschke, D. *The Boolean closures of the deterministic and nondeterministic context-free languages.* GI Jahrestagung, LNCS 1, 1973, pp. 113–121.

17. Wotschke, D. *A characterization of Boolean closures of families of languages.* Automatentheorie und Formale Sprachen, LNCS 2, 1973, pp. 191–200.

18. Wotschke, D. *Nondeterminism and Boolean operations in pda's.* J. Comput. System Sci. 16 (1978), 456–461.

Context-Sensitive Decision Problems in Groups

Stephen R. Lakin[1] and Richard M. Thomas[2]

[1] Cogent Computing Group, School of Mathematical and Information Sciences,
Coventry University, Coventry, UK
`s.lakin@coventry.ac.uk`
[2] Department of Computer Science, University of Leicester, Leicester, UK
`rmt@mcs.le.ac.uk`

Abstract. There already exist classifications of those groups which have regular, context-free or recursively enumerable word problem. The only remaining step in the Chomsky hierarchy is to consider those groups with a context-sensitive word problem. In this paper we consider this problem and prove some results about these groups. We also establish some results about other context-sensitive decision problems in groups.

1 Word Problems in Groups

There are several intriguing connections between group theory and formal language theory. For example, we can consider groups G whose word problem lies in a particular class $\mathcal{F}$ of languages: if Σ is a finite set and φ is a surjective (monoid) homomorphism from Σ^* onto G, then we define the *word problem* WP of G to be $1\varphi^{-1}$ that is the set of words in Σ^* which represent (via φ) the identity in G. It would appear that whether or not the WP lies in $\mathcal{F}$ depends on the choice of Σ and φ, but it is well known that this is not the case if $\mathcal{F}$ is closed under inverse homomorphism (see [10] for example).

In the situation we have just described, we say that Σ is a *monoid generating set* for G. If Σ is a disjoint union $X \cup X^{-1}$, where $X^{-1} = \{x^{-1} \mid x \in X\}$ is in a (1-1) correspondence with X and $x^{-1}\varphi = (x\varphi)^{-1}$, then we say that X is a *group generating set* for G. We consider group generating sets; we simply refer to such a set X as a *generating set* and write $G = \langle X \rangle$. For brevity, we sometimes refer to a word in Σ^* as being a *word over* X. We will only be concerned with *finitely generated* groups G, i.e. groups G where there is a finite set X with $G = \langle X \rangle$.

We may define a *presentation* of a group G to be an expression of the form $\langle X \mid R \rangle$, where X is a generating set for G and R is a set of *relations* which define the structure of G. Each relation is of the form $u_i = v_i$ for words u_i, v_i over X, where the notation $u = v$ means that u and v represent the same element of the group; if two words w and w' represent the same element of G, then this fact will be a consequence of the relations in R (and the normal group laws). G is said to be *finitely presented* if X and R can be chosen to be finite; a *free* group is one with a presentation in which R is empty.

It is a natural question as to which groups have word problems in a particular class $\mathcal{F}$ of languages. For the class of regular languages, Anisimov [1] showed that

C.S. Calude, E. Calude and M.J. Dinneen (Eds.): DLT 2004, LNCS 3340, pp. 296–307, 2004.
© Springer-Verlag Berlin Heidelberg 2004

a group has a regular WP if and only if it is finite. Muller and Schupp [21] proved that a group G has context-free WP if and only if G has a free subgroup of finite index. In fact, they had an extra condition on their group, namely that the group was "accessible", but Dunwoody [7] removed the need for this hypothesis by showing that all finitely-presented groups are accessible (it was already known [2] that groups with a context-free WP are finitely presented). As far as the class of recursively enumerable languages is concerned, the Higman embedding theorem [11] gives us that a finitely generated group G has a recursively enumerable WP if and only if G is a subgroup of a finitely presented group. There is also a characterization of groups with a recursive WP in [5].

Given all this, we have classifications for groups with WP in three of the classes of languages in the Chomsky hierarchy. The only remaining case is that of the class $\mathcal{CS}$ of context-sensitive languages (i.e. those languages which can be recognized in non-deterministic linear space). The problem of giving a precise classification of groups with WP in $\mathcal{CS}$ seems to be very difficult and we present some partial results here. We will also consider the class $\mathcal{DCS}$ of deterministic context-sensitive languages, namely those which can be decided in deterministic linear space. It is still an open question as to whether $\mathcal{DCS}$ and $\mathcal{CS}$ coincide, although it is known [12, 26] that $\mathcal{CS}$ is closed under complementation.

2   Other Decision Problems in Groups

We also consider other decision problems in groups. For example, in a group G, two elements g_1 and g_2 are said to be *conjugate* if there exists $h \in G$ with $h^{-1}g_1 h = g_2$, in which case we write $g_1 \sim_G g_2$. The *conjugacy problem* CP is the question as to whether two given words represent conjugate elements of G.

Another natural problem to consider is the *generalised word problem* GWP of a group G with generators X with respect to a subgroup H, i.e. the question if we can determine whether a word over X lies in H. We will also consider the *effective generalized word problem* EGWP, where not only do we wish to see if a word w lies in the subgroup H, but (if so) also determine which particular element of the subgroup w represents and produce a representative (in the generators of H) for w. For a survey of decision problems in groups, see [19, 20].

As with the other classes of languages mentioned here, $\mathcal{CS}$ and $\mathcal{DCS}$ are closed under inverse homomorphism, and so the WP being (deterministic) context-sensitive is independent of the choice of generating set. We can similarly deduce the same results for the CP and GWP with respect to any subgroup; we will see that the ability to choose a particular generating set can be very convenient.

3   Examples

What sort of groups have a context-sensitive decision problem? Let us give some examples. We begin with free groups. A word w in a free group F can be *reduced* by systematically removing all subwords of the form xx^{-1} or $x^{-1}x$ in w. Two reduced words are equivalent in F if and only if they are identical as words; it immediately follows that the WP for F lies in $\mathcal{DCS}$.

We now turn to the CP. A word w in F is said to be *cyclically reduced* if it is reduced and the first symbol of w is not the inverse of the last symbol. If w is not cyclically reduced, we can conjugate by the last symbol to remove both the first and last symbols; note that, if $u \sim_F v$, then any conjugate of u is also conjugate to any conjugate of v (this is all standard knowledge). Hence the CP reduces to solving the problem for cyclically reduced words. If u and v are cyclically reduced, then they are conjugate if and only if u is a cyclic permutation of v. We can clearly check this in linear space simply by comparing all possible cyclic permutations of u with v; so the CP for a free group lies in $\mathcal{DCS}$.

The GWP in a free group is also deterministic context-sensitive. The classical method of solving this is via *Nielsen reductions*; see [16, 17] for example. Once we have a Nielsen reduced set for the subgroup, it is relatively easy to test membership; from [4], we see that all of this can be done in linear space.

The case of free groups is a special case of that of *linear groups* (those groups isomorphic to matrix groups). It was proved by Lipton and Zalcstein [15] that all linear groups have WP solvable in deterministic logspace, and thus clearly solvable in linear space; so all linear groups have WP in $\mathcal{DCS}$. However, the CP and GWP for such groups may be unsolvable (see Proposition 2 below).

Another important example is the class of *automatic groups*; see [8] for example. This is a wide class which has attracted a great deal of interest. Shapiro showed [27] that any finitely generated subgroup of an automatic group has deterministic context-sensitive WP. (The converse is false; see [14].) The question of the complexity of the CP for automatic groups is still open; it is not even known if such groups have solvable CP. The GWP is unsolvable for automatic groups (see Proposition 3 below).

Given a class of groups with WP in $\mathcal{CS}$ or $\mathcal{DCS}$, we can sometimes generate further examples by taking finitely generated subgroups given the following result (see [14, 23] for example):

Proposition 1. *Let $\mathcal{F}$ be a class of languages closed under inverse homomorphisms and intersection with regular sets; then the class of groups whose WP lies in $\mathcal{F}$ is closed under taking finitely generated subgroups.*

One can easily extend Proposition 1 to cover the GWP as well. Since both $\mathcal{CS}$ and $\mathcal{DCS}$ satisfy the hypothesis of Proposition 1, we have that the classes of groups whose WP or GWP lies in $\mathcal{CS}$ or $\mathcal{DCS}$ are all closed under taking finitely generated subgroups. Unfortunately, this result does not carry over to the situation with regard to the CP. In fact, there are groups with solvable CP (i.e. the CP lies in the class of recursive languages, which is closed under inverse homomorphism and intersection with regular languages) which have subgroups with unsolvable CP (see [6] for example).

4   Direct Products

In this section we investigate whether our decision problems being (deterministic) context-sensitive is preserved under taking direct products. We first conclude that this is the case for the WP:

Theorem 1. *Let G and H be groups and suppose G and H both have (deterministic) context-sensitive WP; then the direct product $K = G \times H$ also has (deterministic) context-sensitive WP.*

Proof. Let X and Y be generating sets for G and H respectively and consider the generating set $Z = X \cup Y$ for K. A word u over Z is equivalent in K to a word $u_g u_h$ where u_g is a word over X and u_h a word over Y, and $u_g u_h$ can be obtained simply by rearranging the symbols in u. Now u represents the identity if and only if u_g and u_h both represent the identity. We simply create $u_g u_h$ and then test u_g and u_h for equivalence to the identity using the (deterministic) context-sensitive algorithms for G and H; this all operates in linear space.   □

Let us now turn to the conjugacy problem.

Theorem 2. *Let G and H be groups, and suppose that G and H both have (deterministic) context-sensitive CP; then the direct product $K = G \times H$ also has (deterministic) context-sensitive CP.*

Proof. Continuing with the notation of Theorem 1, suppose u and v are the input words; then u and v are equivalent in K to words $u_g u_h$ and $v_g v_h$ as before. Now $u \sim_K v$ if and only if $u_g \sim_G v_g$ and $u_h \sim_H v_h$. Our algorithm is very similar to the WP algorithm: we create the words u_g, u_h, v_g, v_h and apply the relevant algorithms for solving the CP in G and H to (u_g, v_g) and (u_h, v_h).   □

We note that the property of having GWP in $\mathcal{CS}$ or $\mathcal{DCS}$ is not preserved under taking direct products. For example, let F be a free group on two generators. We saw in Section 3 that F has GWP in $\mathcal{DCS}$, but, as in Proposition 3 below, the GWP for $F \times F$ is unsolvable.

5   Relationship Between the Decision Problems

It is natural to ask how these decision problems interact with each other (both in general and in the context-sensitive case). We start with an elementary observation. The WP (is w equivalent to 1?) is a special case of the CP (is w conjugate to 1?) and the GWP (is w an element of $\langle 1 \rangle$?); so a group with CP or GWP solvable within some resource bound also has its WP solvable within that bound.

It is possible for all three of these decision problems to be deterministic context-sensitive (as in the case of free groups). It is also possible for none of the problems to be decidable via context-sensitive algorithms as there are groups with unsolvable WP; however, the WP may be genuinely easier than both the CP and the GWP as the following result shows:

Proposition 2. *There exists a linear group with deterministic context-sensitive WP and unsolvable GWP and CP.*

Proof. From Theorem 5.2 of [20], the CP and GWP are unsolvable for some finitely generated subgroups of the matrix group $SL(4, \mathbb{Z})$; however, as we saw above, linear groups have deterministic context-sensitive WP.   □

We also have the following:

Proposition 3. *There exists an automatic group G with deterministic context-sensitive WP and CP and unsolvable GWP.*

Proof. Let F be a free group on two generators, and form the group $G = F \times F$; then G is automatic [8]. From [18], G has a finitely generated subgroup L such that the GWP for L in G is unsolvable. However F has deterministic context-sensitive WP and CP, and thus so does G from Theorems 1 and 2.   □

As far as the remaining possibility, namely the existence of a group with deterministic context-sensitive GWP (and hence WP), but unsolvable CP, is concerned, it appears to be still open whether there even exists a group with solvable GWP but unsolvable CP. If such a group does exist, it is natural to ask whether there is a group with deterministic context-sensitive GWP but unsolvable CP.

6   Amalgamated Free Products of Groups

We have considered direct products of groups; we now consider what happens if we try to combine two groups with a (deterministic) context-sensitive decision problem via an amalgamated free product. Let us first explain what this means.

Suppose we have groups G and H with subgroups $K = \langle k_1, \ldots, k_r \rangle$ and $L = \langle l_1, \ldots, l_r \rangle$ respectively, and also that there is an isomorphism $\varphi : K \to L$ where $\varphi : k_i \mapsto l_i$ for each i. The *amalgamated free product* $G *_{K,L} H$ is the quotient of the free product $G * H$ obtained by equating k_i to l_i for each i. It is a standard result that G and H both embed naturally into $G *_{K,L} H$ and we can then consider K as a common subgroup of G and H (and write $G *_K H$).

Let $X = \langle k_1, \ldots, k_r, g_1, \ldots, g_s \rangle$ and $Y = \langle k_1, \ldots, k_r, h_1, \ldots, h_t \rangle$ be generating sets for the factors G and H respectively, where the g_i and h_j do not lie in K (of course, words over either of these sets of elements may lie in K), and then let Z be the generating set $X \cup Y$ for G. We define a word u over Z to be in *reduced form* if $u = u_K u_1 u_2 \ldots u_n$, where u_K is a (possibly empty) word over $\{k_1, \ldots, k_r\}$, $u_i \notin K$ for all $i \geqslant 1$, and u_i and u_{i+1} do not lie in the same factor for all $i \geqslant 1$. The presence of the subword u_K here allows us to consider words in K without changing our definition (simply by having $n = 0$). However, if $n \geqslant 1$, then we can incorporate u_K into u_1 and simply consider $u = u_1 u_2 \ldots u_n$, which is perhaps more natural. It is known (see [17] for example) that such a word with $n \geqslant 1$ does not represent the identity. We define a word u over Z to be *cyclically reduced* if it is in a reduced form and if u_1 and u_n are not in the same factor (except in the case where $n = 1$). Every word is equivalent to a reduced word and, if $u_K u_1 u_2 \ldots u_m$ and $v_K v_1 v_2 \ldots v_n$ are reduced words equivalent to the same word u, then we must have $m = n$.

We have the following important result (see [17] for example):

Theorem 3 (Solitar's Theorem). *Every element of $G *_K H$ is conjugate to a cyclically reduced element of $G *_K H$. In addition, suppose that u is a cyclically reduced element of $G *_K H$; then:*

1. *if u is conjugate to an element of K, then u lies in some factor;*
2. *if u is conjugate to an element v in some factor but u is not in a conjugate of K, then u and v lie in the same factor and are conjugate therein;*
3. *if u is conjugate to a cyclically reduced element $v = v_1 \ldots v_m$ with $m \geqslant 2$ and v_i in a different factor to v_{i+1} for all i, then u can be obtained by cyclically permuting $v_1 \ldots v_m$ and conjugating by an element of K.*

Let G, H and K be as above. We cannot come up with such a wide-ranging statement for the amalgamated free product as we did for the direct product. The problem is that we need to be able to determine whether or not a word actually represents an element of K, i.e. determine the EGWP for G and H with respect to K. For example, suppose we had a word $u = u_g u_h$ where u_g is a word in the g_i and u_h is a word in the h_i. We may have a situation where neither u_g nor u_h represents the identity, but u_g is equivalent to some element $k \in K$ and u_h is equivalent to k^{-1}, and hence u is equivalent to the identity; the situation is therefore more complicated than that in direct products. However, we can produce at least a partial result; first we need to be able to reduce words:

Proposition 4. *Suppose G and H are groups with a common subgroup K and there exist (deterministic) context-sensitive procedures to decide the EGWP for K in G and H; then, for any word u in $G *_K H$, there exists a (deterministic) context-sensitive procedure to produce a reduced word equivalent to u.*

Proof. This result is fairly obvious since the context-sensitive languages are those decidable in linear space. To summarise the strategy, let

$$G = \langle k_1, \ldots, k_r, g_1, \ldots, g_s \rangle \text{ and } H = \langle k_1, \ldots, k_r, h_1, \ldots, h_t \rangle$$

as usual, where the k_i generate K, and consider the generating set

$$\{k_1, \ldots, k_r, g_1, \ldots, g_s, h_1, \ldots, h_t\}$$

for $G *_K H$.

We go through our word u and consider each (maximal) subword v which is a string of symbols in one of the groups G or H (i.e. a combination of g_i and k_i, or a combination of h_i and k_i). We use our algorithm for the EGWP in the appropriate group to see if v lies in K; if so, then we produce a word $\overline{v}$ in the k_i representing v and replace the subword v of u by $\overline{v}$. We systematically continue our algorithm on the new word.

We continue in this fashion until we either terminate with the empty word (which is obviously the reduced form corresponding to the identity) or a non-empty word in K, or a situation where we cannot make any more deletions or substitutions. This leaves us with a word in reduced form.

This procedure has produced a reduced form for u; note that the word is not always decreasing in length since, when we replace a subword by a representative in K, this may increase the length of the subword by a linear factor C (the procedures for the EGWP are context-sensitive). However, these replacements are bounded by a linear factor and so we have a context-sensitive procedure.

Note that, if G and H have deterministic context-sensitive EGWP with respect to K, then this algorithm is entirely deterministic. $\qquad \square$

We have the following result as a simple consequence:

Theorem 4. *Let G, H and K be groups as above. Suppose G and H both have (deterministic) context-sensitive WP and there exist (deterministic) context-sensitive algorithms to decide the EGWP for G and H with respect to K; then $G *_K H$ also has (deterministic) context-sensitive WP.*

Proof. We produce the reduced form u' of our input word u via the linearly bounded algorithm of Proposition 4. Now u cannot represent the identity unless u' lies in one of the factors; if it does, then we check to see if it is indeed equivalent to the identity in the appropriate group using the (deterministic) context-sensitive algorithm for the WP of that group.   □

As an immediate corollary to Theorem 4 we have the following:

Corollary 1. *Let G and H be groups and suppose that G and H have (deterministic) context-sensitive WP; then the free product $G * H$ also has (deterministic) context-sensitive WP.*

Proof. This result follows immediately from Theorem 4 by taking $K = \{1\}$; our (deterministic) context-sensitive algorithm for determining if a word u lies in K is simply to run the appropriate WP solver on u.   □

In general, it is a difficult question to ask which groups with (deterministic) context-sensitive WP also have (deterministic) context-sensitive EGWP with respect to the appropriate amalgamated subgroup; however, we have the following:

Proposition 5. *Let G be a group with (deterministic) context-sensitive WP and suppose that K is a finite subgroup of G; then G has (deterministic) context-sensitive EGWP with respect to K.*

Proof. Let $G = \langle X \rangle$ and let $k_1, \ldots, k_n$ be words over X representing the elements of K. Suppose we are given a word u over X; we systematically test the words $u^{-1}k_i$ in the WP solver for G. If we accept a word $u^{-1}k_p$, then u lies in K with representative k_p; if none of the words $u^{-1}k_i$ are accepted by the WP solver then u cannot equal any of the k_i, and hence u does not lie in K.

Since the enumeration of the elements of K merely gives us finitely many fixed representatives, this algorithm operates in linear space. If the WP for G is deterministic context-sensitive then this procedure is also deterministic.   □

Given Theorem 4 and Proposition 5, we can immediately deduce the following:

Corollary 2. *Suppose G and H are groups with a common finite subgroup K. If G and H have (deterministic) context-sensitive WP, then $G *_K H$ also has (deterministic) context-sensitive WP.*

Let us now move on to the conjugacy problem. Unfortunately the situation here is somewhat more restrictive, in that it is possible to have two groups with (deterministic) context-sensitive CP and (deterministic) context-sensitive

EGWP with respect to an amalgamated subgroup, and yet the amalgamated free product with respect to this subgroup does not have (deterministic) context-sensitive CP. An example of this occurs in the second half of the proof of the main theorem of [6]. The general situation appears to be very complicated; however, we do have the following result:

Theorem 5. *Let G, H and K be groups as before. Suppose that G and H both have (deterministic) context-sensitive CP and the amalgamated subgroup K is finite; then $G *_K H$ also has (deterministic) context-sensitive CP.*

Proof. Suppose u and v are our input words. We cyclically reduce u and v, first by putting them in reduced form (as in Proposition 4) and then conjugating if necessary (replacing u or v by a conjugate does not affect the property of conjugacy between them). So we can assume that u and v are cyclically reduced and appeal to Theorem 3.

Suppose u is empty; then, if v is empty, we accept u and v as conjugate and, if v is non-empty, we reject them. The situation is similar if v is empty.

If u is conjugate to an element from the subgroup K, then, by Theorem 3, u lies in some factor. So, for u and v to be conjugate, v must lie in some factor (but not necessarily the same factor as u). But we must have $u \sim k_1$ in the factor containing u, $k_1 \sim k_2$ in the factor containing v, and so on, where each k_i is an element of K. Since K is finite, we can check this using a pre-determined lookup table to list the conjugate pairs in K.

Alternatively, suppose u lies in one of the factors but is not conjugate to an element of K; then, by Theorem 3, v must also lie in the same factor. So we check to see if this is true, and, if not, we reject u and v. If it is true, then we use the (deterministic) context-sensitive algorithm for the CP in the appropriate group to determine whether or not they are conjugate, and accept and reject accordingly. We have an entirely similar procedure if v lies in one of the factors.

In the only remaining case, we must have $u = x_1 \ldots x_n$ and $v = y_1 \ldots y_m$ where $n, m \geqslant 2$. By Theorem 3 we must have $m = n$; so, if this is not the case, then we reject the words. If $m = n$ then we know, by Theorem 3, that u and v are conjugate if and only if we can obtain a word equivalent to u by first of all taking some cyclic permutation of the y_i, and then conjugating by an element of K. Enumerating the finite number of elements of K, we consider each permutation of the y_i, and each possible conjugation by an element of K, and see if we obtain a word equivalent to u. Clearly there are only a finite number of words to check.

This algorithm is clearly context-sensitive and, as usual, if our routines are deterministic, then the whole procedure is entirely deterministic.   □

As an immediate consequence of Theorem 5 we have the following:

Corollary 3. *Let G and H be groups, and suppose G and H have (deterministic) context-sensitive CP; then the free product $G * H$ also has (deterministic) context-sensitive CP.*

These ideas allow us to show that other classes of groups have a context-sensitive WP. As an example, we consider a class of one-relator groups as follows:

Proposition 6. *Any one-relator group G which has a presentation of the form $\langle a_1, \ldots, a_m, b_1, \ldots, b_n \mid u_a = u_b \rangle$, where u_a is a word over the a_i and u_b is a word over the b_i, has deterministic context-sensitive WP.*

Proof. Suppose we are given a group G satisfying our hypothesis. Let F_A and F_B be the free groups on the a_i and b_i respectively. Note that G is the amalgamated free product of F_A and F_B with respect to the cyclic subgroups $C_1 = \langle u_a \rangle$ and $C_2 = \langle u_b \rangle$. As F_A and F_B are free groups, the EGWP of F_A with respect to C_1 and F_B with respect to C_2 are deterministic context-sensitive. The result follows from Theorem 4.   □

One important class of groups satisfying the above condition, is the class of *surface groups*. These groups occur when we consider connections between surfaces and combinatorial group theory; for further information see [28] for example. An important point is that there is a basic classification of all closed finite surfaces, either as an orientable surface (of some genus n), a non-orientable surface (of some genus n) or a sphere. The surface groups are the fundamental groups of these surfaces and have the following presentations:

$$\text{orientable, genus } n : \langle x_1, y_1, \ldots, x_n, y_n \mid x_1 y_1 x_1^{-1} y_1^{-1} \ldots x_n y_n x_n^{-1} y_n^{-1} = 1 \rangle;$$
$$\text{non-orientable, genus } n : \langle x_1, \ldots, x_n \mid x_1^2 \ldots x_n^2 = 1 \rangle;$$
$$\text{sphere} : \{1\}.$$

Theorem 6. *All surface groups have deterministic context-sensitive WP.*

Proof. Any surface group is either isomorphic to the fundamental group of a sphere (in which case the group is trivial and the WP is obviously solvable) or else is a one-relator group satisying the hypothesis of Proposition 6.   □

7   Reduced and Irreducible Word Problems

The reduced and irreducible word problems were introduced by Haring-Smith in [9]. Suppose $G = \langle X \rangle$ and let $W = W_X(G)$ denote the WP of G with respect to X. The *reduced word problem* RWP is the set $R_X(G)$ of non-empty words w in W such that no non-empty proper prefix of w lies in W. Similarly the *irreducible word problem* IWP is the set $I_X(G)$ of non-empty words w in W such that no proper non-empty subword of w lies in W. We have the following result [24, 25]:

Proposition 7. *Let $G = \langle X \rangle$; then:*

1. $W_X(G) = R_X(G)^*$.
2. $W_X(G) = I(I_X(G))$ *where $I(L)$ represents the insertion closure of L.*

For an account of insertion closures, see [13]. We now have the following:

Proposition 8. *Let $G = \langle X \rangle$; then the following are equivalent:*

1. *G has (deterministic) context-sensitive WP with respect to X.*

2. G has (deterministic) context-sensitive RWP with respect to X.

3. G has (deterministic) context-sensitive IWP with respect to X.

Proof. This is fairly straightforward given the characterization in Proposition 7. We give the details as an example of the sort of approach often taken to solve problems in linear space.

Suppose that G has (deterministic) context-sensitive WP and let u be a word over X. Let us consider the RWP. First we test whether u represents the identity using the (deterministic) context-sensitive algorithm for the WP for G. If it does not, then u does not lie in the RWP and we reject it. If it does, then we test all proper non-empty prefixes of u. If any of them represent the identity, then we reject u; otherwise, if we have tested all such prefixes without finding one representing the identity, then we accept u. The algorithm for the IWP is entirely similar, but, instead of testing prefixes, we now test every non-empty subword. These two algorithms are clearly (deterministic) context-sensitive procedures, and so we have shown that $1 \Rightarrow 2$ and $1 \Rightarrow 3$.

Conversely, suppose that G has (deterministic) context-sensitive RWP and we are given a word u over X. From Proposition 7 we have that $W_X(G) = R_X(G)^*$. So, if u represents the identity, then some non-empty prefix of u (including possibly u itself) must lie in $R_X(G)$. We simply test each prefix, in turn, with our (deterministic) context-sensitive algorithm for the RWP to see if it lies in $R_X(G)$. If we find no such prefix in $R_X(G)$, then we reject u. If we find a prefix in $R_X(G)$, then we delete this prefix and start our algorithm again on the resulting shorter word. We must eventually terminate, either with rejection or by reducing u to the empty word, in which case we accept u.

The algorithm for the WP, when we have a (deterministic) context-sensitive IWP is similar. Given an input u we simply search through all non-empty subwords of u and delete a subword if we find one in $I_X(G)$; we then continue our algorithm on the resulting shorter word. If we cannot find such a subword then u cannot lie in the insertion closure of $I_X(G)$ and we must reject it; we accept u if we eventually terminate with the empty word. Again these algorithms are clearly (deterministic) context-sensitive and this proves that $2 \Rightarrow 1$ and $3 \Rightarrow 1$, showing the equivalence of the three conditions. $\square$

In the above, we were careful to specify a particular generating set, since we needed this to be able to use the results of Proposition 7. However, Proposition 8 allows us to deduce that the property of having a (deterministic) context-sensitive RWP or IWP is independent of the choice of generating set:

Corollary 4. *Suppose $G = \langle X \rangle = \langle Y \rangle$; then:*

1. if G has (deterministic) context-sensitive RWP with respect to X, then G also has (deterministic) context-sensitive RWP with respect to Y.

2. if G has (deterministic) context-sensitive IWP with respect to X, then G also has (deterministic) context-sensitive IWP with respect to Y.

Proof. Suppose G has (deterministic) context-sensitive RWP with respect to X. Then G has (deterministic) context-sensitive WP with respect to X by Proposition 8, and thus G has (deterministic) context-sensitive WP with respect to Y

(see Proposition 1 and the comments following it). Hence, using Proposition 8 again, G has (deterministic) context-sensitive RWP with respect to Y.

The situation with regard to the IWP is entirely similar.   □

Hence we are able to talk about the RWP or IWP being (deterministic) context-sensitive without worrying about which generating set is being considered. Note, however, that the proof of Corollary 4 uses Proposition 8, which requires linear space, and Corollary 4 does not generalize to arbitrary families of languages closed under inverse homomorphism.

8   Conclusion

In this paper we have studied groups with context-sensitive decision problems. It is clear that there is still a great deal more to do if we are to classify such groups; we are dealing with a wide class of groups with many interesting properties. We hope that our results provide a first step on this road to classification.

One point to consider involves the issue of determinism and non-determinism. It is still unknown whether or not linear space is equivalent to non-deterministic linear space in general. One interesting question is motivated by the result of Muller and Schupp in [22] that, if a group has context-free WP, it also has deterministic context-free WP, even though the classes of context-free and deterministic context-free languages are distinct. (In fact, by [3] the WP of such a group is even an NTS language.) One might ask whether a similar result holds for context-sensitive languages, that is, does every group with a context-sensitive WP also have a deterministic context-sensitive WP?

We should also mention that we have focused on context-sensitive languages as the one remaining case in the Chomsky hierarchy (as far as word problems are concerned). In fact, our results could be couched in terms of (deterministic) algorithms that can be performed in space $O(f(n))$ for any complexity function $f(n)$ such that $O(n) \subseteq O(f(n))$.

Acknowledgements. The constructive comments of the referees were much appreciated. The second author would also like to thank Hilary Craig for all her help and encouragement.

References

1. A. V. Anisimov, Group languages, *Kibernetika* **4** (1971), 18–24.
2. A. V. Anisimov, Some algorithmic problems for groups and context-free languages, *Kibernetika* **8** (1972), 4–11.
3. J. M. Autebert, L. Boasson & G. Sénizergues, Groups and NTS languages, *J. Comp. System Sci.* **35** (1987), 243–267.
4. J. Avenhaus & K. Madlener, The Nielsen reduction and P-complete problems in free groups, *Theoret. Comp. Sci.* **32** (1984), 61–76.
5. W. W. Boone & G. Higman, An algebraic characterization of groups with a solvable word problem, *J. Australian Math. Soc.* **18** (1974), 41–53.

6. D. J. Collins & C. F. Miller III, The conjugacy problem and subgroups of finite index, *Proc. London Math. Soc.* **34** (1977), 535–556.

7. M. J. Dunwoody, The accessibility of finitely presented groups, *Invent. Math.* **81** (1985), 449–457.

8. D. B. A. Epstein, J. W. Cannon, D. F. Holt, S. V. F. Levy, M. S. Paterson & W. P. Thurston, *Word processing in groups* (Jones and Bartlett, 1991).

9. R. H. Haring-Smith, Groups and simple languages, *Trans. Amer. Math. Soc.* **279** (1983), 337–356

10. T. Herbst & R. M. Thomas, Group presentations, formal languages and characterizations of one-counter groups, *Theoret. Comp. Sci.* **112** (1993), 187–213.

11. G. Higman, Subgroups of finitely presented groups, *Proc. Roy. Soc. Ser. A* **262** (1961), 455–475.

12. N. Immerman, Nondeterministic space is closed under complementation, *SIAM J. Comput.* **17** (1988), 935–938.

13. M. Ito, L. Kari & G. Thierren, Insertion and deletion closure of languages, *Theoret. Comp. Sci.* **183** (1997), 3–19.

14. S. R. Lakin, *Context-sensitive decision problems in groups* (PhD thesis, University of Leicester, 2002).

15. R. J. Lipton & Y. Zalcstein, Word problems solvable in logspace, *J. Assoc. Comput. Mach.* **24** (1977), 522–526.

16. R. C. Lyndon & P. E. Schupp *Combinatorial group theory* (Springer-Verlag, 1977).

17. W. Magnus, A. Karrass & D. Solitar *Combinatorial group theory* (Dover, 1976).

18. K. A. Mihailova, The occurrence problem for direct products of groups, *Dokl. Akad. Nauk SSSR* **119** (1958), 1103–1105 .

19. C. F. Miller III *On group-theoretic decision problems and their classification* (Princeton University Press, 1971).

20. C. F. Miller III, Decision problems in groups - surveys and reflections, *in* G. Baumslag & C. F. Miller III (eds.), *Algorithms and classification in combinatorial group theory* (MSRI Publications **23**, Springer-Verlag, 1982), 1–59.

21. D. E. Muller & P. E. Schupp, Groups, the theory of ends, and context-free languages, *J. Comp. System Sci.* **26** (1983), 295–310.

22. D. E. Muller & P. E. Schupp, The theory of ends, pushdown automata and second-order logic, *Theoret. Comp. Sci.* **37** (1985), 51–75.

23. C. E. Röver, D. F. Holt, S. E. Rees & R. M. Thomas, Groups with a context-free co-word problem, *J. London Math. Soc.*, to appear.

24. D. W. Parkes & R. M. Thomas, Groups with context-free reduced word problem, *Comm. Algebra* **30** (2002), 3143–3156.

25. D. W. Parkes & R. M. Thomas, Reduced and irreducible word problems of groups (*MCS Technical Report* **1999/4**, University of Leicester, 1999).

26. R. Szelepcsényi, The method of forcing for nondeterministic automata, *Bull. EATCS* **33** (1987), 96–100.

27. M. Shapiro, A note on context-sensitive languages and word problems, *Internat. J. Algebra Comput.* **4** (1994), 493–497.

28. J. C. Stillwell, *Classical topology and combinatorial group theory* (Springer-Verlag, 1993).

Decidability and Complexity in Automatic Monoids[*]

Markus Lohrey

Lehrstuhl für Informatik I, RWTH Aachen, Germany
lohrey@i1.informatik.rwth-aachen.de

Abstract. We prove several complexity and decidability results for automatic monoids: (i) there exists an automatic monoid with a P-complete word problem, (ii) there exists an automatic monoid such that the first-order theory of the corresponding Cayley-graph is not elementary decidable, and (iii) there exists an automatic monoid such that reachability in the corresponding Cayley-graph is undecidable. Moreover, we show that for every hyperbolic group the word problem belongs to LOGCFL, which improves a result of Cai [4].

1   Introduction

Automatic groups attracted a lot of attention in combinatorial group theory during the last 15 years, see e.g. the textbook [11]. Roughly speaking, a finitely generated group $\mathcal{G}$, generated by the finite set Γ, is automatic, if the elements of $\mathcal{G}$ can be represented by words from a regular language over Γ, and the multiplication with a generator on the right can be recognized by a synchronized 2-tape automaton. This concept easily yields a quadratic time algorithm for the word problem of an automatic group.

It is straight forward to extend the definition of an automatic group to the monoid case; this leads to the class of *automatic monoids*, see e.g. [6, 13, 16, 26]. In the present paper, we study the complexity and decidability of basic algorithmic questions in automatic monoids. In Section 4 we consider the complexity of the word problem for automatic monoids. Analogously to the group case, it is easy to show that for every automatic monoid the word problem can be solved in quadratic time. Here, we prove that there exists a fixed automatic monoid with a P-complete word problem. Thus, unless P = NC, where NC is the class of all problems that can be solved in polylogarithmic time using a polynomial amount of hardware, there exist automatic monoids for which the word problem cannot be efficiently parallelized. Whether there exists an automatic *group* with a P-complete word problem was asked for the first time by Cai [4]. This problem remains open.

---

[*] This work was partly done while the author was at FMI, University of Stuttgart, Germany.

An important subclass of the class of automatic groups is the class of *hyperbolic groups*, which are defined via a geometric hyperbolicity condition on the Cayley-graph. In [4], Cai has shown that for every hyperbolic group the word problem belongs to the parallel complexity class NC^2. Cai also asked, whether the upper bound of NC^2 can be improved. Using known results from formal language theory, we show in Section 4 that the word problem for every hyperbolic group belongs to the complexity class $LOGCFL \subseteq NC^2$. LOGCFL is the class of all problems that are logspace reducible to a context-free language [32]. We also present a class of automatic monoids, namely monoids that can be presented by finite, terminating, confluent, and left-basic semi-Thue systems [29], for which the complexity of the word problem captures the class LOGDCFL (the logspace closure of the *deterministic* context-free languages).

In Section 5 we study *Cayley-graphs* of automatic monoids. The Cayley-graph of a finitely generated monoid $\mathcal{M}$ wrt. a finite generating set Γ is a Γ-labeled directed graph with node set $\mathcal{M}$ and an a-labeled edge from a node x to a node y if $y = xa$ in $\mathcal{M}$. Cayley-graphs of groups are a fundamental tool in combinatorial group theory [23] and serve as a link to other fields like topology, graph theory, and automata theory, see, e.g., [24, 25]. Results on the geometric structure of Cayley-graphs of automatic monoids can be found in [30, 31]. Here we consider Cayley-graphs from a logical point of view, see [20, 21] for previous results in this direction. More precisely, we consider the first-order theory of the Cayley-graph of an automatic monoid $\mathcal{M}$. This theory contains all true statements of the Cayley-graph that result from atomic statements of the form "there is an a-labeled edge between two nodes" using Boolean connectives and quantification over nodes. From the definition of an automatic monoid it follows immediately that the Cayley-graph of an automatic monoid is an automatic graph in the sense of [1, 18]; hence, by a result from [18], its first-order theory is decidable. This allows to verify non-trivial properties for automatic monoids, like for instance right-cancellativity. Here, we prove that there exists an automatic monoid such that the first-order theory of the corresponding Cayley-graph is not elementary decidable. This result sharpens a corresponding statement for general automatic graphs [1]. We remark that, using a result from [22], the Cayley-graph of a right-cancellative automatic monoid has an elementarily decidable first-order theory. Finally we prove that there exists an automatic monoid $\mathcal{M}$ such that reachability in the Cayley-graph (i.e., the question whether for given monoid elements u and v there exists $x \in \mathcal{M}$ with $u = vx$ in $\mathcal{M}$) is undecidable.

2   Monoids and Word Problems

More details and references concerning the material in this section can be found in [3]. In the following, let Γ be always a *finite* alphabet of symbols. A semi-Thue system R over Γ is a (not necessarily finite) set $R \subseteq \Gamma^* \times \Gamma^*$; its elements are called rules. A rule (s, t) will be also written as $s \to t$. W.l.o.g. we may assume that every symbol from Γ appears in a rule of R; thus, Γ is given uniquely by R. Let $\mathrm{dom}(R) = \{\ell \mid \exists r : (\ell, r) \in R\}$ and $\mathrm{ran}(R) = \{r \mid \exists \ell : (\ell, r) \in R\}$. We define the binary relation $\to_R$ on Γ^* by: $x \to_R y$ if $\exists u, v \in \Gamma^* \exists (s,t) \in R : x = usv$ and $y = utv$. Let $\overset{*}{\leftrightarrow}_R$ by the smallest equivalence relation on Γ^* containing $\to_R$; it is a congruence wrt. the concatenation of words and called the *Thue-congruence* associated with R. Hence, we can define the quotient monoid $\Gamma^*/\overset{*}{\leftrightarrow}_R$, which is briefly denoted by Γ^*/R. Let $\pi_R : \Gamma^* \to \Gamma^*/R$ be the canonical surjective monoid homomorphism that maps a word $w \in \Gamma^*$ to its equivalence class wrt. $\overset{*}{\leftrightarrow}_R$. A monoid $\mathcal{M}$ is *finitely generated* if it is isomorphic to a monoid of the form Γ^*/R. In this case, we also say that $\mathcal{M}$ *is finitely generated by* Γ. If in addition to Γ also R is finite, then $\mathcal{M}$ is a *finitely presented monoid*. The *word problem of* $\mathcal{M} \simeq \Gamma^*/R$ *wrt.* R is the set $\{(u,v) \in \Gamma^* \times \Gamma^* \mid \pi_R(u) = \pi_R(v)\}$; it is undecidable in general. If a monoid $\mathcal{M}$ is isomorphic to both Γ^*/R and Σ^*/S for semi-Thue systems R and S, then the word problem of $\mathcal{M}$ wrt. R is logspace-reducible to the word problem of $\mathcal{M}$ wrt. S. Hence, since we are only interested in the decidability (resp. complexity) status of word problems, it makes sense to speak just of the word problem of $\mathcal{M}$.

The semi-Thue system R is *terminating* if there does not exist an infinite chain $s_1 \to_R s_2 \to_R s_3 \to_R \cdots$ in Γ^*. The set of *irreducible words* wrt. R is $\mathrm{IRR}(R) = \{s \in \Gamma^* \mid \neg \exists t \in \Gamma^* : s \to_R t\}$. The system R is *confluent* (resp. *locally confluent*) if for all $s, t, u \in \Gamma^*$ with $s \overset{*}{\to}_R t$ and $s \overset{*}{\to}_R u$ (resp. $s \to_R t$ and $s \to_R u$) there exists $w \in \Gamma^*$ with $t \overset{*}{\to}_R w$ and $u \overset{*}{\to}_R w$. If R is terminating, then by Newman's lemma R is confluent if and only if R is locally confluent. Using *critical pairs* [3] which result from overlapping left-hand sides of R, local confluence is decidable for finite terminating semi-Thue systems. The system R is *length-reducing* if $|s| > |t|$ for all $(s,t) \in R$, where $|w|$ is the length of a word w. The system R is called *length-lexicographic* if there exists a linear order $\succ$ on the alphabet Γ such that for every rule $(s,t) \in R$ either $|s| > |t|$ or ($|s| = |t|$ and there are $u, v, w \in \Gamma^*$ and $a, b \in \Gamma$ such that $s = uav$, $t = ubw$, and $a \succ b$). Clearly, every length-lexicographic semi-Thue system is terminating. In the case when R is terminating and confluent, then every word s has a unique *normal form* $\mathrm{NF}_R(s) \in \mathrm{IRR}(R)$ such that $s \overset{*}{\to}_R \mathrm{NF}_R(s)$ and moreover, the function $\pi_R \restriction \mathrm{IRR}(R)$ (i.e., π_R restricted to $\mathrm{IRR}(R)$) is bijective. Thus, if moreover R is finite, then the word problem of Γ^*/R is decidable: $\pi_R(s) = \pi_R(t)$ if and only if $\mathrm{NF}_R(s) = \mathrm{NF}_R(t)$.

3   Automatic Monoids

Automatic monoids were investigated for instance in [6, 13, 14, 16, 26]. They generalize automatic groups, see [11]. Let us fix a finite alphabet Γ. Let $\# \notin \Gamma$ be an additional padding symbol and let $\Gamma_\# = \Gamma \cup \{\#\}$. We define two encodings $\nu_\ell, \nu_r : \Gamma^* \times \Gamma^* \to (\Gamma_\# \times \Gamma_\#)^*$ as follows: Let $u, v \in \Gamma^*$ and let $k = \max\{|u|, |v|\}$. Define $w = u\#^{k-|u|}$, $x = v\#^{k-|v|}$, $y = \#^{k-|u|}u$, and $z = \#^{k-|v|}v$. Let $w[i]$ denote the i-th symbol of w and similarly for x, y, and z. Then

$$\nu_r(u,v) = (w[1], x[1]) \cdots (w[k], x[k]) \text{ and } \nu_\ell(u,v) = (y[1], z[1]) \cdots (y[k], z[k]).$$

For instance, $\nu_r(aba, bbabb) = (a, b)(b, b)(a, a)(\#, b)(\#, b)$ and $\nu_\ell(aba, bbabb) = (\#, b)(\#, b)(a, a)(b, b)(a, b)$. In the following let $\alpha, \beta \in \{\ell, r\}$.

A relation $R \subseteq \Gamma^* \times \Gamma^*$ is called α-automatic if the language $\{\nu_\alpha(u, v) \mid (u, v) \in R\}$ is a regular language over the alphabet $\Gamma_\# \times \Gamma_\#$. The following simple lemma will turn out to be useful. Its simple proof is left to the reader. A relation $R \subseteq \Gamma^* \times \Gamma^*$ has *bounded length-difference* if there exists a constant γ such that for all $(u, v) \in R$, $|(|u| - |v|)| \leq \gamma$.

Lemma 1. *Let* $R, S \subseteq \Gamma^* \times \Gamma^*$ *have bounded length-difference. Then* R *is* ℓ-*automatic if and only if* R *is* r-*automatic. Moreover, if* R *and* S *are* α-*automatic, then* $R \cdot S = \{(st, uv) \mid (s, u) \in R, (t, v) \in S\}$ *is* α-*automatic as well.*

Let $\mathcal{M}$ be a monoid. A triple (Γ, R, L) is an $\alpha\beta$-*automatic presentation for* $\mathcal{M}$ if: (i) R is a semi-Thue system over the finite alphabet Γ such that $\mathcal{M} \simeq \Gamma^*/R$, (ii) $L \subseteq \Gamma^*$ is a regular language such that $\pi_R{\restriction}L$ maps L surjectively to $\mathcal{M}$, (iii) the relation $\{(u, v) \in L \times L \mid \pi_R(u) = \pi_R(v)\}$ is α-automatic, and (iv) if $\beta = \ell$ (resp. $\beta = r$), then the relation $\{(u, v) \in L \times L \mid \pi_R(au) = \pi_R(v)\}$ (resp. $\{(u, v) \in L \times L \mid \pi_R(ua) = \pi_R(v)\}$) is α-automatic for every $a \in \Gamma$. The monoid $\mathcal{M}$ is $\alpha\beta$-*automatic* if there exists an $\alpha\beta$-automatic presentation for $\mathcal{M}$. Thus, we have four different basic notions of automaticity. Whereas for groups all these four variants are equivalent [13] (which allows to speak of automatic groups), one obtains 15 different notions of automaticity for monoids by combining the four basic variants of $\alpha\beta$-automaticity [13, 14]. For our lower bounds we will mostly work with the strongest possible notion of automaticity, i.e., simultaneous $\alpha\beta$-automaticity for all $\alpha, \beta \in \{\ell, r\}$ (which includes the notion of biautomaticity from the theory of automatic groups, see [11]). Note that a $\alpha\beta$-automatic monoid is by definition finitely generated. Various classes of semi-Thue systems that present automatic monoids can be found in [26].

4   Complexity of the Word Problem

The word problem for an automatic group can be solved in quadratic time [11]. Moreover, the same algorithm also works for $\alpha\beta$-automatic monoids [6]. Here we will show that P is also a lower bound for the monoid case.

Theorem 1. *There is a finite, length-lexicographic, and confluent semi-Thue system* $R \subseteq \Gamma^* \times \Gamma^*$ *such that the word problem for* Γ^*/R *is P-complete and* $(\Gamma, R, \mathrm{IRR}(R))$ *is an* $\alpha\beta$-*automatic presentation for* Γ^*/R *for all* $\alpha, \beta \in \{\ell, r\}$.

Proof. We start with a fixed deterministic Turing machine S that accepts a P-complete language. Let $p(n)$ be a polynomial such that S terminates on an input $w \in L(S)$ after exactly $p(|w|)$ steps (this exact time bound can be easily enforced). We may assume that the tape is restricted to size $p(|w|)$. It is straight forward to simulate S by a new deterministic Turing machine T that operates in a sequence of complete left/right sweeps over the whole tape (of size $p(|w|)$). During a right sweep, the head runs from the left tape end to the right tape end in a sequence of right moves. When reaching the right tape end, the head turns back and starts a left sweep. Let Σ be the tape alphabet of T, Q be the set of states, q_0 be the initial state, and q_f be the final state. With $\mathbb{c} \in \Sigma$ we denote the blank symbol. We write $qa \Rightarrow_T bp$ ($aq \Rightarrow_T pb$), in case T writes b, moves right (left), and enters state p, when reading a in state q. The machine T terminates (and accepts its input) if and only if it finally reaches the final state q_f. Thus, T cannot make any transitions out of q_f. Moreover, we may assume that the tape is blank and that the tape head is scanning the first cell when T terminates in state q_f. Define $\Gamma = \Sigma \cup \overline{\Sigma} \cup Q \cup \overline{Q} \cup \{\$, \overline{\$}\}$, where $\overline{\Sigma} = \{\overline{a} \mid a \in \Sigma\}$ is a disjoint copy of Σ and similarly for $\overline{Q}$. Let R be the following semi-Thue system over Γ:

$$
\begin{array}{ll}
qa \to \overline{b}p \ \text{ if } qa \Rightarrow_T bp & \overline{a}\,\overline{q} \to \overline{p}b \ \text{ if } aq \Rightarrow_T pb \\
q\$ \to \overline{q} \quad \text{ for all } q \in Q & \overline{\$}\,\overline{q} \to q \quad \text{ for all } q \in Q
\end{array}
$$

R is length-lexicographic and confluent (because T is deterministic). Next, let $w \in \Sigma^*$ be an arbitrary input for T and let $m = p(|w|)$. Then w is accepted by T if and only if $\overline{\$}^m q_0 w \mathbb{c}^{m-|w|} \$^m \xrightarrow{*}_R q_f \mathbb{c}^m$ if and only if $\overline{\$}^m q_0 w \mathbb{c}^{m-|w|} \$^m \xleftrightarrow{*}_R q_f \mathbb{c}^m$. Thus, the word problem for Γ^*/R is P-hard.

Next, we show that for all $\alpha, \beta \in \{\ell, r\}$, $(\Gamma, R, \mathrm{IRR}(R))$ is an $\alpha\beta$-automatic presentation for Γ^*/R (then in particular, the word problem for Γ^*/R belongs to P). Due to the symmetry of R, we can restrict to $\beta = \ell$. Thus, we have to show that the relation $E_c = \{(u, v) \in \mathrm{IRR}(R) \times \mathrm{IRR}(R) \mid cu \xrightarrow{*}_R v\}$ is α-automatic for all $c \in \Gamma$ and $\alpha \in \{\ell, r\}$. Note that all relations that appear in the following consideration have bounded length-difference. This allows to make use of Lemma 1. First, note that the following relations are α-automatic:

$$
A_q = \{(u, \overline{v}p) \mid p \in Q, u \in \Sigma^*, \overline{v} \in \overline{\Sigma}^*, qu \xrightarrow{*}_R \overline{v}p\}
$$
$$
B_q = \{(\overline{u}, \overline{p}v) \mid \overline{p} \in \overline{Q}, \overline{u} \in \overline{\Sigma}^*, v \in \Sigma^*, \overline{u}\,\overline{q} \xrightarrow{*}_R \overline{p}v\}
$$

The relation A_q (resp. B_q) describes a single right (resp. left) sweep over the whole tape started in state q, which is just a rational transduction. Since α-automatic relations are closed under composition, the relation

$$
C_q = \{(u\$, \overline{p}v) \mid \overline{p} \in \overline{Q}, u, v \in \Sigma^*, qu\$ \xrightarrow{*}_R \overline{p}v\}
$$

is α-automatic as well. Now the α-automaticity of the relations E_c for $c \in \Gamma$ follows easily: For $c \in \overline{Q} \cup \Sigma \cup \{\$\}$ we have $E_c = \{(u, cu) \mid u \in \mathrm{IRR}(R)\}$, which is clearly α-automatic. For $c = \overline{a} \in \overline{\Sigma}$ and $c = q \in Q$, respectively, we have:

$$
E_{\overline{a}} = \{(u, \overline{a}u) \mid u \in \mathrm{IRR}(R), u \notin \overline{Q}\Gamma^*\} \cup
$$
$$
\{(\overline{q}u, \overline{p}bu) \mid u \in \mathrm{IRR}(R), \overline{q}, \overline{p} \in \overline{Q}, b \in \Sigma, (\overline{a}\,\overline{q}, \overline{p}b) \in R\}
$$
$$
E_q = \{(uw, vw) \mid (u, v) \in A_q, w \in \mathrm{IRR}(R), w \notin (\Sigma \cup \{\$\})\Gamma^*\} \cup
$$
$$
\{(uw, vw) \mid (u, v) \in C_q, w \in \mathrm{IRR}(R)\}.
$$

Finally, $E_{\overline{\$}} = \{(u, \overline{\$}u) \mid u \in \mathrm{IRR}(R), u \notin \overline{Q}\Gamma^*\} \cup \bigcup_{q \in Q}\{(\overline{q}u, v) \mid (u, v) \in E_q\}$. This concludes the proof of the α-automaticity of the relations E_c. $\qquad \square$

Corollary 1. *There exists a fixed finitely presented monoid with a P-complete word problem, which is simultaneously $\alpha\beta$-automatic for all $\alpha, \beta \in \{\ell, r\}$.*

It is open, whether there exists an automatic group (or even cancellative automatic monoid) with a P-complete word problem. An important subclass of the class of automatic groups is the class of hyperbolic groups, which are defined via a geometric hyperbolicity condition on the Cayley-graph. The precise definition is not important for the purpose of this paper. In [4], Cai has shown that for every hyperbolic group the word problem belongs to the parallel complexity class NC^2, which is the class of all problems that can be recognized by a polynomial size family of Boolean circuits of depth $O(\log^2(n))$, where only Boolean gates of fan-in at most 2 are allowed. Cai also asked, whether the upper bound of NC^2 can be improved. Using known results from formal language theory, we will show that for every hyperbolic group the word problem belongs to LOGCFL $\subseteq NC^2$, which is the class of all problems that are logspace reducible to a context-free language [32]. For alternative characterizations of LOGCFL see [27,33].

Theorem 2. *The word problem for every fixed hyperbolic group is in LOGCFL.*

Proof. By [8], a group $\mathcal{G}$ is hyperbolic if and only if $\mathcal{G} \cong \Gamma^*/R$, where R is finite, length-reducing, and $L := \{s \in \Gamma^* \mid s \xrightarrow{*}_R \varepsilon\} = \{s \in \Gamma^* \mid s \xleftrightarrow{*}_R \varepsilon\}$. Since $\mathcal{G}$ is a group, the word problem for $\mathcal{G}$ is logspace reducible to L. Since R is length-reducing, L is growing context-sensitive, i.e., it can be generated by a grammar, where every production is strictly length-increasing. Since every fixed growing context-sensitive language belongs to LOGCFL [9], the theorem follows.   □

In [10,15], hyperbolic groups were generalized to *hyperbolic monoids*. It is not clear whether Theorem 2 can be extended to hyperbolic monoids. It is also open, whether the upper bound of LOGCFL from Theorem 2 can be further improved, for instance to LOGDCFL, which is the class of all problems that are logspace reducible to a deterministic context-free language [32]. For another class of automatic monoids, we can precisely characterize the complexity of the word problem using LOGDCFL: A semi-Thue system R over the alphabet Γ is called *left-basic* [29] if: (i) if $\ell \in \text{dom}(R)$, $r \in \text{ran}(R)$ and $r = u\ell v$ then $u = v = \varepsilon$ and (ii) if $\ell \in \text{dom}(R)$, $r \in \text{ran}(R)$, $ur = \ell v$, and $|\ell| > |u|$, then $v = \varepsilon$. Condition (i) means that a right-hand side does not strictly contain a left-hand side. Condition (ii) means that the following kind of overlapping is not allowed:

u	$r \in \text{ran}(R)$	
$\ell \in \text{dom}(R)$		$v \neq \varepsilon$

Let us define the suffix-rewrite relation $\twoheadrightarrow_R$ by $s \twoheadrightarrow_R t$ if and only if $s = u\ell$ and $t = ur$ for some $u \in \Gamma^*$ and $(\ell, r) \in R$. The following lemma is obvious:

Lemma 2. *If R is left-basic, then for every $s \in \text{IRR}(R)$ and $a \in \Gamma$ we have $sa \xrightarrow{*}_R t$ if and only if $sa \xrightarrow{*}_R t$.*

Left-basic semi-Thue systems generalize monadic semi-Thue systems. Systems that are finite, monadic, and confluent present monoids that are simultaneously rr- and $\ell\ell$-automatic, but in general neither $r\ell$- nor ℓr-automatic [26]. Using arguments similar to those from [26], we can show that for a finite, terminating, confluent, and left-basic semi-Thue system R over an alphabet Γ, the monoid Γ^*/R is rr-automatic.

Theorem 3. *The following problem is in LOGDCFL:*

  INPUT: A finite, terminating, confluent, and left-basic semi-Thue system R over an alphabet Γ, and two words $s, t \in \Gamma^*$

  QUESTION: $s \overset{*}{\leftrightarrow}_R t$?

  Moreover, there exists a finite, length-reducing, confluent, and left-basic semi-Thue system R over an alphabet Γ such that the word problem for Γ^*/R is LOGDCFL-complete.

Proof. Note that the upper bound in the first statement holds in a uniform setting, i.e., the semi-Thue system is part of the input. In order to prove this upper bound, we will use a machine-based characterization of LOGDCFL: A logspace bounded deterministic AuxPDA is a deterministic pushdown automaton that has an auxiliary read-write tape of size $\mathcal{O}(\log(n))$ (where n is the input size). A problem belongs to LOGDCFL if and only if it can be decided by a logspace bounded deterministic AuxPDA that moreover works in polynomial time [32].

  Now, let us assume that the input consists of a tuple (Γ, R, s, t), where R is a finite, terminating, confluent, and left-basic semi-Thue system over the alphabet Γ and $s, t \in \Gamma^*$. Let n be the length of the binary coding of this input. We will construct a logspace bounded deterministic AuxPDA that checks in polynomial time, whether $\mathrm{NF}_R(s) = \mathrm{NF}_R(t)$. For this, we will first show how to calculate $\mathrm{NF}_R(s)$ on a deterministic AuxPDA in logspace and polynomial time. The basic idea of how to do this appeared many times in the literature, see e.g. [3–Thm. 4.2.7]. The only slight complication in our situation results from the fact that the semi-Thue system R belongs to the input. To overcome this, we need the logspace bounded auxiliary store of our AuxPDA. The correctness of the following procedure follows from Lemma 2. Our algorithm for computing $\mathrm{NF}_R(s)$ works in stages. At the beginning of a stage the pushdown contains a word from $\mathrm{IRR}(R)$ and the auxiliary store contains a pointer to a position i in the input word s. Note that a symbol $a \in \Gamma$ can be represented as a bit string of length $\mathcal{O}(\log(n))$, thus the pushdown content is a sequence of blocks of length $\mathcal{O}(\log(n))$, where every block represents a symbol from Γ. The stage begins by pushing the i-th symbol of s onto the pushdown (which is a bit string of length $\mathcal{O}(\log(n))$) and incrementing the pointer to position $i + 1$ in s. Now we have to check whether the pushdown content is of the form $\Gamma^*\mathrm{dom}(R)$. For this we have to scan every left-hand side of R using a second pointer to the input. Every $\ell \in \mathrm{dom}(R)$ is scanned in reverse order and thereby compared with the top of the push-down. During this phase, symbols are popped from the pushdown. If it turns out that the left-hand side that is currently scanned is not a suffix of the pushdown content, then these symbols must be "repushed". This can be done, since the suffix of the pushdown content that was popped so far is a suffix of the currently scanned left-hand side $\ell \in \text{dom}(R)$, which is still available on the read-only input tape. If a left-hand side ℓ is found on top of the pushdown, then the corresponding right-hand side is pushed on the pushdown and we try to find again a left-hand side on top of the pushdown. If finally no left-hand side matches a suffix of the pushdown content, then we know that the pushdown content belongs to $\text{IRR}(R)$ and we can proceed with the next stage. Finally, if the first pointer has reached the end of the input word s (or more precisely points to the first position following s), then the pushdown content equals $\text{NF}_R(s)$.

Claim: In the above procedure, after the i-th stage the pushdown has length at most $i \cdot \alpha$, where $\alpha = \max(\{1\} \cup \{|r| \mid r \in \text{ran}(R)\})$. Moreover, every stage needs only polynomial time.

The first statement can be shown by induction on i. Since R is left-basic, it follows that if w is the pushdown content at the end of the $(i-1)$-th stage, then the pushdown content at the end of the i-th stage either belongs to $w\Gamma$ or is of the form ur for some $r \in \text{ran}(R)$ and some prefix u of w. Moreover, the i-th stage simulates at most $|w| \cdot |R|$ rewrite steps of R.

In order to check whether $\text{NF}_R(s) = \text{NF}_R(t)$, we have to solve one more problem: If we would calculate $\text{NF}_R(t)$ in the same way as above, then the pushdown would finally contain the word $\text{NF}_R(s)\text{NF}_R(t)$. But now there seems to be no way of checking, whether $\text{NF}_R(s) = \text{NF}_R(t)$. Thus, we have to apply another strategy. Note that for a fixed binary coded number $1 \leq i \leq \alpha \cdot |s|$, it is easy to modify our algorithm for calculating $\text{NF}_R(s)$ such that some specified auxiliary storage cell S contains always the i-th symbol of the pushdown content (or some special symbol if the pushdown is shorter than i). For this we have to store the length of the pushdown, for which we need only space $\mathcal{O}(\log(n))$. Moreover, also S only needs space $\mathcal{O}(\log(n))$. Thus, at the end of our modified algorithm for computing $\text{NF}_R(s)$, S contains the symbol $\text{NF}_R(s)[i]$ (the i-th symbol of $\text{NF}_R(s)$) or some special symbol in case $|\text{NF}_R(s)| < i$. Next, we flush the pushdown and repeat the same procedure with the other input word t and the same i, using another storage cell T. In this way we can check, whether $\text{NF}_R(s)[i] = \text{NF}_R(t)[i]$. Finally, we repeat this step for every $1 \leq i \leq \max\{\alpha \cdot |s|, \alpha \cdot |t|\}$. The latter bound is the maximal pushdown-length that may occur, which follows from the above claim. Note that also i needs only space $\mathcal{O}(\log(n))$. This concludes the description of our LOGDCFL-algorithm.

It remains to construct a finite, length-reducing, confluent, and left-basic semi-Thue system R such that the corresponding word problem is LOGDCFL-hard. In [32], Sudborough has shown that there exists a fixed deterministic context-free language $L \subseteq \Sigma^*$ with a LOGDCFL-complete membership problem. Let $\mathcal{A} = (Q, \Delta, \Sigma, \delta, q_0, \bot)$ be a deterministic pushdown automaton with $L = L(\mathcal{A})$, where Q is the set of states, $q_0 \in Q$ is the initial state, Δ is the pushdown alphabet, $\bot \in \Delta$ is the bottom symbol, and $\delta : \Delta \times Q \times \Sigma \to \Delta^* \times Q$ is the transition function. By [32–Lem. 7] we may assume that $\mathcal{A}$ makes no ε-moves and that $\mathcal{A}$ accepts L by empty store in state q_0. Let $m = \max\{|\gamma| \mid \delta(A, q, a) = (\gamma, p), q, p \in Q, A \in \Delta, a \in \Sigma\}$; thus, m is the maximal length of a sequence that is pushed on the pushdown in one step. Let $\# \notin \Delta \cup Q \cup \Sigma$ be an additional symbol and let $\Gamma = \Delta \cup Q \cup \Sigma \cup \{\#\}$. Define the semi-Thue system R by $R = \{Aqa^m\# \to \gamma p \mid \delta(A, q, a) = (\gamma, p)\}$; it is length-reducing, confluent, and left-basic. Moreover, if $h : \Sigma^* \to (\Sigma \cup \{\#\})^*$ denotes the homomorphism defined by $h(a) = a^m\#$, which can be computed in logspace, then $w \in L$ if and only if $\bot q_0 h(w) \xrightarrow{*}_R q_0$ if and only if $\bot q_0 h(w) \xleftrightarrow{*}_R q_0$.   □

5   Cayley-Graphs

Let $\mathcal{M}$ be a monoid, which is finitely generated by Γ, and let $\circ$ denote the monoid operation of $\mathcal{M}$. The *right Cayley-graph* of $\mathcal{M}$ wrt. Γ is the Γ-labeled directed graph $\mathcal{C}(\mathcal{M}, \Gamma) = (\mathcal{M}, (\{(u, v) \mid u \circ a = v\})_{a \in \Gamma})$. Thus, edges are defined via multiplication with generators on the right. The graph that is defined analogously via multiplication with generators on the left is called the *left Cayley-graph* of $\mathcal{M}$ wrt. Γ. In the following, we will always refer to the *right* Cayley-graph when just speaking of the Cayley-graph. Cayley-graphs were mainly investigated for groups, in particular they play an important role in combinatorial group theory [23] (see also the survey of Schupp [28]). Combinatorial properties of Cayley-graphs of monoids are studied in [17]. In [30, 31], Cayley-graphs of automatic monoids are investigated. The work of Calbrix and Knapik on Thue-specifications [5, 19] covers Cayley-graphs of monoids that are presented by terminating and confluent semi-Thue systems as a special case.

In [21], an investigation of Cayley-graphs from a logical point of view was initiated. For a given Cayley-graph $\mathcal{C} = (\mathcal{M}, (E_a)_{a \in \Gamma})$ we consider first-order formulas over the structure $\mathcal{C}$. Atomic formulas are of the form $x = y$ and $E_a(x, y)$, (there is an a-labeled edge from x to y) where x and y are variables that range over the monoid $\mathcal{M}$. Instead of $(x, y) \in E_a$ we write $x \circ a = y$, or briefly $xa = y$. First-order formulas are built from atomic formulas using Boolean connectives and quantifications over variables. The notion of a free variable is defined as usual. A first-order formula without free variables is called a *first-order sentence*. For a first-order sentence φ, we write $\mathcal{C} \models \varphi$ if φ evaluates to true in $\mathcal{C}$. The *first-order theory* of the Cayley-graph $\mathcal{C}$, denoted by $\mathrm{FOTh}(\mathcal{C})$, is the set of all first-order sentences φ such that $\mathcal{C} \models \varphi$. For a detailed introduction into first-order logic over arbitrary structures see [12].

If the monoid $\mathcal{M}$ is finitely generated both by Γ and Σ, then $\mathrm{FOTh}(\mathcal{C}(\mathcal{M}, \Gamma))$ is logspace reducible to $\mathrm{FOTh}(\mathcal{C}(\mathcal{M}, \Sigma))$ and vice versa [20]. Thus, analogously to word problems, the decidability (resp. complexity) status of the first-order theory of a Cayley-graph does not depend on the chosen set of generators. ¿From the definition of an αr-automatic monoid $\mathcal{M}$ it follows immediately that $\mathcal{C}(\mathcal{M}, \Gamma)$ is an automatic graph in the sense of [1, 18] (but the converse is even false for groups, see e.g. [2]). Thus, since every automatic graph has a decidable first-order theory [18], $\mathrm{FOTh}(\mathcal{C}(\mathcal{M}, \Gamma))$ is decidable in case $\mathcal{M}$ is αr-automatic ($\alpha = \ell$ or $\alpha = r$). If $\mathcal{M}$ is an $\alpha \ell$-automatic monoid ($\alpha = \ell$ or $\alpha = r$), then the first-order theory of the left Cayley-graph of $\mathcal{M}$ is decidable.

A problem is elementary decidable if it can be solved in time $\mathcal{O}(2^{\cdot^{\cdot^{2^n}}})$, where the height of this tower of exponents is constant. By [1], there exists an automatic graph with a nonelementary first-order theory. This complexity is already realized by Cayley-graphs of automatic monoids:

Theorem 4. *There is a finite, length-lexicographic, and confluent semi-Thue system $R \subseteq \Gamma^* \times \Gamma^*$ such that $(\Gamma, R, \mathrm{IRR}(R))$ is an $\alpha\beta$-automatic presentation for Γ^*/R for all $\alpha, \beta \in \{\ell, r\}$ and $\mathrm{FOTh}(\mathcal{C}(\Gamma^*/R, \Gamma))$ is nonelementary.*

Proof. Let $\Gamma = \{a, b, \bar{a}, \bar{b}, \$_1, \$_2, \$_a\}$ and let the semi-Thue system R over Γ consist of the following rules, where $c \in \{a, b\}$:

$c\$_1 \to \$_1 c$	$c\$_2 \to \$_2 c$	$\bar{a}\$_a \to a$	$c\$_a \to \$_a c$
$\bar{c}\$_1 \to c$	$\bar{c}\$_2 \to \$_1\bar{c}$	$\bar{b}\$_a \to \$_a\bar{b}$	

R is length-lexicographic and confluent. Arguments similar to those from the proof of Theorem 1 show that $(\Gamma, R, \mathrm{IRR}(R))$ is an $\alpha\beta$-automatic presentation of $\mathcal{M} = \Gamma^*/R$. Let $\mathcal{C} = \mathcal{C}(\mathcal{M}, \Gamma)$. It remains to show that $\mathrm{FOTh}(\mathcal{C})$ is not elementary decidable. For this we reduce the *first-order theory of finite words* over $\{a, b\}$ to $\mathrm{FOTh}(\mathcal{C})$. The former theory is defined as follows: A word $w = a_1 a_2 \cdots a_n \in \{a, b\}^*$ of length n is identified with the relational structure $S_w = (\{1, \ldots, n\}, <, Q_a)$, where $<$ is the usual order on natural numbers and Q_a is the unary predicate $\{i \in \{1, \ldots, n\} \mid a_i = a\}$. Then the first-order theory of finite words over $\{a, b\}$ consists of all first-order sentences ϕ that are built up from the atomic formulas $x < y$ and $Q_a(x)$ such that $S_w \models \phi$ for every word $w \in \{a, b\}^*$. It is known that the first-order theory of finite words is decidable but not elementary, see e.g. [7–Example 8.1] for a simplified proof.

For our reduction first notice that $\mathrm{IRR}(R) = \{\$_1, \$_2, \$_a\}^*\{a, b, \bar{a}, \bar{b}\}^*$. Hence, the latter set can be identified with the monoid $\mathcal{M}$. For $x \in \mathrm{IRR}(R)$ we have $x \in \{\$_1, \$_2, \$_a\}^*\{a, b\}^*$ if and only if $x\$_2\$_1 \neq x\$_1\$_1$ in $\mathcal{M}$. This allows us to represent all words from $\{a, b\}^*$ in $\mathcal{C}$. The fact that a word $w \in \{a, b\}^*$ is represented by infinitely many nodes of $\mathcal{C}$, namely by all elements from $\{\$_1, \$_2, \$_a\}^* w$ does not cause any problems; it is only important that every word $w \in \{a, b\}^*$ is represented at least once. In the sequel let us fix $x = vw$ with $v \in \{\$_1, \$_2, \$_a\}^*$ and $w \in \{a, b\}^*$. The set of all positions within the word w is in one-to-one correspondence with the set of all y such that $y\$_1 = x$ in $\mathcal{M}$: the latter holds if and only if $\exists w_1, w_2 \in \{a, b\}^* \exists c \in \{a, b\} : w = w_1 c w_2$ and $y = v w_1 \bar{c} w_2$. Thus, we can quantify over positions of the word w by quantifying in $\mathcal{C}$ over all those nodes y such that $y\$_1 = x$ in $\mathcal{M}$. Next, assume that $y = v w_1 \bar{c} w_2$ and $w = w_1 c w_2$, i.e., y represents the position $|w_1| + 1$ of w. Then $c = a$ if and only if $y\$_a = x$ in $\mathcal{M}$; thus we can express that a position is labeled with the symbol a. It remains to express that a position is smaller than another one. Assume that $y = v w_1 \bar{c} w_2$, $y' = v w_1' \bar{d} w_2'$, $w_1 c w_2 = w_1' d w_2' = w$, and $w_1 \neq w_1'$, i.e., the two positions represented by y and y' are different. Then $|w_1| < |w_1'|$ if and only if $\exists z \in \mathcal{M} : z\$_1 = y \wedge z\$_2 = y'$ in $\mathcal{M}$.

From the preceding discussion it follows that for every first-order sentence ψ over the signature $(<, Q_a)$ we can construct in polynomial time a first-order formula $\phi(x)$ over the Cayley-graph $\mathcal{C}$ such that ψ belongs to the first-order theory of finite words if and only if $\mathcal{C} \models \forall x : \phi(x)$. This proves the theorem. □

Corollary 2. *There exists a finitely presented monoid $\mathcal{M}$ such that $\mathcal{M}$ is simultaneously $\alpha\beta$-automatic for all $\alpha, \beta \in \{\ell, r\}$ and $\mathrm{FOTh}(\mathcal{C}(\mathcal{M}, \Gamma))$ is not elementary decidable.*

Since the word problem of an automatic group can be solved in time $\mathcal{O}(n^2)$, the results from [20] imply that the nonelementary lower bound from Corollary 2 cannot be realized by an automatic group. This fact even holds for automatic monoids of *finite geometric type*: A finitely generated monoid $\mathcal{M}$ has finite geometric type if for some (and hence every) finite generating set Γ, the Cayley-graph $\mathcal{C}(\mathcal{M}, \Gamma)$ has bounded degree [30], i.e., the number of neighbors of any node is bounded by a fixed constant. Every *right-cancellative monoid* has finite geometric type, but for instance the bicyclic monoid $\{a, b\}^*/\{(ab, \varepsilon)\}$ is not right-cancellative but has finite geometric type. Since the Cayley-graph of an αr-automatic monoid of finite geometric type is an automatic graph of bounded degree, and the first-order theory of every automatic graph of bounded degree belongs to $\mathrm{DSPACE}(2^{2^{2^{\mathcal{O}(n)}}})$ [22], we obtain:

Theorem 5. *Let $\mathcal{M}$ be an αr-automatic monoid ($\alpha \in \{r, \ell\}$) of finite geometric type. Then $\mathrm{FOTh}(\mathcal{C}(\mathcal{M}, \Gamma))$ belongs to $\mathrm{DSPACE}(2^{2^{2^{\mathcal{O}(n)}}})$.*

We conclude this paper with an undecidability result for automatic monoids. Note that for an αr-automatic monoid $\mathcal{M}$ ($\alpha \in \{r, \ell\}$) it is decidable whether for given $u, v \in \mathcal{M}$ there exists $x \in \mathcal{M}$ such that $xu = v$ in $\mathcal{M}$, because this is a first-order property of the Cayley-graph. On the other hand, the reverse question ($\exists x : ux = v$, i.e., reachability in the Cayley-graph) is undecidable in general:

Theorem 6. *There exists a finitely presented monoid $\mathcal{M}$ that is simultaneously ℓr- and rr-automatic such that for given $u, v \in \mathcal{M}$ it is undecidable whether $\exists x \in \mathcal{M} : ux = v$ in $\mathcal{M}$.*

The proof of this result uses the same techniques as the proof of Theorem 1.

References

1. A. Blumensath and E. Grädel. Automatic structures. In *Proc. LICS'2000*, pages 51–62. IEEE Computer Society Press, 2000.
2. A. Blumensath and E. Grädel. Finite presentations of infinite structures: Automata and interpretations. In *Proc. CiAD 2002*, 2002.
3. R. V. Book and F. Otto. *String–Rewriting Systems*. Springer, 1993.
4. J.-Y. Cai. Parallel computation over hyperbolic groups. In *Proc. STOC 92*, pages 106–115. ACM Press, 1992.

5. H. Calbrix and T. Knapik. A string-rewriting characterization of Muller and Schupp's context-free graphs. In *Proc. FSTTCS 1998*, LNCS 1530, pages 331–342. Springer, 1998.

6. C. M. Campbell, E. F. Robertson, N. Ruškuc, and R. M. Thomas. Automatic semigroups. *Theor. Comput. Sci.*, 250(1-2):365–391, 2001.

7. K. J. Compton and C. W. Henson. A uniform method for proving lower bounds on the computational complexity of logical theories. *Ann. Pure Appl. Logic*, 48:1–79, 1990.

8. M. Coornaert, T. Delzant, and A. Papadopoulos. *Géométrie et théorie des groupes*. Number 1441 in Lecture Notes in Mathematics. Springer, 1990.

9. E. Dahlhaus and M. K. Warmuth. Membership for growing context-sensitive grammars is polynomial. *J. Comput. Syst. Sci.*, 33:456–472, 1986.

10. A. Duncan and R. H. Gilman. Word hyperbolic semigroups. *Math. Proc. Cambridge Philos. Soc.*, 136:513–524, 2004.

11. D. B. A. Epstein, J. W. Cannon, D. F. Holt, S. V. F. Levy, M. S. Paterson, and W. P. Thurston. *Word processing in groups*. Jones and Bartlett, Boston, 1992.

12. W. Hodges. *Model Theory*. Cambridge University Press, 1993.

13. M. Hoffmann. *Automatic semigroups*. PhD thesis, University of Leicester, Department of Mathematics and Computer Science, 2000.

14. M. Hoffmann and R. M. Thomas. Notions of automaticity in semigroups. *Semigroup Forum*, 66(3):337–367, 2003.

15. M. Hoffmann and D. Kuske and F. Otto and R. M. Thomas. Some relatives of automatic and hyperbolic groups. In *Workshop on Semigroups, algorithms, automata and languages 2001*, pages 379–406. World Scientific, 2002.

16. J. F. P. Hudson. Regular rewrite systems and automatic structures. In *Semigroups, Automata and Languages*, pages 145–152. World Scientific, 1998.

17. A. V. Kelarev and S. J. Quinn. A combinatorial property and Cayley graphs of semigroups. *Semigroup Forum*, 66(1):89–96, 2003.

18. B. Khoussainov and A. Nerode. Automatic presentations of structures. In *LCC: International Workshop on Logic and Computational Complexity*, LNCS 960, pages 367–392, Springer 1994.

19. T. Knapik and H. Calbrix. Thue specifications and their monadic second-order properties. *Fundam. Inform.*, 39:305–325, 1999.

20. D. Kuske and M. Lohrey. Logical aspects of Cayley-graphs: the group case. to appear in *Ann. Pure Appl. Logic*.

21. D. Kuske and M. Lohrey. Decidable theories of Cayley-graphs. In *Proc. STACS 2003*, LNCS 2607, pages 463–474. Springer, 2003.

22. M. Lohrey. Automatic structures of bounded degree. In *Proc. LPAR 2003*, LNAI 2850, pages 344–358, Springer 2003.

23. R. C. Lyndon and P. E. Schupp. *Combinatorial Group Theory*. Springer, 1977.

24. D. E. Muller and P. E. Schupp. Groups, the theory of ends, and context-free languages. *J. Comput. Syst. Sci.*, 26:295–310, 1983.

25. D. E. Muller and P. E. Schupp. The theory of ends, pushdown automata, and second-order logic. *Theor. Comput. Sci.*, 37(1):51–75, 1985.

26. F. Otto and N. Ruškuc. Confluent monadic string-rewriting systems and automatic structures. *J. Autom. Lang. Comb.*, 6(3):375–388, 2001.

27. W. L. Ruzzo. Tree–size bounded alternation. *J. Comput. Syst. Sci.*, 21:218–235, 1980.

28. P. E. Schupp. Groups and graphs: Groups acting on trees, ends, and cancellation diagrams. *Math. Intell.*, 1:205–222, 1979.

29. G. Sénizergues. Formal languages and word-rewriting. In *Term Rewriting, French Spring School of Theoretical Computer Science*, LNCS 909, pages 75–94. Springer, 1993.

30. P. V. Silva and B. Steinberg. A geometric characterization of automatic monoids. Technical Report CMUP 2000-03, University of Porto, 2001.

31. P. V. Silva and B. Steinberg. Extensions and submonoids of automatic monoids. *Theor. Comput. Sci.*, 289:727–754, 2002.

32. I. H. Sudborough. On the tape complexity of deterministic context–free languages. *J. Assoc. Comput. Mach.*, 25(3):405–414, 1978.

33. H. Venkateswaran. Properties that characterize LOGCFL. *J. Comput. Syst. Sci.*, 43:380–404, 1991.

Relating Tree Series Transducers
and Weighted Tree Automata

Andreas Maletti[*]

Technische Universität Dresden, Fakultät Informatik,
Dresden, Germany
maletti@tcs.inf.tu-dresden.de

Abstract. In this paper we implement bottom-up tree series transducers (tst) over the semiring $\mathcal{A}$ with the help of bottom-up weighted
tree automata (wta) over an extension of $\mathcal{A}$. Therefore we firstly introduce bottom-up DM-monoid weighted tree automata (DM-wta), which
essentially are wta using an operation symbol of a DM-monoid instead of
a semiring element as transition weight. Secondly, we show that DM-wta
are indeed a generalization of tst (using pure substitution). Thirdly, given
a DM-wta we construct a semiring $\mathcal{A}$ along with a wta such that the wta
computes a formal representation of the semantics of the DM-wta.

Finally, we demonstrate the applicability of our presentation result
by deriving a pumping lemma for deterministic tst as well as deterministic DM-wta from a pumping lemma for deterministic wta.

1   Introduction

In formal language theory several different accepting and transducing devices
were intensively studied [13]. A classical folklore result shows how to implement
generalized sequential machines (cf., e.g., [1]) on weighted automata [14, 5, 11]
with the help of the particular semiring $(\mathfrak{P}(\Sigma^*), \cup, \circ)$ of languages over the
alphabet Σ. Naturally, this semiring is not commutative, notwithstanding the
representation allows us to transfer results obtained for weighted automata to
generalized sequential machines. In this sense, the study of arbitrary weighted automata subsumes the study of generalized sequential machines.

We translate the above representation result to tree languages (cf., e.g., [4]),
i.e., we show how to implement bottom-up tree transducers [12, 15] on bottom-up
weighted tree automata (wta) [3, 9]. More generally, we even unearth a relationship between bottom-up tree series transducers (tst) [7, 8] using pure substitution
and wta. Therefore we first introduce bottom-up DM-monoid weighted tree automata (DM-wta), which essentially are wta where the weight of a transition is
an operation symbol of a DM-monoid [9] instead of a semiring element. These devices can easily simulate both wta and tst by a proper choice of the DM-monoid
(cf. Proposition 5). Next we devise a monoid $\mathcal{A}$ which is capable of emulating

---

[*] Financially supported by the German Research Foundation (DFG, GK 334/3)

C.S. Calude, E. Calude and M.J. Dinneen (Eds.): DLT 2004, LNCS 3340, pp. 321–333, 2004.

322   A. Maletti

the effect of the operation symbols of a DM-monoid $\mathcal{D}$ (cf. Theorem 6). Then we extend $\mathcal{A}$ to a semiring using the addition of a semiring $\mathcal{B}$ for which $\mathcal{D}$ is a semimodule (cf. Theorem 8). In this way we obtain an abstract addition (of $\mathcal{B}$), which allows us to perform the concrete addition (of $\mathcal{D}$) later. Thereby we obtain a representation result, in which a tst or a DM-wta is presented as wta, which computes a formal representation of the semantics of the tst or DM-wta.

For a tst M over a completely idempotent semiring $\mathcal{A}$, e.g., all tree transducers, we can refine the constructed semiring with the help of a congruence relation such that the factor semiring uses (an extension of) the concrete addition of $\mathcal{A}$ (cf. Theorem 10). Then one can construct a wta such that it computes the same tree series as M. Finally, we note that the construction of the semiring preserves many beneficial properties (concerning the addition) of the original DM-monoid.

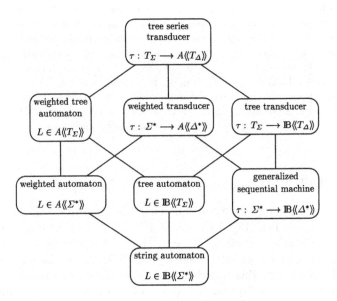

Fig. 1. Generalization hierarchy

Hence the study of wta subsumes the study of tst over completely idempotent semirings. In fact, the subsumption also holds for deterministic devices, i.e., the study of deterministic wta subsumes the study of deterministic tst or DM-wta. To illustrate the applicability of the relationship we transfer a pumping lemma [2] for deterministic finite wta to both tst and DM-wta. This is possible, because the semiring addition is irrelevant for deterministic wta and the determinism property is preserved by the constructions. This yields that for a given tst or DM-wta M we can construct a wta M' such that $\|M'\| = \|M\|$. Hence the pumping lemma for wta can readily be transfered to tst and DM-wta.

2 Preliminaries

The set $\{0, 1, 2, \ldots\}$ of all non-negative integers is denoted by $\mathbb{N}$ and we let $\mathbb{N}_+ = \mathbb{N} \setminus \{0\}$. In the following let $k, n \in \mathbb{N}$. The interval $[k, n]$ abbreviates $\{i \in \mathbb{N} \mid k \leq i \leq n\}$ and we use $[n]$ to stand for $[1, n]$. The set of all subsets of a set A is denoted by $\mathfrak{P}(A)$ and the set of all (total) mappings $f : A \longrightarrow B$ is denoted by B^A as customary. Finally, the set of all words over A is displayed as A^*, the length of a word $w \in A^*$ is denoted by $|w|$, and $\cdot$ is used to denote concatenation as well as to delimit subwords.

2.1 Trees and Substitutions

A non-empty set Σ equipped with a mapping $\mathrm{rk}_\Sigma : \Sigma \longrightarrow \mathbb{N}$ is called an *operator alphabet*. The set $\Sigma_k = \{\sigma \in \Sigma \mid \mathrm{rk}_\Sigma(\sigma) = k\}$ denotes the set of operators of arity k. Given a set V, the set $T_\Sigma(V)$ of (finite, labeled, and ordered) Σ-trees indexed by V is the smallest set T such that $\Sigma_0 \cup V \subseteq T$ and for every $k \in \mathbb{N}_+$, $\sigma \in \Sigma_k$, and $t_1, \ldots, t_k \in T$ also $\sigma(t_1, \ldots, t_k) \in T$. The notation T_Σ abbreviates $T_\Sigma(\emptyset)$. The mapping $\mathrm{pos} : T_\Sigma(V) \longrightarrow \mathfrak{P}(\mathbb{N}^*)$ is defined for every $v \in V$, $\sigma \in \Sigma_k$, and $t_1, \ldots, t_k \in T_\Sigma(V)$ by

$$\mathrm{pos}(v) = \{\varepsilon\} \qquad \mathrm{pos}(\sigma(t_1, \ldots, t_k)) = \{\varepsilon\} \cup \{i \cdot w_i \mid i \in [k], w_i \in \mathrm{pos}(t_i)\} \ .$$

Moreover, $\mathrm{height}(t) = 1 + \max\{|w| \mid w \in \mathrm{pos}(t)\}$ for every $t \in T_\Sigma(V)$. The *label* of t at $w \in \mathrm{pos}(t)$ is denoted by $\mathrm{lab}_t(w)$, i.e.,

$$\mathrm{lab}_v(\varepsilon) = v \qquad \mathrm{lab}_{\sigma(t_1, \ldots, t_k)}(w) = \begin{cases} \sigma & \text{, if } w = \varepsilon \\ \mathrm{lab}_{t_i}(w_i) & \text{, if } w = i \cdot w_i \text{ with } i \in [k] \end{cases} .$$

For convenience, we assume a countably infinite set $X = \{x_i \mid i \in \mathbb{N}_+\}$ of formal variables and its finite subsets $X_n = \{x_i \mid i \in [n]\}$. A Σ-tree $t \in T_\Sigma(X_n)$ is in the set $\widehat{T}_\Sigma(X_n)$, if and only if every $x \in X_n$ occurs exactly once in t. Given $t \in T_\Sigma(X_n)$ and $t'_1, \ldots, t'_n \in T_\Sigma(V)$, the expression $t[t'_1, \ldots, t'_n]$ denotes the *(parallel) tree substitution* of t'_i for every occurrence of x_i in t, i.e., $x_i[t'_1, \ldots, t'_n] = t'_i$ for every $i \in [n]$ and

$$\sigma(t_1, \ldots, t_k)[t'_1, \ldots, t'_n] = \sigma(t_1[t'_1, \ldots, t'_n], \ldots, t_k[t'_1, \ldots, t'_n])$$

for every $k \in \mathbb{N}$, $\sigma \in \Sigma_k$, and $t_1, \ldots, t_k \in T_\Sigma(X_n)$. Let $t \in \widehat{T}_\Sigma(X_n)$ with $n \geq 1$ and $t' \in \widehat{T}_\Sigma(X_k)$. The *non-identifying tree substitution* of t' into t, denoted by $t \langle\!| t' |\!\rangle$, yields a tree of $\widehat{T}_\Sigma(X_{k+n-1})$ which is defined by

$$t \langle\!| t' |\!\rangle = t[t', x_{k+1}, \ldots, x_{k+n-1}] \ .$$

This way no variable of t' is identified with a variable of t. To complete the definition we let $t \langle\!| t' |\!\rangle = t$ whenever $t \in T_\Sigma$, i.e., $n = 0$. One can compare this with the classical lambda-calculus, where (except for reordering of the arguments)

$$(\lambda x_1 \ldots x_n.t)(\lambda x_1 \ldots x_k.t') \Rightarrow \lambda x_1 \ldots x_{k+n-1}.t \langle\!| t' |\!\rangle \ .$$

2.2   Algebraic Structures

Given a carrier set A, an operator alphabet Ω, and a family $I = (I_k)_{k \in \mathbb{N}}$ of mappings $I_k : \Omega_k \longrightarrow A^{(A^k)}$ interpreting the symbols as operations on A, the triple (A, Ω, I) is called an *(abstract) Ω-algebra*. The algebra (T_Ω, Ω, I) where $I_k(\omega) = \overline{\omega}$ for every $k \in \mathbb{N}$, $\omega \in \Omega_k$, and $\overline{\omega}(t_1, \ldots, t_k) = \omega(t_1, \ldots, t_k)$ for every $t_1, \ldots, t_k \in T_\Omega$ is called the *initial (term) Ω-algebra*. In the sequel we often do not differentiate between the symbol and the actual operation. Usually the context will provide sufficient information as to clarify which meaning is intended. Further we occasionally omit the operator alphabet and instead list the operators and identify nullary operators with elements of A.

Monoids are algebraic structures $\mathcal{A} = (A, \otimes)$ with carrier set A, an *associative* operation $\otimes : A^2 \longrightarrow A$, i.e., $a_1 \otimes (a_2 \otimes a_3) = (a_1 \otimes a_2) \otimes a_3$ for every $a_1, a_2, a_3 \in A$, and a *neutral element* $1 \in A$, i.e., $1 \otimes a = a = a \otimes 1$ for every $a \in A$. The neutral element is unique and denoted by 0_A or 1_A in the sequel. The monoid is said to be *commutative*, if $a_1 \otimes a_2 = a_2 \otimes a_1$ for every $a_1, a_2 \in A$, and it is said to be *idempotent*, if $a = a \otimes a$ for every $a \in A$. A commutative monoid is called *complete*, if it is possible to define an (infinitary) operation $\bigotimes$ such that the following two additional axioms hold for all index sets I, J and all families $(a_i)_{i \in I}$ of monoid elements.

(i) $\bigotimes_{i \in \{j\}} a_i = a_j$ and $\bigotimes_{i \in \{j_1, j_2\}} a_i = a_{j_1} \otimes a_{j_2}$ for $j_1 \neq j_2$.

(ii) $\bigotimes_{j \in J} \bigotimes_{i \in I_j} a_i = \bigotimes_{i \in I} a_i$, if $\bigcup_{j \in J} I_j = I$ and $I_{j_1} \cap I_{j_2} = \emptyset$ for $j_1 \neq j_2$.

The relation $\sqsubseteq \subseteq A^2$ is defined by $a_1 \sqsubseteq a_2$ if and only if there exists $a \in A$ such that $a_1 \otimes a = a_2$. If $\sqsubseteq$ is a partial order, then $\mathcal{A}$ is said to be *naturally ordered*. Finally, a naturally ordered and complete monoid is *continuous*, if for every $a \in A$, index set I, and family $(a_i)_{i \in I}$ of elements $a_i \in A$

$$\bigotimes_{i \in E} a_i \sqsubseteq a \text{ for all finite } E \subseteq I \iff \bigotimes_{i \in I} a_i \sqsubseteq a \ .$$

Note that an idempotent monoid is continuous, if and only if it is *completely idempotent*, i.e., it is complete and for every non-empty index set I and element $a \in A$ we have that $\bigotimes_{i \in I} a = a$.

Algebraic structures $\mathcal{A} = (A, \oplus, \odot)$ made of two monoids $(A, \oplus)$ and $(A, \odot)$ with neutral elements 0_A and 1_A, respectively, of which the former monoid is commutative and the latter monoid has 0_A as an *absorbing element*, i.e., $a \odot 0_A = 0_A = 0_A \odot a$ for every $a \in A$, are called *semirings (with one and absorbing zero)*, if the monoids are connected via the distributivity laws, i.e., $a_1 \odot (a_2 \oplus a_3) = (a_1 \odot a_2) \oplus (a_1 \odot a_3)$ and $(a_1 \oplus a_2) \odot a_3 = (a_1 \odot a_3) \oplus (a_2 \odot a_3)$ for every $a_1, a_2, a_3 \in A$. The semiring $\mathcal{A}$ is called *(additively) idempotent*, if $(A, \oplus)$ is idempotent. Finally, a *complete* semiring consists of a complete monoid $(A, \oplus)$ and satisfies the additional constraint that for every index set I, $a \in A$, and family $(a_i)_{i \in I}$ of semiring elements

$$\bigoplus_{i \in I}(a \odot a_i) = a \odot \bigoplus_{i \in I} a_i \quad \text{and} \quad \bigoplus_{i \in I}(a_i \odot a) = \left(\bigoplus_{i \in I} a_i\right) \odot a \ .$$

Let $\mathcal{B} = (B, +)$ be a commutative monoid, $\mathcal{A} = (A, \oplus, \odot)$ be a semiring, and $\cdot : A \times B \longrightarrow B$ be a mapping. Then $\mathcal{B}$ is called a *(left) $\mathcal{A}$-semimodule (via $\cdot$)*, if the conditions (i)-(iii) hold for all $a, a_1, a_2 \in A$ and all $b, b_1, b_2 \in B$.

(i) $a \cdot 0_{\mathcal{B}} = 0_{\mathcal{B}}$ and $1_{\mathcal{A}} \cdot b = b$.
(ii) $(a_1 \odot a_2) \cdot b = a_1 \cdot (a_2 \cdot b)$.
(iii) $a \cdot (b_1 + b_2) = (a \cdot b_1) + (a \cdot b_2)$ and $(a_1 \oplus a_2) \cdot b = (a_1 \cdot b) + (a_2 \cdot b)$.

Given that $\mathcal{B}$ and $\mathcal{A}$ are complete, $\mathcal{B}$ is called a *complete $\mathcal{A}$-semimodule*, if for every family $(b_i)_{i \in I}$ of monoid elements and family $(a_i)_{i \in I}$ of semiring elements the additional axioms (iv) and (v) hold.

(iv) $a \cdot \sum_{i \in I} b_i = \sum_{i \in I} (a \cdot b_i)$
(v) $(\bigoplus_{i \in I} a_i) \cdot b = \sum_{i \in I} (a_i \cdot b)$

Clearly each commutative monoid $\mathcal{B} = (B, +)$ is an $\mathbb{N}$-semimodule, where the semiring of non-negative integers is given by $(\mathbb{N}, +, \cdot)$, using the mixed operation $\cdot : \mathbb{N} \times B \longrightarrow B$ defined as $n \cdot b = \sum_{i \in [n]} b$ for every $n \in \mathbb{N}$ and $b \in B$. Note that $\sum_{i \in [0]} b = 0_{\mathcal{B}}$. Similarly, every commutative and continuous monoid is a complete $\mathbb{N}_{\infty}$-semimodule (cf. [9]), where $\mathbb{N}_{\infty} = (\mathbb{N} \cup \{+\infty\}, +, \cdot)$. Furthermore, any idempotent and commutative monoid $\mathcal{B}$ is a $\mathbb{B}$-semimodule where $\mathbb{B} = (\{0, 1\}, \vee, \wedge)$ is the *boolean semiring*, and $\mathcal{B}$ is a complete $\mathbb{B}$-semimodule, if $\mathcal{B}$ additionally is completely idempotent (cf. [9]).

Let (D, Ω) be an Ω-algebra. The algebraic structure $\mathcal{D} = (D, +, \Omega)$ is called a *distributive multi-operator monoid* (DM-monoid) [9], if $(D, +)$ is a commutative monoid with neutral element $0_{\mathcal{D}}$ and for every $k \in \mathbb{N}$, $\omega \in \Omega_k$, $i \in [k]$, and $d, d_1, \ldots, d_k \in D$

(i) $\omega(d_1, \ldots, d_{i-1}, 0_{\mathcal{D}}, d_{i+1}, \ldots, d_k) = 0_{\mathcal{D}}$,
(ii) $\omega(d_1, \ldots, d_{i-1}, d + d_i, d_{i+1}, \ldots, d_k) = \omega(d_1, \ldots, d, \ldots, d_k) + \omega(d_1, \ldots, d_k)$.

For $\mathcal{D}$ to be *complete* we demand that $(D, +)$ is complete and for every $k \in \mathbb{N}$, $\omega \in \Omega_k$, index sets $I_1, \ldots, I_k$, and family $(d_i)_{i \in I_j}$ of monoid elements for every $j \in [k]$ the equality

$$\omega\left(\sum_{i_1 \in I_1} d_{i_1}, \ldots, \sum_{i_k \in I_k} d_{i_k}\right) = \sum_{i_1 \in I_1} \cdots \sum_{i_k \in I_k} \omega(d_{i_1}, \ldots, d_{i_k})$$

is satisfied. Finally, $\mathcal{D}$ is *continuous*, if $\mathcal{D}$ is complete and $(D, +)$ is continuous.

The DM-monoid $\mathcal{D}$ is said to be an *$\mathcal{A}$-semimodule* for some commutative semiring $\mathcal{A} = (A, \oplus, \odot)$, if $(D, +)$ is an $\mathcal{A}$-semimodule and for every $k \in \mathbb{N}$, $\omega \in \Omega_k$, $a \in A$, $i \in [k]$, and $d_1, \ldots, d_k \in D$ the equality

$$\omega(d_1, \ldots, d_{i-1}, a \cdot d_i, d_{i+1}, \ldots, d_k) = a \cdot \omega(d_1, \ldots, d_k)$$

holds. The DM-monoid $\mathcal{D}$ is a *complete $\mathcal{A}$-semimodule*, if both $\mathcal{A}$ and $\mathcal{D}$ are by itself complete and for every $a \in A$, $d \in D$, index set I, and family $(a_i)_{i \in I}$ and $(d_i)_{i \in I}$ of semiring and monoid elements, respectively, we have

$$\left(\bigoplus_{i \in I} a_i\right) \cdot d = \sum_{i \in I} (a_i \cdot d) \qquad \text{and} \qquad a \cdot \sum_{i \in I} d_i = \sum_{i \in I} (a \cdot d_i) \ .$$

Clearly, every DM-monoid is an $\mathbb{N}$-semimodule.

2.3   Formal Power Series and Tree Series Substitution

Any mapping $\varphi : T \longrightarrow A$ from a set T into a commutative monoid $\mathcal{A} = (A, \oplus)$ is also called *(formal) power series*. The set of all power series is denoted by $A\langle\!\langle T \rangle\!\rangle$. We write (φ, t) instead of $\varphi(t)$ for $\varphi \in A\langle\!\langle T \rangle\!\rangle$ and $t \in T$. The *sum* $\varphi_1 \oplus \varphi_2$ of two power series $\varphi_1, \varphi_2 \in A\langle\!\langle T \rangle\!\rangle$ is defined pointwise by $(\varphi_1 \oplus \varphi_2, t) = (\varphi_1, t) \oplus (\varphi_2, t)$ for every $t \in T$. The *support* $\mathrm{supp}(\varphi)$ of φ is defined by

$$\mathrm{supp}(\varphi) = \{\, t \in T \mid (\varphi, t) \neq 0_A \,\} \ .$$

If the support of φ is finite, then φ is said to be a *polynomial*. The power series with empty support is denoted by $\widetilde{0}_A$.

In case $T = T_\Sigma(V)$ for some ranked alphabet Σ and set V, then φ is also called *(formal) tree series*. Let $\mathcal{A} = (A, \oplus, \odot)$ now be a complete semiring and let $n \in \mathbb{N}$, $\varphi \in A\langle\!\langle T_\Sigma(X_n) \rangle\!\rangle$, and $\psi_1, \dots, \psi_n \in A\langle\!\langle T_\Sigma \rangle\!\rangle$. We define the *tree series substitution* of $(\psi_1, \dots, \psi_n)$ into φ, denoted by $\varphi \longleftarrow (\psi_1, \dots, \psi_n)$, as

$$\varphi \longleftarrow (\psi_1, \dots, \psi_n) = \bigoplus_{\substack{t \in T_\Sigma(X_n), \\ t_1, \dots, t_n \in T_\Sigma}} \left((\varphi, t) \odot \bigodot_{i \in [n]} (\psi_i, t_i) \right) t[t_1, \dots, t_n] \ .$$

Note that the order in the product is given by the order $1 < \cdots < n$ of the indices. Furthermore, note that irrespective of the number of occurrences of x_i the coefficient (ψ_i, t_i) is taken into account exactly once, even if x_i does not appear at all in t. This notion of substitution is called *pure IO-substitution* [7]. Other notions of substitution, like o-IO-substitution [8] and OI-substitution [10], have been defined, but in this paper we will exclusively deal with pure IO-substitution.

2.4   Tree Automata and Tree Series Transducers

Let I and J be sets. An $(I \times J)$-*matrix* over a set S is a mapping $M : I \times J \longrightarrow S$. The set of all $(I \times J)$-matrices is denoted by $S^{I \times J}$ and the (i, j)-entry with $i \in I$ and $j \in J$ of a matrix $M \in S^{I \times J}$ is usually denoted by $M_{i,j}$ instead of $M(i, j)$. Let Σ be an operator alphabet, I be a non-empty set, and $\mathcal{A} = (A, \oplus)$ be a commutative monoid. Every family $\mu = (\mu_k)_{k \in \mathbb{N}}$ of mappings $\mu_k : \Sigma_k \longrightarrow A^{I \times I^k}$ is called *tree representation* over Σ, I, and $\mathcal{A}$. A *deterministic* tree representation additionally fulfills the restriction that for every $\sigma \in \Sigma_k$ and $i_1, \dots, i_k \in I$ there exists at most one $i \in I$ such that $\mu_k(\sigma)_{i, (i_1, \dots, i_k)} \neq 0_A$.

A *(bottom-up) weighted tree automaton* (wta) is a system $M = (I, \Sigma, \mathcal{A}, F, \mu)$ comprising of a set I of *states*, a finite input ranked alphabet Σ, a semiring $\mathcal{A} = (A, \oplus, \odot)$, a vector $F \in A^I$ of *final weights*, and a tree representation μ over Σ, I, and $\mathcal{A}$. If I is infinite, then $\mathcal{A}$ must be complete, otherwise M is called *finite*. Moreover, M is *deterministic*, if μ is deterministic. Let $\boldsymbol{\mu} = (\boldsymbol{\mu}_k(\sigma))_{k \in \mathbb{N}, \sigma \in \Sigma_k}$ where $\boldsymbol{\mu}_k(\sigma) : (A^I)^k \longrightarrow A^I$ is defined componentwise for every $i \in I$ and $V_1, \dots, V_k \in A^I$ by

$$\boldsymbol{\mu}_k(\sigma)(V_1, \dots, V_k)_i = \bigoplus_{i_1, \dots, i_k \in I} \mu_k(\sigma)_{i, (i_1, \dots, i_k)} \odot (V_1)_{i_1} \odot \cdots \odot (V_k)_{i_k} \ .$$

Let $h_\mu : T_\Sigma \longrightarrow A^I$ be the unique homomorphism from (T_Σ, Σ) to (A^I, μ). The tree series $\|M\| \in A\langle\langle T_\Sigma \rangle\rangle$ recognized by M is $(\|M\|, t) = \bigoplus_{i \in I} F_i \odot h_\mu(t)_i$ for every $t \in T_\Sigma$.

A *(bottom-up) tree series transducer* (tst) M is a system $(I, \Sigma, \Delta, \mathcal{A}, F, \mu)$ in which I is a set of *states*, Σ and Δ are finite input and output ranked alphabets, respectively, $\mathcal{A} = (A, \oplus, \odot)$ is a semiring, $F \in A\langle\langle T_\Delta(X_1) \rangle\rangle^I$ is a vector of *final outputs*, and μ is a tree representation over Σ, I, and $A\langle\langle T_\Delta(X) \rangle\rangle$ such that $\mu_k(\sigma) \in A\langle\langle T_\Delta(X_k) \rangle\rangle^{I \times I^k}$ for every $k \in \mathbb{N}$ and $\sigma \in \Sigma_k$. If I is finite and each tree series in the range of $\mu_k(\sigma)$ is a polynomial, then M is called *finite*, otherwise $\mathcal{A}$ must be complete. Finite tst over the Boolean semiring $\mathbb{B}$ are also called *tree transducers*. The tst M is *deterministic*, if μ is deterministic. Let $\mu = (\mu_k(\sigma))_{k \in \mathbb{N}, \sigma \in \Sigma_k}$ where $\mu_k(\sigma) : (A\langle\langle T_\Delta \rangle\rangle^I)^k \longrightarrow A\langle\langle T_\Delta \rangle\rangle^I$ is defined componentwise for every $i \in I$ and $V_1, \ldots, V_k \in A\langle\langle T_\Delta \rangle\rangle^I$ by

$$\mu_k(\sigma)(V_1, \ldots, V_k)_i = \bigoplus_{i_1, \ldots, i_k \in I} \mu_k(\sigma)_{i, (i_1, \ldots, i_k)} \longleftarrow ((V_1)_{i_1}, \ldots, (V_k)_{i_k}) \ .$$

Let $h_\mu : T_\Sigma \longrightarrow A\langle\langle T_\Delta \rangle\rangle^I$ be the unique homomorphism from the initial Σ-algebra (T_Σ, Σ) to $(A\langle\langle T_\Delta \rangle\rangle^I, \mu)$. For every $t \in T_\Sigma$ the *tree-to-tree-series transformation* (t-ts transformation) $\|M\| : T_\Sigma \longrightarrow A\langle\langle T_\Delta \rangle\rangle$ computed by M is $(\|M\|, t) = \bigoplus_{i \in I} F_i \longleftarrow (h_\mu(t)_i)$.

3   Establishing the Relationship

Inspired by the automaton definition of [9] we define DM-monoid weighted tree automata (DM-wta). Roughly speaking, to each transition of a DM-wta an operation symbol of a DM-monoid is associated.

Definition 1. *A DM-monoid weighted tree automaton (DM-wta) is a system $M = (I, \Sigma, \mathcal{D}, F, \mu)$, where*

- I *is a non-empty set of* states,
- Σ *is a finite operator alphabet of* input symbols,
- $\mathcal{D} = (D, +, \Omega)$ *is a DM-monoid,*
- $F \in (\Omega_1)^I$ *is the final weight vector, and*
- $\mu = (\mu_k)_{k \in \mathbb{N}}$ *is a tree representation over I, Σ, and Ω.*

If I is infinite, then $\mathcal{D}$ must be complete. Otherwise, M is called finite. *Finally, M is* deterministic, *if μ is deterministic.*

Unless stated otherwise let $M = (I, \Sigma, \mathcal{D}, F, \mu)$ be a DM-wta over the DM-monoid $\mathcal{D} = (D, +, \Omega)$. In the following let $k \in \mathbb{N}$, $\sigma \in \Sigma_k$, $i \in I$, and $t = \sigma(t_1, \ldots, t_k) \in T_\Sigma$. Moreover, all function arguments range over their respective domains. Next we define two semantics, namely initial algebra semantics [16] and a semantics based on runs. In the latter the weight of a run is obtained by combining the weights obtained for the direct subtrees with the help of the operation symbol associated to the topmost transition. Nondeterminism is taken care of by adding the weights of all runs on a given input tree.

Definition 2. *Let* $\mu = (\mu_k(\sigma))_{k\in\mathbb{N}, \sigma\in\Sigma_k}$ *where* $\mu_k(\sigma) : (D^I)^k \longrightarrow D^I$ *is defined componentwise for every* $i \in I$ *by*

$$\mu_k(\sigma)(V_1, \ldots, V_k)_i = \sum_{i_1, \ldots, i_k \in I} \mu_k(\sigma)_{i,(i_1, \ldots, i_k)}((V_1)_{i_1}, \ldots, (V_k)_{i_k}) .$$

Let $h_\mu : T_\Sigma \longrightarrow D^I$ *be the unique homomorphism from* (T_Σ, Σ) *to* (D^I, μ). *The tree series recognized by* M *is defined as* $(\|M\|, t) = \sum_{i\in I} F_i(h_\mu(t)_i)$.

Definition 3. *A run on* $t \in T_\Sigma$ *is a mapping* $r : \text{pos}(t) \longrightarrow I$. *The set of all runs on* t *is denoted by* $R(t)$. *The weight of* r *is defined by the mapping* $\text{wt}_r : \text{pos}(t) \longrightarrow D$ *which is defined for* $w \in \text{pos}(t)$ *with* $\text{lab}_t(w) \in \Sigma_k$ *by*

$$\text{wt}_r(w) = \mu_k(\text{lab}_t(w))_{r(w),(r(w\cdot 1), \ldots, r(w\cdot k))}(\text{wt}_r(w\cdot 1), \ldots, \text{wt}_r(w\cdot k)) .$$

The run-based semantics *of* M *is* $(|M|, t) = \sum_{r\in R(t)} F_{r(\varepsilon)}(\text{wt}_r(\varepsilon))$.

The next proposition states that the initial algebra semantics coincides with the run-based semantics, which is mainly due to the distributivity of the DM-monoid. Intuitively speaking, this reflects the property that nondeterminism can equivalently either be handled locally (initial algebra semantics) or globally (run-based semantics).

Proposition 4. *For every DM-wta* M *we have* $\|M\| = |M|$.

The next proposition demonstrates how powerful DM-wta are. In fact, every wta and every tst can be simulated by a DM-wta.

Proposition 5. *Let* M_1 *be a wta and* M_2 *be a tst.*
(i) There exists a DM-wta M *such that* $\|M\| = \|M_1\|$.
(ii) There exists a DM-wta M *such that* $\|M\| = \|M_2\|$.

Proof. Since it is clear (cf. [7]), how to simulate a wta with the help of a tst, we only show Statement (ii). Let $M_2 = (I_2, \Sigma, \Delta, \mathcal{A}, F_2, \mu_2)$ be a tst,

$$\Omega = \{ \underline{\varphi}_k \mid k \in \mathbb{N}, \varphi \in A\langle\langle T_\Delta(X_k)\rangle\rangle \} ,$$

and let $\underline{\varphi}_k : A\langle\langle T_\Delta\rangle\rangle^k \longrightarrow A\langle\langle T_\Delta\rangle\rangle$ be defined as

$$\underline{\varphi}_k(\psi_1, \ldots, \psi_k) = \varphi \longleftarrow (\psi_1, \ldots, \psi_k) .$$

Then, by [9, 7], $\mathcal{D} = (A\langle\langle T_\Delta\rangle\rangle, \oplus, \Omega)$ is a DM-monoid, which is complete whenever $\mathcal{A}$ is. Hence we let $M = (I_2, \Sigma, \mathcal{D}, F, \mu)$ with $F_i = \underline{F_2(i)}_1$ and for every $i, i_1, \ldots, i_k \in I_2$ we set $\mu_k(\sigma)_{i,(i_1, \ldots, i_k)} = \underline{(\mu_2)_k(\sigma)_{i,(i_1, \ldots, i_k)}}_k$.

Note that in both statements of Proposition 5, M can be constructed to be deterministic, whenever the input device, i.e., M_1 or M_2, is deterministic. Let $\mathcal{D} = (D, \Omega)$ be an Ω-algebra. In the following ω ranges over Ω_k. We denote by ΩX the set of all terms $\{ \overline{\omega}(x_1, \ldots, x_k) \mid \omega \in \Omega_k \}$. We can define a monoid which simulates the algebra $\mathcal{D}$ as follows. Recall that we use overlining, if we want to refer to the term obtained by top-concatenation of the overlined symbol with its arguments.

Theorem 6. *For every Ω-algebra (D, Ω) there exists a monoid $(B, \leftarrow)$ such that $D \cup \Omega X \subseteq B$ and for all $d_1, \ldots, d_k \in D$*

$$\omega(d_1, \ldots, d_k) = \overline{\omega}(x_1, \ldots, x_k) \leftarrow d_1 \leftarrow \cdots \leftarrow d_k .$$

Proof. Assume that $\Omega \cap D = \emptyset$ and let $\Omega' = \Omega \cup D$, where the elements of D are treated as nullary symbols. Firstly, we define a mapping $h : T_{\Omega'}(X) \longrightarrow T_{\Omega'}(X)$ for every $v \in D \cup X$ as follows.

$$h(v) = v$$

$$h(\overline{\omega}(t_1, \ldots, t_k)) = \begin{cases} \omega(h(t_1), \ldots, h(t_k)) & , \text{ if } h(t_1), \ldots, h(t_k) \in D \\ \overline{\omega}(h(t_1), \ldots, h(t_k)) & , \text{ otherwise} \end{cases}$$

Note that $h(t) \in \widehat{T_{\Omega'}}(X_n)$ whenever $t \in \widehat{T_{\Omega'}}(X_n)$. Secondly, let

$$B = D^* \cup \bigcup_{n \in \mathbb{N}_+} D^* \cdot \widehat{T_{\Omega'}}(X_n) .$$

Next we define the operation $\leftarrow : B^2 \longrightarrow B$ for every $w \in D^*$, $b \in B$, $t \in \widehat{T_{\Omega'}}(X_n)$, and $t' \in D \cup \widehat{T_{\Omega'}}(X_n)$ by

$$w \leftarrow b = w \cdot b$$
$$w \cdot t \leftarrow \varepsilon = w \cdot t$$
$$w \cdot t \leftarrow t' \cdot b = w \cdot (h(t \langle\!| t' |\!\rangle)) \leftarrow b .$$

Roughly speaking, one can understand $\leftarrow$ as function composition where the arguments are lambda-terms and the evaluation (which is done via h) is call-by-value. Next we would like to extend this monoid to a semiring by introducing the addition of the DM-monoid. However, the addition should also be able to sum up terms, hence we first use an abstract addition coming from a semiring for which the DM-monoid is a complete semimodule.

Let $\mathcal{A} = (A, \oplus, \odot)$ be a semiring. We lift the operation $\leftarrow : B^2 \longrightarrow B$ to an operation $\leftarrow : A\langle\!\langle B \rangle\!\rangle^2 \longrightarrow A\langle\!\langle B \rangle\!\rangle$ by

$$\psi_1 \leftarrow \psi_2 = \bigoplus_{b_1, b_2 \in B} ((\psi_1, b_1) \odot (\psi_2, b_2)) (b_1 \leftarrow b_2) .$$

Let the monoid $\mathcal{D} = (D, +)$ be a complete $\mathcal{A}$-semimodule. Then we define the *sum of a series* $\varphi \in A\langle\!\langle D \rangle\!\rangle$ (summed in D) by the mapping $\sum : A\langle\!\langle D \rangle\!\rangle \longrightarrow D$ with $\sum \varphi = \sum_{d \in D} (\varphi, d) \cdot d$. For a vector $V \in A\langle\!\langle D \rangle\!\rangle^I$ we let $(\sum V)_i = \sum V_i$. By convenience we identify the series $1_A d$ with d.

Proposition 7. *Let the DM-monoid $\mathcal{D} = (D, +, \Omega)$ be a complete $(A, \oplus, \odot)$-semimodule and $\varphi_1, \ldots, \varphi_k \in A\langle\!\langle D \rangle\!\rangle$. Then*

(i) $\sum(\bigoplus_{i \in I} \varphi_i) = \sum_{i \in I} \sum \varphi_i$ for every family $(\varphi_i)_{i \in I}$ of series and
(ii) $\omega(\sum \varphi_1, \ldots, \sum \varphi_k) = \sum(\overline{\omega}(x_1, \ldots, x_k) \leftarrow \varphi_1 \leftarrow \cdots \leftarrow \varphi_k)$.

Thus we can construct a semiring with the following properties.

Theorem 8. *For every continuous DM-monoid $\mathcal{D} = (D, +, \Omega)$ there exists a semiring $(C, \oplus, \leftarrow)$ such that $D \cup \Omega X \subseteq C$ and for all $d_1, \ldots, d_k \in D$:*

(i) $\omega(d_1, \ldots, d_k) = \overline{\omega}(x_1, \ldots, x_k) \leftarrow d_1 \leftarrow \cdots \leftarrow d_k,$
(ii) $\sum (\bigoplus_{i \in I} d_i) = \sum_{i \in I} d_i.$

Proof. Let $\mathcal{A} = (A, \oplus, \odot)$ be a semiring such that $\mathcal{D}$ is a complete $\mathcal{A}$-semimodule. For example, $\mathcal{A}$ can always be chosen to be $\mathbb{N}_\infty$. By Theorem 6 there exists a monoid $(B, \leftarrow)$ such that Statement (i) holds. Consequently, let $C = A\langle\!\langle B \rangle\!\rangle$ and $\leftarrow : C^2 \longrightarrow C$ be the extension of $\leftarrow$ on B. Clearly, $(C, \oplus, \leftarrow)$ is a semiring and by Theorem 6 and Proposition 7 the Statements (i) and (ii) hold.

The semiring $(A\langle\!\langle B \rangle\!\rangle, \oplus, \leftarrow)$ constructed in Theorem 8 will be denoted by $G_A(\mathcal{D})$ in the sequel. We note that $G_A(\mathcal{D})$ is complete, because $\mathcal{A}$ is complete (cf. [9]). Hence we are ready to state the first main representation theorem.

Theorem 9. *Let $M_1 = (I_1, \Sigma, \mathcal{D}, F_1, \mu_1)$ be a DM-wta and M_2 be a tst.*

 - *There exists a wta $M = (I_1, \Sigma, G_A(\mathcal{D}), F, \mu)$ such that $\|M_1\| = \sum \|M\|$.*
 - *There exists a wta M such that $\|M_2\| = \sum \|M\|$.*

Proof. The second statement follows from the first and Proposition 5, so it remains to prove the first statement. Let $F_i = \overline{(F_1)_i}(x_1)$ and

$$\mu_k(\sigma)_{i, (i_1, \ldots, i_k)} = \overline{(\mu_1)_k(\sigma)_{i, (i_1, \ldots, i_k)}}(x_1, \ldots, x_k) .$$

Note that again M can be chosen to be deterministic, whenever the input device is deterministic. The main reason for the remaining summation is the fact that we do not know how to define sums like $\overline{\omega}(x_1, \ldots, x_k) + \overline{\omega}'(x_1, \ldots, x_k)$ for $\omega, \omega' \in \Omega_k$. Hence, we finally consider tst, because there we know more about the operations of Ω.

Theorem 10. *Let $\mathcal{A}$ be a completely idempotent semiring and let M_1 be a tst over $\mathcal{A}$. There exists a wta M such that $\|M\| = \|M_1\|$.*

The last theorem admits a trivial corollary.

Corollary 11. *For every bottom-up tree transducer M_1 there exists a wta M such that $\|M\| = \|M_1\|$.*

4   Pumping Lemmata

In this section we would like to demonstrate how to make use of the representation theorem derived in the previous section (Theorem 9). Unfortunately, very few results exist for weighted tree automata over arbitrary semirings (in particular: non-commutative semirings). However, in [2] a pumping lemma for deterministic finite wta is presented and we would like to translate this result to deterministic finite tst and deterministic finite DM-wta.

In this section, let $\mathcal{A} = (A, \oplus, \odot)$ be a semiring and $\mathcal{D} = (D, +, \Omega)$ be a DM-monoid. Let $\mathcal{L}^d_\Sigma(\mathcal{A})$ be the class of deterministically recognizable tree series, i.e., for every $L \in \mathcal{L}^d_\Sigma(\mathcal{A})$ there exists a deterministic finite wta $M = (I, \Sigma, \mathcal{A}, F, \mu)$ such that $L = \|M\|$. Similarly, let $\mathcal{T}^d_{\Sigma, \Delta}(\mathcal{A})$ be the class of deterministically computable t-ts transformations, i.e., for every $\tau \in \mathcal{T}^d_{\Sigma, \Delta}(\mathcal{A})$ there exists a deterministic finite tst $M = (I, \Sigma, \Delta, \mathcal{A}, F, \mu)$ such that $\tau = \|M\|$. Finally, let $\mathcal{L}^d_\Sigma(\mathcal{D})$ be the class of deterministically recognizable DM-monoid tree series, i.e., for every $L \in \mathcal{L}^d_\Sigma(\mathcal{D})$ there exists a deterministic finite DM-wta $M = (I, \Sigma, \mathcal{D}, F, \mu)$ such that $L = \|M\|$.

Firstly, we state the original corollary of [2].

Corollary 12 (Corollary 5.8 of [2]). *Let $L \in \mathcal{L}^d_\Sigma(\mathcal{A})$. There exists $m \in \mathbb{N}$ such that for every tree $t \in \mathrm{supp}(L)$ with $\mathrm{height}(t) \geq m + 1$ there exist trees $C, C' \in \widehat{T_\Sigma}(X_1)$ and $t' \in T_\Sigma$, and semiring elements $a, a', b, b', d \in A$ such that*

- $t = C[C'[t']]$,
- $\mathrm{height}(C[t']) \leq m + 1$ and $C \neq x_1$, and
- $(L, C'[C^n[t']]) = a' \odot a^n \odot d \odot b^n \odot b$ for every $n \in \mathbb{N}$.

We have already noted that the determinism and finiteness properties are preserved by all our constructions, so given a deterministic finite DM-wta M_1, we can construct a deterministic finite wta M such that $\sum \|M\| = \|M_1\|$ (cf. Theorem 9). Since the addition of the semiring is irrelevant for deterministic devices, we actually obtain $\|M\| = \|M_1\|$. Now we can apply the pumping lemma (Corollary 12) to this wta and thereby obtain a pumping lemma for tree series of $\mathcal{L}^d_\Sigma(\mathcal{D})$.

Theorem 13. *Let $L \in \mathcal{L}^d_\Sigma(\mathcal{D})$ and $\Omega' = \Omega \cup D$. There exists $m \in \mathbb{N}$ such that for every $t \in \mathrm{supp}(L)$ with $\mathrm{height}(t) \geq m + 1$ there exist $C, C' \in \widehat{T_\Sigma}(X_1)$, $t' \in T_\Sigma$, and $a, a' \in \widehat{T_{\Omega'}}(X_1)$, and $d \in D$ such that*

- $t = C[C'[t']]$,
- $\mathrm{height}(C[t']) \leq m + 1$ and $C \neq x_1$, and
- $(L, C'[C^n[t']]) = a' \leftarrow a^n \leftarrow d$ for every $n \in \mathbb{N}$.

Proof. The statement follows from Corollary 5.8 of [2] and Theorem 9.

With the help of Proposition 5 we can also obtain a pumping lemma for deterministic finite tst in the very same manner.

Theorem 14. *Let $\tau \in \mathcal{T}^d_{\Sigma, \Delta}(\mathcal{A})$ be a t-ts transformation. There exists $m \in \mathbb{N}$ such that for every tree $t \in \mathrm{supp}(T)$ with $\mathrm{height}(t) \geq m + 1$ there exist trees $C, C' \in \widehat{T_\Sigma}(X_1)$, $t' \in T_\Sigma$, and $a, a' \in A\langle\langle T_\Delta(X_1) \rangle\rangle$, and $c \in A\langle\langle T_\Delta \rangle\rangle$ such that*

- $t = C[C'[t']]$,
- $\text{height}(C[t']) \le m + 1$ and $C \ne x_1$, and
- $(\tau, C'[C^n[t']]) = a' \leftarrow a^n \leftarrow c$ for every $n \in \mathbb{N}$.

Proof. The statement is an immediate consequence of Proposition 5 and Theorem 13.

Finally, if we instantiate the previous theorem to the Boolean semiring, then we obtain the classical pumping lemma for deterministic bottom-up tree transducers (cf. [6]).

Acknowledgements. The author would like to thank the anonymous referees for their valuable suggestions, which improved the readability of the paper.

References

1. J. Berstel. *Transductions and Context-Free Languages.* Teubner, Stuttgart, 1979.
2. Björn Borchardt. A pumping lemma and decidability problems for recognizable tree series. *Acta Cybernetica*, 2004. to appear.
3. Symeon Bozapalidis. Equational elements in additive algebras. *Theory of Computing Systems*, 32:1–33, 1999.
4. H. Comon, M. Dauchet, R. Gilleron, F. Jacquemard, D. Lugiez, S. Tison, and M. Tommasi. Tree automata – techniques and applications. Available on: http://www.grappa.univ-lille3.fr/tata, 1997.
5. Samuel Eilenberg. *Automata, Languages, and Machines – Volume A*, volume 59 of *Pure and Applied Mathematics*. Academic Press, 1974.
6. Joost Engelfriet. Some open questions and recent results on tree transducers and tree languages. In R. V. Book, editor, *Formal Language Theory – Perspectives and Open Problems*, pages 241–286. Academic Press, 1980.
7. Joost Engelfriet, Zoltán Fülöp, and Heiko Vogler. Bottom-up and top-down tree series transformations. *Journal of Automata, Languages and Combinatorics*, 7(1):11–70, 2002.
8. Zoltán Fülöp and Heiko Vogler. Tree series transformations that respect copying. *Theory of Computing Systems*, 36:247–293, 2003.
9. Werner Kuich. Formal power series over trees. In Symeon Bozapalidis, editor, *Proceedings of 3rd DLT 1997*, pages 61–101. Aristotle University of Thessaloniki, 1997.
10. Werner Kuich. Tree transducers and formal tree series. *Acta Cybernetica*, 14:135–149, 1999.
11. Werner Kuich and Arto Salomaa. *Semirings, Automata, Languages*, volume 5 of *EATCS Monographs on Theoretical Computer Science*. Springer, 1986.
12. W. C. Rounds. Mappings and grammars on trees. *Mathematical Systems Theory*, 4:257–287, 1970.
13. Grzegorz Rozenberg and Arto Salomaa, editors. *Handbook of Formal Languages*, volume 1. Springer, 1997.
14. M. P. Schützenberger. Certain elementary families of automata. In *Proceedings of Symposium on Mathematical Theory of Automata*, pages 139–153. Polytechnic Institute of Brooklyn, 1962.

15. J. W. Thatcher. Generalized[2] sequential machine maps. *Journal of Computer and System Sciences*, 4:339–367, 1970.

16. J. W. Thatcher, E. G. Wagner, and J. B. Wright. Initial algebra semantics and continuous algebra. *Journal of the ACM*, 24:68–95, 1977.

An NP-Complete Fragment of LTL

Anca Muscholl[1,*] and Igor Walukiewicz[2,*]

[1] LIAFA, Université Paris, Paris, France
muscholl@liafa.jussieu.fr
[2] LaBRI, Université Bordeaux-1, France
igw@labri.fr

Abstract. A fragment of linear time temporal logic (LTL) is presented. It is proved that the satisfiability problem for this fragment is NP-complete. The fragment is larger than previously known NP-complete fragments. It is obtained by prohibiting the use of until operator and requiring to use only next operators indexed by a letter.

1 Introduction

Linear time temporal logic (LTL) is a well-studied and broadly used formalism for reasoning about events in time. It is equivalent to first-order logic over finite and infinite words [6]. The operators of the logic correspond to well-known semi-groups which gives a starting point of the successful classification research [13]. LTL is used to formulate properties of finite or infinite words. Such a formalization permits to do model-checking – verify if the given model has the given property. It turns out that, for LTL and its fragments, in almost all cases the model-checking problem is equivalent to a satisfiability-checking problem. This is why the satisfiability problem for LTL and its fragments is so well-studied. It is well-known that the problem for whole LTL is PSPACE-complete [11]. It is known also [11] that the fragment using only the "sometimes in the future" modality, denoted F, as well as the fragment using only the "next" modality, denoted X, have NP-complete satisfiability problems. Nevertheless, the fragment when both F and X are allowed is PSPACE-complete. This is a decidable fragment of LTL [1, 13]. We show that restricting the next operator X to operators X_a ($a \in \Sigma$) that enforce the current letter to be a, we get a fragment with the satisfiability problem in NP.

Thus, this paper shows that the PSPACE-completeness of the $F + X$ fragment is in some sense an accident due to some syntactic conventions. A very common approach to formalization of LTL is to have propositions in the logic and to consider a model to be a sequence of valuations of propositions. Another approach is to consider models to be words over a given alphabet and to have next modalities indexed by the letters, i.e. $X_a \varphi$ says that the first letter of the model is a and after cutting a the rest of the model satisfies φ. Of course it is very easy

---

* Work supported by the ACI Sécurité Informatique VERSYDIS.

C.S. Calude, E. Calude and M.J. Dinneen (Eds.): DLT 2004, LNCS 3340, pp. 334–344, 2004.
© Springer-Verlag Berlin Heidelberg 2004

to translate between the two conventions but the fragments that look natural in one convention do not necessary do so in the other. In particular consider the next operator X. Having operators X_a we can express X as $X\varphi \equiv \bigvee_{a\in\Sigma} X_a\varphi$, where Σ is the alphabet. An important point about this translation is that it induces an exponential blow-up. We show that it is having X as a primitive in the language that is the source of PSPACE-hardness. We prove that the fragment of LTL without until operator and using X_a operators instead of X is NP-complete.

Related Work. We have mentioned already above the classic results on PSPACE-completeness of the full logic and NP-completeness of the fragments only with F and only with X, [11]. Matched with the PSPACE-completeness of $F + X$ fragment, these results were considered sharp and the later work has concentrated mostly on extensions of LTL [7, 4, 3, 8]. Nevertheless the question about the fragment considered here was posed by the second author [12]. Recently the search of "easy" fragments of LTL has regained some interest [5, 2, 10]. The main motivation is to understand why the model-checkers (or the satisfiability-checkers) behave relatively well in practice despite the PSPACE lower bound. The fragments considered in recent papers put restrictions on the nesting of operators and on the number of propositions used [2].

2   Preliminaries

We will use Σ for a finite alphabet, the letters of which will be denoted by $a, b, c, \ldots$ As usual Σ^* denotes the set of finite and Σ^ω the set of infinite words over Σ. We use $u, v, w, \ldots$ to range over words.

Let $A \subseteq \Sigma^*$ be a finite set of words. The size of A is the sum of the lengths $|v|$ of all $v \in A$. We write $\Sigma^{\leq n}$ for the set of words of length $\leq n$.

Definition 1. *The set of subLTL formulas over an alphabet Σ is defined by the following grammar:*

$$\varphi ::= \mathsf{tt} \mid \mathsf{ff} \mid X_b\varphi \mid F\varphi \mid G\varphi \mid \varphi_1 \vee \varphi_2 \mid \varphi_1 \wedge \varphi_2$$

where subscript b ranges over Σ and X_b, F and G are called "next", "finally" and "globally" modalities.

For a non-empty word $v = a_1 \cdots a_k$ we write for short $X_v\phi$ instead of the formula $X_{a_1} \ldots X_{a_k}\varphi$.

The models are infinite words $v \in \Sigma^\omega$. The semantic is standard so we recall just the most important clauses:

- $v \models X_a\varphi$ if v can be factorized as av' and $v' \models \varphi$;
- $v \models F\varphi$ if there is a factorization uw of v, where $u \in \Sigma^*$, such that $w \models \varphi$;
- $v \models G\varphi$ if for all factorizations uw of v with $u \in \Sigma^*$, we have $w \models \varphi$.

Observe that there is no negation in the syntax. This is because we can define the negation of a formula using the equivalence rules $\neg(F\varphi) = G(\neg\varphi)$ and $\neg(X_a\varphi) = X_a(\neg\varphi) \vee \bigvee_{b\neq a} X_b\mathsf{tt}$. Note that these rules increase the size of the formula by a linear factor only.

In this paper we prove:

Theorem 1. *The satisfiability problem for subLTL is NP-complete.*

Let us compare subLTL with linear time temporal logic with propositional constants, that we call PTL here. In PTL instead of an alphabet we have a set of propositional constants $Prop = \{P, Q, \dots\}$. The formulas are built from propositions and their negations using the modalities X, F and G. There is also an until operator but we do not need it for our discussion here. The models are infinite sequences of valuations of propositions. When interested in satisfiability of a given formula φ one can restrict to the set of propositions that appear in the formula, call it $Prop_\varphi$. This way a model can be coded as a word over the finite alphabet $\Delta = 2^{Prop_\varphi}$. Given this, the semantics is the best explained by the translation to LTL:

- P is translated to $\bigvee\{X_a\text{tt} \mid a \in \Delta, \ P \text{ true in } a\}$ (recall that letters are valuations),
- $X\varphi = \bigvee_{a \in \Delta} X_a \varphi$

The rest of the clauses being identities.

Having definitions of both subLTL and PTL we can make the comparisons. First, observe that the fragment of PTL without X corresponds to the fragment of subLTL where after X_a we can put only tt. Next, observe that the translation of $X\varphi$ induces an exponential blowup. For example a formula $X^n\text{tt}$ (X n-times followed by tt) is translated to a formula of exponential size. Finally, observe that subLTL can express more properties than PTL without X. A simple example is $G(X_a\text{tt} \Rightarrow X_{ab}\text{tt})$ which states that after each a there is b. This property is not expressible in PTL without X if we have more than two letters. Another interesting formula is $G(X_{ab}\text{tt} \vee X_{ba}\text{tt})$. This formula has only the words $(ab)^\omega$ and $(ba)^\omega$ as models. This indicates that constructing a model for a subLTL formula may require a bit of combinatorics on words as the phenomena of interplay between different prefixes start to occur.

3   The Lower-Bound

Showing NP-hardness of the satisfiability problem for subLTL is quite straightforward.

We reduce SAT. Given a propositional formula α over variables $x_1, \dots, x_n$ we consider models over the letters $b, a_1, \dots, a_n$ were b will be used to fill the "empty spaces". A valuation of the variables $x_1, \dots, x_n$ will be encoded by a word in such a way that x_i is true iff a_i occurs in the word. Let φ_α be a formula obtained by replacing each occurrence of x_i in α by $FX_{a_i}\text{tt}$. Then there is a valuation satisfying α iff there is a word which is a model for φ_α.

In this reduction the alphabet is not fixed. Nevertheless it is quite straightforward to modify the reduction so that it works also for a two letter alphabet. For example, one can code each letter a_i as a word ba^ib. We omit the details.

4   The Upper-Bound

To show that the satisfiability problem for subLTL is in NP we will prove the small model property. The algorithm will be then to guess the model, of the form uv^ω for $u, v \in \Sigma^*$ and to check if the formula holds. This latter task can be done in polynomial time [11]. As a digression let us mention that the precise complexity of this problem is not known [9].

Hence, our goal in this section is the following theorem:

Theorem 2. *Every satisfiable formula φ of subLTL has a model of the form uv^ω with $|u| + |v|$ polynomial in the size of φ.*

The proof will be split into two subsections. In the first we will consider periodic words, i.e., ones of the form v^ω. We will show that if φ has a model v^ω then there is short word w such that w^ω is also a model of φ. In the second subsection we consider the case of ultimately periodic words, i.e., of the form uv^ω and show how to shorten u. Putting the two together we will obtain a small model for any satisfiable formula.

4.1   Periodic Words

We will first characterize the models of a subLTL formula that are periodic infinite words, i.e., words of the form v^ω for some $v \in \Sigma^*$.

Let Swords(w) be the set of finite factors of w.

Definition 2. *For $A, B \subseteq \Sigma^*$ and $p \in \Sigma^*$ we say that $w \in \Sigma^* \cup \Sigma^\omega$ is an (A, B, p)-word if*

- *p is a prefix of w,*
- *$A \subseteq Swords(w)$,*
- *$B \cap Swords(w) = \emptyset$.*

For the proof of the proposition below it is important to note that for any word $v \in \Sigma^*$ and any factorization $v = xy$ we have that $\text{Swords}(v^\omega) = \text{Swords}((yx)^\omega)$.

Proposition 1. *Let ϕ be a subLTL formula of size n. Then there exists a set $\mathcal{T}(\phi)$ of triples (A, B, p), where $A, B \subseteq \Sigma^{\leq n}$ are of polynomial size in n and $p \in \Sigma^{\leq n}$, such that for any word $v \in \Sigma^*$:*

$$v^\omega \models \phi \quad \text{iff} \quad v^\omega \text{ is an } (A, B, p)\text{-word for some } (A, B, p) \in \mathcal{T}(\phi).$$

Proof. We show the assertion by induction on the given formula ϕ.

1. We have $\mathcal{T}(\text{tt}) = \{(\emptyset, \emptyset, \lambda)\}$ and $\mathcal{T}(\text{ff}) = \{(\emptyset, \Sigma, \lambda)\}$.
2. Suppose $\phi = \phi_1 \wedge \phi_2$. We define $\mathcal{T}(\phi)$ as the set of triples (A, B, p) constructed as follows. For every two triples $(A_1, B_1, p_1) \in \mathcal{T}(\phi_1)$ and $(A_2, B_2, p_2) \in \mathcal{T}(\phi_2)$ with $p_1 \leq p_2$ ($p_2 \leq p_1$ respectively) we let $A = A_1 \cup A_2$, $B = B_1 \cup B_2$ and $p = p_2$ ($p = p_1$ respectively). It is easy to check that v^ω is a (A_i, B_i, p_i)-word for $i = 1, 2$ if and only if it is a (A, B, p)-word.

3. Suppose $\phi = X_a \psi$. We define $T(\phi) = \{(A, B, ap) \mid (A, B, p) \in T(\psi)\}$. We have $v^\omega \models \phi$ if and only if $v = aw$ and $(wa)^\omega \models \psi$. By induction, this happens if and only if $(wa)^\omega$ is an (A, B, p)-word for some $(A, B, p) \in T(\psi)$. But this is the case if and only if $(aw)^\omega$ is an (A, B, ap)-word.

4. Suppose $\phi = F\psi$. We define $T(\phi) = \{(A \cup \{p\}, B, \lambda) \mid (A, B, p) \in T(\psi)\}$. We have $v^\omega \models F\psi$ if and only if there exists some factorization $v = wx$ with $(xw)^\omega \models \psi$. By induction hypothesis this is equivalent to $(xw)^\omega$ being an (A, B, p)-word for some triple $(A, B, p) \in T(\psi)$. It is now easy to see that v^ω is an $(A \cup \{p\}, B, \lambda)$-word iff there is a factorization xw of v with $(wx)^\omega$ being an (A, B, p)-word.

5. Let $\phi = G\psi$. For each subset $\{(A_0, B_0, p_0), \ldots, (A_k, B_k, p_k)\} \subseteq T(\psi)$ with a distinguished element (A_0, B_0, p_0) we add the tuple (A, B, p_0) to $T(\phi)$ where

$$A = \bigcup_{i=0,\ldots,k} A_i, \qquad B = \bigcup_{i=0,\ldots,k} B_i \cup Y$$

and Y is the set of minimal words that are neither prefixes nor contain as a prefix any of the words $p_0, \ldots, p_k$. It is easy to see that Y is of the size polynomial in n. A word belongs to Y if it is of the form va with v a prefix of one of the words $p_0, \ldots, p_n$ and va neither a prefix of any of these words nor containing any of them.

Suppose that $v^\omega \models G\psi$. For every factorization xw of v we know, by the induction hypothesis, that $(wx)^\omega$ is a (A_w, B_w, p_w)-word for some (A_w, B_w, p_w) in $T(\psi)$. Let $(A, B, p_v) \in T(\phi)$ be the triple constructed as above from the set $\{(A_w, B_w, p_w) \mid w \text{ suffix of } v\}$. A direct verification shows that v^ω is a (A, B, p_v)-word. For example, let us show that $\mathrm{Swords}(v^\omega) \cap B = \emptyset$. Directly from the definition we have that $\mathrm{Swords}(v^\omega) \cap B_w$ for every w a suffix of v. For the set Y defined as above we have $\mathrm{Swords}(v^\omega) \cap Y = \emptyset$ because all suffixes of v^ω have some p_w as a prefix.

For the opposite direction suppose that v is an (A, B, p)-word constructed from some set $\{(A_0, B_0, p_0), \ldots, (A_k, B_k, p_k)\} \subseteq T(\psi)$. Take any factorization xw of v. We want to show that $(wx)^\omega \models \psi$. Because of the set Y, as defined above, the word $(wx)^\omega$ has some p_i as a prefix. From the definition we know that $A_i \subseteq A$ and $B_i \subseteq B$. Hence $(wx)^\omega$ is a (A_i, B_i, p_i)-word and consequently $(wx)^\omega \models \psi$.

$\square$

Example 1. Consider the formula $\phi = G\psi$, where $\psi = X_{ab}\mathtt{tt} \vee X_{ba}\mathtt{tt}$ and $\Sigma = \{a, b\}$. We have $T(\psi) = \{(\emptyset, \emptyset, ab), (\emptyset, \emptyset, ba)\}$. The construction above yields $T(\phi) = \{(\emptyset, \{b, aa\}, ab), (\emptyset, \{a, bb\}, ba), (\emptyset, \{aa, bb\}, ab), (\emptyset, \{aa, bb\}, ba)\}$. Clearly, only for the last two triples of $T(\phi)$ there can be a solution.

The next two lemmas show that finite (periodic, respectively.) (A, B, p)-words can be chosen of polynomial length.

Lemma 1. *Let $A, B \subseteq \Sigma^*$ and $p \in \Sigma^*$. If there is a finite (A, B, p)-word then there is one of size polynomial in the sizes of A, B and p.*

Proof. We construct a (deterministic) finite automaton $\mathcal{A}$ accepting (A, B, p)-words. Then we show that it accepts a short word.

As a preparation observe that we may assume that no word in A is a factor of some other word in A; if it is the case then we simply delete the smaller one. Similarly for B but here we can delete the bigger one.

The states of $\mathcal{A}$ will be triples (u, v, r) where u is a prefix of a word in A, v is a prefix of a word in B and r is a prefix of p. Intuitively such a state will say that u and v are suffixes of the word being read and they are the longest possible suffixes for the words in A and B respectively. The last component r is used for testing that the word starts with p. The initial state will be $(\lambda, \lambda, \lambda)$.

The transitions of $\mathcal{A}$ are deterministic. We have

$$(u, v, r) \xrightarrow{a} (u', v', r')$$

when

- either $u' = ua$, or if ua is not a prefix of a word in A then u' is the longest suffix of ua that is a prefix of word from A.
- for v' we have exactly the same rule but with respect to B.
- either $r = p = r'$, or $r' = ra$ if ra is a prefix of p.

A state (u, v, r) is *rejecting* if $v \in B$. It is a u-state if its first component is u.

Our first claim is that a word w is an (A, B, p)-word iff $\mathcal{A}$ has a run on this word that does not visit a rejecting state, passes through a u-state for every $u \in A$ and ends in a state where the last component is p (called a p-state). This follows from the observation that if there is a B-factor v in w then after reading the last letter of v the automaton $\mathcal{A}$ enters in the rejecting state. If $u \in A$ is a factor of w, then after reading u the state of $\mathcal{A}$ is a u-state.

It remains to see that there is a short (A, B, p)-word if there is one at all. Consider an (A, B, p)-word u and an accepting run $\rho = (s_0, s_1, \ldots, s_m)$ of $\mathcal{A}$ on $u = a_1 \cdots a_m$. Thus, s_m is p-state and for each $u \in A$ there is some position j such that s_j is a u-state. Let us fix for each $u \in A$ such a position $j(u)$ and let $J = \{j(u) \mid u \in A\}$. We can delete now any loop contained between two consecutive positions in J, and the run obtained is still accepting. The length of the run is at most $O(|A||\mathcal{A}|)$, hence polynomial in $|A|, |B|, |p|$. $\qquad\square$

Lemma 2. *If there is a periodic (A, B, p)-word then there is one of the form s^ω with $|s|$ polynomial in the sizes of A, B and p.*

Proof. The construction is as in the previous lemma but now we need to start the automaton in some state of the form (u, v, λ) and to require that it reaches the state (u, v, p).

Suppose that we have a run from (u, v, λ) to (u, v, p) for some u, v, and let s be the word defining this run. Then s defines also a run from (u, v, p) back to itself, thus s^ω is the desired periodic (A, B, p)-word. By the same argument as in the lemma before we can also see that there is always a run of polynomial length, if any. For the other direction, suppose that s^ω is a periodic (A, B, p)-word. We can consider the run of $\mathcal{A}$ on this word starting in $(\lambda, \lambda, \lambda)$. Since s^ω

is an (A, B, p)-word, this run never blocks. Hence, there must be two indices $i < j$ such that after reading s^i and s^j the automaton is in the same state, say (u, v, p). But then, there is a run of automaton $\mathcal{A}$ from (u, v, λ) to (u, v, p) on s^{j-i}. $\qquad\square$

4.2   Ultimately Periodic Words

We turn now to the case where the model of ϕ is ultimately periodic, i.e., of the form uv^ω for some $u, v \in \Sigma^*$. For the rest of the section, n denotes the length of the given formula φ.

Lemma 3. *Any subLTL formula ϕ using only the operators X_a $(a \in \Sigma)$ is equivalent to a disjunction of the form $\bigvee_{v \in V} X_v \mathrm{tt}$, for some set $V \subseteq \Sigma^{\leq |\phi|}$ of at most $|\phi|$ words.*

Proof. We show the assertion by induction on the given formula ϕ. For $\phi = X_a\psi$ we suppose that ψ is equivalent to $\bigvee_{v \in V} X_v \mathrm{tt}$, hence ϕ is equivalent to $\bigvee_{v \in V'} X_v \mathrm{tt}$ for $V' = aV$. Let now ϕ_i be equivalent to $\bigvee_{v \in V_i} X_v \mathrm{tt}$ for $i = 1, 2$. Then $\phi_1 \wedge \phi_2$ is equivalent to $\bigvee_{v \in V} X_v \mathrm{tt}$, for V defined as follows: a word $v \in V_i$ belongs to V if and only if there exists $v' \in V_j$, $j \neq i$, such that $v' \leq v$. $\qquad\square$

An *F-formula* is a formula that begins with F. Similarly for X and *G-formulas*. The set $\mathrm{El}(\phi)$ of *elementary subformulas* of ϕ is the set of those subformulas of φ that are either F- or G-formulas.

For any word we write $w[i, j]$ for the factor $a_i \ldots a_j$. For an infinite word $w = a_1 a_2 \ldots$ we write $w[j, \infty]$ for the suffix $a_j a_{j+1} \ldots$ We use $w(i)$ to denote a_i, the i-th letter a_i.

For the rest of the section we fix a model uv^ω of a formula φ. With each position $i \leq |u|$ in the word uv^ω we associate the set of subformulas:

$$S_i = \{\psi \in \mathrm{El}(\varphi) \mid u[i, |u|] v^\omega \models \psi\}$$

Remark 1. If a formula $G\alpha$ is in S_i then it is in all S_j for $j \geq i$. Analogously, if $F\alpha \in S_i$ then $F\alpha \in S_j$ for all $j \leq i$.

A position $i \leq |u|$ is called *important* if there is a formula $F\alpha$ in $S_i \setminus S_{i+1}$ or there is a formula $G\alpha$ in $S_{i+1} \setminus S_i$. Let VIP be the set of important positions in u. Clearly the number of important positions is bounded by $n = |\varphi|$.

We will show how to reduce distances between consecutive important positions, in order to obtain a short word u. From now on we fix two consecutive important positions $i, j \in$ VIP. This means that $S_{i+1} = \cdots = S_j$ contain the same F- and G-subformulas and $S_i \neq S_{i+1}$.

For a subformula ψ of φ let $\widehat{\psi}$ be a formula obtained by substituting tt for every F- or G-subformula of ψ appearing in S_j and ff for all other F or G subformulas. By Lemma 3, for every subformula ψ there exists a polynomial-size set of words $V_\psi \subseteq \Sigma^{\leq n}$ such that $\widehat{\psi}$ is $\bigvee_{v \in V_\psi} X_v \mathrm{tt}$ (in consequence, if $V_\psi = \emptyset$ then $\widehat{\psi} = \mathrm{ff}$ and if $V_\psi = \{\lambda\}$ then $\widehat{\psi} = \mathrm{tt}$).

Example 2. $X_a\widehat{G(X_b\text{tt})}$ is $X_a\text{tt}$ or $X_a\text{ff}$ depending on $G(X_b\text{tt})$ being in S_j or not. Moreover $\widehat{G(X_a\widehat{G(X_b\text{tt})})}$ is just tt or ff. (Both hats range over whole formulas).

Our goal is to replace $u[i+1, j]$ in uv^ω by a short word so that the result is still a model of φ. In order to do this we will use (Y, p, s)-words, that we define in the following. Let $Y \subseteq \Sigma^*$ and $p, s \in \Sigma^*$. A finite word w is a (Y, p, s)-word, if it starts with p, finishes with s and for all positions $1 \leq k \leq |w| - |s|$, the suffix $w[k, |w|]$ starts with a word from Y.

Lemma 4. *If there exists some (Y, p, s) word, then there exists some of length polynomial in the sizes of Y, p, s.*

Proof. Consider some (Y, p, s)-word w and a position $k \leq |w| - |s|$. By definition, there exists some $l > k$ such that $w[k, l] \in Y$. With k we associate the position $r(k)$ defined by $r(k) = \max\{l' \geq l \mid \exists k' \leq k : w[k', l'] \in Y\}$. That is, $r(k)$ is the rightmost end of a word in Y that begins at the left of k. By definition, $w[l+1, r(k)]$ is a suffix (possibly empty) of a word in Y. Thus, there are at most $|Y|^2$ different words $w[k, r(k)]$.

Suppose now that $|p| < k < k' \leq |w| - |s|$ are such that $w[k, r(k)] = w[k', r(k')]$. Obviously, the word $w[1, k]w[k' + 1, |w|]$ obtained by cutting out $w[k + 1, k']$ is still a (Y, p, s)-word. Thus, by the above remark we know that there exists a (Y, p, s)-word of length at most $|Y|^2 + |ps|$.   □

For each $G\alpha \in S_j$ let V_α be the set of words obtained by Lemma 3 such that $\widehat{\alpha} = \bigvee_{v \in V_\alpha} X_v\text{tt}$. Applying again Lemma 3 we obtain a set Y such that $\bigwedge_{G\alpha \in S_j} \widehat{\alpha} = \bigvee_{v \in Y} X_v\text{tt}$. Note that Y is of polynomial size and contains only words of length at most n, since $Y \subseteq \bigcup_{G\alpha \in S_j} V_\alpha$. Moreover, let $p = u[i+1, i+n]$, $s = u[j - n + 1, j]$ be the prefix and suffix, respectively., of length n of $u[i, j]$.

The next lemma follows immediately from the definition of Y:

Lemma 5. *Let $i, j \in VIP$ be consecutive important positions with $j - i > n$, and let Y, p, s be defined as above. Then the word $u[i + 1, j]$ is a (Y, p, s)-word. Moreover, each (Y, p, s)-word w starts with $u[i+1, i+n]$, finishes with $u[j-n+1, j]$ and for all $k \leq |w| - n$ the word $w[k, |w|]$ has a word from V_α as a prefix for all subformulas $G\alpha \in S_j$.*

Our goal is to show that $wu[j + 1, |u|]v^\omega \models S_{i+1}$ for any (Y, p, s)-word w, as this will imply $u[1, i]wu[j + 1, |u|]v^\omega \models \varphi$. To do this we will need a definition and several lemmas. Consider a formula α and let

$$\gamma_\alpha = \bigwedge\{\delta \in S_{i+1} \mid \delta \text{ is a } F\text{- or } G\text{-subformula of } \alpha\}$$

Hence γ_α is the conjunction of all F- and G-formulas that appear in $S_{i+1} = S_j$ and that are subformulas of α. By Lemma 3 the formula $\widehat{\alpha}$ is equivalent to $\bigvee_{r \in R} X_r\text{tt}$. We define $\widetilde{\alpha}$ to be the formula $\bigvee_{r \in R} \bigwedge_{s \leq r} X_s\gamma_\alpha$. The definition of $\widetilde{\alpha}$ is important because it gives an underapproximation of α that is easy to work with.

Lemma 6. *If α is a subformula of the initial formula then $\widetilde{\alpha} \Rightarrow \alpha$ holds.*

Proof. By induction on α:

If α is an F- or G-formula then $\widehat{\alpha}$ is either tt or ff and $\widetilde{\alpha}$ is either γ_α or ff, respectively. We have that $\widetilde{\alpha} \Rightarrow \alpha$ in this case.

If $\alpha = X_a\beta$ then $\gamma_\alpha \Rightarrow \gamma_\beta$. Let $\widehat{\beta} = \bigvee_{r \in R} X_r$tt, then $\widetilde{\beta} = \bigvee_{r \in R} \bigwedge_{s \leq r} X_s\gamma_\beta$. We have that $\widetilde{\alpha} = \bigvee_{ar \in aR} \bigwedge_{s \leq ar} X_s\gamma_\alpha = \bigvee_{r \in R}(\gamma_\alpha \wedge \bigwedge_{s \leq r} X_a X_s\gamma_\alpha) = \gamma_\alpha \wedge X_a(\bigvee_{r \in R} \bigwedge_{s \leq r} X_s\gamma_\alpha) \Rightarrow X_a(\bigvee_{r \in R} \bigwedge_{s \leq r} X_s\gamma_\beta) = X_a\widehat{\beta} \Rightarrow X_a\beta = \alpha$.

For $\alpha = \beta_0 \vee \beta_1$ the argument is straightforward. It remains to consider the case when α is of the form $\beta_0 \wedge \beta_1$. We have that $\widehat{\beta_0}$ is of the form $\bigvee_{r \in R_0} X_r$tt and $\widehat{\beta_1}$ is $\bigvee_{r \in R_1} X_r$tt. Now by construction $\widehat{\beta_0 \wedge \beta_1}$ is $\bigvee_{s \in S} X_s$tt where $s \in S$ if there is $i \in \{0,1\}$ with $s \in R_i$ and some prefix of s in R_{1-i}. We have by definition that $\widetilde{\beta_0 \wedge \beta_1}$ is $\bigvee_{s \in S} \bigwedge_{w \leq s} X_w\gamma_\alpha$. It is easy to check that this implies $\bigvee_{s \in R_i} \bigwedge_{w \leq s} X_w\gamma_\alpha$ for $i \in \{0,1\}$, hence also $\widetilde{\beta_0 \wedge \beta_1}$. $\square$

We have now all ingredients to show that $wu[j+1, |u|]v^\omega \models S_{i+1}$ for any (Y, p, s)-word w. To shorten the notation we will write z for the suffix $u[j+1, |u|]v^\omega$. The proof goes through two lemmas because we need to consider G-formulas separately.

Lemma 7. *Let w be a (Y, p, s)-word. For every G-formula $G\alpha \in S_{i+1}$ and every non-empty suffix w' of w we have $w'z \models G\alpha$.*

Proof. The proof is by induction on the size of $G\alpha$. Take a formula $G\alpha \in S_{i+1}$ and any suffix w' of w. We want to show that $w'z \models \alpha$. Since $z \models G\alpha$, this suffices for showing that $w'z \models G\alpha$.

If $|w'| \leq n$ then w' is a suffix of $u[j-n+1, j]$ and we have $w'z \models G\alpha$ by definition of S_{i+1}. Hence also $w'z \models \alpha$.

If $|w'| > n$ then we know by Lemma 5 that w' starts with a word from V_α, say $r \in V_\alpha$. We will show that $w'z \models \widetilde{\alpha}$. Consider now the conjunct $\bigwedge_{s \leq r} X_s\gamma_\alpha$ of $\widetilde{\alpha}$, where γ_α is, as before, the conjunction of all F and G-subformulas of α from S_{i+1}. We know that $|w'| > n$ and $|r| \leq n$, hence we can use the induction hypothesis and we obtain that $w'z \models \bigwedge_{s \leq r} X_s\gamma_\alpha$. But then we have $w'z \models \widetilde{\alpha}$ which implies $w'z \models \alpha$ by Lemma 6. $\square$

Lemma 8. *If w is a (Y, p, s)-word then $wz \models S_{i+1}$.*

Proof. For formulas $F\alpha \in S_{i+1}$ we know that $u(j)z \models S_j = S_{i+1}$ and we are done as w ends with the letter $u(j)$. The case of G-formulas follows from the previous lemma.

It remains to consider an X-formula α. The first observation is that $wz \models \widehat{\alpha}$. This is because $u[i+1, j]z \models \alpha$, the size of α is not bigger than n, and w starts with $u[i+1, i+n]$. Now, by the same reasoning as in the previous lemma we get that $wz \models \widetilde{\alpha}$. Finally, by Lemma 6 we have $wz \models \alpha$. $\square$

We obtain:

Proposition 2. *If $uv^\omega \models \varphi$ and $i, j \in VIP$ are successive important positions then there is a word w of size polynomial in $n = |\varphi|$ such that $u[1, i] \, w \, u[j + 1, |u|]v^\omega \models \varphi$.*

Proof. From Lemma 8 we know that $wz \models S_{i+1}$. By induction on $k = i, \dots, 1$ it is easy to see that $u[k, i] \, w \, [j + 1, |u|]v^\omega \models S_k$. Lemma 4 gives the bound on the length of w. □

Using Proposition 2 repetitively we can shorten the size of u. From the previous subsection, Proposition 1 and Lemma 2, we know that we can shorten v. This proves the small model theorem, Theorem 2.

5   Conclusions

We have show the small model property for subLTL which implies that the satisfiability problem for the logic is NP-complete. This indicates that the temporal logic formalism based on word models is more subtle than the one based on propositions and on sequences of valuations. This also indicates that the X operator of LTL may be a source of complexity. On the other hand it seems that X_a operators are easier algorithmically, but the proofs become much more involved because some combinatorics on words becomes necessary. As further work we would like to investigate global trace logics with X_a instead of X. The hope being to have a reasonable such logic with the complexity lower than EXPSPACE. The other question is the complexity of the model-checking problem for subLTL with respect to a single path uv^ω. This question has been asked for general LTL in [9]. For general LTL it can be solved in polynomial time, but no good lower bound is known.

References

1. J. Cohen, D. Perrin, and J.-E. Pin. On the expressive power of temporal logic. *Journal of Computer and System Sciences*, 46(3):271–294, 1993.
2. S. Demri and Ph. Schnoebelen. The complexity of propositional linear temporal logics in simple cases. *Information and Computation*, 174(1):84–103, 2002.
3. C. Dixon, M. Fisher, and M. Reynolds. Execution and proof in a horn-clause temporal logic. In H. Barringer, M. Fisher, D. Gabbay, and G. Gough, editors, *Advances in Temporal Logic*, volume 16 of *Applied Logic Series*, pages 413–433. Kluwer Academic, 2000.
4. E. A. Emerson and C. Lei. Modalities for model-checking: Branching-time logic strikes back. *Science of Computer Programming*, 8(3):275–306, 1987.
5. K. Etessami, M. Vardi, and Th. Wilke. First-order logic with two variables and unary temporal logic. In *LICS'97 - 12th Annual IEEE Symposium on Logic in Computer Science, Warsaw, Poland*, pages 228–235, 1997.
6. J. Kamp. *Tense Logic and the Theory of Linear Order*. PhD thesis, University of California, Los Angeles, 1968.

7. O. Lichtenstein and A. Pnueli. Checking that finite state concurrent programs satisfy their linear specification. In *POPL'85 - 12th Annual ACM Symposium on Principles of Programming Languages, New Orleans, Louisiana*, pages 97–107, 1985.
8. N. Markey. Past is for free: On the complexity of verifying linear temporal properties with past. *Acta Informatica*, 40(6-7):431–458, 2004.
9. N. Markey and Ph. Schnoebelen. Model checking a path. In *CONCUR'03 - Concurrency Theory, 14th International Conference, Marseille, France*, number 2761 in Lecture Notes in Computer Science, pages 248–262. Springer, 2003.
10. Ph. Schnoebelen. The complexity of temporal logic model checking. *Advances in Modal Logic*, 4:437–459, 2003.
11. A. P. Sistla and E. M. Clarke. The complexity of propositional linear temporal logic. *Journal of the ACM*, 32(3):733–749, 1985.
12. I. Walukiewicz. Difficult configurations – on the complexity of LTrL. In *ICALP'98 - Automata, Languages and Programming, 25th International Colloquium, Aalborg (Denmark)*, number 1443 in Lecture Notes in Computer Science, pages 140–151, 1998.
13. Th. Wilke. Classifying discrete temporal properties. In *STACS'99, 16th Annual Symposium on Theoretical Aspects of Computer Science, Trier (Germany)*, number 1563 in Lecture Notes in Computer Science, pages 32–46, 1999.

From Post Systems to the Reachability Problems for Matrix Semigroups and Multicounter Automata

Igor Potapov*

Department of Computer Science, University of Liverpool, Liverpool, U.K.
igor@csc.liv.ac.uk

Abstract. The main result of this paper is the reduction of PCP(n) to the vector reachability problem for a matrix semigroup generated by n 4×4 integral matrices. It follows that the vector reachability problem is undecidable for a semigroup generated by 7 integral matrices of dimension 4. The question whether the vector reachability problem is decidable for $n=2$ and $n=3$ remains open. Also we show that proposed technique can be applied to Post's tag-systems. As a result we define new classes of counter automata that lie on the border between decidability and undecidability.

1   Introduction

In this paper we show the connection between decision problems for Post systems and the reachability problems for matrix semigroups and counter automata.

We start from the vector reachability problem for a matrix semigroup, which is a generalisation of the orbit problem [12]. The vector reachability problem is formulated as follows: *"Let S be a given finitely generated semigroup of $n \times n$ matrices from $\mathbb{Q}^{n \times n}$ and vectors $\bar{x}, \bar{y}$ from $\mathbb{Q}^n$. Decide whether there is a matrix $M \in S$ such that $M \cdot \bar{x} = \bar{y}$."*

In fact, the problem has close relation to the membership problem in matrix semigroups [1, 7, 14] and the reachability in linear iterative maps. It is equivalent to the following reachability problem for non-deterministic linear maps: "Given two vectors u and v in n-dimensional vector space over $\mathbb{Q}$ and a set A of linear transformations. Determine whether exists a sequence of transformations from A such that maps v to u". In algebraic terms the vector reachability problem can be expressed as a problem of determining whether it is possible to get a vector u by an action of matrix semigroup on the initial vector v.

In case where a semigroup generated by one matrix the vector reachability is decidable in polynomial time [12]. It is shown in [14] that the vector reachability problem is decidable for the case of row-monomial matrix semigroups over semigroup $\mathcal{S}$, where $\mathcal{S}$ is an arbitrary finitely generated *commutative* matrix

---

* Work partially supported by NAL/00684/G Nuffield Foundation grant.

C.S. Calude, E. Calude and M.J. Dinneen (Eds.): DLT 2004, LNCS 3340, pp. 345–356, 2004.

semigroup over an algebraic number field F. In particular the vector reachability problem is decidable for a case of row-monomial matrix semigroups over $\mathbb{Q}$ or $\mathbb{C}$. In general case the problem is also decidable for $n = 1$ since it can be reduced to the solution of the system of linear Diophantine equations. Since the membership problem for 3×3 matrix semigroup is undecidable it is also easy to see that the vector reachability problem is undecidable for any $n \geq 9$ by reduction of membership problem of dimension 3 to the vector reachability problem of dimension 9 [7].

As a main result of this paper we prove that the vector reachability problem is undecidable for any $n \geq 4$. In particular the undecidability result is based on the reduction the Post correspondence problem to the 4-dimensional vector reachability problem. The question whether the problem is decidable for n=2 and n=3 is still open.

As an extension of proposed reduction we apply a similar technique to the Post's tag-systems to show their connection to reachability problems for counter automata and to identify some classes of automata with decidable and undecidable reachability problems.

This paper is organised as follows. Next section contains preliminaries. In the Section 3 we show a matrix interpretation of Post correspondence problem and the main result. Another application of the developed technique is presented in Section 4, where we convert tag-systems to four counter automata. The paper ends with some conclusions and open problems.

2   Preliminaries

In what follows we use traditional denotations $\mathbb{N}, \mathbb{Z}, \mathbb{Q}$ and $\mathbb{Q}^+$ for the sets of naturals (non-negative integers), integers, rationals and non-negative rationals, respectively. A *semigroup* is a pair $(S, \cdot)$, where S is a set and $\cdot$ is an associative binary operation on S. A semigroup $(S, \cdot)$ is generated by a set A of its elements iff every element of S is a finite product $a_{i_1} \cdot a_{i_2} \cdot \ldots \cdot a_{i_k}$ where $a_{i_j} \in A$. The set of $n \times n$ matrices over rationals (integers) is denoted by $\mathbb{Q}^{n \times n}$ ($\mathbb{Z}^{n \times n}$). It is clear that the identity element for a semigroup $(\mathbb{Q}^{n \times n}, \cdot)$ or for $(\mathbb{Z}^{n \times n}, \cdot)$ is the identity matrix that we denote by E_n (or E).

We denote an empty word by ϵ. The concatenation of two strings w and v is a string obtained by appending the symbols of v to the right end of w, that is, if $w = a_1 a_2 \ldots a_n$ and $v = b_1 b_2 \ldots b_n$ then the concatenation of w and v, denoted by $w \cdot v$ or wv, is $a_1 a_2 \ldots a_n b_1 b_2 \ldots b_n$. The reverse of a string is obtained by writing the symbols in reverse order; if w is a string as shown above, then its reverse w^{-1} is $a_n \ldots a_2 a_1$.

2.1   The Stern-Brocot Tree

The Stern-Brocot tree is an elegant way for constructing the set of all nonnegative fractions $\frac{m}{n}$ where m and n are relatively prime [8].

Take two rational numbers, $\frac{a}{b}$ and $\frac{c}{d}$, and insert between them a third value, called the mediant, equal to $\frac{a+c}{b+d}$. Now with three numbers in hand, construct mediant between the first and second and between second and third, so that the next level of the tree has five numbers.

The canonical version of the Stern-Brocot tree starts with two irreducible fractions $\frac{0}{1}$, $\frac{1}{0}$ representing zero and infinity. Actually $\frac{0}{1}$ is a fraction while $\frac{1}{0}$ is not. However, we use them to describe a way to get all possible positive fractions arranged in a binary tree form as illustrated on Figure 1.

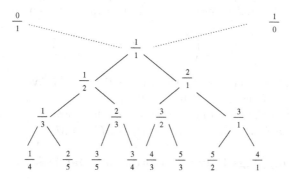

Fig. 1. The Stern-Brocot Tree

So, for example, from $\frac{0}{1}$ and $\frac{1}{0}$ we get $\frac{1}{1}$. The mediant of $\frac{0}{1}$ and $\frac{1}{1}$ is $\frac{1}{2}$ while the mediant of $\frac{1}{1}$ and $\frac{1}{0}$ is $\frac{2}{1}$. On the next stage of the construction, we form four new fractions: $\frac{1}{3}$ from $\frac{0}{1}$ and $\frac{1}{2}$, $\frac{2}{3}$ from $\frac{1}{2}$ and $\frac{1}{1}$, $\frac{3}{2}$ from $\frac{1}{1}$ and $\frac{2}{1}$, and, finally, the mediant of $\frac{2}{1}$ and $\frac{1}{0}$ which is $\frac{3}{1}$. Continuing this way we get an infinite tree known as the Stern-Brocot tree [1].

If we specify the position of a fraction in the tree as a path consisting of L(eft) an R(ight) moves along the tree starting from the top (fraction), and also define matrices

$$M_L = \begin{pmatrix} 1 & 1 \\ 0 & 1 \end{pmatrix}, M_R = \begin{pmatrix} 1 & 0 \\ 1 & 1 \end{pmatrix}$$

then product of the matrices corresponding to the path (or binary word) is matrix $\begin{pmatrix} n & n' \\ m & m' \end{pmatrix}$ whose entries are numerators and denominators of parent fractions $\frac{m}{n}$ and $\frac{m'}{n'}$. For example, the path leading to fraction $\frac{3}{5}$ is LRL. The corresponding matrix product is

$$M_L \times M_R \times M_L = \begin{pmatrix} 1 & 1 \\ 0 & 1 \end{pmatrix} \times \begin{pmatrix} 1 & 0 \\ 1 & 1 \end{pmatrix} \times \begin{pmatrix} 1 & 1 \\ 0 & 1 \end{pmatrix} = \begin{pmatrix} 2 & 3 \\ 1 & 2 \end{pmatrix}$$

and the parents of $\frac{3}{5}$ are $\frac{1}{2}$ and $\frac{2}{3}$.

---

[1] It was discovered independently by the German mathematician Moriz Stern (1858) and by the French clock maker Achille Brocot (1860).

Let us consider a binary word $w=w_1 \cdot \ldots \cdot w_n$, that corresponds to some path in the Stern-Brocot tree. In case of matrix representation the empty word corresponds to the identity matrix and the binary word w corresponds to a matrix $M_w = M_{w_1} \times \ldots \times M_{w_n}$.

So let w be a binary word and M_w be its matrix representation. For example, in order to append character L to the word w (right concatenation) we have to multiply matrix M_w by M_L from the right, so the word $w' = w \cdot L$ corresponds to $M_w \times M_L$. Similarly we can add character R to the head of the word w (left concatenation), so the word $w' = R \cdot w$ corresponds to $M_R \times M_w$.

The left and the right contractions can be expressed using inverse matrices:

$$M_L^{-1} = \begin{pmatrix} 1 & -1 \\ 0 & 1 \end{pmatrix}, M_R^{-1} = \begin{pmatrix} 1 & 0 \\ -1 & 1 \end{pmatrix}.$$

Let the head character of the word $w' = R \cdot w$ be R. We can delete a head character R if we multiply M_R^{-1} by $M_{w'}$ then $M_R^{-1} \times M_{w'} = M_R^{-1} \times M_R \times M_w = M_w$. In order to delete the tail character L of the word $w' = w \cdot L$ we multiply $M_{w'}$ by M_L^{-1}, so $M_{w'} \times M_L^{-1} = M_w \times M_L \times M_L^{-1} = M_w$.

2.2   Two Mappings Between Words and Matrices

Now we derive two mappings ψ and ϕ from the Stern-Brocot number system. First, let us consider the mapping ψ between $\{L, R\}^*$ and 2×2 matrices:

$$\psi : \epsilon \mapsto \begin{pmatrix} 1 & 0 \\ 0 & 1 \end{pmatrix} = E \quad \psi : L \mapsto \begin{pmatrix} 1 & 1 \\ 0 & 1 \end{pmatrix} = M_L \quad \psi : R \mapsto \begin{pmatrix} 1 & 0 \\ 1 & 1 \end{pmatrix} = M_R$$

$$\psi : w_1 \cdot \ldots \cdot w_r \mapsto M_{w_1} \times \ldots \times M_{w_r}.$$

It follows from the properties of Stern-Brocot number system [8, 10] that the mapping ψ is an isomorphism between $\{L, R\}^*$ and elements of matrix semigroup generated by 2×2 matrices $\begin{pmatrix} 1 & 1 \\ 0 & 1 \end{pmatrix}$ and $\begin{pmatrix} 1 & 0 \\ 1 & 1 \end{pmatrix}$. Since for every matrix with nonzero determinant there is only one unique inverse matrix we can also define a similar mapping ϕ. It can be defined using inverse matrices of the semigroup generator. Mapping ϕ is also an isomorphism between $\{L, R\}^*$ and elements of matrix semigroup generated by 2×2 matrices $\{M_L^{-1}, M_R^{-1}\}$:

$$\phi : \epsilon \mapsto \begin{pmatrix} 1 & 0 \\ 0 & 1 \end{pmatrix} = E \quad \phi : L \mapsto \begin{pmatrix} 1 & -1 \\ 0 & 1 \end{pmatrix} = M_L^{-1} \quad \phi : R \mapsto \begin{pmatrix} 1 & 0 \\ -1 & 1 \end{pmatrix} = M_R^{-1}$$

$$\phi : w_1 \cdot \ldots \cdot w_r \mapsto M_{w_1}^{-1} \times \ldots \times M_{w_r}^{-1}.$$

Note, that these two mappings from $\{L, R\}^*$ to matrices are injective, the mappings from $w \in \{L, R\}^+$ to $w^{-1} \in \{L, R\}^+$ and from $\psi(u)$ to $\phi(u^{-1})$ are bijective (see Figure 2). There are many interesting properties of mappings ϕ and ψ, but we state here only some of them that will be used in the paper.

Proposition 1. *Given a word* $w \in \{L, R\}^+$ *and* $w' \in \{L, R\}^*$.

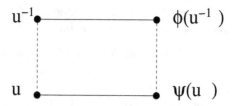

Fig. 2. Mappings between words and matrices

- Let $\psi(w) = \begin{pmatrix} e & f \\ g & h \end{pmatrix}$ then $e, h > 0$ and $f, g \geq 0$;

- Let $\phi(w) = \begin{pmatrix} a & b \\ c & d \end{pmatrix}$ then $a, d > 0$ and $b, c \leq 0$;

- Let $\phi(w' \cdot L) = \begin{pmatrix} a & b \\ c & d \end{pmatrix}$ then $a \leq |b|$; Let $\phi(w' \cdot R) = \begin{pmatrix} a & b \\ c & d \end{pmatrix}$ then $d \leq |c|$;

- Let $\psi(R \cdot w') = \begin{pmatrix} e & f \\ g & h \end{pmatrix}$ then $e \leq g$; Let $\psi(L \cdot w') = \begin{pmatrix} e & f \\ g & h \end{pmatrix}$ then $h \leq f$.

Proof. These properties hold for a case of four initial matrices $\phi(L)$, $\phi(R)$, $\psi(L)$, $\psi(R)$. The straightforward check of the matrix operations

$$\begin{pmatrix} a & b \\ c & d \end{pmatrix} \times \phi(L), \begin{pmatrix} a & b \\ c & d \end{pmatrix} \times \phi(R), \psi(L) \times \begin{pmatrix} e & f \\ g & h \end{pmatrix}, \psi(R) \times \begin{pmatrix} e & f \\ g & h \end{pmatrix}$$

and easy induction on the length of words w and w' shows the above properties.

Lemma 1. *Given two words* $u, v \in X^*$, $u = v$ *iff* $\phi(u^{-1}) = (\psi(v))^{-1}$.

Proof. $\Rightarrow$ Let $u = v$ then $\phi(u^{-1}) \times \psi(v)$ corresponds to a product of matrices of even length of the form: $A_n \times \ldots \times A_1 \times B_1 \times \ldots \times B_n$ where $A_i \in \{M_L^{-1}, M_R^{-1}\}$, $B_j \in \{M_L, M_R\}$, $i, j \in 1..n$ and for each symmetric pair of matrices A_k, B_k we have $A_k = B_k^{-1}$. ¿From it follows that $A_n \times \ldots \times A_1 \times B_1 \times \ldots \times B_n = A_n \times \ldots \times A_2 \times E \times B_2 \times \ldots \times B_n = \ldots = A_n \times E \times B_n = E$. The same idea of proof works for $\psi(v) \times \phi(u^{-1})$.

$\Leftarrow$ Let $\phi(u^{-1}) = (\psi(v))^{-1}$ then $\psi(v)$ is the unique inverse matrix for $\phi(u^{-1})$ that should be equal to $\psi(u)$, since $\phi(u^{-1}) = (\psi(u))^{-1}$. By injectivity of mapping from $\{L, R\}^*$ to matrices we have that $u = v$.

3   Post Correspondence Problem and Its Matrix Interpretation

Post correspondence problem (in short, PCP) is formulated as follows: Given a finite alphabet X and a finite sequence of pairs of words in X^*: $(u_1, v_1), \ldots, (u_k, v_k)$. Is there a finite sequence of indexes $\{i_j\}$ with $\{i_j \in \{1..k\}\}$, such that

$$u_{i_1} \cdot \ldots \cdot u_{i_n} = v_{i_1} \cdot \ldots \cdot v_{i_n}?$$

PCP(n) denotes the same problem with a sequence of n pairs. Without loss of generality we assume that the alphabet X is binary.

Lemma 2. *Given a finite binary alphabet X and a finite sequence of pairs of words in X^*:*

$$(u_1, v_1), \ldots, (u_k, v_k)$$

and a finite sequence of indexes $\{i_j\}$ with $\{i_j \in \{1..k\}\}$. The word $u = u_{i_1} \cdot \ldots \cdot u_{i_n}$ is equal to the word $v = v_{i_1} \cdot \ldots \cdot v_{i_n}$ if and only if

$$\phi(u^{-1}) \times \psi(v) = E.$$

Proof. The proposition ffollows from Lemma 1 and the fact that matrices $\psi(w)$ and $\phi(w)$ have the inverse matrices (elements) for any word $w \in \{L, R\}^*$.

Theorem 1. *PCP(n) can be reduced to the vector reachability problem for a semigroup generated by n 4×4-matrices.*

Proof. Given a sequence of pairs of words in a binary alphabet $A^* = \{L, R\}$: $(u_1, v_1), \ldots, (u_n, v_n)$. Let us construct the sequence of pairs of 2×2 matrices using two mappings ϕ and ψ: $(\phi(u_1), \psi(v_1)), \ldots, (\phi(u_n), \psi(v_n))$.

Instead of equation $u = v$ we would like to consider a concatenation of two words $u^{-1} \cdot v$ that is a palindrome in case where $u = v$. Now we show a matrix interpretation of this concatenation. We associate 2×2 matrix C with a word w of the form $u^{-1} \cdot v$. Initially C is an identity matrix corresponding to an empty word. The extension of a word w by a new pair of words (u_r, v_r) (i.e. that gives us $w' = u_r^{-1} \cdot w \cdot v_r$) corresponds to the following matrix multiplication

$$C_{w'} = C_{u_r^{-1} \cdot w \cdot v_r} = \phi(u_r^{-1}) \times C_w \times \psi(v_r) \tag{1}$$

Let us rewrite the operation (1) in more details.

$$\begin{pmatrix} c_{w'}^{11} & c_{w'}^{12} \\ c_{w'}^{21} & c_{w'}^{22} \end{pmatrix} = \begin{pmatrix} u^{11} & u^{12} \\ u^{21} & u^{22} \end{pmatrix} \times \begin{pmatrix} c_w^{11} & c_w^{12} \\ c_w^{21} & c_w^{22} \end{pmatrix} \times \begin{pmatrix} v^{11} & v^{12} \\ v^{21} & v^{22} \end{pmatrix} \tag{1'}$$

According to the Lemma 2 $u = u_{i_1} \cdot \ldots \cdot u_{i_n} = v_{i_1} \cdot \ldots \cdot v_{i_n} = v$ for a finite sequence of indexes $\{i_j\}$ with $\{i_j \in \{1..k\}\}$ if and only if $\phi(u^{-1}) \times \psi(v)$ is equal to the identity matrix. So the question of the word equality can be reduced to the problem of finding a sequence of pairwise matrix multiplications that gives us the identity matrix.

Now we show that it is possible to avoid pairwise matrix multiplications by increasing the dimension from 2 to 4. Actually we represent matrices C_w and $C_{w'}$ from (1') as 4×1 vectors and we unite every pair of matrices $\phi(u_r^{-1})$ and $\psi(v_r)$ into 4×4 joint matrix $M_{u_r^{-1}, v_r}$ in the following way:

$$\begin{pmatrix} c_{w'}^{11} \\ c_{w'}^{12} \\ c_{w'}^{21} \\ c_{w'}^{22} \end{pmatrix} = \underbrace{\begin{pmatrix} u^{11} \cdot v^{11} & u^{11} \cdot v^{21} & u^{12} \cdot v^{11} & u^{12} \cdot v^{21} \\ u^{11} \cdot v^{12} & u^{11} \cdot v^{22} & u^{12} \cdot v^{12} & u^{12} \cdot v^{22} \\ u^{21} \cdot v^{11} & u^{21} \cdot v^{21} & u^{21} \cdot v^{11} & u^{22} \cdot v^{21} \\ u^{21} \cdot v^{12} & u^{21} \cdot v^{22} & u^{21} \cdot v^{12} & u^{22} \cdot v^{22} \end{pmatrix}}_{M_{u_r^{-1}, v_r}} \cdot \begin{pmatrix} c_w^{11} \\ c_w^{12} \\ c_w^{21} \\ c_w^{22} \end{pmatrix} \tag{2}$$

Note that the equation (1') is equivalent to the equation (2) in sense that the expression for computing values of $c_{w'}^{11}$, $c_{w'}^{12}$, $c_{w'}^{21}$ and $c_{w'}^{22}$ in (2) coincide with the corresponding values in (1').

Thus for every pair of words (u_r, v_r) we construct the matrix $M_{u_r^{-1}, v_r}$. Now PCP(n) can be reduced to the following vector reachability problem: Given a matrix semigroup S generated by set of matrices $\{M_{u_1^{-1}, v_1}, \ldots, M_{u_n^{-1}, v_n}\}$. Decide whether there is a matrix $M \in S$ such that

$$\begin{pmatrix} 1 \\ 0 \\ 1 \\ 0 \end{pmatrix} = M \cdot \begin{pmatrix} 1 \\ 0 \\ 1 \\ 0 \end{pmatrix}.$$

It follows that if we can solve the above problem then we can solve the PCP(n).

Matiyasevich and Senizergues proved in [16] that the PCP(7) is undecidable. Thus the following corollary of the Theorem 1 holds.

Corollary 1. *The vector reachability problem is undecidable for a semigroup generated by 7 matrices of dimension ≥ 4.*

4 Post's Tag-Systems

In this section we consider tag-systems, which is another interesting family of systems proposed by Post. We plan to show that problems about tag systems can be converted to the reachability problems in counter automata using the techniques and two mappings ψ and ϕ presented in Sections 2 and 3.

A tag system is set of rules that specifies a fixed number of elements to be removed from the beginning of a sequence and a set of elements to be appended ("tagged" onto the end) based on the elements that were removed from the beginning. Tag systems have a Turing machine-like halting problem for deciding based on an arbitrarily given initial sequence whether repeated application of the rules leads to a word of length smaller than the number of elements removed from the beginning [17]. Wang [19] also considered a sort of opposite to a tag system that he dubbed a lag system. Lag systems allow dependence on more than just the first element, but remove only the first element. The tag and lag systems can be represented as a communicating finite state machine [6] that interacts with one FIFO channel.

4.1 CFSM with FIFO Channel

Definition 1. *A Communication Finite State Machine (CFSM) is a finite transition system given by a 4-tuple $C = (Q, q_0, \Sigma, \delta)$ where : Q is a finite set of states, $q_0 \in Q$ is the initial state, Σ is a finite alphabet, and δ is a transition function such that $\delta \subseteq Q \times (\{+, -\} \times \Sigma) \times Q$.*

Note that C sees $(+, a)$ and $(-, a)$ as single symbols, which we henceforth write as $+a$ and $-a$ respectively. The label $-a$ denotes the emission of a in the channel, and $+a$ denotes the reception of a from the channel (Figure 3).

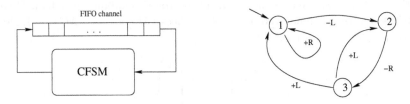

Fig. 3. The communicating finite state machine model

The FIFO channel F, that contains the word w, can be changed only by two modes: either to send another message a to the channel and then F will store the word $w \cdot a$, or if F contains a word $w' = a \cdot w$, to receive a message a, then F will store a word w.

Now for any CFSM with a finite alphabet $\{L, R\}$ we can construct an equivalent model of a matrix transition system T with the same set of states and transitions but with the different transition function and the set of labels. The configuration of T will be represented by pair (M, s), where M is an integral matrix $\begin{pmatrix} n & n' \\ m & m' \end{pmatrix}$ and s is a state such that $s \in Q$.

Note, that the operation of concatenation can be performed on any word. However the contraction is a conditional operation since we need to know the head or the tail character of a word in the FIFO channel. The Proposition 1 states that we can identify the head or the tail symbol of a word by checking linear inequalities on matrix elements.

The transition system T can be constructed by changing the labels of CFSM in compliance with the following rules:

- L : to substitute all labels $-L$ by label $\times M_L$

- R : to substitute all labels $-L$ by label $\times M_R$

+ L : to substitute all labels $+L$ by guard $(m < n)$ and label $M_L^{-1} \times$

+ R : to substitute all labels $+R$ by guard $(m \geq n)$ and label $M_R^{-1} \times$

Here the label $\times M_a$ denotes the right multiplication and corresponds to the emission of a symbol a in the channel, and $M_a^{-1} \times$ denotes the left multiplication and corresponds to the reception of a from the channel.

The dynamics of the matrix transition system can be defined as follows. Let the current configuration of T be a pair $[M, s]$. The transition system T can reach the configuration $[M \times M_L, s']$ from (M, s) by a transition $\times M_L$ or the configuration $[M \times M_R, s']$ by a transition $\times M_R$ form state s to state s'.

Also the transition system T could have transition with pre and post conditions. For example the configuration $[M_L^{-1} \times M, s']$ can be reached from (M, s) by a transition $n > m | M_L^{-1} \times$ from state s to state s' if the precondition $n > m$ is true.

It is easy to see that we can go from the transition system T to counters automaton by introducing the individual counters for each matrix element. Let us rewrite matrix multiplications as a set of equations with four counters c_1, c_2, c_3, c_4 to keep track of values n, n', m and m', respectively .

$$\times M_R \text{ corresponds to } \begin{pmatrix} c_1 & c_2 \\ c_3 & c_4 \end{pmatrix} \times \begin{pmatrix} 1 & 0 \\ 1 & 1 \end{pmatrix} = \begin{pmatrix} c_1 + c_2 & c_2 \\ c_3 + c_4 & c_4 \end{pmatrix}$$

$$\times M_L \text{ corresponds to } \begin{pmatrix} c_1 & c_2 \\ c_3 & c_4 \end{pmatrix} \times \begin{pmatrix} 1 & 1 \\ 0 & 1 \end{pmatrix} = \begin{pmatrix} c_1 & c_1 + c_2 \\ c_3 & c_3 + c_4 \end{pmatrix}$$

$$M_R^{-1} \times \text{ corresponds to } \begin{pmatrix} 1 & 0 \\ -1 & 1 \end{pmatrix} \times \begin{pmatrix} c_1 & c_2 \\ c_3 & c_4 \end{pmatrix} = \begin{pmatrix} c_1 & c_2 \\ c_3 - c_1 & c_4 - c_2 \end{pmatrix}$$

$$M_L^{-1} \times \text{ corresponds to } \begin{pmatrix} 1 & -1 \\ 0 & 1 \end{pmatrix} \times \begin{pmatrix} c_1 & c_2 \\ c_3 & c_4 \end{pmatrix} = \begin{pmatrix} c_1 - c_3 & c_2 - c_4 \\ c_3 & c_4 \end{pmatrix}$$

Now we have finished reduction of CFSM with one FIFO into a special form of counter automaton. Since the system of CFSM with FIFO can model a Turing Machine [19, 9] the following theorem holds:

Theorem 2. *Let A be a 4-counter automata where each transition is from the set of unconditional jumps:*

$$\begin{cases} c_1 := c_1 + c_2 \\ c_3 := c_3 + c_4 \end{cases}, \quad \begin{cases} c_2 := c_1 + c_2 \\ c_4 := c_3 + c_4 \end{cases}$$

or from the set of conditional jumps:

$$\text{if } (c_1 > c_3) \text{ then } \begin{cases} c_1 := c_1 - c_3 \\ c_2 := c_2 - c_4 \end{cases};$$

$$\text{if } (c_1 \le c_3) \text{ then } \begin{cases} c_3 := c_3 - c_1 \\ c_4 := c_4 - c_2 \end{cases}.$$

Then A can simulate a Turing Machine.

The most popular universal model of computations is a model of two register (counter) machine or so called Minsky machine. In that model we can independently increment or decrement each counter and check them for zero. There are many other models of 2 and 3 counter machines that are universal as well but that have different basic operations such as *reset* $(c_i := 0)$, *transfer* $(c_i := c_i + c_j)$, different tests like checking the equality or inequality between counters $(c_i = c_j$ or $c_i < c_j)$.

For example the reachability problem for a model of two counter automata with four operations such as $\{+1, -1, reset, transfer\}$ is decidable. The similar weak automaton (without testing for zero) with three counters have undecidable reachability and boundedness problems [4]. The model of two counter machine with only one test for zero and one reset (or transfer operation) has effectively computable semilinear reachability set [5]. So the idea balancing between the size of dimension and power of basic operations gives a number of classes on the border between decidability and undecidability.

The class of automata from Theorem 2 is another class of such system, where in dimension 4 we use *transfer* operation, *reverse transfer* operation ($c_i := c_i - c_j$) and comparison between only two counters. Moreover, as it will be shown in the next subsection, there is a resembling class of automata with decidable reachability properties.

4.2   CFSM with LIFO Channel

The CFSM with a LIFO channel or stack is a pushdown automata. The only difference from tag system is that we write or read to/from one side of the channel. In such case the LIFO channel can be represented as a product of matrices that could be extended only from one side. So we use the following labels $M_R\times$, $M_L\times$, $M_R^{-1}\times$ and $M_L^{-1}\times$ for the matrix transition system and other guards from Proposition 1 to check the first symbol in the stack.

The LIFO channel L, that contains the word w, can be changed only by two modes: either to send another message a to the channel and then L will store the word $a \cdot w$, or if L contains a word $w' = a \cdot w$, to receive a message a, then L will store a word w. The label $M_a\times$ denotes the right matrix multiplication and corresponds to the push a into the stack or LIFO channel and $M_a^{-1}\times$ corresponds to the pop an a off the top of the stack. We use the same method that converts a CFSM into the counter automaton. again. Note that the 4-counter automata from Theorem 3 is very similar to the automata from Theorem 2, but the status of reachability problem is opposite.

Theorem 3. *Let A is a 4-counter automata where each transition is from the set of unconditional jumps:*

$$\begin{cases} c_1 := c_1 + c_3 \\ c_2 := c_2 + c_4 \end{cases}, \quad \begin{cases} c_3 := c_1 + c_3 \\ c_4 := c_2 + c_4 \end{cases}$$

or from the set of conditional jumps:

$$\text{if } (c_1 > c_3) \text{ then } \begin{cases} c_1 := c_1 - c_3 \\ c_2 := c_2 - c_4 \end{cases};$$

$$\text{if } (c_1 \le c_3) \text{ then } \begin{cases} c_3 := c_3 - c_1 \\ c_4 := c_4 - c_2 \end{cases}.$$

The counters automata A is equivalent to a pushdown automata and therefore has decidable reachability problem.

It is possible to apply the same methods to different tag-systems [19, 9, 13] to derive a variety of multicounter automata with different structural constraints that have decidable or undecidable properties.

5   Conclusion

The main result of this paper is the reduction of PCP(n) to the vector reachability problem with 4×4 matrix semigroup with n generators. We leave the question whether the problem is decidable or undecidable for n=2 and n=3 as an open problem. We have also shown that this technique can be applied for other Post-systems and can be used for further investigation about the border between decidability and undecidability in the class of multicounter automata.

Author is very grateful to anonymous referees for their careful review and helpful suggestions on an earlier version of the paper and also to Alexei Lisitsa for many discussions and valuable comments about this work.

References

1. J. Bestel and J. Karhumaki. Combinatorics on Words - A Tutorial, Bulletin of the EATCS, February 2003, 178 - 228
2. V. Blondel and J. Tsitsiklis. A survey of computational complexity results in systems and control. Automatica, 36, 2000, 1249-1274
3. H. Comon and Y. Jurski. Multiple counters automata, safety analysis and Presburger arithmetic, Research Report LSV-98-1, Mar.,1998.
4. C. Dudourd, A. Finkel and Ph. Schnoebelen. Resets nets between decidability and undecidability. In Proc. 25th Int. Coll. Automata, Languages and Programming, LNCS 1443, 1998, 103-115
5. A. Finkel and G. Sutre. Decidability of reachability problems for classes of two counters automata. 17th Ann. Symp. Theoretical Aspects of Computer Science (STACS'2000), Lille, France, Feb. 2000, volume 1770 of Lecture Notes in Computer Science, 2000, 346-357
6. Alain Finkel, S. Purushothaman Iyer, G. Sutre. Well-abstracted transition systems: application to FIFO automata. Inf. Comput. 181(1), 2003, 1-31
7. S. Gaubert and R. Katz. Reachability problems for products of matrices in semirings. arXiv:math.OC/0310028, v1. 2003, 1-21
8. R. Graham, D. Knuth, O. Patashnik. Concrete Mathematics. 2nd edition, Addison-Wesley, 1994
9. M. Gouda, E. Manning, Y. Yu. On the progress of communication between two finite state machines. Information and control, 63, 1984, 200-216
10. Brian Hayes. On the Teeth of Wheels. American Scientist, Volume 88, Number 4, July-August, 2000, 296-300
11. O. H. Ibarra, J. Su, T. Bultan, Z. Dang, and R. A. Kemmerer. Counter Machines: Decidable Properties and Applications to Verification Problems. MFCS, LNCS 1893, 2000, 426-435
12. R. Kannan and R. J. Lipton. Polynomial-time algorithm for the orbit problem, Journal of the ACM (JACM), Volume 33 , Issue 4, 1986, 808 - 821
13. O. Kurganskyy, I. Potapov. On the bound of algorithmic resolvability of correctness problems of automaton interaction through communication channels. *Cybernetics and System Analysis*, 3, 1999, 49-57
14. A. Lisitsa, I. Potapov. Membership and reachability problems for row-monomial transformations. MFCS, 2004, 623-634
15. Maurice Mergenstern. Frontier between decidability and undecidability: a survey. Theoretical Computer Science, v.231, 2000, 217-251

16. Yu. V. Matiyasevich, G. Senizergues. Decision problems for semi-Thue systems with a few rules, Proc. LICS'96, 1996, 523-531
17. M. Minsky. Computation: Finite and Infinite Machines. Englewood Cliffs, N.J.: Prentice-Hall, 1967
18. C. Moore. Unpredictability and Undecidability in Dynamical Systems, *Physical Review Letters*, Vol.64, N. 20, 1990, 2354-2357
19. H. Wang. Tag Systems and Lag Systems. Math. Ann. 152, 1963, 65-74

Words Avoiding $\frac{7}{3}$-Powers and the Thue–Morse Morphism

Narad Rampersad

School of Computer Science, University of Waterloo, Waterloo, ON, Canada
nrampersad@math.uwaterloo.ca

Abstract. In 1982, Séébold showed that the only overlap-free binary words that are the fixed points of non-identity morphisms are the Thue–Morse word and its complement. We strengthen Séébold's result by showing that the same result holds if the term 'overlap-free' is replaced with '$\frac{7}{3}$-power-free'. Furthermore, the number $\frac{7}{3}$ is best possible.

1 Introduction

In 1912, Thue [15] gave a construction of an infinite overlap-free word over a binary alphabet. Since then the properties of infinite overlap-free words have been studied extensively (see, for example, the survey by Séébold [12]). Thue [15] also gave a complete characterization of the bi-infinite overlap-free binary words (see also [6]), and Fife [5] gave a characterization of the (one-sided) infinite overlap-free binary words. Séébold [11, 13] showed that the Thue–Morse word and its complement are the only infinite overlap-free binary words that can be obtained by iteration of a morphism. Another proof of this fact was later given by Berstel and Séébold [3]. We show that this result can be strengthened somewhat.

In this paper we are particularly concerned with $\frac{7}{3}$-powers. Several results previously known for overlap-free binary words have recently been shown to be true for $\frac{7}{3}$-power-free binary words as well. For example, Restivo and Salemi's factorization theorem for overlap-free binary words [10] was recently shown to be true for $\frac{7}{3}$-power-free binary words by Karhumäki and Shallit [7]. Similarly, Shur [14] gave a characterization of the bi-infinite $\frac{7}{3}$-power-free binary words that is analogous to that given by Thue for the bi-infinite overlap-free words. In fact, Shur showed that these two sets of words are equal. Furthermore, Shur showed that the number $\frac{7}{3}$ is best possible; *i.e.*, the result no longer holds if the number $\frac{7}{3}$ is replaced by a larger number.

The number $\frac{7}{3}$ has been shown to be a threshold for other properties as well. Karhumäki and Shallit [7] also showed that the threshold between polynomial growth and exponential growth for binary words is $\frac{7}{3}$; *i.e.*, for $2 < \alpha \le \frac{7}{3}$, there are polynomially many binary words of length n that avoid α-powers, but for $\alpha > \frac{7}{3}$, there are exponentially many binary words of length n that avoid α-powers.

Kolpakov, Kucherov, and Tarannikov [8] showed that $\frac{7}{3}$ is also a threshold for the minimal letter density in binary words; *i.e.*, for $2 < \alpha \le \frac{7}{3}$, the minimal

C.S. Calude, E. Calude and M.J. Dinneen (Eds.): DLT 2004, LNCS 3340, pp. 357–367, 2004.
© Springer-Verlag Berlin Heidelberg 2004

letter density in binary words avoiding α-powers is $\frac{1}{2}$, but for $\alpha > \frac{7}{3}$, the minimal letter density in binary words avoiding α-powers is less than $\frac{1}{2}$.

The goal of this paper is to generalize Séébold's result by showing that the Thue–Morse word and its complement are the only infinite $\frac{7}{3}$-power-free binary words that can be obtained by iteration of a morphism. At first glance, it may seem that this is an immediate consequence of Shur's result; however, this is not necessarily so, as there are infinite $\frac{7}{3}$-power-free binary words that cannot be extended to the left to form bi-infinite $\frac{7}{3}$-power-free binary words. For example, if we denote the complement of the Thue–Morse word by $\mu^\omega(1)$ (see Section 2), the infinite binary word $001001\mu^\omega(1)$ has been shown by Allouche, Currie, and Shallit [1] to be the lexicographically least infinite overlap-free binary word; however, it cannot be extended to the left to form a $\frac{7}{3}$-power-free word: prepending a 0 creates the cube 000, and prepending a 1 creates the $\frac{7}{3}$-power 1001001.

In general, results regarding infinite words are often more difficult to obtain than the analogous results for bi-infinite words; for example, note that a characterization of the bi-infinite overlap-free words was known to Thue [15] in 1912, whereas a characterization of the infinite overlap-free words was only given much later by Fife [5] in 1980.

2   Definitions and Notation

Let Σ be a finite, non-empty set called an *alphabet*. We denote the set of all finite words over the alphabet Σ by Σ^*. We also write Σ^+ to denote the set $\Sigma^* - \{\epsilon\}$, where ϵ is the empty word. Let Σ_k denote the alphabet $\{0, 1, \ldots, k-1\}$. Throughout this paper we will work exclusively with the binary alphabet Σ_2.

Let $\mathbb{N}$ denote the set $\{0, 1, 2, \ldots\}$. An *infinite word* is a map from $\mathbb{N}$ to Σ, and a *bi-infinite* word is a map from $\mathbb{Z}$ to Σ. The set of all infinite words over the alphabet Σ is denoted Σ^ω. We also write Σ^∞ to denote the set $\Sigma^* \cup \Sigma^\omega$.

A map $h : \Sigma^* \to \Delta^*$ is called a *morphism* if h satisfies $h(xy) = h(x)h(y)$ for all $x, y \in \Sigma^*$. A morphism may be defined simply by specifying its action on Σ. A morphism $h : \Sigma^* \to \Sigma^*$ such that $h(a) = ax$ for some $a \in \Sigma$ is said to be *prolongable on a*; we may then repeatedly iterate h to obtain the *fixed point* $h^\omega(a) = axh(x)h^2(x)h^3(x)\cdots$.

An *overlap* is a word of the form $axaxa$, where $a \in \Sigma$ and $x \in \Sigma^*$. A word w' is called a subword of $w \in \Sigma^\infty$ if there exist $u \in \Sigma^*$ and $v \in \Sigma^\infty$ such that $w = uw'v$. We say a word w is *overlap-free* (or *avoids overlaps*) if no subword of w is an overlap.

Let μ be the *Thue–Morse morphism*; i.e., the morphism defined by $\mu(0) = 01$ and $\mu(1) = 10$. It is well-known [9, 15] that the *Thue–Morse word*, $\mu^\omega(0)$, is overlap-free. The complement of the Thue–Morse word, given by $\mu^\omega(1)$, is also overlap-free.

We also need the notion of a *fractional power*, which was first introduced by Dejean [4]. Let α be a rational number such that $\alpha \geq 1$. An α-*power* is a word of the form $x^n x'$, where $x, x' \in \Sigma^*$, and x' is a prefix of x with $n + |x'|/|x| = \alpha$. We say a word w is α-*power-free* (or *avoids α-powers*) if no subword of w is an

β-power for any rational $\beta \geq \alpha$; otherwise, we say w *contains an* α-*power*. Note that a word is overlap-free if and only if it is $(2 + \epsilon)$-power-free for all $\epsilon > 0$; for example, an overlap-free word is necessarily $\frac{7}{3}$-power-free.

3 Preliminary Lemmata

We will need the following result due to Shur [14].

Theorem 1 (Shur). *Let* $w \in \Sigma_2^*$, *and let* $\alpha > 2$ *be a real number. Then* w *is* α-*power-free iff* $\mu(w)$ *is* α-*power-free.*

We will also make frequent use of the following result due to Karhumäki and Shallit [7]. This theorem is a generalization of a similar factorization theorem for overlap-free words due to Restivo and Salemi [10].

Theorem 2 (Karhumäki and Shallit). *Let* $x \in \Sigma_2^*$ *be a word avoiding* α-*powers, with* $2 < \alpha \leq \frac{7}{3}$. *Then there exist* u, v, y *with* $u, v \in \{\epsilon, 0, 1, 00, 11\}$ *and a word* $y \in \Sigma_2^*$ *avoiding* α-*powers, such that* $x = u\mu(y)v$.

Next, we will establish a few lemmata. Lemma 1 is analogous to a similar lemma for overlap-free words given in Allouche and Shallit [2, Lemma 1.7.6]. (This result for overlap-free words was also stated without formal proof by Berstel and Séébold [3].)

Lemma 1. *Let* $w \in \Sigma_2^*$ *be a* $\frac{7}{3}$-*power-free word with* $|w| \geq 52$. *Then* w *contains* $\mu^3(0) = 01101001$ *and* $\mu^3(1) = 10010110$ *as subwords.*

Proof. Since w is $\frac{7}{3}$-power-free, by Theorem 2 we can write

$$w = u\mu(y)v , \tag{1}$$

where y is $\frac{7}{3}$-power-free and $|y| \geq 24$. Similarly, we can write

$$y = u'\mu(y')v' , \tag{2}$$

where y' is $\frac{7}{3}$-power-free and $|y'| \geq 10$. Again, we can write

$$y' = u''\mu(y'')v'' , \tag{3}$$

where y'' is $\frac{7}{3}$-power-free and $|y''| \geq 3$. From (1)–(3), we get

$$w = u\mu(u'\mu(u''\mu(y'')v'')v')v$$
$$= u\mu(u')\mu^2(u'')\mu^3(y'')\mu^2(v'')\mu(v')v ,$$

where $u, u', u'', v, v', v'' \in \{\epsilon, 0, 1, 00, 11\}$. Since y'' is $\frac{7}{3}$-power-free and $|y''| \geq 3$, y'' contains both 0 and 1, and so $\mu^3(y'')$, and consequently w, contains both $\mu^3(0) = 01101001$ and $\mu^3(1) = 10010110$ as subwords as required. $\qquad\square$

Lemma 2. *Let w' be a subword of $w \in \Sigma_2^*$, where w' is either of the form $abb\mu(w'')$ or $\mu(w'')bba$ for some $a, b \in \Sigma_2$ and $w'' \in \Sigma_2^*$. Suppose also that $a \neq b$ and $|w''| \geq 2$. Then w contains a $\frac{7}{3}$-power.*

Proof. Suppose $ab = 10$ and $w' = 100\mu(w'')$ (the other cases follow similarly). The word $\mu(w'')$ may not begin with a 0 as that would create the cube 000. Hence we have $w' = 10010\mu(w''')$ for some $w''' \in \Sigma_2^*$. If $\mu(w''')$ begins with 01, then w' contains the $\frac{7}{3}$-power 1001001. If $\mu(w''')$ begins with 10, then w' contains the $\frac{5}{2}$-power 01010. Hence, w contains a $\frac{7}{3}$-power. $\qquad\square$

Lemma 3. *For $i, j \in \mathbb{N}$, let w be a $\frac{7}{3}$-power-free word over Σ_2 such that $|w| = (7 + 2j)2^i - 1$. Let a be an element of Σ_2. Then waw contains a $\frac{7}{3}$-power x, where $|x| \leq 7 \cdot 2^i$.*

Proof. Suppose $a = 1$ (the case $a = 0$ follows similarly). The proof is by induction on i. For the base case we have $i = 0$. Hence, $|w| \geq 6$ and $|w|$ is even. If w either begins or ends with 11, then $w1w$ contains the cube 111, and the result follows. Suppose then that w neither begins nor ends with 11. By explicitly examining all 13 words of length six that avoid $\frac{7}{3}$-powers and neither begin nor end with 11, we see that all such words of length at least six can be written in the form $pbbq$, where $p, q \in \Sigma_2^+$ and $b \in \Sigma_2$. Hence, $w1w$ must have at least one subword with prefix bb and suffix bb. Moreover, since $|w|$ is even, there must exist such a subword where the prefix bb and the suffix bb each begin at positions of different parity in $w1w$. Let x be a smallest such subword such that $w1w$ neither begins nor ends with x. Suppose $b = 0$ (the case $b = 1$ follows similarly). Then either $x = 00100$ or x contains a subword 01010 or 10101. Hence, $w1w$ contains one of the subwords 01010, 10101, or 1001001 as required.

Let us now assume that the lemma holds for all i', where $0 < i' < i$. Since w avoids $\frac{7}{3}$-powers, and since $|w| \geq 7$, by Theorem 2 we can write $w = u\mu(w')v$, where $u, v \in \{\epsilon, 0, 1, 00, 11\}$ and $w' \in \Sigma_2^*$ is $\frac{7}{3}$-power-free. By applying a case analysis similar to that used in Cases (1)–(4) of the proof of Theorem 3, we can eliminate all but three cases: $(u, v) \in \{(\epsilon, \epsilon), (\epsilon, 0), (0, \epsilon)\}$.

Case 1: $(u, v) = (\epsilon, \epsilon)$. In this case $w = \mu(w')$. This is clearly not possible, since for $i > 0$, $|w| = (7 + 2j)2^i - 1$ is odd.

Case 2: $(u, v) = (\epsilon, 0)$. Then $w = \mu(w')0$ and $w1w = \mu(w')01\mu(w')0 = \mu(w'0w')0$. If $|w| = (7 + 2j)2^i - 1$, we see that $|w'| = (7 + 2j)2^{i-1} - 1$. Hence, if $i' = i - 1$, we may apply the inductive assumption to $w'0w'$. We thus obtain that $w'0w'$ contains a $\frac{7}{3}$-power x', where $|x'| \leq 7 \cdot 2^{i-1}$, and so $w1w$ must contain a $\frac{7}{3}$-power $x = \mu(x')$, where $|x| \leq 7 \cdot 2^i$.

Case 3: $(u, v) = (0, \epsilon)$. This case is handled similarly to the previous case, and we omit the details.

By induction then, we have that waw contains a $\frac{7}{3}$-power x, where $|x| \leq 7 \cdot 2^i$. $\qquad\square$

Lemma 4. *For $i \in \mathbb{N}$, let w be a $\frac{7}{3}$-power-free word over Σ_2 such that $|w| = 5 \cdot 2^i - 1$. Let a be an element of Σ_2. Then waw contains a $\frac{7}{3}$-power x, where $|x| \le 5 \cdot 2^i$.*

Proof. The proof is analogous to that of Lemma 3 and we omit the details.   □

Lemma 5. *For $i, j \in \mathbb{Z}^+$, let w and s be $\frac{7}{3}$-power-free words over Σ_2 such that $|w| = 2^{i+1} - 1$ or $|w| = 3 \cdot 2^i - 1$, and $|s| = 2^{j+1} - 1$ or $|s| = 3 \cdot 2^j - 1$. Assume also that $|s| \ge |w|$. Let a be an element of Σ_2. Then $sawawas$ contains a $\frac{7}{3}$-power.*

Proof. Suppose $a = 1$ (the case $a = 0$ follows similarly). The proof is by induction on i. For the base case we have $i = 1$ and either $|w| = 3$ or $|w| = 5$. An easy computation suffices to verify that for all w with $|w| = 3$ or $|w| = 5$, and all $a, b \in \Sigma_2^2$, $a1w1w1b$ contains a $\frac{7}{3}$-power.

Let us now assume that the lemma holds for all i', where $1 < i' < i$. Since w avoids $\frac{7}{3}$-powers, and since $|w| \ge 7$, by Theorem 2 we can write $w = u\mu(w')v$, where $u, v \in \{\epsilon, 0, 1, 00, 11\}$ and $w' \in \Sigma_2^*$ is $\frac{7}{3}$-power-free. Similarly, we can write $s = u'\mu(s')v'$, where $u', v' \in \{\epsilon, 0, 1, 00, 11\}$ and $s' \in \Sigma_2^*$ is $\frac{7}{3}$-power-free. By applying a case analysis similar to that used in Cases (1)–(4) of the proof of Theorem 3 below, we can eliminate all but three cases: $(u, v, u', v') \in \{(\epsilon, \epsilon, \epsilon, \epsilon), (\epsilon, 0, \epsilon, 0), (0, \epsilon, 0, \epsilon)\}$.

Case 1: $(u, v, u', v') = (\epsilon, \epsilon, \epsilon, \epsilon)$. In this case $w = \mu(w')$. This is clearly not possible, since for $i > 1$, both $|w| = 2^{i+1} - 1$ and $|w| = 3 \cdot 2^i - 1$ are odd.

Case 2: $(u, v, u', v') = (\epsilon, 0, \epsilon, 0)$. Then $w = \mu(w')0$, $s = \mu(s')0$, and

$$s1w1w1s = \mu(s')01\mu(w')01\mu(w')01\mu(s')0 = \mu(s'0w'0w'0s')0 \ .$$

If $|w| = 2^{i+1} - 1$ or $|w| = 3 \cdot 2^i - 1$, we see that $|w'| = 2^i - 1$ or $|w| = 3 \cdot 2^{i-1} - 1$. Similarly, if $|s| = 2^{j+1} - 1$ or $|s| = 3 \cdot 2^j - 1$, we see that $|s'| = 2^j - 1$ or $|s| = 3 \cdot 2^{j-1} - 1$. Hence, if $i' = i - 1$, we may apply the inductive assumption to $s'0w'0w'0s'$. We thus obtain that $s'0w'0w'0s'$ contains a $\frac{7}{3}$-power x', and so $s1w1w1s$ must contain a $\frac{7}{3}$-power $x = \mu(x')$.

Case 3: $(u, v, u', v') = (0, \epsilon, 0, \epsilon)$. This case is handled similarly to the previous case, and we omit the details.

By induction then, we have that $sawawas$ contains a $\frac{7}{3}$-power.   □

Lemma 6. *Let n be a positive integer. Then n can be written in the form $2^i - 1$, $3 \cdot 2^i - 1$, $5 \cdot 2^i - 1$, or $(7 + 2j)2^i - 1$ for some $i, j \in \mathbb{N}$.*

Proof. If $n = 1$ then $n = 2^1 - 1$ as required. Suppose then that $n > 1$. Then we may write $n - 1 = m2^i$, where m is odd and $i \in \mathbb{N}$. But for any odd positive integer m, either $m \in \{1, 3, 5\}$, or m is of the form $7 + 2j$ for some $j \in \mathbb{N}$, and the result follows.   □

4   Main Theorem

Let $h : \Sigma^* \to \Sigma^*$ be a morphism. We say that h is *non-erasing* if, for all $a \in \Sigma$, $h(a) \neq \epsilon$. Let E be the morphism defined by $E(0) = 1$ and $E(1) = 0$. The following theorem is analogous to a result regarding overlap-free words due to Berstel and Séébold [3].

Theorem 3. *Let $h : \Sigma_2^* \to \Sigma_2^*$ be a non-erasing morphism. If $h(01101001)$ is $\frac{7}{3}$-power-free, then there exists an integer $k \geq 0$ such that either $h = \mu^k$ or $h = E \circ \mu^k$.*

Proof. Let $h(0) = x$ and $h(1) = x'$ with $|x|, |x'| \geq 1$. The proof is by induction on $|x| + |x'|$. If $|x| < 7$ and $|x'| < 7$, then a quick computation suffices to verify that if $h(01101001)$ is $\frac{7}{3}$-power-free, then either $h = \mu^k$ or $h = E \circ \mu^k$, where $k \in \{0, 1, 2\}$. Let us assume then, without loss of generality, that $|x| \geq |x'|$ and $|x| \geq 7$. The word x must avoid $\frac{7}{3}$-powers, and so, by Theorem 2, we can write $x = u\mu(y)v$, where $u, v \in \{\epsilon, 0, 1, 00, 11\}$ and $y \in \Sigma_2^*$. We will consider all 25 choices for (u, v).

Case 1: $(u, v) \in \{(0, 00), (00, 0), (00, 00), (1, 11), (11, 1), (11, 11)\}$. Suppose (u, v) $= (0, 00)$. Then $h(00) = 0\mu(y)000\mu(y)00$ contains the cube 000, contrary to the assumptions of the theorem. The argument for the other choices for (u, v) follows similarly.

Case 2: $(u, v) \in \{(0, 11), (00, 1), (00, 11), (1, 00), (11, 0), (11, 00)\}$. For any of these choices for (u, v), $h(00) = u\mu(y)vu\mu(y)v$ contains a subword of the form $abb\mu(y)$ or $\mu(y)bba$ for some $a, b \in \Sigma_2$, where $a \neq b$. Since $|x| \geq 7$, $|y| \geq 2$, and so by Lemma 2 we have that $h(00)$ contains a $\frac{7}{3}$-power, contrary to the assumptions of the theorem.

Case 3: $(u, v) \in \{(\epsilon, 0), (0, \epsilon), (\epsilon, 1), (1, \epsilon)\}$. Suppose $(u, v) = (0, \epsilon)$. Then $h(00) = 0\mu(y)0\mu(y)$. We have two subcases.

Case 3a: $\mu(y)$ begins with 01 or ends with 10. Then by Lemma 2, $h(00)$ contains a $\frac{7}{3}$-power, contrary to the assumptions of the theorem.

Case 3b: $\mu(y)$ begins with 10 and ends with 01. Then $h(00) = 0\mu(y')$ $01010\mu(y'')$ contains the $\frac{5}{2}$-power 01010, contrary to the assumptions of the theorem.

The argument for the other choices for (u, v) follows similarly.

Case 4: $(u, v) \in \{(\epsilon, 00), (0, 0), (00, \epsilon), (\epsilon, 11), (1, 1), (11, \epsilon)\}$. Suppose $(u, v) = (00, \epsilon)$. Then $h(00) = 00\mu(y)00\mu(y)$. The word $\mu(y)$ may not begin with a 0 as that would create the cube 000. We have then that $h(00) = 00\mu(y)0010\mu(y')$ for some $y' \in \Sigma_2^*$. By Lemma 2, $h(00)$ contains a $\frac{7}{3}$-power, contrary to the assumptions of the theorem. The argument for the other choices for (u, v) follows similarly.

Case 5: $(u, v) \in \{(0, 1), (1, 0)\}$. Suppose $(u, v) = (0, 1)$. By Lemma 6, the following three subcases suffice to cover all possibilities for $|y|$.

Case 5a: $|y| = (7 + 2j)2^i - 1$ for some $i, j \in \mathbb{N}$. We have $h(00) = 0\mu(y)10\mu(y)1$ $= 0\mu(y1y)1$. By Lemma 3, $y1y$ contains a $\frac{7}{3}$-power. The word $h(00)$ must then contain a $\frac{7}{3}$-power, contrary to the assumptions of the theorem.

Case 5b: $|y| = 5 \cdot 2^i - 1$ for some $i \in \mathbb{N}$. Again we have $h(00) = 0\mu(y)10\mu(y)1 = 0\mu(y1y)1$. By Lemma 4, $y1y$ contains a $\frac{7}{3}$-power. The word $h(00)$ must then contain a $\frac{7}{3}$-power, contrary to the assumptions of the theorem.

Case 5c: $|y| = 2^i - 1$ or $|y| = 3 \cdot 2^i - 1$ for some $i \in \mathbb{N}$. We have two subcases.

Case 5c.i: $|x'| < 7$. We have $h(0110) = 0\mu(y)1x'x'0\mu(y)1$. The only $x' \in \Sigma_2^*$ where $|x'| < 7$ and $1x'x'0$ does not contain a $\frac{7}{3}$-power is

$$x' \in \{10, 0110, 1001, 011010, 100110, 101001\} \ .$$

However, each of these words either begins or ends with 10, and so we have that $h(0110)$ contains a subword of the form $100\mu(y)$ or $\mu(y)110$. Hence, by Lemma 2 we have that $h(0110)$ contains a $\frac{7}{3}$-power, contrary to the assumptions of the theorem.

Case 5c.ii: $|x'| \geq 7$. By Theorem 2, we can write $x' = u'\mu(z)v'$, where $u', v' \in \{\epsilon, 0, 1, 00, 11\}$ and $z \in \Sigma_2^*$ is $\frac{7}{3}$-power-free. Applying the preceding case analysis to x' allows us to eliminate all but three subcases.

Case 5c.ii.A: $(u', v') = (0, 1)$. We have

$$h(0110) = 0\mu(y)10\mu(z)10\mu(z)10\mu(y)1 = 0\mu(y1z1z1y)1 \ .$$

Moreover, by applying to x' the same reasoning used in Case 5a and Case 5b, we have $|z| = 2^j - 1$ or $|z| = 3 \cdot 2^j - 1$ for some $j \in \mathbb{N}$, and so by Lemma 5, $y1z1z1y$ contains a $\frac{7}{3}$-power. The word $h(0110)$ must then contain a $\frac{7}{3}$-power, contrary to the assumptions of the theorem.

Case 5c.ii.B: $(u', v') = (1, 0)$. Then $h(01) = 0\mu(y)11\mu(z)0$. The word $\mu(z)$ may not begin with a 1 as that would create the cube 111. We have then that $h(01) = 0\mu(y)1101\mu(z')0$ for some $z' \in \Sigma_2^*$. By Lemma 2, $h(01)$ contains a $\frac{7}{3}$-power, contrary to the assumptions of the theorem.

Case 5c.ii.C: $(u', v') = (\epsilon, \epsilon)$. Then $h(01) = 0\mu(y)1\mu(z)$. We have two subcases.

- $\mu(z)$ begins with 01. Then $h(01) = 0\mu(y)101\mu(z')$ for some $z' \in \Sigma_2^*$. The word $\mu(y)$ may not end in 10 as that would create the $\frac{5}{2}$-power 10101. Hence $h(01) = 0\mu(y')01101\mu(z')$ for some $y' \in \Sigma_2^*$. If $\mu(z')$ begins with 10, then $h(01)$ contains the $\frac{7}{3}$-power 0110110. If $\mu(z')$ begins with 01, then $h(01)$ contains the $\frac{5}{2}$-power 10101. Either situation contradicts the assumptions of the theorem.

- $\mu(z)$ begins with 10. Then $h(01) = 0\mu(y)110\mu(z')$ for some $z' \in \Sigma_2^*$. By Lemma 2, $h(01)$ contains a $\frac{7}{3}$-power, contrary to the assumptions of the theorem.

The argument for the other choice for (u, v) follows similarly.

Case 6: $(u, v) = (\epsilon, \epsilon)$. In this case we have $x = \mu(y)$.

All cases except $x = \mu(y)$ lead to a contradiction. The same reasoning applied to x' gives $x' = \mu(y')$ for some $y' \in \Sigma_2^*$. Let the morphism h' be defined by $h'(0) = y$ and $h'(1) = y'$. Then $h = \mu \circ h'$, and by Theorem 1, $h'(01101001)$ is $\frac{7}{3}$-power-free. Moreover, $|y| < |x|$ and $|y'| < |x'|$. Also note that for the preceding case analysis it sufficed to consider the following words only: $h(00)$, $h(01)$, $h(10)$, $h(11)$, $h(0110)$, $h(1001)$, and $h(01101001)$. However, 00, 01, 10, 11, 0110, and 1001 are all subwords of 01101001. Hence, the induction hypothesis can be applied, and we have that either $h' = \mu^k$ or $h' = E \circ \mu^k$. Since $E \circ \mu = \mu \circ E$, the result follows.   □

We now establish the following corollary.

Corollary 1. *Let* $h : \Sigma_2^* \to \Sigma_2^*$ *be a morphism such that* $h(01) \neq \epsilon$. *Then the following statements are equivalent.*

(a) *The morphism* h *is non-erasing, and* $h(01101001)$ *is* $\frac{7}{3}$-*power-free.*
(b) *There exists* $k \geq 0$ *such that* $h = \mu^k$ *or* $h = E \circ \mu^k$.
(c) *The morphism* h *maps any infinite* $\frac{7}{3}$-*power-free word to an infinite* $\frac{7}{3}$-*power-free word.*
(d) *There exists an infinite* $\frac{7}{3}$-*power-free word whose image under* h *is* $\frac{7}{3}$-*power-free.*

Proof.

(a) $\Longrightarrow$ (b) was proved in Theorem 3.
(b) $\Longrightarrow$ (c) follows from Theorem 1 via König's Infinity Lemma.
(c) $\Longrightarrow$ (d) : We need only exhibit an infinite $\frac{7}{3}$-power-free word: the Thue–Morse word, $\mu^\omega(0)$, is overlap-free and so is $\frac{7}{3}$-power-free.
(d) $\Longrightarrow$ (a) : Let **w** be an infinite $\frac{7}{3}$-power-free word whose image under h is $\frac{7}{3}$-power-free. By Lemma 1, **w** must contain 01101001, and so $h(01101001)$ is $\frac{7}{3}$-power-free.

To see that h is non-erasing, note that if $h(0) = \epsilon$, then since $h(01) \neq \epsilon$, $h(1) \neq \epsilon$. But then $h(01101001) = h(1)^4$ is not $\frac{7}{3}$-power-free, contrary to what we have just shown. Similarly, $h(1) \neq \epsilon$, and so h is non-erasing.   □

Let $h : \Sigma_2^* \to \Sigma_2^*$ be a morphism. We say that h is the *identity morphism* if $h(0) = 0$ and $h(1) = 1$. The following corollary gives the main result.

Corollary 2. *An infinite* $\frac{7}{3}$-*power-free binary word is a fixed point of a non-identity morphism if and only if it is equal to the Thue–Morse word,* $\mu^\omega(0)$, *or its complement,* $\mu^\omega(1)$.

Proof. Let $h : \Sigma_2^* \to \Sigma_2^*$ be a non-identity morphism, and let us assume that h has a fixed point that avoids $\frac{7}{3}$-powers. Then h maps an infinite $\frac{7}{3}$-power-free word to an infinite $\frac{7}{3}$-power-free word, and so, by Corollary 1, h is of the form μ^k or $E \circ \mu^k$ for some $k \geq 0$. Since h has a fixed point, it is not of the form $E \circ \mu^k$, and since h is not the identity morphism, $h = \mu^k$ for some $k \geq 1$. But the only fixed points of μ^k are $\mu^\omega(0)$ and $\mu^\omega(1)$, and the result follows.   □

5   The Constant $\frac{7}{3}$ Is Best Possible

It remains to show that the constant $\frac{7}{3}$ given in Corollary 2 is best possible; *i.e.*, Corollary 2 would fail to be true if $\frac{7}{3}$ were replaced by any larger rational number. To show this, it suffices to exhibit an infinite binary word $\mathbf{w}$ that avoids $(\frac{7}{3} + \epsilon)$-powers for all $\epsilon > 0$, such that $\mathbf{w}$ is the fixed point of a morphism $h : \Sigma_2^* \to \Sigma_2^*$, where h is not of the form μ^k for any $k \geq 0$. Kolpakov *et al.* [8] have already given an example of such a word. Their example was the fixed point of a 21-uniform morphism; we will give a similar solution using a 19-uniform morphism.

For rational α, we say that a word w *avoids* α^+*-powers* if w avoids $(\alpha + \epsilon)$-powers for all $\epsilon > 0$.

Let $h : \Sigma_2^* \to \Sigma_2^*$ be the morphism defined by

$$h(0) = 0110100110110010110$$
$$h(1) = 1001011001001101001 \ .$$

Since $|h(0)| = |h(1)| = 19$, h is not of the form μ^k for any $k \geq 0$. We will show that the fixed point $h^\omega(0)$ avoids $\frac{7}{3}^+$-powers by using a technique similar to that given by Karhumäki and Shallit [7]. We first state the following lemma, which may be easily verified computationally.

Lemma 7. *(a) Suppose $h(ab) = th(c)u$ for some letters $a, b, c \in \Sigma_2$ and words $t, u \in \Sigma_2^*$. Then this inclusion is trivial (that is, $t = \epsilon$ or $u = \epsilon$).*
(b) Suppose there exist letters $a, b, c \in \Sigma_2$ and words $s, t, u, v \in \Sigma_2^*$ such that $h(a) = st$, $h(b) = uv$, and $h(c) = sv$. Then either $a = c$ or $b = c$.

Theorem 4. *The fixed point $h^\omega(0)$ avoids $\frac{7}{3}^+$-powers.*

Proof. The proof is by contradiction. Let $w \in \Sigma_2^*$ avoid $\frac{7}{3}^+$-powers, and suppose that $h(w)$ contains a $\frac{7}{3}^+$-power. Then we may write $h(w) = xyyy'z$ for some $x, z \in \Sigma_2^*$ and $y, y' \in \Sigma_2^+$, where y' is a prefix of y, and $|y'|/|y| > \frac{1}{3}$. Let us assume further that w is a shortest such string, so that $0 \leq |x|, |z| < 19$. We will consider two cases.

Case 1 : $|y| \leq 38$. In this case we have $|w| \leq 6$. Checking all 20 words $w \in \Sigma_2^6$ that avoid $\frac{7}{3}^+$-powers, we see that, contrary to our assumption, $h(w)$ avoids $\frac{7}{3}^+$-powers in every case.

Case 2 : $|y| > 38$. Noting that if $h(w)$ contains a $\frac{7}{3}^+$-power, it must contain a square, we may apply a standard argument (see [7] for an example) to show that Lemma 7 implies that $h(w)$ can be written in the following form:

$$h(w) = A_1 A_2 \ldots A_j A_{j+1} A_{j+2} \ldots A_{2j} A_{2j+1} A_{2j+2} \ldots A_{n-1} A_n' A_n'' \ ,$$

for some j, where

$$A_i = h(a_i) \quad \text{for} \quad i = 1, 2, \ldots, n \quad \text{and} \quad a_i \in \Sigma_2$$
$$A_n = A'_n A''_n$$
$$y = A_1 A_2 \ldots A_j$$
$$\quad = A_{j+1} A_{j+2} \ldots A_{2j}$$
$$y' = A_{2j+1} A_{2j+2} \ldots A_{n-1} A'_n$$
$$z = A''_n .$$

Since y' is a prefix of y, and since $|y'|/|y| > \frac{1}{3}$, A'_n must be a prefix of A_k, where $k = \lfloor \frac{j}{3} \rfloor + 1$. However, noting that for any $a \in \Sigma_2$, any prefix of $h(a)$ suffices to uniquely determine a, we may conclude that $A_k = A_n$. Hence, we may write

$$h(w) = A_1 A_2 \ldots A_{k-1} A_k \ldots A_j A_{j+1} A_{j+2} \ldots A_{j+k-1} A_{j+k} \ldots A_{2j}$$
$$A_{2j+1} A_{2j+2} \ldots A_{n-1} A_n ,$$

where

$$y = A_1 A_2 \ldots A_{k-1} A_k \ldots A_j$$
$$\quad = A_{j+1} A_{j+2} \ldots A_{j+k-1} A_{j+k} \ldots A_{2j}$$
$$y' z = A_{2j+1} A_{2j+2} \ldots A_{n-1} A_n$$
$$\quad = A_1 A_2 \ldots A_{k-1} A_k .$$

We thus have

$$w = (a_1 a_2 \ldots a_j)^2 a_1 a_2 \ldots a_k ,$$

where $k = \lfloor \frac{j}{3} \rfloor + 1$. Hence, w is a $\frac{7}{3}^+$-power, contrary to our assumption. The result now follows.   □

Theorem 4 thus implies that the constant $\frac{7}{3}$ given in Corollary 2 is best possible.

Acknowledgements

The author would like to thank Jeffrey Shallit for suggesting the problem, as well as for several other suggestions, such as the example $001001\mu^\omega(1)$ given in the introduction, and for pointing out the applicability of the proof technique used in Section 5.

References

1. J.-P. Allouche, J. Currie, J. Shallit, "Extremal infinite overlap-free binary words", *Electron. J. Combin.* **5** (1998), #R27.

2. J.-P. Allouche, J. Shallit, *Automatic Sequences: Theory, Applications, Generalizations*, Cambridge University Press, 2003.
3. J. Berstel, P. Séébold, "A characterization of overlap-free morphisms", *Discrete Appl. Math.* **46** (1993), 275–281.
4. F. Dejean, "Sur un théorème de Thue", *J. Combin. Theory Ser. A* **13** (1972), 90–99.
5. E. Fife, "Binary sequences which contain no BBb", *Trans. Amer. Math. Soc.* **261** (1980), 115–136.
6. W. Gottschalk, G. Hedlund, "A characterization of the Morse minimal set", *Proc. Amer. Math. Soc.* **15** (1964), 70–74.
7. J. Karhumäki, J. Shallit, "Polynomial versus exponential growth in repetition-free binary words", *J. Combin. Theory Ser. A* **104** (2004), 335–347.
8. R. Kolpakov, G. Kucherov, Y. Tarannikov, "On repetition-free binary words of minimal density", WORDS (Rouen, 1997), *Theoret. Comput. Sci.* **218** (1999), 161–175.
9. M. Morse, G. Hedlund, "Unending chess, symbolic dynamics, and a problem in semi-groups", *Duke Math. J.* **11** (1944), 1–7.
10. A. Restivo, S. Salemi, "Overlap free words on two symbols". In M. Nivat, D. Perrin, eds., *Automata on Infinite Words*, Vol. 192 of *Lecture Notes in Computer Science*, pp. 198–206, Springer-Verlag, 1984.
11. P. Séébold, "Morphismes itérés, mot de Morse et mot de Fibonacci", *C. R. Acad. Sci. Paris Sér. I Math.* **295** (1982), 439–441.
12. P. Séébold, "Overlap-free sequences". In M. Nivat, D. Perrin, eds., *Automata on Infinite Words*, Vol. 192 of *Lecture Notes in Computer Science*, pp. 207–215, Springer-Verlag, 1984.
13. P. Séébold, "Sequences generated by infinitely iterated morphisms", *Discrete Appl. Math.* **11** (1985) 255–264.
14. A.M. Shur, "The structure of the set of cube-free $\mathbb{Z}$-words in a two-letter alphabet" (Russian), *Izv. Ross. Akad. Nauk Ser. Mat.* **64** (2000), 201–224. English translation in *Izv. Math.* **64** (2000), 847–871.
15. A. Thue, "Über die gegenseitige Lage gleicher Teile gewisser Zeichenreihen", *Kra. Vidensk. Selsk. Skrifter. I. Math. Nat. Kl.* **1** (1912), 1–67.

On the Equivalence Problem for
E-Pattern Languages Over Small Alphabets

Daniel Reidenbach*

Fachbereich Informatik, Technische Universität Kaiserslautern,
Kaiserslautern, Germany
reidenba@informatik.uni-kl.de

Abstract. We contribute new facets to the discussion on the equivalence problem for E-pattern languages (also referred to as extended or
erasing pattern languages). This fundamental open question asks for the
existence of a computable function that, given any pair of patterns, decides whether or not they generate the same language. Our main result
disproves Ohlebusch and Ukkonen's conjecture (*Theoretical Computer
Science* 186, 1997) on the equivalence problem; the respective argumentation, that largely deals with the nondeterminism of pattern languages,
is restricted to terminal alphabets with at most four distinct letters.

1   Introduction

Patterns—finite strings that consist of variables and terminal symbols—are compact and "natural" devices for the definition of numerous regular and nonregular
formal languages. A pattern generates a word by a uniform substitution of the
variables with arbitrary strings of terminal symbols, and, accordingly, its language is the set of all words that can be obtained by suchlike morphisms. For
instance, the language generated by the pattern $\alpha = x_1 x_1 \, \mathsf{a} \, \mathsf{b} \, x_2$ (with variables
x_1, x_2 and terminals a, b) includes all words where the prefix consists of two
occurrences of the same string, followed by the string ab and concluded by an ar-
bitrary suffix. Thus, the language of α contains, e.g., the words $w_1 = \mathsf{b} \, \mathsf{a} \, \mathsf{b} \, \mathsf{a} \, \mathsf{a} \, \mathsf{b} \, \mathsf{a}$
and $w_2 = \mathsf{a} \, \mathsf{b} \, \mathsf{b} \, \mathsf{b}$, whereas $v_1 = \mathsf{b} \, \mathsf{b} \, \mathsf{b} \, \mathsf{b} \, \mathsf{a}$ and $v_2 = \mathsf{b} \, \mathsf{a} \, \mathsf{a} \, \mathsf{b} \, \mathsf{b}$ are not covered by α.

The investigation of patterns in strings—initiated by Thue [16, 17]—may be
seen as a classical topic in the research on word monoids and combinatorics of
words (cf., e.g., [2], a survey is given in [3]). Contrary to this, the definition
of *pattern languages* as described above—introduced by Angluin [1]—originally
has been motivated by considerations on algorithmic language learning within
the scope of *inductive inference*. Since then, however, the properties of pattern
languages have been intensively studied from a language theoretical point of view
as well, e.g. by Jiang, Kinber, Salomaa, Salomaa, Yu [7, 8]; for a survey see [10].
These examinations reveal that the characteristics of languages generated by a
definition which disallows the substitution of variables with the empty word—as

---

* Supported by the Deutsche Forschungsgemeinschaft (DFG), Grant Wi 1638/1-3.

C.S. Calude, E. Calude and M.J. Dinneen (Eds.): DLT 2004, LNCS 3340, pp. 368–380, 2004.

given by Angluin—and of those produced by a definition allowing the empty substitution (as applied when generating w_2 in our example) differ significantly. Languages of the latter type have been introduced by Shinohara [15]; they are called *extended, erasing*, or simply *E*-pattern languages.

In spite of the wide range of profound examinations, a number of fundamental properties of E-pattern languages is still unresolved; one of the best-known open problems among these is the decidability of the *equivalence*, i.e. the question on the existence of a total computable function that, given any pair of patterns, decides whether or not they generate the same language. This problem, that for Angluin's pattern languages has a trivial answer in the affirmative, has been tackled several times (cf. [5, 7, 8, 4, 11, 13]), contributing a number of positive results on subclasses, properties, conjectures, and conditions, but no comprehensive answer. Consequently, the anticipation of a positive outcome, as expressed in [8], so far could be neither verified nor refuted.

The current state of knowledge on E-pattern languages reveals that several of their properties strongly depend on the size of the terminal alphabet. For instance, the subclass generated by terminal-free patterns is learnable *if and only if* the alphabet is not binary (cf. [12]), whereas the full class is learnable for unary alphabets, but not for those with two, three or four letters (cf. [13]). Consequently, and particularly for small alphabets, E-pattern languages show a variety of (frequently fairly surprising) discontinuities. This phenomenon is brought about by the fact that especially those words over only a few distinct letters tend to be *ambiguous*, i.e. one and the same pattern can generate such a word by different substitutions. The influence of this nondeterminism of E-pattern languages—that is even more complex provided that the patterns do not consist of variables only—on several open problems is not completely understood yet, and therefore most corresponding partial results are restricted to those cases where ambiguity of words is somewhat easy to grasp (cf., e.g., [7, 8, 11]).

These observations establish the background of the present paper, that provides new insight into the consequences of nondeterminism of pattern languages. We apply our approach to the prevailing conjecture on the equivalence problem for E-pattern languages—given by Ohlebusch and Ukkonen [11]—according to which, for terminal alphabets with at least three distinct letters, two arbitrary patterns α and β generate the same language if and only if there exist terminal-preserving morphisms ϕ and ψ such that $\phi(\alpha) = \beta$ and $\psi(\beta) = \alpha$. This conjecture, that we recently have claimed to be incorrect for alphabets with exactly three letters (cf. [13]), in the present paper is disproven for alphabets of size 4.

2   Preliminaries

We now proceed formally. For notions and preliminary results not given in this paper we refer to [14] or, if appropriate, to the respective referenced literature.

$\mathbb{N}$ is the set of natural numbers, $\{0, 1, 2, \dots\}$. A *word* is a finite string of symbols. For an arbitrary set A of symbols, A^+ denotes the set of all non-empty words over A and A^* the set of all (empty and non-empty) words over A. Any set $L \subseteq A^*$ is a *language* over an alphabet A. We designate the *empty* word as ε. For the word that results from the n-fold concatenation of a letter a or of a word w we write a^n or $(w)^n$, respectively. $|\cdot|$ denotes the size of a set or the length of a word, respectively, and $|w|_\mathsf{a}$ the frequency of a letter a in a word w.

The following notion allows to address certain parts of a word w over an alphabet A: If w contains n, $n \geq 1$, occurrences of a subword u then for every i, $1 \leq i \leq n$, $u\langle i \rangle$ is the ith occurrence (from the left) of u in w. For that case, the subword $[w/u\langle i \rangle]$ is the prefix of w up to (but not including) the leftmost letter of $u\langle i \rangle$ and the subword $[u\langle i \rangle \backslash w]$ is the suffix of w beginning with the first letter that is to the right of $u\langle i \rangle$. Moreover, for every word w that contains at least i occurrences of a subword u, j occurrences of subword v and that satisfies $w = w_1 u\langle i \rangle w_2 v\langle j \rangle w_3$ with $w_1, w_2, w_3 \in A^*$, we use $[u\langle i \rangle \backslash w/v\langle j \rangle]$ as an abbreviation for $[u\langle i \rangle \backslash [w/v\langle j \rangle]]$. Thus, for appropriate u, v, w, the specified subwords satisfy $w = [w/u\langle i \rangle] \, u\langle i \rangle \, [u\langle i \rangle \backslash w]$ or $w = [w/u\langle i \rangle] \, u\langle i \rangle \, [u\langle i \rangle \backslash w/v\langle j \rangle] \, v\langle j \rangle \, [v\langle j \rangle \backslash w]$, respectively; e.g., with $w = \mathsf{abcabb}$, $u = \mathsf{a}$ and $v = \mathsf{ab}$, the definition leads to $[w/u\langle 2 \rangle] = \mathsf{abc}$, $[u\langle 2 \rangle \backslash w] = \mathsf{bb}$, and $[u\langle 1 \rangle \backslash w/v\langle 2 \rangle] = \mathsf{bc}$.

We proceed with the pattern specific terminology. Σ is a finite alphabet of *terminal* symbols and $X = \{x_1, x_2, x_3, \dots\}$ an infinite set of *variables*, $\Sigma \cap X = \emptyset$. Henceforth, we use lower case letters in typewriter font, e.g. $\mathsf{a}, \mathsf{b}, \mathsf{c}$, as terminal symbols exclusively, and terminal words are named as u, v, or w.

A *pattern* is a non-empty word over $\Sigma \cup X$, a *terminal-free pattern* is a non-empty word over X; naming patterns we use lower case letters from the beginning of the Greek alphabet such as α, β, γ. $\text{var}(\alpha)$ denotes the set of all variables of a pattern α. We write Pat_Σ for the set of all patterns over the union of X and a specific alphabet Σ or, if there is no need to emphasise the terminal alphabet, Pat for short.

Following [4] we designate two patterns α, β as *similar* if and only if $\alpha = \alpha_0 u_1 \alpha_1 u_2 \dots \alpha_{m-1} u_m \alpha_m$ and $\beta = \beta_0 u_1 \beta_1 u_2 \dots \beta_{m-1} u_m \beta_m$ with $m \in \mathbb{N}$, $\alpha_i, \beta_i \in X^+$ for $1 \leq i < m$, $\alpha_0, \beta_0, \alpha_m, \beta_m \in X^*$ and $u_i \in \Sigma^+$ for $i \leq m$; in other words, we call patterns similar if their terminal substrings are identical and occur in the same order in the patterns.

A morphism $\phi : (\Sigma \cup X)^* \longrightarrow (\Sigma \cup X)^*$ is *terminal-preserving* if and only if, for every $\mathsf{a} \in \Sigma$, $\phi(\mathsf{a}) = \mathsf{a}$. If additionally, for a terminal-preserving morphism ϕ and for all $x_i \in X$, $\phi(x_i) \in X^*$ then we call ϕ *similarity-preserving*. We say that patterns α, β are *(morphically) coincident* if there exist similarity-preserving morphisms ϕ and ψ such that $\phi(\alpha) = \beta$ and $\psi(\beta) = \alpha$; we call them *(morphically) semi-coincident* if there is either such a ϕ or such a ψ, and, for the case that there is neither such a ϕ nor such a ψ, they are designated as *(morphically) incoincident*.

A terminal-preserving morphism σ is a *substitution* if and only if, for every $x_i \in X$, $\sigma(x_i) \in \Sigma^*$. The *E-pattern language* $L_\Sigma(\alpha)$ of a pattern α is defined as the set of all $w \in \Sigma^*$ such that $\sigma(\alpha) = w$ for some substitution σ. For any word $w = \sigma(\alpha)$ we say that σ *generates* w, and for any language $L = L_\Sigma(\alpha)$ we say that α *generates* L. If Σ is understood then we denote the E-pattern language of a pattern α simply as $L(\alpha)$. We use ePAT as an abbreviation for the full class of E-pattern languages (or ePAT_Σ if the corresponding alphabet is of interest).

We designate a pattern α as *succinct* if and only if $|\alpha| \leq |\beta|$ for all patterns β with $L(\beta) = L(\alpha)$. The pattern $\beta = x_1 x_2 x_1 x_2$, e.g., generates the same language as $\alpha = x_1 x_1$, and therefore β is not succinct; α is succinct as $L(x_1) \neq L(\alpha)$.

According to [9] we denote a word w as *ambiguous* (in respect of a pattern α) if and only if it can be generated by several substitutions of α, i.e. there exist substitutions σ and σ', $\sigma(x_i) \neq \sigma'(x_i)$ for some $x_i \in \text{var}(\alpha)$, such that $\sigma(\alpha) = w = \sigma'(\alpha)$. Correspondingly, we call a word w *unambiguous* (in respect of α) if and only if there is exactly one substitution σ with $\sigma(\alpha) = w$. The word $w_1 = \text{aaba}$, for instance, is ambiguous in respect of $\alpha = x_1 \text{a} \, x_2$ since it can be generated by, e.g., σ and σ' with $\sigma(x_1) = \text{a}$, $\sigma(x_2) = \text{ba}$ and $\sigma'(x_1) = \varepsilon$, $\sigma'(x_2) = \text{aba}$. The example word $w_2 = \text{ba}$ is unambiguous in respect of α.

We now proceed with some decidability problems on E-pattern languages: Let ePAT* be any set of E-pattern languages. We say that the *inclusion problem* for ePAT* is *decidable* if and only if there exists a computable function which, given two arbitrary patterns α, β with $L(\alpha), L(\beta) \in$ ePAT*, decides whether or not $L(\alpha) \subseteq L(\beta)$. Accordingly, the *equivalence problem* is decidable if and only if there exists another computable function which for every pair of patterns α, β with $L(\alpha), L(\beta) \in$ ePAT* decides whether or not $L(\alpha) = L(\beta)$. Obviously, the decidability of the inclusion implies the decidability of the equivalence. As mentioned in Section 1, the decidability of the equivalence problem for ePAT has not been resolved yet, but there is a number of positive results on subclasses given in [11]. The inclusion problem is known to be undecidable (cf. [8]). Under certain circumstances, however, the inclusion problem is decidable; this results from the following fact:

Fact 1 ([11]). *Let Σ be an alphabet and α, β two arbitrary similar patterns such that Σ contains two distinct letters not occurring in α and β. Then $L_\Sigma(\beta) \subseteq L_\Sigma(\alpha)$ iff there exists a similarity-preserving morphism ϕ such that $\phi(\alpha) = \beta$.*

In particular, Fact 1 implies the decidability of the inclusion problem for the class of terminal-free E-pattern languages if $|\Sigma| \geq 2$ (proven in [5] and [8]).

The following theorem shows that any consideration on the equivalence problem can be restricted to similar patterns. Therefore, Fact 1 implies the decidability of the equivalence for all pairs of patterns if one of the two patterns does not contain at least two distinct letters of the alphabet.

Fact 2 ([5] and [7]). *Let Σ be an alphabet, $|\Sigma| \geq 3$, and let $\alpha, \beta \in \text{Pat}_\Sigma$. If $L_\Sigma(\alpha) = L_\Sigma(\beta)$ then α and β are similar.*

Moreover, Fact 2 suggests a possible approach to the equivalence problem, that has been addressed by [4] and [11]: Obviously, the equivalence of E-pattern languages is decidable provided that Fact 1 holds for all similar patterns (and not only for those satisfying the additional condition).

We conclude this section with a definition that originates in [11] and that is motivated by the facts stated above: Let Σ be an alphabet and define $\Sigma' := \Sigma \cup \{\text{a}\}$ for an arbitrary $\text{a} \notin \Sigma$. We say that the *equivalence for* ePAT_Σ *is preserved under alphabet extension* if and only if, for every pair $\alpha, \beta \in \mathrm{Pat}_\Sigma$, $L_\Sigma(\alpha) = L_\Sigma(\beta)$ implies $L_{\Sigma'}(\alpha) = L_{\Sigma'}(\beta)$ and vice versa.

3 On Ohlebusch and Ukkonen's Conjecture

The equivalence problem for E-pattern languages has first been examined in [5] and [7] and later in [8], [4], and [11]. The latter authors give a procedure that for every pattern computes a shortest *normal form*. They conjecture that, for alphabets with at least three letters, two patterns generate the same language if and only if their normal forms are the same, and the authors paraphrase their conjecture as follows:

Conjecture 1 ([11]). For an alphabet Σ, $|\Sigma| \geq 3$, and patterns $\alpha_1, \alpha_2 \in \mathrm{Pat}_\Sigma$, $L_\Sigma(\alpha_1) = L_\Sigma(\alpha_2)$ if and only if α_1 and α_2 are morphically coincident.

Furthermore, as a consequence of Fact 1 and Fact 2, the authors annotate that the equivalence problem is decidable for $|\Sigma| \geq 3$ if the equivalence for ePAT_Σ is preserved under alphabet extension (cf. [11], Open Question 2).

The choice of alphabet size 3 as a lower bound in Conjecture 1 might be caused by the following observations: The patterns $\alpha_1 = x_1\,\mathsf{a}\,x_2\,\mathsf{b}\,x_3$ and $\alpha_2 = x_1\,\mathsf{a}\,\mathsf{b}\,x_2$, for instance, generate the same language if $\Sigma = \{\mathsf{a}, \mathsf{b}\}$ (although they are semi-coincident) since for every word in $\{\sigma_i(\alpha_1) \mid \sigma_i(x_2) \neq \varepsilon\}$ a second substitution σ_i' can be given with $\sigma_i'(\alpha_1) = \sigma_i(\alpha_1)$ and $\sigma'(x_2) = \varepsilon$. Thus, this specific ambiguity of words, that is caused by the small alphabet and by the composition of variables and terminal symbols in α_1, brings about the equivalence of $L_\Sigma(\alpha_1)$ and $L_\Sigma(\alpha_2)$. Contrary to this, for $|\Sigma| \geq 3$, the existence of analogue examples seems to be rather implausible since for every variable in a pattern at least one occurrence can be chosen for assigning a substitution that contains a letter which differs from the terminal symbols to the left and to the right of the variable (cf., e.g., the proof of Fact 2 as given in [7]). Consequently, Conjecture 1 suggests that such patterns do not exist for alphabets containing at least three letters.

As a by-product of learning theoretical studies, [13] anticipates that—at least for alphabets with *exactly* three letters—this conjecture is incorrect. More precisely, the paper claims that, for $\Sigma := \{\mathsf{a}, \mathsf{b}, \mathsf{c}\}$, $\Sigma' := \Sigma \cup \{\mathsf{d}\}$ and

$$\tilde{\alpha}_{\mathrm{abc},1} := x_1\,\mathsf{a}\,x_2\,x_3^2\,x_4^2\,x_5^2\,x_6^2\,x_7^2\,x_8\,\mathsf{b}\,x_9\,\mathsf{a}\,x_2\,x_{10}^2\,x_4^2\,x_5^2\,x_6^2\,x_{11}^2\,x_8\,\mathsf{b}\,x_{12},$$

$$\tilde{\alpha}_{\mathrm{abc},2} := x_1\,\mathsf{a}\,x_2\,x_3^2\,x_4^2\,x_7^2\,x_8\,\mathsf{b}\,x_9\,\mathsf{a}\,x_2\,x_{10}^2\,x_4^2\,x_{11}^2\,x_8\,\mathsf{b}\,x_{12},$$

$L_\Sigma(\tilde{\alpha}_{\mathrm{abc},1}) = L_\Sigma(\tilde{\alpha}_{\mathrm{abc},2})$, but $L_{\Sigma'}(\tilde{\alpha}_{\mathrm{abc},1}) \supset L_{\Sigma'}(\tilde{\alpha}_{\mathrm{abc},2})$, and $\tilde{\alpha}_{\mathrm{abc},1}$ and $\tilde{\alpha}_{\mathrm{abc},2}$ are semi-coincident.

The present paper actually dis*proves* Conjecture 1; to this end, however, we regard different alphabets, namely those of size 4. Thus, we establish an additional result to that in [13]. Besides, the chosen alphabet size is by far more challenging, and therefore it requires a significantly more elaborate and instructive reasoning. Hence, our presumably unexpected main result reads as follows:

Theorem 1. *Let Σ be an alphabet, $|\Sigma| = 4$. Then the equivalence for ePAT$_\Sigma$ is not preserved under alphabet extension.*

Theorem 2. *Let Σ be an alphabet, $|\Sigma| = 4$. Then there exist morphically incoincident patterns $\alpha_1, \alpha_2 \in$ Pat$_\Sigma$ such that $L_\Sigma(\alpha_1) = L_\Sigma(\alpha_2)$.*

Referring to these statements we can conclude that for "small" alphabets (i.e. for those with at most four distinct letters) the equivalence of E-pattern languages has some common properties, which nicely contrast with the expectations (potentially) involved in Fact 2, Conjecture 1, and Theorem 7.2 of [8]:

Corollary 1. *Let Σ be an alphabet, $|\Sigma| \leq 4$. Then the equivalence for ePAT$_\Sigma$ is not preserved under alphabet extension.*

Corollary 2. *Let Σ be an alphabet, $|\Sigma| \leq 4$. Then there exist morphically incoincident or semi-coincident patterns $\alpha_1, \alpha_2 \in$ Pat$_\Sigma$ such that $L_\Sigma(\alpha_1) = L_\Sigma(\alpha_2)$.*

The proof of Theorem 1 and Theorem 2, that for the given patterns $\tilde{\alpha}_{abc,1}$ and $\tilde{\alpha}_{abc,2}$ can be adapted to the case $|\Sigma| = 3$ with little effort, is accomplished in Section 3.1. Its underlying principle follows the course indicated above: We compose two sophisticated incoincident example patterns—each of them consisting of 82 variables and terminals—and identify "decisive" words in their languages. Then we precisely examine the ambiguity of these words and reveal that all of them can be generated by substitutions assigning the empty word to at least one among two specific variables; thereby we can conclude that both patterns generate the same language. In other words, we analyse the nondeterminism of E-pattern languages, that has been the subject, e.g., of [9]. However, the prevailing point of view in literature does not exactly meet our requirements as it investigates the ambiguity of pattern *languages*, i.e. the maximum ambiguity among all words in the language, whereas we ask for the existence of particular alternative substitutions for selected words. Thus, our method is rather related to the research on *equality sets* (cf. [6]).

In spite of the extensive argumentation required even for a single alphabet, we expect our method to be useful for future examinations of Conjecture 1 with regard to different alphabet sizes as well. Moreover, we suggest that the (supposably necessary) complexity of our example patterns explains the lack of comprehensive results on the equivalence problem so far, and we consider the subsequent section to provide an insight into the extraordinary combinatorial depth of E-pattern languages.

Obviously, with the present state of knowledge on the subject, the given results do not imply the non-decidability of the equivalence problem for ePAT$_\Sigma$ with $|\Sigma| = 4$. They show, first, that the expected lower bound in terms of alphabet size for a uniform behaviour of E-pattern languages concerning the decidability of the equivalence—as expressed in Conjecture 1—needs to be redetermined (provided such a bound exists at all). Second, they suggest that any decision procedure for $|\Sigma| = 4$ (if any) presumably needs to be more elaborate than that given in [11]—which, by the way, still might be applicable to $|\Sigma| \geq 5$. Additional remarks on suchlike aspects are given in Section 3.2.

3.1   Proof of the Main Results

The present section contains four lemmata. Lemma 1 and Lemma 4 prove Theorem 1; the argumentation on Theorem 2 is accomplished by Lemma 1 again and, additionally, Lemma 3—which, in turn, utilises Lemma 2.

We begin with the example patterns that constitute the core of our reasoning:

Definition 1 (First Version). *The patterns $\tilde{\alpha}_{\mathrm{abcd},1}$ and $\tilde{\alpha}_{\mathrm{abcd},2}$ are given by*

$$\tilde{\alpha}_{\mathrm{abcd},1} := x_1 \; \mathsf{a} \; x_2 \, x_3^2 \, x_4^2 \, x_5^2 \, x_6^2 \, x_7 \, \mathsf{b} \; x_8 \; \mathsf{a} \; x_2 \, x_9^2 \, x_4^2 \, x_5^2 \, x_{10}^2 \, x_7 \, \mathsf{b} \; x_{11}$$
$$\mathsf{c} \; x_{12} \, x_{13}^2 \, x_{14}^2 \, x_{15}^2 \, x_{16}^2 \, x_{17} \, \mathsf{d} \; x_{18} \; \mathsf{c} \; x_{12} \, x_{19}^2 \, x_{14}^2 \, x_{15}^2 \, x_{20}^2 \, x_{17} \, \mathsf{d} \; x_{21}$$
$$x_{14}^2 \, x_{15}^2 \, x_{14}^2 \, x_{15}^2 \, x_{14}^2 \, x_{15}^2 \, x_{22} \, x_4^2 \, x_5^2 \, x_4^2 \, x_5^2 \, x_4^2 \, x_5^2 \, x_{23} \, x_4 \, x_{14} \, x_{24}$$
$$\tilde{\alpha}_{\mathrm{abcd},2} := x_1 \; \mathsf{a} \; x_2 \, x_3^2 \, x_4^2 \, x_5^2 \, x_6^2 \, x_7 \, \mathsf{b} \; x_8 \; \mathsf{a} \; x_2 \, x_9^2 \, x_4^2 \, x_5^2 \, x_{10}^2 \, x_7 \, \mathsf{b} \; x_{11}$$
$$\mathsf{c} \; x_{12} \, x_{13}^2 \, x_{14}^2 \, x_{15}^2 \, x_{16}^2 \, x_{17} \, \mathsf{d} \; x_{18} \; \mathsf{c} \; x_{12} \, x_{19}^2 \, x_{14}^2 \, x_{15}^2 \, x_{20}^2 \, x_{17} \, \mathsf{d} \; x_{21}$$
$$x_{14}^2 \, x_{15}^2 \, x_{14}^2 \, x_{15}^2 \, x_{14}^2 \, x_{15}^2 \, x_{22} \, x_4^2 \, x_5^2 \, x_4^2 \, x_5^2 \, x_4^2 \, x_5^2 \, x_{23} \, x_{14} \, x_4 \, x_{24}.$$

Since $\tilde{\alpha}_{\mathrm{abcd},1}$ and $\tilde{\alpha}_{\mathrm{abcd},2}$ might be regarded as fairly intricate we give a second version of Definition 1 revealing the structure of the patterns:

Definition 1 (Second Version). *Consider the patterns*

$$\gamma_1 := x_4^2 \, x_5^2,$$
$$\gamma_2 := x_{14}^2 \, x_{15}^2,$$
$$\beta_1 := x_2 \, x_3^2 \, \gamma_1 \, x_6^2 \, x_7,$$
$$\beta_1' := x_2 \, x_9^2 \, \gamma_1 \, x_{10}^2 \, x_7,$$
$$\beta_2 := x_{12} \, x_{13}^2 \, \gamma_2 \, x_{16}^2 \, x_{17},$$
$$\beta_2' := x_{12} \, x_{19}^2 \, \gamma_2 \, x_{20}^2 \, x_{17}$$
$$\hat{\alpha}_1 := x_1 \; \mathsf{a} \; \beta_1 \; \mathsf{b} \; x_8 \; \mathsf{a} \; \beta_1' \; \mathsf{b} \; x_{11} \; \mathsf{c} \; \beta_2 \; \mathsf{d} \; x_{18} \; \mathsf{c} \; \beta_2' \; \mathsf{d} \; x_{21},$$
$$\hat{\alpha}_2 := (\gamma_2)^3 \, x_{22} \, (\gamma_1)^3.$$

Then $\tilde{\alpha}_{\mathrm{abcd},1} := \hat{\alpha}_1 \, \hat{\alpha}_2 \, x_{23} \, x_4 \, x_{14} \, x_{24}$ and $\tilde{\alpha}_{\mathrm{abcd},2} := \hat{\alpha}_1 \, \hat{\alpha}_2 \, x_{23} \, x_{14} \, x_4 \, x_{24}$.

In order to facilitate the understanding of our reasoning we give some brief informal explanatory remarks before proceeding with the actual proof of Theorems 1 and 2: Evidently, concerning the question whether or not $L(\tilde{\alpha}_{\mathrm{abcd},1})$ and $L(\tilde{\alpha}_{\mathrm{abcd},2})$ are different, only those words are of interest that are generated by a substitution which is not empty for both x_4 and x_{14}, as the order of the last occurrences of these variables is the only difference between the patterns. Therefore, the components of $\tilde{\alpha}_{\mathrm{abcd},1}$ and $\tilde{\alpha}_{\mathrm{abcd},2}$ are tailor-made for ensuring the ambiguity of all words generated by a substitution σ that satisfies $\sigma(x_4) \neq \varepsilon \neq \sigma(x_{14})$.

With regard to the subpatterns of $\tilde{\alpha}_{\mathrm{abcd},1}$ and $\tilde{\alpha}_{\mathrm{abcd},2}$, we first examine the kernels, i.e. γ_1 and γ_2. Obviously, for any substitution σ, if $\sigma(\gamma_1)$ or $\sigma(\gamma_2)$ do not contain at least two different letters then $\sigma(\gamma_1)$ or $\sigma(\gamma_2)$, respectively, are ambiguous and can be generated simply by x_5 or x_{15}, respectively; thus, x_4 or x_{14} can be substituted empty. In our formal argumentation, we utilise this fact only for $\sigma(\gamma_1) \in \{\mathsf{c}\}^* \cup \{\mathsf{d}\}^*$ or $\sigma(\gamma_2) \in \{\mathsf{a}\}^* \cup \{\mathsf{b}\}^*$. The other cases are covered by the ambiguity of $\sigma(x_1 \, \mathsf{a} \, \beta_1 \, \mathsf{b} \, x_8)$ (resp. $\sigma(x_{11} \, \mathsf{c} \, \beta_2 \, \mathsf{d} \, x_{18})$) whenever $\sigma(\gamma_1)$ (resp. $\sigma(\gamma_2)$) contains—possibly among others—the letters a or b (resp. c or d), leading again to an optional empty substitution for x_4 (resp. x_{14}). Thus, σ only can generate a decisive word if $\sigma(\gamma_1)$ consists of c and d and $\sigma(\gamma_2)$ of a and b. Such a choice of a substitution utilising letters that are distinguishable from the terminals to the left and to the right of the corresponding variable subword in the pattern probably is the most natural option and is used frequently (see, e.g., proof on Theorem 7.2 in [8]). However, for that case, $\sigma(\hat{\alpha}_2) = w_0 \, \mathsf{a} \, \mathsf{b} \, w_1 \, \mathsf{a} \, \mathsf{b} \, w_2 \, \mathsf{c} \, \mathsf{d} \, w_3 \, \mathsf{c} \, \mathsf{d} \, w_4$ for some words w_i, $i \leq 4$. Consequently, $\sigma(\tilde{\alpha}_{\mathsf{abcd},1})$ and $\sigma(\tilde{\alpha}_{\mathsf{abcd},2})$ can be generated by the (sub-)pattern $x_1 \, \mathsf{a}\mathsf{b} \, x_8 \, \mathsf{a}\mathsf{b} \, x_{11} \, \mathsf{c}\mathsf{d} \, x_{18} \, \mathsf{c}\mathsf{d} \, x_{21}$, and therefore x_4 and x_{14} can be substituted empty again. The variables with single occurrences, such as x_8 and x_{23}, are used to compensate the side effects of the empty substitution of x_4 or x_{14}; the modified repetitions of β_1 (as β_1') and β_2 (as β_2') and, particularly, those variables that distinguish β_1 from β_1' (e.g. x_3) and β_2 from β_2' (e.g. x_{13}) guarantee that $\tilde{\alpha}_{\mathsf{abcd},1}$ and $\tilde{\alpha}_{\mathsf{abcd},2}$ are incoincident. The latter point, one of the statements of Theorem 2, is discussed in Lemma 3.

As the ambiguity of decisive words affects $\tilde{\alpha}_{\mathsf{abcd},1}$ and $\tilde{\alpha}_{\mathsf{abcd},2}$ in the same way, the stated phenomenon allows us to prove the following, crucial lemma:

Lemma 1. *Let $\Sigma_1 = \{\mathsf{a}, \mathsf{b}, \mathsf{c}, \mathsf{d}\}$. Then $L_{\Sigma_1}(\tilde{\alpha}_{\mathsf{abcd},1}) = L_{\Sigma_1}(\tilde{\alpha}_{\mathsf{abcd},2})$.*

Proof. We first prove $L_{\Sigma_1}(\tilde{\alpha}_{\mathsf{abcd},1}) \subseteq L_{\Sigma_1}(\tilde{\alpha}_{\mathsf{abcd},2})$. Hence, let σ be an arbitrary substitution that is applicable to $\tilde{\alpha}_{\mathsf{abcd},1}$. We show that there exists a substitution σ' such that $\sigma'(\tilde{\alpha}_{\mathsf{abcd},2}) = \sigma(\tilde{\alpha}_{\mathsf{abcd},1})$. To this end, we refer to $\tilde{\alpha}_{\mathsf{abcd},1}$ and $\tilde{\alpha}_{\mathsf{abcd},2}$ as declared in the second version of Definition 1 and regard the following cases— that evidently can be restricted to a consideration of $\sigma(\gamma_1)$ and $\sigma(\gamma_2)$:

<u>Case 1</u> $\sigma(\gamma_1) \in \{\mathsf{a}, \mathsf{b}, \mathsf{c}, \mathsf{d}\}^+ \setminus \{\mathsf{b}, \mathsf{c}, \mathsf{d}\}^+$:

Define $\quad \sigma'(x_1) := \sigma(x_1 \, \mathsf{a} \, x_2 \, x_3^2) \, [\sigma(\gamma_1)/\,\mathsf{a}\langle 1\rangle]$,

$\qquad \sigma'(x_2) := [\mathsf{a}\langle 1\rangle \setminus \sigma(\gamma_1)]$,

$\qquad \sigma'(x_8) := \sigma(x_8 \, \mathsf{a} \, x_2 \, x_9^2) \, [\sigma(\gamma_1)/\,\mathsf{a}\langle 1\rangle]$,

$\qquad \sigma'(x_{22}) := \sigma(x_{22} \, (\gamma_1)^3)$,

$\qquad \sigma'(x_{23}) := \sigma(x_{23} \, x_4)$,

$\qquad \sigma'(x_j) := \sigma(x_j), \; x_j \in \mathrm{var}(\beta_2 \, \beta_2') \cup \{x_6, x_7, x_{10}, x_{11}, x_{18}, x_{21}, x_{24}\}$,

$\qquad \sigma'(x_j) := \varepsilon, \; x_j \in \mathrm{var}(\gamma_1) \cup \{x_3, x_9\}$.

<u>Case 2</u> $\sigma(\gamma_1) \in \{\mathsf{b}, \mathsf{c}, \mathsf{d}\}^+ \setminus \{\mathsf{c}, \mathsf{d}\}^+$:

Define $\quad \sigma'(x_7) := [\sigma(\gamma_1)/\,\mathsf{b}\langle 1\rangle]$,

$\qquad \sigma'(x_8) := [\mathsf{b}\langle 1\rangle \setminus \sigma(\gamma_1)] \, \sigma(x_6^2 \, x_7 \, \mathsf{b} \, x_8)$,

$\qquad \sigma'(x_{11}) := [\mathsf{b}\langle 1\rangle \setminus \sigma(\gamma_1)] \, \sigma(x_{10}^2 \, x_7 \, \mathsf{b} \, x_{11})$,

$\qquad \sigma'(x_{22}) := \sigma(x_{22} \, (\gamma_1)^3)$,

$\qquad \sigma'(x_{23}) := \sigma(x_{23} \, x_4)$,

$\qquad \sigma'(x_j) := \sigma(x_j), \; x_j \in \mathrm{var}(\beta_2 \, \beta_2') \cup \{x_1, x_2, x_3, x_9, x_{18}, x_{21}, x_{24}\}$,

$\qquad \sigma'(x_j) := \varepsilon, \; x_j \in \mathrm{var}(\gamma_1) \cup \{x_6, x_{10}\}$.

<u>Case 3</u> $\sigma(\gamma_1) \in \{c\}^* \cup \{d\}^*$:

Define $\sigma'(x_4) := \varepsilon,$

$\sigma'(x_5) := \sigma(x_4\, x_5),$

$\sigma'(x_{23}) := \sigma(x_{23}\, x_4),$

$\sigma'(x_j) := \sigma(x_j),\ x_j \in \mathrm{var}(\tilde{\alpha}_{\mathrm{abcd},2}) \setminus (\mathrm{var}(\gamma_1) \cup \{x_{23}\}).$

<u>Case 4</u> $\sigma(\gamma_1) \in \{c,d\}^+ \setminus (\{c\}^+ \cup \{d\}^+)$ and $\sigma(\gamma_2) \in \{a,b,c,d\}^+ \setminus \{a,b,d\}^+$:

Define $\sigma'(x_{11}) := \sigma(x_{11}\ c\ x_{12}\, x_{13}^2)\, [\sigma(\gamma_2)/\, c\langle 1\rangle],$

$\sigma'(x_{12}) := [c\langle 1\rangle \setminus \sigma(\gamma_2)],$

$\sigma'(x_{18}) := \sigma(x_{18}\ c\ x_{12}\, x_{19}^2)\, [\sigma(\gamma_2)/\, c\langle 1\rangle],$

$\sigma'(x_{22}) := \sigma((\gamma_2)^3\, x_{22}),$

$\sigma'(x_{24}) := \sigma(x_{14}\, x_{24}),$

$\sigma'(x_j) := \sigma(x_j),\ x_j \in \mathrm{var}(\beta_1\, \beta_1') \cup \{x_1, x_8, x_{16}, x_{17}, x_{20}, x_{21}, x_{23}\},$

$\sigma'(x_j) := \varepsilon,\ x_j \in \mathrm{var}(\gamma_2) \cup \{x_{13}, x_{19}\}.$

<u>Case 5</u> $\sigma(\gamma_1) \in \{c,d\}^+ \setminus (\{c\}^+ \cup \{d\}^+)$ and $\sigma(\gamma_2) \in \{a,b,d\}^+ \setminus \{a,b\}^+$:

Define $\sigma'(x_{17}) := [\sigma(\gamma_2)/\, d\langle 1\rangle],$

$\sigma'(x_{18}) := [d\langle 1\rangle \setminus \sigma(\gamma_2)]\, \sigma(x_{16}^2\, x_{17}\ d\ x_{18}),$

$\sigma'(x_{21}) := [d\langle 1\rangle \setminus \sigma(\gamma_2)]\, \sigma(x_{20}^2\, x_{17}\ d\ x_{21}),$

$\sigma'(x_{22}) := \sigma((\gamma_2)^3\, x_{22}),$

$\sigma'(x_{24}) := \sigma(x_{14}\, x_{24}),$

$\sigma'(x_j) := \sigma(x_j),\ x_j \in \mathrm{var}(\beta_1\, \beta_1') \cup \{x_1, x_8, x_{11}, x_{12}, x_{13}, x_{19}, x_{23}\},$

$\sigma'(x_j) := \varepsilon,\ x_j \in \mathrm{var}(\gamma_2) \cup \{x_{16}, x_{20}\}.$

<u>Case 6</u> $\sigma(\gamma_1) \in \{c,d\}^+ \setminus (\{c\}^+ \cup \{d\}^+)$ and $\sigma(\gamma_2) \in \{a\}^* \cup \{b\}^*$:

Define $\sigma'(x_{14}) := \varepsilon,$

$\sigma'(x_{15}) := \sigma(x_{14}\, x_{15}),$

$\sigma'(x_{24}) := \sigma(x_{14}\, x_{24}),$

$\sigma'(x_j) := \sigma(x_j),\ x_j \in \mathrm{var}(\tilde{\alpha}_{\mathrm{abcd},2}) \setminus (\mathrm{var}(\gamma_2) \cup \{x_{24}\}).$

<u>Case 7</u> $\sigma(\gamma_1) \in \{c,d\}^+ \setminus (\{c\}^+ \cup \{d\}^+)$ and $\sigma(\gamma_2) \in \{a,b\}^+ \setminus (\{a\}^+ \cup \{b\}^+)$:

Consequently, $\sigma((\gamma_1)^3)$ contains at least two occurrences of the subword c d and $\sigma((\gamma_2)^3)$ contains at least two occurrences of the subword a b. Furthermore, due to the shape of these subwords, their occurrences must be non-overlapping. Therefore σ' can be given as follows:

Define $\sigma'(x_1) := \sigma(\hat{\alpha}_1)\, [\sigma(\hat{\alpha}_2)/\, a\, b\langle 1\rangle],$

$\sigma'(x_8) := [a\, b\langle 1\rangle \setminus \sigma(\hat{\alpha}_2)/\, a\, b\langle 2\rangle],$

$\sigma'(x_{11}) := [a\, b\langle 2\rangle \setminus \sigma(\hat{\alpha}_2)/\, c\, d\langle 1\rangle],$

$\sigma'(x_{18}) := [c\, d\langle 1\rangle \setminus \sigma(\hat{\alpha}_2)/\, c\, d\langle 2\rangle],$

$\sigma'(x_{21}) := [c\, d\langle 2\rangle \setminus \sigma(\hat{\alpha}_2)]\, \sigma(x_{23}\, x_4\, x_{14}\, x_{24}),$

$\sigma'(x_j) := \varepsilon,\ x_j \in \mathrm{var}(\tilde{\alpha}_{\mathrm{abcd},2}) \setminus \{x_1, x_8, x_{11}, x_{18}, x_{21}\}.$

With the annotations on the shape of $\tilde{\alpha}_{\mathrm{abcd},1}$ and $\tilde{\alpha}_{\mathrm{abcd},2}$ in mind, it is obvious that, in every of the seven cases, $\sigma'(\tilde{\alpha}_{\mathrm{abcd},2}) = \sigma(\tilde{\alpha}_{\mathrm{abcd},1})$. Thus, since σ has been chosen arbitrarily and as the cases are exhaustive, $L_{\Sigma_1}(\tilde{\alpha}_{\mathrm{abcd},1}) \subseteq L_{\Sigma_1}(\tilde{\alpha}_{\mathrm{abcd},2})$.

The proof for $L_{\Sigma_1}(\tilde{\alpha}_{\mathrm{abcd},2}) \subseteq L_{\Sigma_1}(\tilde{\alpha}_{\mathrm{abcd},1})$ is similar: In the argumentation given above, it is sufficient to replace $\tilde{\alpha}_{\mathrm{abcd},1}$ by $\tilde{\alpha}_{\mathrm{abcd},2}$ and vice versa and, additionally, to adapt $\sigma'(x_{23})$ and $\sigma'(x_{24})$ in Cases 1-7 in an adequate manner such that it matches the shape of $\tilde{\alpha}_{\mathrm{abcd},1}$. The rest is verbatim the same. $\square$

With Lemma 1, the crucial element of Theorem 1 and Theorem 2 is proven. In the next step we complete the proof of Theorem 2. As a prerequisite thereof, we proceed with an evident lemma that is of great use for the upcoming proof of Lemma 3 and that is a direct consequence of Lemma 1 in [13]:

Lemma 2. *Let α be a terminal-free pattern and let $\phi : X^* \longrightarrow X^*$ be a morphism with $\phi(\alpha) = \alpha$. Then either $\phi(x_j) = x_j$ for every $x_j \in \mathrm{var}(\alpha)$ or there is an $x_{j'} \in \mathrm{var}(\alpha)$ such that $|\phi(x_{j'})| \geq 2$ and $x_{j'} \in \mathrm{var}(\phi(x_{j'}))$.*

We call any $x_{j'}$ satisfying these two conditions an *anchor variable* (in respect of the morphism ϕ).

Now we can prove that there are no similarity-preserving morphisms mapping $\tilde{\alpha}_{\mathrm{abcd},1}$ and $\tilde{\alpha}_{\mathrm{abcd},2}$ onto each other:

Lemma 3. $\tilde{\alpha}_{\mathrm{abcd},1}$ *and* $\tilde{\alpha}_{\mathrm{abcd},2}$ *are morphically incoincident.*

Proof. Assume to the contrary there is a similarity-preserving morphism ϕ with $\phi(\tilde{\alpha}_{\mathrm{abcd},1}) = \tilde{\alpha}_{\mathrm{abcd},2}$ or with $\phi(\tilde{\alpha}_{\mathrm{abcd},2}) = \tilde{\alpha}_{\mathrm{abcd},1}$. Then, obviously, $\phi(x_4) \neq x_4$ or $\phi(x_{14}) \neq x_{14}$. Consequently—since, e.g., β_1 and β_2 occur in $\tilde{\alpha}_{\mathrm{abcd},1}$ as well as in $\tilde{\alpha}_{\mathrm{abcd},2}$ and since necessarily $\phi(\beta_1) = \beta_1$ and $\phi(\beta_2) = \beta_2$—there must be an anchor variable $x_{j'}$ in β_1 or β_2 (cf. Lemma 2).

We start with β_1. First, for $j' \in \{3,4,5,6\}$, $x_{j'}$ being an anchor variable implies that $\phi(x_{j'}^2) = x_k x_{k'} \delta x_k x_{k'} \delta$ with variables $x_k, x_{k'}$ and $\delta \in X^*$, but there is no substring in β_1 that equals the given shape of $\phi(x_{j'})$. Second, because of the necessity of $\phi(\beta_1') = \beta_1'$, x_2 cannot be an anchor variable since $\phi(x_2)$ had to equal both $x_2 x_3 \delta$ and $x_2 x_9 \delta$ for a $\delta \in X^*$. Finally, due to an analogous reason, $j' \neq 7$. Thus, there is no anchor variable in $\mathrm{var}(\beta_1)$. This contradicts $\phi(x_4) \neq x_4$.

With regard to β_2, the argumentation is equivalent, and, consequently, there is no anchor variable in $\mathrm{var}(\beta_2)$. Therefore, the assumption is incorrect. $\square$

With Lemma 1 and Lemma 3, the proof of Theorem 2 is accomplished. Consequently, and referring to Fact 1, it is obvious that, for a terminal alphabet Σ_3 with at least six distinct letters, $L_{\Sigma_3}(\tilde{\alpha}_{\mathrm{abcd},1}) \neq L_{\Sigma_3}(\tilde{\alpha}_{\mathrm{abcd},2})$. Hence, for $|\Sigma| = 4$ or $|\Sigma| = 5$, the equivalence for ePAT_Σ is not preserved under alphabet extension. In order to conclude the proof of Theorem 1, we therefore have to show explicitly that the given example patterns generate different languages for alphabets with exactly five letters:

Lemma 4. *Let $\Sigma_2 \supseteq \{\mathsf{a}, \mathsf{b}, \mathsf{c}, \mathsf{d}, \mathsf{e}\}$. Then $L_{\Sigma_2}(\tilde{\alpha}_{\mathrm{abcd},1}) \neq L_{\Sigma_2}(\tilde{\alpha}_{\mathrm{abcd},2})$.*

Proof. We show that there is a word in $L_{\Sigma_2}(\tilde{\alpha}_{\mathrm{abcd},1}) \setminus L_{\Sigma_2}(\tilde{\alpha}_{\mathrm{abcd},2})$. To this end, we refer to $\tilde{\alpha}_{\mathrm{abcd},1}$ and $\tilde{\alpha}_{\mathrm{abcd},2}$ as declared in the second version of Definition 1 and consider the substitution σ given by

$$\sigma(x_j) := \begin{cases} c\,e^{3j-2}\,c\ c\,e^{3j-1}\,c\ c\,e^{3j}\,c & , \quad x_j \in \mathrm{var}(\beta_1\,\beta_1'), \\ a\,e^{3j-2}\,a\ a\,e^{3j-1}\,a\ a\,e^{3j}\,a & , \quad x_j \in \mathrm{var}(\beta_2\,\beta_2'), \\ \varepsilon & , \quad \text{else.} \end{cases}$$

Then $\sigma(\tilde{\alpha}_{\mathrm{abcd},1})$ has the following suffix generated by $\sigma(\mathrm{d}\ x_{21}\ \hat{\alpha}_2\ x_{23}\ x_4\ x_{14}\ x_{24})$:

$$\mathrm{d}\,((a\,e^{40}\,a\ a\,e^{41}\,a\ a\,e^{42}\,a)^2(a\,e^{43}\,a\ a\,e^{44}\,a\ a\,e^{45}\,a)^2)^3$$
$$((c\,e^{10}\,c\ c\,e^{11}\,c\ c\,e^{12}\,c)^2(c\,e^{13}\,c\ c\,e^{14}\,c\ c\,e^{15}\,c)^2)^3$$
$$c\,e^{10}\,c\ c\,e^{11}\,c\ c\,e^{12}\,c\ a\,e^{40}\,a\ a\,e^{41}\,a\ a\,e^{42}\,a$$

and this is the only occurrence of that subword in $\sigma(\tilde{\alpha}_{\mathrm{abcd},1})$.

Now assume to the contrary there is a substitution σ' with $\sigma'(\tilde{\alpha}_{\mathrm{abcd},2}) = \sigma(\tilde{\alpha}_{\mathrm{abcd},1})$. As, due to $\sigma(x_j) \in \{a, c, e\}^*$ for all $x_j \in \mathrm{var}(\tilde{\alpha}_{\mathrm{abcd},1})$, the letters b and d each occur exactly twice in $\sigma(\tilde{\alpha}_{\mathrm{abcd},1})$ we may conclude that $\sigma'(\beta) = \sigma(\beta)$ for $\beta \in \{\beta_1, \beta_1', \beta_2, \beta_2'\}$. Therefore—and since, according to Theorem 3 of [12], the patterns $\beta_1\,\beta_1'$ and $\beta_2\,\beta_2'$ are succinct—Lemma 1 of [12] is applicable, which shows that in the given case necessarily $\sigma'(x_j) = v_0\ c\,c\,e^{3j-1}\,c\,c\ v_1$, $v_0, v_1 \in \Sigma^*$, for all $x_j \in \mathrm{var}(\beta_1\,\beta_1')$ and $\sigma'(x_j) = v_2\ a\,a\,e^{3j-1}\,a\,a\ v_3$, $v_2, v_3 \in \Sigma^*$, for all $x_j \in \mathrm{var}(\beta_2\,\beta_2')$. Consequently, $\sigma'(x_{23}\,x_{14}\,x_4\,x_{24}) = v_4\ a\,a\,e^{41}\,a\,a\,w\,c\,c\,e^{11}\,c\,c\ v_5$, $v_4, v_5 \in \Sigma^*$, for some $w \in \{a, c, e\}^*$. However, for every occurrence of this subword in $\sigma(\tilde{\alpha}_{\mathrm{abcd},1})$—or, more precisely, in $\sigma(\hat{\alpha}_2)$—we have $w = v_6\ a\,e^{44}\,a\ v_7$, $v_6, v_7 \in \Sigma^*$ (see suffix of $\sigma(\tilde{\alpha}_{\mathrm{abcd},1})$ as depicted above). Thus, we may conclude $\sigma'(x_{15}) \neq v_8\ a\,e^{44}\,a\ v_9$, $v_8, v_9 \in \Sigma^*$, since the frequency of the subword $a\,e^{44}\,a$ in $\sigma(\tilde{\alpha}_{\mathrm{abcd},1})$ equals $|\tilde{\alpha}_{\mathrm{abcd},2}|_{x_{15}}$ and since at least one occurrence of $a\,e^{44}\,a$— in fact, it is even all six occurrences in $\sigma(\hat{\alpha}_2)$—is contained in $\sigma'(x_{14}\,x_4)$. This contradicts the claim $\sigma'(x_j) = v_2\ a\ a\,e^{3j-1}\,a\ a\ v_3$ for all $x_j \in \mathrm{var}(\beta_2\,\beta_2')$.

Consequently, there is no substitution σ' with $\sigma'(\tilde{\alpha}_{\mathrm{abcd},2}) = \sigma(\tilde{\alpha}_{\mathrm{abcd},1})$. $\square$

Thus, with Lemma 1 and Lemma 4, Theorem 1 is proven. Moreover, the proof of Lemma 4 shows that our way of composing example patterns cannot directly be used for the transition between $|\Sigma| = 5$ and $|\Sigma| = 6$. The argumentation on Lemma 1 is based on the fact that every substitution either matches the "easier" Cases 1 - 6 or exactly reconstructs the terminal substring of the pattern (see Case 7). We are uncertain whether these substitutions can be avoided for all patterns—and not only for our examples—in case of $|\Sigma| \geq 5$.

3.2   Some Notes

The proof of Lemma 4 can be extended canonically such that in addition to $L_{\Sigma_2}(\tilde{\alpha}_{\mathrm{abcd},2}) \not\supseteq L_{\Sigma_2}(\tilde{\alpha}_{\mathrm{abcd},1})$ the opposite direction $L_{\Sigma_2}(\tilde{\alpha}_{\mathrm{abcd},1}) \not\supseteq L_{\Sigma_2}(\tilde{\alpha}_{\mathrm{abcd},2})$ is shown. Consequently, both languages are incomparable, and it seems as if, for $|\Sigma| = 4$ and $|\Sigma'| > 4$, there is no pair of patterns $\alpha, \beta \in \mathrm{Pat}_\Sigma$ such that $L_\Sigma(\alpha) = L_\Sigma(\beta)$ and $L_{\Sigma'}(\alpha) \subset L_{\Sigma'}(\beta)$. In contrast to this, for smaller alphabets there are patterns that possess such a feature, for instance

- $\alpha = x_1^2$ and $\beta = x_1^2 x_2^2$ for the transition $|\Sigma| = 1$ vs. $|\Sigma| = 2$,
- $\alpha = x_1 \, \mathsf{a} \, \mathsf{b} \, x_2$ and $\beta = x_1 \, \mathsf{a} \, x_2 \, \mathsf{b} \, x_3$ for the transition $|\Sigma| = 2$ vs. $|\Sigma| = 3$, and
- $\alpha = \tilde{\alpha}_{\mathsf{abc},2}$ and $\beta = \tilde{\alpha}_{\mathsf{abc},1}$ for the transition $|\Sigma| = 3$ vs. $|\Sigma| = 4$.

In this context, we conjecture that, for an alphabet Σ with four letters and morphically semi-coincident patterns $\alpha, \beta \in \mathrm{Pat}_\Sigma$, necessarily $L_\Sigma(\alpha) \neq L_\Sigma(\beta)$. Particularly with regard to Theorem 2, we consider this fairly counter-intuitive.

We conclude this paper with a hint on a potential problem concerning any common normal form for $\tilde{\alpha}_{\mathsf{abcd},1}$ and $\tilde{\alpha}_{\mathsf{abcd},2}$: We conjecture that both patterns are succinct for all alphabets with at least four letters. If this is correct then, for $|\Sigma| = 4$, not only the concrete algorithm in [11] has to fail (as shown in Theorem 2), but any suchlike approach as there are E-pattern languages that presumably do not have a "natural" unique shortest normal form.

Acknowledgements. The author wishes to thank Rolf Wiehagen and a referee for their helpful comments on a draft of this paper.

References

1. D. Angluin. Finding patterns common to a set of strings. *J. Comput. Syst. Sci.*, 21:46–62, 1980.
2. D.R. Bean, A. Ehrenfeucht, and G.F. McNulty. Avoidable patterns in strings of symbols. *Pacific J. Math.*, 85:261–294, 1979.
3. C. Choffrut and J. Karhumäki. Combinatorics of words. In G. Rozenberg and A. Salomaa, editors, *Handbook of Formal Languages*, volume 1, chapter 6, pages 329–438. Springer, 1997.
4. G. Dány and Z. Fülöp. A note on the equivalence problem of E-patterns. *Inf. Process. Lett.*, 57:125–128, 1996.
5. G. Filè. The relation of two patterns with comparable language. In *Proc. STACS 1988*, volume 294 of *LNCS*, pages 184–192, 1988.
6. T. Harju and J. Karhumäki. Morphisms. In G. Rozenberg and A. Salomaa, editors, *Handbook of Formal Languages*, volume 1, chapter 7, pages 439–510. Springer, 1997.
7. T. Jiang, E. Kinber, A. Salomaa, K. Salomaa, and S. Yu. Pattern languages with and without erasing. *Int. J. Comput. Math.*, 50:147–163, 1994.
8. T. Jiang, A. Salomaa, K. Salomaa, and S. Yu. Decision problems for patterns. *J. Comput. Syst. Sci.*, 50:53–63, 1995.
9. A. Mateescu and A. Salomaa. Finite degrees of ambiguity in pattern languages. *RAIRO Inform. théor.*, 28(3–4):233–253, 1994.
10. A. Mateescu and A. Salomaa. Patterns. In G. Rozenberg and A. Salomaa, editors, *Handbook of Formal Languages*, volume 1, chapter 4.6, pages 230–242. Springer, 1997.
11. E. Ohlebusch and E. Ukkonen. On the equivalence problem for E-pattern languages. *Theor. Comp. Sci.*, 186:231–248, 1997.
12. D. Reidenbach. A discontinuity in pattern inference. In *Proc. STACS 2004*, volume 2996 of *LNCS*, pages 129–140, 2004.
13. D. Reidenbach. On the learnability of E-pattern languages over small alphabets. In *Proc. COLT 2004*, volume 3120 of *LNAI*, pages 140–154, 2004.
14. G. Rozenberg and A. Salomaa. *Handbook of Formal Languages*, volume 1. Springer, Berlin, 1997.

15. T. Shinohara. Polynomial time inference of extended regular pattern languages. In *Proc. RIMS Symp.*, volume 147 of *LNCS*, pages 115–127, 1982.

16. A. Thue. Über unendliche Zeichenreihen. *Kra. Vidensk. Selsk. Skrifter. I Mat. Nat. Kl.*, 7, 1906.

17. A. Thue. Über die gegenseitige Lage gleicher Teile gewisser Zeichenreihen. *Kra. Vidensk. Selsk. Skrifter. I Mat. Nat. Kl.*, 1, 1912.

Complementation of Rational Sets on Countable Scattered Linear Orderings

Chloé Rispal[1] and Olivier Carton[2]

[1] IGM, Université de Marne-la-Vallée, Marne-la-Vallée, France
chloe.rispal@univ-mlv.fr
[2] LIAFA, Université Paris 7, Paris, France
Olivier.Carton@liafa.jussieu.fr

Abstract. In a preceding paper (Bruyère and Carton, automata on linear orderings, MFCS'01), automata have been introduced for words indexed by linear orderings. These automata are a generalization of automata for finite, infinite, bi-infinite and even transfinite words studied by Büchi. Kleene's theorem has been generalized to these words. We prove that rational sets of words on countable scattered linear ordering are closed under complementation using an algebraic approach.

1 Introduction

In his seminal paper [12], Kleene showed that automata on finite words and regular expressions have the same expressive power. Since then, this result has been extended to many classes of structures like infinite words [6, 15], bi-infinite words [10, 16], transfinite words [8, 1], traces, trees, pictures.

In [4], automata accepting linear-ordered structures have been introduced with corresponding rational expressions. These linear structures include finite words, infinite, transfinite words and their mirrors. These automata are usual automata on finite words, extended with limit transitions. A Kleene-like theorem was proved for words on countable scattered linear orderings. Recall that an ordering is scattered if it does not contain a dense subordering isomorphic to $\mathbb{Q}$.

For many structures, the class of rational sets is closed under many operations like substitutions, inverse substitutions and boolean operations. As for boolean operations, the closure under union and intersection are almost always easy to get. The closure under complementation is often much more difficult to prove. This property is important both from the practical and the theoretical point of view. It means that the class of rational sets forms an effective boolean algebra. It is used whenever some logic is translated into automata. For instance, in both proofs of the decidability of the monadic second-order theory of the integers by Büchi [7] and the decidability of the monadic second-order theory of the infinite binary tree by Rabin [19], the closure under complementation of automata is the key property.

C.S. Calude, E. Calude and M.J. Dinneen (Eds.): DLT 2004, LNCS 3340, pp. 381–392, 2004.

In [4], the closure under complementation was left as an open problem. In this paper, we solve that problem in a positive way. We show that the complement of a rational set of words on countable scattered linear orderings is also rational.

The classical method to get an automaton for the complement of a set of finite words accepted by an automaton $\mathcal{A}$ is through determinization. It is already non-trivial that the complement of a rational set of infinite words is also rational. The determinization method cannot be easily extended to infinite words. In his seminal paper [7], Büchi used another approach based on a congruence on finite words and Ramsey's theorem. This method is somehow related to our algebraic approach. McNaughton extended the determinization method to infinite words [13] proving that any Büchi automaton is equivalent to a deterministic Muller automaton. Büchi pushed further this method and extended it to transfinite words [8]. It is then very complex. In [3], the algebraic approach was used to give another proof of the closure under complementation for transfinite words. In [9], we have already proved the result for words on countable scattered linear orderings of finite ranks. The determinization method cannot be applied because any automaton is not equivalent to a deterministic one. In that paper, we extended the method used by Büchi in [7] using an additional induction on the rank. Since ranks of countable scattered linear orderings range over all countable ordinals, this approach is not suitable for words on all these orderings. In this paper, we prove the whole result for all countable scattered linear orderings using an algebraic approach. We define a generalization of semigroups, called ◇-semigroups. We show that, when finite, these ◇-semigroups are equivalent to automata. We also show that, by analogy with the case of finite words, a canonical ◇-semigroup, called the syntactic ◇-semigroup, can be associated with any rational set X. It has the property of being the smallest ◇-semigroup recognizing X. A continuation of this paper would be to extend the equivalence between star-free sets, first order logic and aperiodic semigroups [22, 14, 2] and also between rational sets and the monadic second order theory.

Both hypotheses that the orderings are scattered and countable are really necessary. Büchi already pointed out that rational sets of transfinite words of length greater that ω_1 (the least non-countable ordinal) are not closed under complement. It can be proved that the set of words on scattered linear orderings is not rational as a subset of words on all linear orderings although its complement is rational.

Our proof of the complementation closure is effective. Given an automaton $\mathcal{A}$, it gives another automaton $\mathcal{B}$ that accepts words that are not accepted by $\mathcal{A}$. It gives another proof of the decidability of the equivalence of these automata [5].

This paper is organized as follows. Definitions concerning linear orderings and rational sets are first recalled in Sections 2 and 3. Then, Section 4 introduces the algebraic structure of ◇-semigroup. The proof of equivalence between finite ◇-semigroups and automata is sketched in both directions in Sections 5 and 6. Finally, the syntactic ◇-semigroup corresponding to a rational set is defined in Section 7.

2   Words on Linear Orderings

This section recalls basic definitions on linear orderings but the reader is referred to [21] for a complete introduction. Hausdorff's characterization of countable scattered linear orderings is given and words indexed by linear orderings are introduced.

Let J be a set equipped with an order $<$. The ordering J is *linear* if for any j and k in J, either $j < k$ or $k < j$. Let A be a finite alphabet. A *word* $x = (a_j)_{j \in J}$ indexed by a linear ordering J is a function from J to A. J is called the *length* of x. For instance ω is the length of right-infinite words $a_0 a_1 ...$ and ζ is the length of bi-infinite words $... a_{-1} a_0 a_1 ...$.

2.1   Product of Words Indexed by Linear Orderings

For any linear ordering J, we denote by $-J$ the backward linear ordering that is the set J equipped with the reverse ordering. For instance, $-\omega$ is the linear ordering of negative integers.

The sum $J + K$ of two linear orderings is the set $J \cup K$ equipped with the ordering $<$ extending the orderings of J and K by setting $j < k$ for any $j \in J$ and $k \in K$. Formally, the *sum* $\sum_{j \in J} K_j$ is the set of all pairs (k, j) such that $k \in K_j$ equipped with the ordering defined by $(k_1, j_1) < (k_2, j_2)$ if and only if $j_1 < j_2$ or $(j_1 = j_2$ and $k_1 < k_2$ in $K_{j_1})$.

The sum of linear orderings helps to define the products of words. Let J be a linear ordering and let $(x_j)_{j \in J}$ be words of respective length K_j for any $j \in J$. The word $x = \prod_{j \in J} x_j$ obtained by concatenation of the words x_j with respect to the ordering on J is of length $L = \sum_{j \in J} K_j$. For instance, if for any $j \in \omega$, we denote by $x_j = a^{\omega^j}$, then $x = \prod_{j \in \omega} x_j$ is the word $x = a^{\omega^\omega}$ of length $\sum_{j \in \omega} \omega^j = \omega^\omega$.

The sequence $(x_j)_{j \in J}$ of words is called a J-*factorization* of the word $x = \prod_{j \in J} x_j$.

2.2   Scattered Linear Orderings

A linear ordering J is *dense* if for any j and k in J such that $j < k$, there exists an element i of J such that $j < i < k$. It is *scattered* if it contains no dense subordering. The ordering ω of natural integers and the ordering ζ of relative integers are scattered. More generally, ordinals are scattered orderings. We denote by $\mathcal{N}$ the subclass of finite linear orderings, $\mathcal{O}$ the class of countable ordinals and of $\mathcal{S}$ the class of countable scattered linear orderings. The following characterization of scattered orderings is due to Hausdorff.

Theorem 1. *[Hausdorff [11]] A countable linear ordering J is scattered if and only if J belongs to $\bigcup_{\alpha \in \mathcal{O}} V_\alpha$ where the classes V_α are inductively defined by:*

1. $V_0 = \{\mathbf{0}, \mathbf{1}\}$

2. $V_\alpha = \{\sum_{j \in J} K_j \mid J \in \mathcal{N} \cup \{\omega, -\omega, \zeta\} \text{ and } K_j \in \bigcup_{\beta < \alpha} V_\beta\}.$

where $\mathbf{0}$ and $\mathbf{1}$ are respectively the orderings with zero and one element.

In order to simplify the proofs, we use slightly different inductive classes: For any $\alpha \in \mathcal{O}$, the class W_α is defined by : $W_\alpha = \{\sum_{j \in J} K_j \mid J \in \mathcal{N} \text{ and } K_j \in V_\alpha\}.$
The inclusions $V_\alpha \subset W_\alpha \subset V_{\alpha+1}$ hold for any ordinal α thus, using Theorem 1, scattered linear orderings can be defined from the classes W_α by: $\mathcal{S} = \bigcup_{\alpha \in \mathcal{O}} W_\alpha.$
The *rank* of a linear ordering J is the smallest ordinal α such that $J \in W_\alpha$. We denote by $A^\diamond$ the set of all words over A indexed by countable scattered linear orderings.

3   Rational Sets of Words on Linear Orderings

Bruyère and Carton have introduced rational expressions and automata for words indexed by countable scattered linear orderings. They have proved that a set of words is rational if and only if it is accepted by a finite automaton extending Kleene's theorem. This section shortly recalls definitions of rational operations and automata but the reader is referred to [4] for more details.

3.1   Rational Expressions

Let A be a finite alphabet. The set $Rat(A^\diamond)$ of rational sets of words over A indexed by countable scattered linear orderings is the smallest set containing $\{a\}$ for any $a \in A$ and closed under the following rational operations defined for any subsets X and Y of $A^\diamond$ by :

$$X + Y = \{z \mid z \in X \cup Y\}$$
$$X \cdot Y = \{x \cdot y \mid x \in X, y \in Y\} \qquad X^* = \{\prod_{j=1}^{n} x_j \mid n \in \mathcal{N}, x_j \in X\}$$
$$X^\omega = \{\prod_{j \in \omega} x_j \mid x_j \in X\} \qquad X^{-\omega} = \{\prod_{j \in -\omega} x_j \mid x_j \in X\}$$
$$X^\# = \{\prod_{j \in \alpha} x_j \mid \alpha \in \mathcal{O}, x_j \in X\} \, X^{-\#} = \{\prod_{j \in -\alpha} x_j \mid \alpha \in \mathcal{O}, x_j \in X\}$$
$$X \diamond Y = \{\prod_{j \in J \cup \hat{J}^*} z_j \mid J \in \mathcal{S} \setminus \emptyset, z_j \in X \text{ if } j \in J \text{ and } z_j \in Y \text{ if } j \in \hat{J}^*\} \text{ where}$$
$$\hat{J}^* = \hat{J} \setminus \{(\emptyset, J), (J, \emptyset)\}.$$

The notation $\hat{J}$ is defined in the next section.

3.2   Automata on Linear Orderings

An automaton on linear orderings is a classical finite automaton with additional limit transitions of the form $P \longrightarrow q$ or $q \longrightarrow P$ where P is a set of states.

Definition 1. *An automaton $\mathcal{A} = (Q, A, E, I, F)$ on linear orderings is defined by a finite set of states Q, a finite alphabet A, a set of transitions $E \subseteq (Q \times A \times$*

$Q) \cup (\mathcal{P}(Q) \times Q) \cup (Q \times \mathcal{P}(Q))$ *and initial and final sets of states* $I \subseteq Q$ *and* $F \subseteq Q$.

The definition of paths is based on the notion of cut that we explain now: Let x be a word indexed by an ordering $J \in \mathcal{S}$. To any two-factorization $x = yz$ of x, one can associate a partition of J into two intervals (K, L) such that $|y| = K$ and $|z| = L$. Such a partition is called a *cut* of J. The set $\hat{J} = \{(K, L) | K \cup L = J \wedge \forall k \in K, \forall l \in L, k < l\}$ is the set of cuts of the ordering J. Then, a path labelled x is a function from the set $\hat{J}$ into the set of states. As the set $\hat{J}$ is naturally equipped with the ordering $(K_1, L_1) < (K_2, L_2)$ if and only if $K_1 \subset K_2$, a path labelled by a word of length J is a word over Q of length $\hat{J}$.

Let $\gamma = (q_c)_{c \in \hat{J}}$ be a word of length $\hat{J}$ over Q, the limit sets of states of γ at a given cut c of $\hat{J}$ are defined by:

$$\lim_{c^-} \gamma = \{q \in Q | \forall c' < c, \exists c'' \ \ c' < c'' < c \text{ and } q = q_{c''}\}$$

$$\lim_{c^+} \gamma = \{q \in Q | \forall c' > c, \exists c'' \ \ c < c'' < c' \text{ and } q = q_{c''}\}$$

Definition 2. *Let* $\mathcal{A} = (Q, A, E, I, F)$ *be an automaton on linear orderings and let* $x = (a_j)_{j \in J}$ *be a word of length* J *on* A. *A path* γ *of label* x *in* $\mathcal{A}$ *is a word* $\gamma = (q_c)_{c \in \hat{J}}$ *of length* $\hat{J}$ *over* Q *such that for any* $(K, L) \in \hat{J}$:

- *If there exists* $l \in L$ *such that* $(K \cup \{l\}, L \setminus \{l\}) \in \hat{J}$
  then $q_{(K,L)} \xrightarrow{a_l} q_{(K \cup \{l\}, L \setminus \{l\})} \in E$ *else* $\lim_{(K,L)^-} \gamma \to q_{(K,L)} \in E$.
- *If there exists* $k \in K$ *such that* $(K \setminus \{k\}, L \cup \{k\}) \in \hat{J}$ *then*
  $q_{(K \setminus \{k\}, L \cup \{k\})} \xrightarrow{a_k} q_{(K,L)} \in E$ *else* $q_{(K,L)} \to \lim_{(K,L)^+} \gamma \in E$.

Thus, if a cut has a predecessor or a successor, usual transitions are used, otherwise the path uses limit transitions.

As $\hat{J}$ has the least element $(\emptyset, J)$ and the greatest element $(J, \emptyset)$ for any linear ordering J, a path has always a first and a last state. A word is *accepted* by an automata if it is the label of a path leading from an initial state to a final state. We denote by $p \xRightarrow{x} q$ the existence of a path leading from the state p to the state q of label x.

It has been proved in [4] that automata and rational expressions have the same expressive power.

$$a$$

$$0 \to \{1\}$$

$$0 \qquad \qquad 1 \qquad \qquad \{0,1\} \to 0$$
$$b$$

Fig. 1. Automaton on linear orderings accepting the set $(a^{-\omega}b)^{\#}$

Theorem 2. *[4] A set of words indexed by countable scattered linear orderings is rational if and only if it is accepted by a finite automata.*

4 Algebraic Characterization of Rational Sets

A semigroup is a set S equipped with an associative binary product. The semigroup S in which had been added a neutral element is denoted by S^1. An element $e \in S$ is an *idempotent* if $e^2 = e$ and the set of idempotents of S is denoted by $E(S)$. A pair $(s, e) \in S \times S$ is *right linked* (respectively left linked) if $e \in E(S)$ and $se = s$ (respectively $es = s$). Two right linked pairs (s_1, e_1) and (s_2, e_2) are *conjugated* if there exists $a, b \in S^1$ such that $e_1 = ab$, $e_2 = ba$, $s_1a = s_2$ and $s_2b = s_1$. The conjugacy relation is an equivalence relation on right linked pairs [17].

4.1 ⬦-Semigroups

The product of semigroups is generalized to recognize sets of words indexed by countable scattered linear orderings. A ⬦-semigroup is a generalization of a usual semigroup. The product of a sequence indexed by any scattered ordering is defined.

Definition 3. *A ⬦-semigroup is a set S equipped with product $\pi : S^\diamond \longrightarrow S$ which maps any word of countable scattered linear length over S to an element of S.*

- *for any element s of S, $\pi(s) = s$.*
- *for any word x over S of countable scattered linear length and for any factorization $x = \prod_{j \in J} x_j$ where $J \in \mathcal{S}$,*

$$\pi(x) = \pi(\prod_{j \in J} \pi(x_j))$$

The latter condition is a generalization of associativity.

For instance, the set $A^\diamond$ equipped with the concatenation is a ⬦-semigroup.

Example 1. The set $S = \{0, 1\}$ equipped with the product π defined for any $u \in S^\diamond$ by $\pi(u) = 0$ if u has at least one occurence of the letter 0 and $\pi(u) = 1$ otherwise is a ⬦-semigroup.

For any two elements s and t of a ⬦-semigroup (S, π), the finite product $\pi(st)$ is merely denoted by st.

A *sub-⬦-semigroup* T of a ⬦-semigroup S is a subset of S closed under product. A *morphism of ⬦-semigroup* is an application which preserves the product. A *congruence* of ⬦-semigroup is an equivalence relation $\sim$ stable under product: If $s_j \sim t_j$ for any $j \in J$, then $\pi(\prod_{j \in J} s_j) \sim \pi(\prod_{j \in J} t_j)$. The set $S/\sim$ is a ⬦-semigroup. A ⬦-semigroup T is a *quotient* of a ⬦-semigroup S if there exists an onto morphism from S to T. A ⬦-semigroup T *divides* S if T is the quotient of a sub-⬦-semigroup of S.

4.2   Finite ◇-Semigroups

A ◇-semigroup (S, π) is said to be finite if S is finite. Even when S is finite, the function π is not easy to describe because the product of any sequence has to be given. It turns out that the function π can be described using a semigroup structure on S with two additional functions (called τ and $-\tau$) from S to S. This gives a finite description of the function π. The functions τ and $-\tau$ are the counterpart of limit transitions of automata. This finite description is based on the next Lemma which follows directly from Ramsey's Theorem [20].

Let $x = \prod_{i \in \omega} x_i$ an ω-factorization. Another factorization $x = \prod_{i \in \omega} y_i$ is called a *superfactorization* if there is a sequence $(k_i)_{i \in \omega}$ of integers such that $y_0 = x_0 \ldots x_{k_0}$ and $y_i = x_{k_{i-1}+1} \ldots x_{k_i}$ for all $i \geq 1$.

Lemma 1. *Let* $\varphi : A^\diamond \longrightarrow S$ *be a morphism into a finite ◇-semigroup. For any factorization* $x = \prod_{i \in \omega} x_i$, *there exists a superfactorization* $x = \prod_{i \in \omega} y_i$ *and a right linked pair* $(s, e) \in S \times E(S)$ *such that* $\varphi(y_0) = s$ *and* $\varphi(y_i) = e$ *for any* $i > 0$.

Such a factorization is called a *ramseyan factorization*, see Theorem 3.2 in [18].

Definition 4. *Let* S *be a semigroup. A function* $\tau : S \longrightarrow S$ *(respectively* $-\tau : S \longrightarrow S$) *is* compatible to the right *with* S *(respectively* to the left) *if and only if for any* s, t *in* S *and any integer* n *the following properties hold:* $s(ts)^\tau = (st)^\tau$ *and* $(s^n)^\tau = s^\tau$ *(respectively* $(st)^{-\tau}s = (ts)^{-\tau}$ *and* $(s^n)^{-\tau} = s^{-\tau}$).

The product of a finite ◇-semigroup S can be finitely described by functions compatible to the right and to the left with S.

Theorem 3. *Let* (S, π) *be a finite ◇-semigroup. The binary product defined for any* s, t *in* S *by* $s \cdot t = \pi(st)$ *naturally endows a structure of semigroup and the functions* τ *and* $-\tau$ *respectively defined by* $s^\tau = \pi(s^\omega)$ *and* $s^{-\tau} = \pi(s^{-\omega})$ *are respectively compatible to the right and to the left with* S.

Conversely, let S *be a finite semigroup and let* τ *and* $-\tau$ *be functions respectively compatible to the right and to the left with* S. *Then* S *can be uniquely endowed with a structure of ◇-semigroup* (S, π) *such that* $s^\tau = \pi(s^\omega)$ *and* $s^{-\tau} = \pi(s^{-\omega})$.

The first part of the theorem follows directly from the associativity of the product π. Conversely, let S be a finite semigroup and let τ and $-\tau$ be functions respectively compatible to the right and to the left with S. The product of a word $x = (s_j)_{j \in J}$ over S of length $J \in \mathcal{S}$ is defined by induction on $\alpha \in \mathcal{O}$ for any $J \in W_\alpha$ by the following way:

Let $J \in W_0$ and let $x \in S^J$. There exists an integer m and $s_1, \ldots, s_m$ in S such that $x = s_1 \ldots s_m$. We set $\pi(x) = s_1 \cdot s_2 \ldots s_m$.

Let $J \in W_\alpha$ where $\alpha > 1$ and let $x \in S^J$. The linear ordering J can be decomposed as a sum $J = \sum_{i \in I} K_i$ where $I \in \mathcal{N} \cup \{\omega, -\omega\}$ and for all $i \in I$, $K_i \in \bigcup_{\beta < \alpha} W_\beta$. There exists a factorization $x = \prod_{i \in I} x_i$ such that for all $i \in I$, $|x_i| = K_i$.

- $J = \{1, \ldots, m\} \in \mathcal{N}$: we set $\pi(x) = \pi(x_1) \ldots \pi(x_m)$.
- $J = \omega$: There exists a superfactorization $x = \prod_{i \in \omega} y_i$ and a right linked pair $(s, e) \in S \times E(S)$ such that $\varphi(y_0) = s$ and $\varphi(y_i) = e$ for any $i > 0$. We set $\pi(x) = se^\tau$.
- $J = -\omega$: Symmetrically to the previous case, we set $\pi(x) = e^{-\tau}s$.

Since two linked pairs associated with two factorizations of a word are conjugated [18], it can be proved by induction on α that π is uniquely defined and associative on $S^\diamond$.

Example 2. The $\diamond$-semigroupe $S = \{0, 1\}$ of Example 1 is defined by the finite product $00 = 01 = 10 = 0$ and $11 = 1$ and by the compatible functions τ and $-\tau$ defined by $0^\tau = 0^{-\tau} = 0$ and $1^\tau = 1^\tau = 1$.

4.3   Recognizability

It is well known that rational sets of finite words are exactly those recognized by finite semigroups. This result is generalized for words indexed by countable scattered linear orderings.

Definition 5. *Let S and T be two $\diamond$-semigroups. The $\diamond$-semigroup T recognizes a subset X of S if and only if there exists a morphism $\varphi : S \longrightarrow T$ and a subset $P \subseteq T$ such that $X = \varphi^{-1}(P)$. A set $X \subseteq A^\diamond$ is recognizable if and only if there exists a **finite** $\diamond$-semigroup recognizing it.*

Example 3. The set $S = \{0, 1\}$ equipped with the product π defined for any $u \in S^\diamond$ by $\pi(u) = 1$ if $u \in 1^\#$ and $\pi(u) = 0$ otherwise is a $\diamond$-semigroup. It is also defined by the finite product $00 = 01 = 10 = 0$ et $11 = 1$ and by the compatible functions τ and $-\tau$ defined by $0^\tau = 0^{-\tau} = 1^{-\tau} = 0$ and $1^\tau = 1$. Define the morphism of $\diamond$-semigroup $\varphi : A^\diamond \longrightarrow S$ by $\varphi(a) = 1$ for any $a \in A$. The set $A^\#$ is recognizable since $A^\# = \varphi^{-1}(\{1\})$.

For any finite alphabet A, $Rec(A^\diamond)$ denotes the set of subsets of $A^\diamond$ recognizable by a finite $\diamond$-semigroup.

Theorem 4. *A set of words indexed by countable scattered linear orderings is rational iff it is recognizable.*

Example 4. The set $X = (ab)^\diamond$ is recognized by the $\diamond$-semigroup $S = \{s, t, e, f, 0\}$ whose product is defined by $st = e$, $ts = f$, $ee = e$, $ff = f$, $es = s$, $ft = t$, $sf = s$, $te = t$, $e^\tau = e$, $e^{-\tau} = e$, $f^\tau = t$, $f^{-\tau} = s$ where any other product is equal to 0. Defining the morphism $\varphi : A^\diamond \to S$ by $\varphi(a) = s$ and $\varphi(b) = t$, we get $X = \varphi^{-1}(e)$.

If X is recognized by a morphism $\varphi : S \longrightarrow T$, the set $A^\diamond \setminus X$ is also recognized by φ since $A^\diamond \setminus X = \varphi^{-1}(S \setminus P)$. Therefore, we obtain following theorem.

Theorem 5. *Rational sets of words on countable scattered linear orderings are closed under complementation.*

Example 5. The set $X = A^*$ is recognized by the $\diamond$-semigroup $S = \{0,1\}$ whose product is defined by $11 = 1$, $01 = 10 = 00 = 0$ and by the compatible functions $0^\tau = 0^{-\tau} = 1^\tau = 1^{-\tau} = 0$. Define the morphism $\varphi : A^\diamond \to S$ by $\varphi(a) = 1$ for any $a \in A$. One gets $X = \varphi^{-1}(1)$ and the complement $A^\diamond \setminus X = (A^\diamond)^\omega A^\diamond + A^\diamond (A^\diamond)^{-\omega} = \varphi^{-1}(0)$.

The next section is devoted to sketches of proof of Theorem 4.

5   From $\diamond$-Semigroups to Automata

Let (S, π) be a finite $\diamond$-semigroup. By Theorem 3, the product π is defined by compatible functions τ and $-\tau$. Let X be a subset of $A^\diamond$ recognized by S. There exists a morphism of $\diamond$-semigroup $\varphi : A^\diamond \longrightarrow S$ and a subset P of S such that $X = \varphi^{-1}(P)$. Since rational sets are closed under finite union, one may suppose that P is a single element $\{p\}$. Let h be the finite substitution which associates to each element s of S the set $\varphi^{-1}(s) \cap A$. Since $X = h(\pi^{-1}(p) \cap \varphi(A)^\diamond)$, it suffices to prove that the set $\pi^{-1}(p)$ of words over S whose product is p, is rational. Recall that the Green's relations are defined from the following preorders:

$$s \leq_\mathcal{R} t \iff \exists a \in S^1, \; s = ta$$
$$s \leq_\mathcal{L} t \iff \exists a \in S^1, \; s = at$$
$$s \leq_\mathcal{J} t \iff \exists a, b \in S^1, \; s = atb$$

For any $\mathcal{K} \in \{\mathcal{R}, \mathcal{L}, \mathcal{J}\}$, $s\mathcal{K}t$ if and only if $s \leq_\mathcal{K} t$ and $t \leq_\mathcal{K} s$. We also denote by $s <_\mathcal{K} t$ iff $s \leq_\mathcal{K} t$ and not $t \leq_\mathcal{K} s$. Recall that the equivalence relation $\mathcal{D} = \mathcal{R}\mathcal{L} = \mathcal{L}\mathcal{R}$ is equal to $\mathcal{J}$ when S is finite.

The proof is by induction on the $\mathcal{D}$-class structure of S. For any $\mathcal{D}$-class D of S, denote by:

$$S_D = \{s \in S \mid \forall p \in D, s \geq_\mathcal{J} p\} \quad \text{and} \quad T_D = \{s \in S \mid \forall p \in D, s >_\mathcal{J} p\}$$

We define an automaton on linear orderings accepting words over S_D and computing the product π of its path's labels in both directions.

Let $\mathcal{A}_D = (Q_D, S_D, E_D)$ be the automaton defined by:

$Q_D = S_D^1 \times S_D^1 \times \mathbb{B}$ is the set of states where $\mathbb{B} = \{0,1\}$

$E_D = \{(s, rt, b) \overset{r}{\longrightarrow} (sr, t, b') \mid b \in \mathbb{B}, b' = (r \in D)\}$

$\quad \cup \; \{\{(s_i, t_i, b_i)\}_{1 \leq i \leq m} \longrightarrow (s, t, b) \mid b \in \mathbb{B}, \exists 1 \leq i \leq m, \; b_i = 1,$
$\quad \exists 1 \leq k \leq m, \; \exists e \in E(D), \; s_k e = s_k, \; et_k = t_k, \; s = s_k e^\tau \text{ and } t_k = e^\tau t\}$

$\quad \cup \; \{(s, t, b) \longrightarrow \{(s_i, t_i, b_i)\}_{1 \leq i \leq m} \mid b \in \mathbb{B}, \exists 1 \leq i \leq m, \; b_i = 1,$
$\quad \exists 1 \leq k \leq m, \; \exists e \in E(D), \; s_k e = s_k, \; et_k = t_k, \; t = e^{-\tau} t_k \text{ and } s_k = s e^{-\tau}\}$

The boolean component of Q_D allows limit transitions only if the label of the path admits a ramseyan factorization associated to an idempotent of D. Since two right linked pairs (s_1, e_1) and (s_2, e_2) of a same $\mathcal{D}$-class are conjugated iff $s_1 \mathcal{R} s_2$, it can be shown by induction on the rank that $\mathcal{A}_D$ computes properly the product π. The words of $S_D^\diamond$ admitting ramseyan factorizations associated to idempotents of $\mathcal{J}$-above $\mathcal{D}$-classes are taken care by a substitution which is rational by induction. Let D be a $\mathcal{D}$-class of S and let f be the rational substitution defined by:

$$f : S_D \longrightarrow Rat(S_D^\diamond)$$
$$s \longrightarrow \begin{cases} \pi^{-1}(s) & \text{if } s \in T_D \\ \{s\} \cup F_s \cup G_s & \text{if } s \in D \end{cases}$$

where for any $s \in D$,

$$F_s = \bigcup_{\substack{s_1, \dots, s_m >_{\mathcal{J}} s, \\ s_1 \dots s_m = s}} \pi^{-1}(s_1) \dots \pi^{-1}(s_m)$$

$$G_s = \bigcup_{\substack{t, e >_{\mathcal{J}} s, \\ te^\tau = s}} \pi^{-1}(t)\pi^{-1}(e)^\omega \cup \bigcup_{\substack{t, e >_{\mathcal{J}} s, \\ e^{-\tau} t = s}} \pi^{-1}(e)^{-\omega}\pi^{-1}(t).$$

If L_s denotes the set of words recognized by the automaton $\mathcal{A}_D$ with the initial state $\{(1, s, 0)\}$ and the final set of states $\{(s, 1, b)| \ b \in \mathbb{B}\}$ for any $s \in S_D$, it can be proved that for any $p \in D$, $f(L_p) = \pi^{-1}(p)$ using another induction on the rank.

6   From Automata to ◇-Semigroups

This proof of the converse is adapted from [3]. Let $\mathcal{A} = (Q, A, E, I, F)$ be an automaton on linear orderings accepting a set $X \subseteq A^\diamond$. The *content* of a path is the set of states occurring in the path and $p \overset{x}{\underset{P}{\Longrightarrow}} q$ denotes a path leading from p to q of label x and of content P. Let $T = \mathcal{P}(Q)$ be the set of all subsets of Q and $K = \mathcal{P}(T)$ be the set of subsets of T. The set K is equipped with the following product and union:

$$kk' = \{t \cup t' \mid t \in k, \ t' \in k'\} \text{ and } k + k' = k \cup k'$$

Let S be the set of all $Q \times Q$ matrices whose entries are in K with product defined by:

$$(m \cdot m')_{q,q'} = \bigcup_{p \in Q} m_{q,p} \cdot m'_{p,q'} = \{t \cup t' \mid \exists p \in Q, \ t \in m_{q,p}, \ t' \in m'_{p,q'}\}$$

The semigroup S is finite and by Theorem 3, it suffices to define compatible functions to endow a structure of ◇-semigroup. Define the function τ by:

$$m^\tau_{q,q'} = \{t \cup \{q'\} \mid \exists t' \subset t, \ \exists p \in Q, \ t \in m^\pi_{q,p}, \ t' \in m^\pi_{p,p} \text{ and } t' \longrightarrow q' \in E\}$$

where π is the smallest integer such that m^π is an idempotent matrix. The function $-\tau$ is defined symmetrically and it can be proved that τ and $-\tau$ are functions respectively compatible on the right and left with S. It remains to define a morphism $\varphi : A^\diamond \longrightarrow S$ recognizing X. For each letter a of A, we define the matrix $m_a = \varphi(a)$ corresponding to the edges of $\mathcal{A}$ labelled by a: The entry (q, q') of m_a is equal to $\{\{q, q'\}\}$ if $q \xrightarrow{a} q' \in E$ or $\emptyset$ otherwise. An induction on the rank would show that for all word $x \in A^\diamond$, $\varphi(x) = m$ where the matrix m memorizes the contents of paths labelled by x:

$$m_{q,q'} = \{l \mid q \underset{l}{\overset{x}{\Longrightarrow}} q'\}$$

A word $x \in A^\diamond$ belongs to X iff $\varphi(x)$ has a (i, f) non-empty entry where i and f are respectively initial and final states. Thus X is recognized by S.

7   Syntactic $\diamond$-Semigroup

Let X be a recognizable subset of $A^\diamond$. Among all $\diamond$-semigroups recognizing X, there exists one which is minimal in the sense of division. It is called the syntactic $\diamond$-semigroup of X and is the first canonical object associated to rational sets on linear orderings. For any $\diamond$-semigroup (S, π) and any set $P \subseteq S$, the equivalence relation $\sim_P$ is defined for any s,t in S by $s \sim_P t$ iff for any integer m:

$$\forall s_1, s_2, \ldots, s_m, t_1, t_2, \ldots, t_m \in S^1 \;,\; \forall \theta_1, \theta_2, \ldots, \theta_{m-1} \in \{\omega, -\omega\} \cup \mathcal{N},$$

$$\pi(s_m(\ldots (s_2(s_1 s t_1)^{\theta_1} t_2)^{\theta_2} \ldots)^{\theta_{m-1}} t_m) \in P$$

$$\Longleftrightarrow \pi(s_m(\ldots (s_2(s_1 t t_1)^{\theta_1} t_2)^{\theta_2} \ldots)^{\theta_{m-1}} t_m) \in P$$

The equivalence relation $\sim_P$ is a congruence of $\diamond$-semigroup. If S finite, then and the quotient $S/\sim_P$ is an effective $\diamond$-semigroup.

If X is a recognizable subset of $A^\diamond$, then the quotient $A^\diamond/\sim_X$ is finite and recognizes X.

Proposition 1. *Let X be a subset of $A^\diamond$. The set X is recognizable if and only if the relation $\sim_X$ is a congruence of $\diamond$-semigroup of finite index.*

For any recognizable subset X of $A^\diamond$, the $\diamond$-semigroup $A^\diamond/\sim_X$ is called the *syntactic semigroup* of X and is denoted by $S(X)$. It is the smallest $\diamond$-semigroup recognizing X in the sense of division.

Proposition 2. *Let X be a recognizable set of $A^\diamond$ and let T be a $\diamond$-semigroup. Then T recognizes X if and only if $S(X)$ divides T.*

In particular, for any recognizable set X, the relation $\sim_X$ is the coarsest congruence such that the quotient $A^\diamond/\sim_X$ recognizes X. From Theorem 4 and Proposition 2, it follows that the syntatic $\diamond$-semigroup of a rational set is finite.

Theorem 6. *A set of words indexed by countable scattered linear orderings is rational iff its syntactic $\diamond$-semigroup is finite.*

References

1. N. Bedon. Finite automata and ordinals. *Theoret. Comput. Sci.*, 156:119–144, 1996.
2. N. Bedon. Star-free sets of words on ordinals. *Inform. Comput.*, 166:93–111, 2001.
3. N. Bedon and O. Carton. An Eilenberg theorem for words on countable ordinals. In Cláudio L. Lucchesi and Arnaldo V. Moura, editors, *Latin'98: Theoretical Informatics*, volume 1380 of *Lect. Notes in Comput. Sci.*, pages 53–64. Springer-Verlag, 1998.
4. V. Bruyère and O. Carton. Automata on linear orderings. In J. Sgall, A. Pultr, and P. Kolman, editors, *MFCS'2001*, volume 2136 of *Lect. Notes in Comput. Sci.*, pages 236–247, 2001. IGM report 2001-12.
5. V. Bruyère, O. Carton, and G. Sénizergues. Tree automata and automata on linear orderings. In Tero Harju and Juhani Karhumäki, editors, *WORDS'2003*, volume 27, pages 222–231, 2003. Turku Center for Computer Science.
6. J. R. Büchi. Weak second-order arithmetic and finite automata. *Z. Math. Logik und grundl. Math.*, 6:66–92, 1960.
7. J. R. Büchi. On a decision method in the restricted second-order arithmetic. In *Proc. Int. Congress Logic, Methodology and Philosophy of science, Berkeley 1960*, pages 1–11. Stanford University Press, 1962.
8. J. R. Büchi. Transfinite automata recursions and weak second order theory of ordinals. In *Proc. Int. Congress Logic, Methodology, and Philosophy of Science, Jerusalem 1964*, pages 2–23. North Holland, 1965.
9. O. Carton and C. Rispal. Complementation of rational sets on scattered linear orderings of finite rank. In *LATIN'04*, 2004.
10. D. Girault-Beauquier. Bilimites de langages reconnaissables. *Theoret. Comput. Sci.*, 33(2–3):335–342, 1984.
11. F. Hausdorff. Set theory. In *Chelsea*, New York, 1957.
12. S. C. Kleene. Representation of events in nerve nets and finite automata. In C.E. Shannon, editor, *Automata studies*, pages 3–41. Princeton university Press, Princeton, 1956.
13. R. McNaughton. Testing and generating infinite sequences by a finite automaton. *Inform. Comput.*, 9:521–530, 1966.
14. R. McNaughton and S. Papert. *Counter free automata*. MIT Press, Cambridge, MA, 1971.
15. D. Muller. Infinite sequences and finite machines. In Proc. of Fourth Annual IEEE Symp., editor, *Switching Theory and Logical Design*, pages 3–16, 1963.
16. M. Nivat and D. Perrin. Ensembles reconnaissables de mots bi-infinis. In *Proceedings of the Fourteenth Annual ACM Symposium on Theory of Computing*, pages 47–59, 1982.
17. J.-P. Pécuchet. Etude syntaxique des parties reconnaissables de mots infinis. volume 226, pages 294–303, 1986.
18. D. Perrin and J.-E. Pin. Infinite words. In Elsevier, editor, *Academic Press*, 2003.
19. M. O. Rabin. Decidability of second-order theories and automata on infinite trees. 141:1–35, 1969.
20. F. D. Ramsey. On a problem of formal logic. *Proc. of the London math. soc.*, 30:338–384, 1929.
21. J. G. Rosenstein. *Linear ordering*. Academic Press, New York, 1982.
22. M.-P. Schützenberger. On finite monoids having only trivial subgroups. *Inform. Control*, 8:190–194, 1965.

On the Hausdorff Measure of
ω-Power Languages

Ludwig Staiger

Martin-Luther-Universität Halle-Wittenberg,
Institut für Informatik, Halle, Germany
staiger@informatik.uni-halle.de

Abstract. We use formal language theory to estimate the Hausdorff
measure of sets of a certain shape in Cantor space. These sets are closely
related to infinite iterated function systems in fractal geometry.

Our results are used to provide a series of simple examples for the
non-coincidence of limit sets and attractors for infinite iterated function
systems.

There are a series of approaches which use means of formal language theory in
order to describe fractals. Another well-known approach for describing fractals
is via iterated function systems (IFS) (see [3]).

In the papers [15, 2, 5] a combination of IFS controlled by finite automata
was introduced for the description of a wider class of fractals. This leads to a
further extension by using arbitrary languages a control structures for IFS (see
[10]).

A different way of generalising IFS was pursued, e.g. in [9, 13, 14] where the
iterated function systems were allowed to contain infinitely many functions.

In this paper we consider infinite iterated function systems (IIFS) in Cantor
space, that is, in a space of infinite words over a finite alphabet. These IIFS are
of a special shape and can be identified with languages of finite words (cf. [20]).
As is known from [13] for IIFS limit sets and their closures (attractors) do not
coincide, in general. We aim at presenting different levels of non-coincidence for
our IIFS in Cantor space using Hausdorff dimension and Hausdorff measure.

It is well-known that the estimation of the Hausdorff dimension and, in particular, of the Hausdorff measure of even rather simply definable sets is already
a complicated task (cf. [6, 7]). It was shown in [19, 16, 10] that results from language theory might facilitate this task. Here we derive results estimating Hausdorff measure for ω-power languages generated by prefix codes and use it to
provide simple examples of languages for the above mentioned non-coincidence
cases. The majority of the examples are also simple in the sense of language theory, that is, they are products of one or two one-turn deterministic one-counter
languages (see [1]).

It should be mentioned that these results are not restricted to the Cantor
space of infinite words, as a direct translation of our results on IIFS to the
unit interval $[0, 1] \subseteq \mathbb{R}$ can be obtained by considering an infinite word $\xi \in
\{0, \ldots, r - 1\}^{\omega}$ as the r-ary expansion $0.\xi$ of a real number. As indicated in

C.S. Calude, E. Calude and M.J. Dinneen (Eds.): DLT 2004, LNCS 3340, pp. 393–405, 2004.
© Springer-Verlag Berlin Heidelberg 2004

[16], this translation generalises easily to unit cubes in d-dimensional space $\mathbb{R}^d$. Moreover, this translation preserves Hausdorff dimension and, up to a certain linear bound, also Hausdorff measure.

1   Notation and Preliminary Results

Next we introduce the notation used throughout the paper. By $\mathbb{N} = \{0, 1, 2, \ldots\}$ we denote the set of natural numbers. Let X be an alphabet of cardinality $|X| = r$. By X^* we denote the set (monoid) of words on X, including the *empty word* e, and X^ω is the set of infinite sequences (ω-words) over X. For $w \in X^*$ and $\eta \in X^* \cup X^\omega$ let $w \cdot \eta$ be their *concatenation*. This concatenation product extends in an obvious way to subsets $W \subseteq X^*$ and $B \subseteq X^* \cup X^\omega$. For a language W let $W^* := \bigcup_{i \in \mathbb{N}} W^i$ be the *submonoid* of X^* generated by W, and by $W^\omega := \{w_1 \cdots w_i \cdots : w_i \in W \setminus \{e\}\}$ we denote the set of infinite strings formed by concatenating words in W. Furthermore $|w|$ is the *length* of the word $w \in X^*$ and $\mathbf{A}(B)$ is the set of all finite prefixes of strings in $B \subseteq X^* \cup X^\omega$. We shall abbreviate $w \in \mathbf{A}(\eta)$ ($\eta \in X^* \cup X^\omega$) by $w \sqsubseteq \eta$.

As usual, a language $V \subseteq X^*$ is called a code provided every word $w \in V^*$ has a unique factorisation into words $v_1, \ldots, v_k \in V$. If for arbitrary $w, v \in V$ the relation $w \sqsubseteq v$ implies $w = v$ the language V is called *prefix-free* or a *prefix code*. Further we denote by $B/w := \{\eta : w \cdot \eta \in B\}$ the *left derivative* of the set $B \subseteq X^* \cup X^\omega$.

For a language $W \subseteq X^*$ let $\mathsf{s}_W : \mathbb{N} \to \mathbb{N}$ where $\mathsf{s}_W(n) := |W \cap X^n|$ be its *structure function*. The *structure generating function* corresponding to s_W is

$$\mathfrak{s}_W(t) := \sum_{i \in \mathbb{N}} \mathsf{s}_W(i) \cdot t^i. \tag{1}$$

$\mathfrak{s}_W$ is a power series with convergence radius $\operatorname{rad} W := \left(\limsup_{n \to \infty} \sqrt[n]{\mathsf{s}_W(n)}\right)^{-1}$. It is convenient to consider $\mathfrak{s}_W$ also as a function mapping $[0, \infty)$ to $[0, \infty) \cup \{\infty\}$.

The convergence radius $\operatorname{rad} W$ is closely related to the entropy of the language, $H_W = \limsup_{n \to \infty} \frac{\log_r(1 + \mathsf{s}_W(n))}{n}$ (cf. [11, 19]).

The parameter $\mathbf{t}_1(W) := \sup\{t : t \geq 0 \wedge \mathfrak{s}_W(t) \leq 1\}$ satisfies $\mathbf{t}_1(W) \leq \operatorname{rad} W^*$. Moreover, it fulfils the following (see [11, 19]).

Lemma 1. $\mathfrak{s}_W(\mathbf{t}_1(W)) = 1$ *iff* $\mathfrak{s}_W(\operatorname{rad} W) \geq 1$, *and, if* $\mathfrak{s}_W(\operatorname{rad} W) < 1$, *then* $\mathbf{t}_1(W) = \operatorname{rad} W = \operatorname{rad} W^*$.

If W *is a code then we have always* $\operatorname{rad} W^* = \mathbf{t}_1(W)$.

We consider the set X^ω as a metric space (Cantor space) (X^ω, ρ) of all ω-words over the alphabet X where the metric ρ is defined as follows.

$$\rho(\xi, \eta) := \inf\{r^{-|w|} : w \sqsubseteq \xi \wedge w \sqsubseteq \eta\} .$$

This space is compact, and the mapping $\phi_w(\xi) := w \cdot \xi$ is a contracting similitude if only $w \neq e$. Thus a language $W \subseteq X^* \setminus \{e\}$ defines a possibly infinite IFS (IIFS) in (X^ω, ρ). Moreover, $\mathcal{C}(F) := \{\xi : \mathbf{A}(\xi) \subseteq \mathbf{A}(F)\}$ is the closure of the set F (smallest closed subset containing F) in (X^ω, ρ).

Next we recall the definition of the Hausdorff dimension of a subset of (X^ω, ρ) (see [19, 23]). For $F \subseteq X^\omega$ and $0 \leq \alpha \leq$ the equation

$$\mathbb{L}_\alpha(F) := \lim_{l \to \infty} \inf\left\{ \sum_{w \in W} r^{-\alpha \cdot |w|} : F \subseteq W \cdot X^\omega \wedge \forall w(w \in W \to |w| \geq l) \right\} \quad (2)$$

defines the α-dimensional metric outer measure on X^ω. $\mathbb{L}_\alpha$ satisfies the following.

Corollary 1. *If $\mathbb{L}_\alpha(F) < \infty$ then $\mathbb{L}_{\alpha+\epsilon}(F) = 0$ for all $\epsilon > 0$.*

Then the *Hausdorff dimension* of F is defined as

$$\dim F := \sup\{\alpha : \alpha = 0 \vee \mathbb{L}_\alpha(F) = \infty\} = \inf\{\alpha : \mathbb{L}_\alpha(F) = 0\}.$$

It should be mentioned that dim is countably stable and shift invariant, that is,

$$\dim \bigcup_{i \in \mathbb{N}} F_i = \sup\{\dim F_i : i \in \mathbb{N}\} \quad \text{and} \quad \dim w \cdot F = \dim F.$$

We list some relations of the Hausdorff dimension and measure for ω-power languages to the properties of the structure generation functions of the corresponding languages (see [19, 16]).

Proposition 1. $\dim W^\omega = -\log_r \operatorname{rad} W^*$

Proposition 2. *If $\alpha = \dim W^\omega$ then $\mathbb{L}_\alpha(W^\omega) \leq 1$.*

If W is a regular language then $0 < \mathbb{L}_\alpha(W^\omega) \leq \mathbb{L}_\alpha(\mathcal{C}(W^\omega)) \leq 1$, and if W is regular and a union of codes then $\mathbb{L}_\alpha(W^\omega) = \mathbb{L}_\alpha(\mathcal{C}(W^\omega))$.

The following direct connections between $\mathfrak{s}_W$ and $\mathbb{L}_\alpha(W^\omega)$ or $\dim W^\omega$ are helpful.

Proposition 3. *1. If $\mathfrak{s}_W(r^{-\alpha}) \leq 1$ then $\alpha \geq \dim W^\omega$.*
2. If $\mathfrak{s}_W(r^{-\alpha}) < 1$ then $\mathbb{L}_\alpha(W^\omega) = 0$.
3. If W is a code and $\mathfrak{s}_W(r^{-\alpha}) > 1$ then $\alpha < \dim W^\omega$.

2   The Hausdorff Measure of ω-Power Languages

As we have seen in Proposition 2 the Hausdorff measure $\mathbb{L}_\alpha(W^\omega)$ varies between 0 and 1 when $\alpha = \dim W^\omega$. In this section we give a more precise estimate, in particular, we derive a formula the measure $\mathbb{L}_\alpha(V^\omega)$ for prefix-free languages V.

To this end we mention the following known properties of the ω-power W^ω.

$$(V \cdot W)^\omega = V \cdot (W \cdot V)^\omega \quad (3)$$
$$(V \cup W)^\omega = (V^* \cdot W)^\omega \cup (V \cup W)^* \cdot V^\omega \quad (4)$$

These properties are called the *rotation* (Eq. (3)) and *union splitting* (Eq. (4)) properties, respectively.

Lemma 2. *Let $w \in \mathbf{A}(V) \setminus V \cdot X \cdot X^*$, that is, $w \sqsubseteq v$ for some $v \in V$ but no $v' \in V$ is a proper prefix of w, and let $W := V \cap w \cdot X^*$ and $\hat{V} := V \setminus W$. Then*

$$V^\omega \cap w \cdot X^\omega = W \cdot V^\omega = W \cdot (\hat{V}^* \cdot W)^\omega \cup W \cdot V^* \cdot \hat{V}^\omega \text{ and} \qquad (5)$$
$$V^\omega / w = (V/w \cdot \hat{V}^* \cdot w)^\omega \cup (V/w) \cdot V^* \cdot \hat{V}^\omega . \qquad (6)$$

Proof. In the first equation the first identity follows from the fact that every $w_1 \cdot w_2 \cdots \in V^\omega$ having $w \sqsubseteq w_1 \cdot w_2 \cdots$ has $w_1 \in W$, and the second one is an application of union splitting (see Eq. (4)) of $(W \cup \hat{V})^\omega$.

The second equation follows from the first one, the rotation property and the observations that $V/w = W/w$ and $w \cdot (V/w) = W$. □

We have the following identity.

Lemma 3. *If $V \subseteq X^*$ is a code, $\alpha \geq \dim V^\omega$, and $w \in \mathbf{A}(V) \setminus V \cdot X \cdot X^*$ then*
$$\mathbb{L}_\alpha(V^\omega/w) = \mathbb{L}_\alpha\Big((V/w \cdot (V \setminus w \cdot X^*)^* \cdot w)^\omega\Big). \text{ In particular, } \mathbb{L}_\alpha(V^\omega/w) \leq 1.$$

Proof. We use $\hat{V} := V \setminus w \cdot X^*$ as in Lemma 2.

Since V is a code, we have $\sum_{v \in V} r^{-\alpha|v|} \leq 1$ for $\alpha \geq \dim V^\omega$. Now $\hat{V} \subset V$ implies $\sum_{v \in \hat{V}} r^{-\alpha|v|} < 1$. Hence $\mathbb{L}_\alpha(\hat{V}^\omega) = 0$ and $\mathbb{L}_\alpha((V/w) \cdot V^* \cdot \hat{V}^\omega) = 0$. Thus, the first assertion follows from Eq. (6), and then the second one from Proposition 2. □

We say that a language $V \subseteq X^*$ satisfies the *countable intersection property* provided V^ω is infinite and the set $w \cdot V^\omega \cap v \cdot V^\omega$ is at most countable for every pair $w, v \in V$, $w \neq v$. It should be noted that every language $V \subseteq X^*$ satisfying the countable intersection property is a code.

Theorem 1. *Assume $V \subseteq X^*$ satisfies the countable intersection property, $\sum_{v \in V} r^{-\alpha|v|} = 1$ for some $\alpha, 0 < \alpha \leq 1$, and $\sum_{w \sqsubseteq v, v \in V} r^{-\alpha|v|} \geq c \cdot r^{-\alpha|w|}$ for some word $w \in \mathbf{A}(V) \setminus V \cdot X \cdot X^*$. Then $\alpha = \dim V^\omega$ and $\mathbb{L}_\alpha(V^\omega) \leq c^{-1}$.*

Proof. $\alpha = \dim V^\omega$ follows from Lemma 1 and Proposition 1.

Set $W := V \cap w \cdot X^*$ and $\hat{V} := V \setminus w \cdot X^*$ as above and observe that $\sum_{v \in \hat{V}} r^{-\alpha|v|} < 1$. As V satisfies the countable intersection property, we have $\mathbb{L}_\alpha(W \cdot V^\omega) = \sum_{v \in W} \mathbb{L}_\alpha(v \cdot V^\omega) = \sum_{w \sqsubseteq v, v \in V} r^{-\alpha|v|} \cdot \mathbb{L}_\alpha(V^\omega)$.

On the other hand, using the identity $W/w = V/w$ we obtain $W \cdot V^\omega = w \cdot (V^\omega/w)$ and from Lemma 3 the inequality $\mathbb{L}_\alpha(W \cdot V^\omega) = r^{-\alpha \cdot |w|} \cdot \mathbb{L}_\alpha(V^\omega/w) \leq r^{-\alpha \cdot |w|}$.

Thus $\mathbb{L}_\alpha(V^\omega) \leq r^{-\alpha \cdot |w|} \cdot \left(\sum_{w \sqsubseteq v, v \in V} r^{-\alpha|v|}\right)^{-1} \geq c^{-1}$. □

Letting the constant c in Theorem 1 tend to infinity (if possible) we obtain the following.

Corollary 2. *Assume $V \subseteq X^*$ satisfies the countable intersection property, $\sum_{v \in V} r^{-\alpha|v|} = 1$ for some $\alpha, 0 < \alpha \leq 1$, and that for all $k \in \mathbb{N}$ there is a word*

$w \in \mathbf{A}(V) \setminus V \cdot X \cdot X^*$ such that $\sum_{w \sqsubseteq v, v \in V} r^{-\alpha|v|} \geq k \cdot r^{-\alpha|w|}$. Then $\alpha = \dim V^\omega$ and $\mathbb{L}_\alpha(V^\omega) = 0$.

A converse to Theorem 1 can be proved for prefix codes.

Theorem 2. Let $V \subseteq X^*$ be a prefix code, $\sum_{v \in V} r^{-\alpha|v|} = 1$ for some $\alpha, 0 < \alpha \leq 1$, and assume that there is a constant $c > 0$ such that $\sum_{w \sqsubseteq v, v \in V} r^{-\alpha|v|} \leq c \cdot r^{-\alpha|w|}$ for all $w \in \mathbf{A}(V)$. Then $\alpha = \dim V^\omega$ and $\mathbb{L}_\alpha(V^\omega) \geq c^{-1}$.

Proof. As above $\sum_{v \in V} r^{-\alpha|v|} = 1$ implies $\dim V^\omega = \alpha$ because V is a code.

Since $\sum_{v \in V} r^{-\alpha|v|} = 1$, in case V is infinite we may choose a sequence of natural numbers $l_n, n \in \mathbb{N}$, such that for $V_n := \{v : v \in V \wedge |v| \leq l_n\}$ we have $p_n := \sum_{v \in V_n} r^{-\alpha|v|} \geq 1 - r^{-(n+1)}$. Then $1 \geq \prod_{i=0}^\infty p_i > 0$. If V is finite, we choose $V_n := V$ for all $n \in \mathbb{N}$.

For technical reasons, we introduce the following concepts depending on the sequence $(l_n)_{n \in \mathbb{N}}$:

$$W := \bigcup_{i=0}^\infty \prod_{n=0}^i V_n \text{ , and} \tag{7}$$

$$l(w) := \min\{i : \exists w'(w \cdot w' \in \prod_{n=0}^i V_n)\} \text{ for } w \in \mathbf{A}(W) \tag{8}$$

Thus $l(w) = i$ for $w \in V_0 \cdots V_i$. In order to use the mass distribution principle ([7–Principle 4.2]) we introduce a set function ν on balls $w \cdot X^\omega$ with $w \in W$:

$$\nu(w \cdot X^\omega) := \prod_{n=0}^{l(w)} \frac{1}{p_n} \cdot r^{-\alpha|w|}$$

Due to the choice of the coefficients p_n we have the identity

$$\sum_{v \in V_{l(w)+1}} \nu(w \cdot v \cdot X^\omega) = \sum_{v \in V_{l(w)+1}} \prod_{n=0}^{l(wv)} \frac{1}{p_n} \cdot r^{-\alpha|wv|}$$

$$= r^{-\alpha|w|} \cdot \prod_{n=0}^{l(w)} \frac{1}{p_n} \cdot \sum_{v \in V_{l(w)+1}} \frac{1}{p_{l(w)+1}} \cdot r^{-\alpha|v|}$$

$$= \left(\prod_{n=0}^{l(w)} \frac{1}{p_n} \right) \cdot r^{-\alpha|w|} = \nu(w \cdot X^\omega)$$

Letting $\nu(u \cdot X^\omega) := 0$ for $u \notin \mathbf{A}(W)$ we observe that ν is extendible to a metric outer measure on X^ω with support $\mathbf{supp}\, \nu = V_0 \cdot V_1 \cdots V_i \cdots \subseteq V^\omega$ and $\nu(\mathbf{supp}\, \nu) = 1$ as follows:

The inclusion $\nu(\mathbf{supp}\,\nu) \cap w \cdot X^\omega \subseteq \bigcup\limits_{\substack{w \sqsubseteq v \\ v \in V_0 \cdots V_{l(w)}}} v \cdot X^\omega \subseteq w \cdot X^\omega$ yields

$$\nu(w \cdot X^\omega) = \sum_{w \sqsubseteq v, v \in V_0 \cdots V_{l(w)}} \nu(v \cdot X^\omega) \qquad \text{for } w \in \mathbf{A}(W) = \mathbf{A}(\mathbf{supp}\,\nu).$$

Therefore, for $F \subseteq w \cdot X^\omega$ with $\operatorname{diam} F = r^{-|w|}$ we have $\nu(w \cdot X^\omega) = 0$ or

$$\nu(F) \le \nu(w \cdot X^\omega) \le \sum_{\substack{w \sqsubseteq v \\ v \in V_0 \cdots V_{l(w)}}} \prod_{n=0}^{l(w)} \frac{1}{p_n} \cdot r^{-\alpha|v|} \le \prod_{n=0}^{\infty} \frac{1}{p_n} \cdot \sum_{\substack{w \sqsubseteq v \\ v \in V_0 \cdots V_{l(w)}}} r^{-\alpha|v|}$$

Now, $w \in \mathbf{A}(W)$ splits uniquely into the product $w = v' \cdot w'$ where $v' \in \prod_{i=1}^{l(w)-1} V_i$ and $w' \in A(V) \setminus V$. Consequently the inequality assumed in the theorem implies

$$\sum_{\substack{w \sqsubseteq v \\ v \in V_0 \cdots V_{l(w)}}} r^{-\alpha|v|} = r^{-\alpha|v'|} \cdot \sum_{\substack{w' \sqsubseteq v \\ v \in V_{l(w)}}} r^{-\alpha|v|} \le c \cdot r^{-\alpha|v'|} \cdot r^{-\alpha|w'|} = c \cdot r^{-\alpha|w|}.$$

Thus, $\nu(F) \le \prod_{n=0}^{\infty} \frac{c}{p_n} \cdot (\operatorname{diam} F)^\alpha$ for all $F \subseteq X^\omega$. So we can apply the mass distribution principle of [7] to obtain $\mathbb{L}_\alpha(V^\omega) \ge \frac{\nu(V^\omega)}{c \cdot \prod_{n=0}^{\infty} p_n^{-1}} = \frac{1}{c} \cdot \prod_{n=0}^{\infty} p_n > 0$.

Since the choice of the sequence $(l_n)_{n \in \mathbb{N}}$ is arbitrary, we can make $\prod_{n=0}^{\infty} p_n$ as close to 1 as possible, and the assertion $L_\alpha(V^\omega) \ge \frac{1}{c}$ follows. $\qquad \square$

Combining Theorems 1 and 2 we obtain the following.

Theorem 3. *Let $V \subseteq X^*$ be a prefix code and $\sum_{v \in V} r^{-\alpha|v|} = 1$ for some $\alpha, 0 < \alpha \le 1$. Then $\alpha = \dim V^\omega$ and $\mathbb{L}_\alpha(V^\omega) = \inf\{\mathfrak{s}_{V/w}(r^{-\alpha})^{-1} : w \in \mathbf{A}(V)\}$.*

Proof. Observe that $\mathfrak{s}_{V/w}(r^{-\alpha}) = r^{\alpha \cdot |w|} \cdot \sum_{w \sqsubseteq v, v \in V} r^{-\alpha|v|}$ whenever $w \in \mathbf{A}(V)$. $\qquad \square$

The next theorem gives a formula for the Hausdorff measure of the ω-power of the product of two prefix codes. The product of two prefix codes is known to be also a prefix code (see [4]).

Theorem 4. *Let $V, W \subseteq X^*$ be prefix codes which satisfy $\dim(W \cdot V)^\omega \ge \max\{\dim W^\omega, \dim V^\omega\}$. Then $\dim(W \cdot V)^\omega = \max\{\dim W^\omega, \dim V^\omega\}$ and $\mathbb{L}_\alpha((W \cdot V)^\omega) = \min\{\mathbb{L}_\alpha(W^\omega), \mathbb{L}_\alpha(V^\omega)\}$ for $\alpha = \dim(W \cdot V)^\omega$.*

Proof. Since the product of W and V is unambiguous, we have $\mathfrak{s}_{W \cdot V}(t) = \mathfrak{s}_W(t) \cdot \mathfrak{s}_V(t)$. Let $\alpha' \ge \max\{\dim W^\omega, \dim V^\omega\}$. This implies $\mathfrak{s}_W(r^{-\alpha'}) \le 1$ and $\mathfrak{s}_V(r^{-\alpha'}) \le 1$ and, consequently, $\mathfrak{s}_{W \cdot V}(r^{-\alpha'}) \le 1$ whence $\alpha' \ge \dim(W \cdot V)^\omega$. This shows $\dim(W \cdot V)^\omega \le \max\{\dim W^\omega, \dim V^\omega\}$, hence the first assertion.

To show the second one we distinguish two cases. If $\mathfrak{s}_V(r^{-\alpha}) < 1$ we have $\mathfrak{s}_{W \cdot V}(r^{-\alpha}) = \mathfrak{s}_W(r^{-\alpha}) \cdot \mathfrak{s}_V(r^{-\alpha}) < 1$ and, consequently, $\mathbb{L}_\alpha(V^\omega) = \mathbb{L}_\alpha((W \cdot V)^\omega) == 0$

If $\mathfrak{s}_V(r^{-\alpha}) = 1$ we use the relation

$$\mathfrak{s}_{W \cdot V/u}(t) = \begin{cases} \mathfrak{s}_{V/v}(t) & \text{, when } u = w \cdot v \text{ with } w \in W \text{ and } v \in \mathbf{A}(V), \\ \mathfrak{s}_{W/u}(t) \cdot \mathfrak{s}_V(t) & \text{, if } u \in \mathbf{A}(W), \end{cases}$$

for $u \in \mathbf{A}(W \cdot V)$. Then Theorem 3 yields the following estimate.

$$\mathbb{L}_\alpha((W \cdot V)^\omega) = \inf\{\mathfrak{s}_{W \cdot V/u}(r^{-\alpha})^{-1} : u \in \mathbf{A}(W \cdot V)\}$$

$$= \min\{\inf\{\frac{1}{\mathfrak{s}_{W/w}(r^{-\alpha})} : w \in \mathbf{A}(W)\}, \inf\{\frac{1}{\mathfrak{s}_{V/v}(r^{-\alpha})} : v \in \mathbf{A}(V)\}\}$$

$$= \min\{\mathbb{L}_\alpha(W^\omega), \mathbb{L}_\alpha(V^\omega)\} \qquad\qquad \square$$

3   Construction of Prefix Codes from Languages

In this section we derive our examples which show that limit sets and their closures (attractors) for IIFS in Cantor space do not coincide. We present different levels of non-coincidence using Hausdorff dimension and Hausdorff measure.

We intend to find simple examples for these levels of non-coincidence. Simplicity here means, on the one hand that our examples are prefix codes, which makes the IIFS simple, and on the other hand, as indicated above, we try to choose them in low classes of the Chomsky hierarchy, preferably linear context-free languages.

3.1   Limit Set and Attractor

The limit set in Cantor space of an IIFS described by a language $L \subseteq X^* \setminus \{e\}$ is L^ω. It is also the largest solution (fixed point) of the equation $F = L \cdot F$ when $F \subseteq X^\omega$ (see [22]). The attractor of the IIFS is $\mathcal{C}(L^\omega)$. Using the ls-limit (or adherence) of [12] (see also [21]) we can describe the difference $\mathcal{C}(L^\omega) \setminus L^\omega$ more precisely.

Set $\mathit{ls}\, L := \{\xi : \xi \in X^\omega \wedge \mathbf{A}(\xi) \subseteq \mathbf{A}(L)\}$, for $L \subseteq X^*$. Then (see [12, 21])

$$\mathcal{C}(L^\omega) = \mathit{ls}\, L^* = L^\omega \cup L^* \cdot \mathit{ls}\, L \qquad\qquad (9)$$

This yields $\dim \mathcal{C}(L^\omega) = \max\{\dim L^\omega, \dim \mathit{ls}\, L\}$. For prefix codes L we have additionally the following identity (see [23]).

$$\mathbb{L}_\alpha(\mathcal{C}(L^\omega)) = \mathbb{L}_\alpha(L^\omega) + \sum\nolimits_{i \in \mathbb{N}} \mathfrak{s}_L(r^{-\alpha})^i \cdot \mathbb{L}_\alpha(\mathit{ls}\, L) \qquad\qquad (10)$$

If $\mathbb{L}_\alpha(L^\omega) > 0$ then $\mathfrak{s}_L(r^{-\alpha}) \geq 1$. Consequently,

$$\mathbb{L}_\alpha(\mathcal{C}(L^\omega)) = \begin{cases} \mathbb{L}_\alpha(L^\omega) & \text{, if } \mathbb{L}_\alpha(\mathit{ls}\, L) = 0 \text{, and} \\ \infty & \text{, otherwise.} \end{cases} \qquad\qquad (11)$$

If $\mathfrak{s}_L(r^{-\alpha'}) < 1$ then $\mathbb{L}_{\alpha'}(L^\omega) = 0$. In this case Eq. (10) shows that $\mathbb{L}_{\alpha'}(\mathcal{C}(L^\omega))$ is zero, non-null finite or infinite according to whether $\mathbb{L}_{\alpha'}(\mathit{ls}\, L)$ is zero, non-null finite or infinite, respectively. The condition $\mathfrak{s}_L(r^{-\alpha'}) < 1$ is fulfilled, in particular, if $\dim L^\omega < \alpha'$.

3.2   The Padding Construction

Let $W \subseteq (X \setminus \{d\})^*$ where d is a letter in X. Define for an injective function $f : \mathbb{N} \to \mathbb{N}$ satisfying $f(n) > n$ when $\mathsf{s}_W(n) > 0$

$$L := \{w \cdot d^{f(|w|)-|w|} : w \in W\} \, . \tag{12}$$

Then L is a prefix code, $\mathsf{s}_L(t) = \sum_{n \geq 0} \mathsf{s}_W(n) \cdot t^{f(n)}$. Because f is injective and $\mathsf{s}_L(i) > 0$ implies that $i = f(j)$ for some $j \in \mathbb{N}$ we have

$$(\operatorname{rad} W)^{\liminf_{n\to\infty} \frac{n}{f(n)}} \geq \operatorname{rad} L = \liminf_{n\to\infty} \frac{1}{f(n)\sqrt{\mathsf{s}_W(n)}} \geq (\operatorname{rad} W)^{\limsup_{n\to\infty} \frac{n}{f(n)}} \, . \tag{13}$$

If $\lim_{n\to\infty} \sqrt[n]{\mathsf{s}_W(n)}$ exists then we have $\operatorname{rad} L = (\operatorname{rad} W)^{\limsup_{n\to\infty} \frac{n}{f(n)}}$.

Since $\mathcal{C}(L^\omega) = L^\omega$ whenever L is finite, we are interested only in infinite languages $W, L \subseteq X^*$. Thus $0 < \frac{1}{|X|-1} \leq \operatorname{rad} W \leq \operatorname{rad} L \leq 1$.

Consider the identity $\mathsf{s}_L(t) = \sum_{n \geq 0} \mathsf{s}_W(n) \cdot t^{f(n)} = \sum_{n \geq 0} \mathsf{s}_W(n) \cdot t^{\gamma \cdot n} \cdot t^{f(n)-\gamma \cdot n}$. Since $0 \leq t \leq \operatorname{rad} L \leq 1$, we get the following relations between the parameters of the languages L and W.

$$\begin{aligned} \mathsf{s}_L(t) &\leq \mathsf{s}_W(t^\gamma) && \text{for } 0 \leq t \leq \sqrt[\gamma]{\operatorname{rad} W} \, , \\ \operatorname{rad} L &\geq \sqrt[\gamma]{\operatorname{rad} W} && \text{and} && \text{if } f(n) \geq \gamma \cdot n \, . \tag{14} \\ \operatorname{rad} L^* &= \mathbf{t}_1(L) \geq \sqrt[\gamma]{\mathbf{t}_1(W)} \end{aligned}$$

Moreover, Eq. (13) implies that $\operatorname{rad} L = \sqrt[\gamma]{\operatorname{rad} W}$ whenever $\gamma = \lim_{n\to\infty} \frac{f(n)}{n}$, and if $f(n) = \gamma \cdot n$ for $\mathsf{s}_W(n) > 0$ we have, in addition, $\mathsf{s}_L(t) = \mathsf{s}_W(t^\gamma)$ and $\mathbf{t}_1(L) = \sqrt[\gamma]{\mathbf{t}_1(W)}$. It should be mentioned, however, that $\gamma = \lim_{n\to\infty} \frac{f(n)}{n}$ does not imply $\mathbf{t}_1(L) = \sqrt[\gamma]{\mathbf{t}_1(W)}$ (see Example 8).

In order to apply Theorem 3 we are interested in connections between $\mathsf{s}_{L/w}$ and $\mathsf{s}_{W/w}$ for $w \in \mathbf{A}(W)$.

Lemma 4. Let $W \subseteq (X \setminus \{d\})^*$, $f(n) \geq \gamma \cdot n$ for $\mathsf{s}_W(n) > 0$. If $w \in \mathbf{A}(W)$ then $\mathsf{s}_{L/w}(t) \leq \mathsf{s}_{W/w}(t^\gamma)$ for $0 \leq t \leq \sqrt[\gamma]{\operatorname{rad} W}$.

If, moreover, W is a regular language then there is a $k \in \mathbb{N}$ such that $\mathsf{s}_{L/w}(t) \leq \mathsf{s}_{W/w}(t^\gamma) \leq t^{-\gamma \cdot k} \cdot \mathsf{s}_W(t^\gamma)$ for $0 \leq t \leq \sqrt[\gamma]{\operatorname{rad} W}$.

If $w \notin \mathbf{A}(W)$ then $\mathsf{s}_{L/w}(t) \leq 1$ for $0 \leq t \leq 1$.

Proof. Let $w \in \mathbf{A}(W)$. We consider the identity $L/w = \bigcup_{wu \in W} u \cdot d^{f(|wu|)-|wu|}$.

From this we obtain

$$\mathsf{s}_{L/w}(t) = \sum_{n \in \mathbb{N}} \mathsf{s}_{W/w}(n) \cdot t^{f(|w|+n)-|w|} = \sum_{n \in \mathbb{N}} \mathsf{s}_{W/w}(n) \cdot t^{\gamma \cdot n} \cdot t^{f(|w|+n)-|w|-\gamma \cdot n} \, ,$$

whence $\mathsf{s}_{L/w}(t) \leq \mathsf{s}_{W/w}(t^\gamma)$ if $f(n) \geq \gamma \cdot n$ for $\mathsf{s}_W(n) > 0$, and the first assertion is proved.

To show the next one, observe that $t^{|w|} \cdot \mathfrak{s}_{W/w}(t') \leq \mathfrak{s}_W(t')$ whenever $0 \leq t'$. If W is regular, there is a constant $k \in \mathbb{N}$ such that for every $w \in X^*$ there is a $\hat{w}, |\hat{w}| \leq k$ with $W/w = W/\hat{w}$.

The last assertion is obvious. $\qquad\qquad\qquad\qquad\qquad\qquad\qquad\qquad\qquad\qquad\square$

With Theorem 3 we obtain the following.

Corollary 3. *If $W \subseteq (X \setminus \{d\})^*$ is a regular language, $f(n) \geq \gamma \cdot n$ for $\mathfrak{s}_W(n) > 0$, $\mathfrak{t}_1(L) \leq \sqrt[\gamma]{\operatorname{rad} W}$ and $\mathfrak{s}_L(\mathfrak{t}_1(L)) = 1$ then $\dim L^\omega = -\log \mathfrak{t}_1(L)$ and $\mathbb{L}_\alpha(L^\omega) > 0$ for $\alpha = \dim L^\omega$.*

If we change the order in the construction of Eq. (12) we obtain for $\tilde{d} \in X$ and $W \subseteq (X \setminus \{\tilde{d}\})^*$

$$\tilde{L} := \{\tilde{d}^{f(|w|)-|w|} \cdot w : w \in W\}, \tag{15}$$

and the results on the structure generating function Eqs. (13) and (14) remain valid. In particular, $\tilde{L}$ is also a prefix code. Moreover we have a lower bound for $\mathfrak{s}_{\tilde{L}/w}$.

Proposition 4. *If $w = \tilde{d}^{f(n)}$ then $\mathfrak{s}_{\tilde{L}/w}(t) \geq \mathfrak{s}_W(n) \cdot t^n$.*

This enables us to apply Theorem 3.

Corollary 4. *Let $W \subseteq (X \setminus \{\tilde{d}\})^*$, $f : \mathbb{N} \to \mathbb{N}$ injective and $f(n) > n$ for $\mathfrak{s}_W(n) > 0$. If $\tilde{L} = \{\tilde{d}^{f(|w|)-|w|} \cdot w : w \in W\}$ and $\mathfrak{t}_1(L) > \operatorname{rad} W$ then $\mathbb{L}_\alpha(\tilde{L}) = 0$ for $\alpha = \dim \tilde{L}$.*

It should be mentioned that for linear functions $f : \mathbb{N} \to \mathbb{N}$, $f(n) = \gamma \cdot n + \delta$ with rational coefficients, and regular languages W the resulting languages L and $\tilde{L}$ are one-turn deterministic one-counter languages. These have rational structure generating functions $\mathfrak{s}_L$ and $\mathfrak{s}_{\tilde{L}}$ respectively, which implies that $\mathfrak{s}_L(\operatorname{rad} L) = \mathfrak{s}_{\tilde{L}}(\operatorname{rad} \tilde{L}) = \infty$ whence $\mathfrak{s}_L(\mathfrak{t}_1(L)) = \mathfrak{s}_{\tilde{L}}(\mathfrak{t}_1(\tilde{L})) = 1$ (see [11]).

3.3   Examples

In this section we give our announced examples. Here we consider the following cases which might appear for $\alpha = \dim L^\omega$ and $\hat{\alpha} = \dim \mathcal{C}(L^\omega)$, $\mathbb{L}_\alpha(L^\omega)$ and $\mathbb{L}_{\hat{\alpha}}(\mathcal{C}(L^\omega))$. The principal possibilities are shown in the figure below. The case $\mathbb{L}_\alpha(L^\omega) = \infty$ is excluded by Proposition 2.

We consider only prefix codes L. In this case Proposition 3.2 and Eq. (10) give some principal limitations.

Proposition 5. *Let $L \subseteq X^*$ be a prefix code and let $0 < \mathbb{L}_{\alpha'}(\mathbf{ls}\,L) < \infty$. If $\mathfrak{s}_L(r^{-\alpha'}) < 1$ then $\mathbb{L}_{\alpha'}(\mathcal{C}(L^\omega)) < \infty$, and if $\mathbb{L}_{\alpha'}(L^\omega) > 0$ then $\mathbb{L}_{\alpha'}(\mathcal{C}(L^\omega)) = \infty$.*

Proof. The first assertion is obvious, and the second follows from Eq. (10) and the fact that $\mathbb{L}_{\alpha'}(L^\omega) > 0$ implies $\mathfrak{s}_L(r^{-\alpha'}) = 1$. $\qquad\qquad\qquad\square$

Thus Case 4 is not possible for prefix codes L, and in order to achieve the identity $\mathbb{L}_{\hat{\alpha}}(\mathcal{C}(L^\omega)) = \infty$ one must have $\mathbb{L}_{\hat{\alpha}}(\mathbf{ls}\,L) = \infty$ or, in case $\alpha = \hat{\alpha}$, $\mathbb{L}_\alpha(\mathbf{ls}\,L) > 0$ and $\mathfrak{s}_L(r^{-\alpha}) = 1$.

fixed point L^ω	attractor $\mathcal{C}(L^\omega)$	**Example**
$\alpha = \dim L^\omega$	$\hat{\alpha} = \dim \mathcal{C}(L^\omega)$	

Dimension $\alpha = \hat{\alpha}$ and $\dim \textbf{\textit{ls}}\, L \leq \alpha$

1. $\mathbb{L}_\alpha(L^\omega) = 0$	$\mathbb{L}_\alpha(\mathcal{C}(L^\omega)) = 0$	Example 1
2. $\mathbb{L}_\alpha(L^\omega) > 0$	$\mathbb{L}_\alpha(\mathcal{C}(L^\omega)) = \mathbb{L}_\alpha(L^\omega)$	Proposition 2
3. $\mathbb{L}_\alpha(L^\omega) = 0$	$0 < \mathbb{L}_\alpha(\mathcal{C}(L^\omega)) < \infty$	Example 8
4. $\mathbb{L}_\alpha(L^\omega) > 0$	$\mathbb{L}_\alpha(L^\omega) < \mathbb{L}_\alpha(\mathcal{C}(L^\omega)) < \infty$ not possible	
5. $\mathbb{L}_\alpha(L^\omega) = 0$	$\mathbb{L}_\alpha(\mathcal{C}(L^\omega)) = \infty$	Example 5
6. $\mathbb{L}_\alpha(L^\omega) > 0$	$\mathbb{L}_\alpha(\mathcal{C}(L^\omega)) = \infty$	Example 2

Dimension $\alpha < \hat{\alpha}$ and $\dim \textbf{\textit{ls}}\, L = \hat{\alpha}$

7. $\mathbb{L}_\alpha(L^\omega) = 0$ $\mathbb{L}_{\hat{\alpha}}(\mathcal{C}(L^\omega)) = 0$		Example 10
8. $\mathbb{L}_\alpha(L^\omega) > 0$ $\mathbb{L}_{\hat{\alpha}}(\mathcal{C}(L^\omega)) = 0$		Example 9
9. $\mathbb{L}_\alpha(L^\omega) = 0$ $0 < \mathbb{L}_{\hat{\alpha}}(\mathcal{C}(L^\omega)) < \infty$		Example 6[1]
10. $\mathbb{L}_\alpha(L^\omega) > 0$ $0 < \mathbb{L}_{\hat{\alpha}}(\mathcal{C}(L^\omega)) < \infty$		Example 3
11. $\mathbb{L}_\alpha(L^\omega) = 0$ $\mathbb{L}_{\hat{\alpha}}(\mathcal{C}(L^\omega)) = \infty$		Example 7
12. $\mathbb{L}_\alpha(L^\omega) > 0$ $\mathbb{L}_{\hat{\alpha}}(\mathcal{C}(L^\omega)) = \infty$		Example 4

Let X consist of the four letters a, b, d and $\tilde{d}$. We arrange our examples according to increasing complexity. All examples, except for Example 8, have $f(n) = \gamma \cdot n$.

In the first seven examples we use the regular languages $W^{(1)} := \{a,b\}^* \setminus \{e\}$, $W^{(2)} := (\{a,b\} \cdot a)^* \setminus \{e\}$ and $W^{(3)} := \{a,b\}^* \cdot \tilde{d} \cdot \{a,b\}^*$ with the parameters:

$$\mathfrak{s}_{W^{(1)}}(t) = \frac{2t}{1-2t} \quad , \; \mathfrak{t}_1(W^{(1)}) = \tfrac{1}{4} \text{ and } \mathbb{L}_{\frac{1}{2}}(\textbf{\textit{ls}}\, W^{(1)}) = 1$$

$$\mathfrak{s}_{W^{(2)}}(t) = \frac{2t^2}{1-2t^2} \quad , \; \mathfrak{t}_1(W^{(2)}) = \tfrac{1}{2} \text{ and } \mathbb{L}_{\frac{1}{4}}(\textbf{\textit{ls}}\, W^{(2)}) = 1$$

$$\mathfrak{s}_{W^{(3)}}(t) = \frac{t}{(1-2t)^2} \, , \; \mathfrak{t}_1(W^{(3)}) = \tfrac{1}{4} \text{ and } \mathbb{L}_{\frac{1}{2}}(\textbf{\textit{ls}}\, W^{(3)}) = \infty$$

The first four examples are linear languages, and in Examples 2, 3 and 4 we use the construction of Eq. (12) and Corollary 3 to show that $\mathbb{L}_\alpha(L^\omega) > 0$.

Example 1. Set $W_1 := W^{(1)}$, $\gamma_1 := 4$ and use the construction of Eq. (15). Eq. (14) shows $\mathfrak{t}_1(L_1) = \frac{1}{\sqrt{2}} > \operatorname{rad} W_1 = \frac{1}{2}$. Since $\textbf{\textit{ls}}\, L_1 = \{\tilde{d}\}^\omega$, we have $\mathbb{L}_\alpha(\mathcal{C}(L_1^\omega)) = \mathbb{L}_\alpha(L_1^\omega)$, and Corollary 4 yields $\mathbb{L}_\alpha(L_1^\omega) = 0$. □

Example 2. We set $W_2 := W^{(1)}$ and $\gamma_2 := 2$. Then $\dim L_2^\omega = -\log_4 \mathfrak{t}_1(L_2) = \frac{1}{2}$ and $\mathfrak{s}_{L_2}(\mathfrak{t}_1(L_2)) = 1$. Now, Proposition 5 implies $\mathbb{L}_\alpha(\mathcal{C}(L_2^\omega)) = \infty$. □

Example 3. We use $W_3 := W^{(1)}$ and $\gamma_3 := 4$. Then $\dim L_3^\omega = \frac{1}{4}$, $\hat{\alpha} = \dim \textbf{\textit{ls}}\, L = \frac{1}{2}$, $\mathfrak{s}_{L_3}(4^{-\hat{\alpha}}) = \mathfrak{s}_{W_3}(\frac{1}{16}) = \frac{1}{7}$ and, finally, $\mathbb{L}_\alpha(\mathcal{C}(L_3^\omega)) = \frac{7}{6}$. □

Example 4. Set $W_4 := W^{(3)}$ and $\gamma_4 := 4$. This yields $\alpha = \dim L_4^\omega = -\log_4 \mathfrak{t}_1(L_4) = \frac{1}{4}$ and $\hat{\alpha} = \dim \textbf{\textit{ls}}\, L_4 = \frac{1}{2}$ and $\mathbb{L}_{\hat{\alpha}}(\textbf{\textit{ls}}\, L_4) = \infty$ (cf Example B of [16]). □

---

[1] An example of a language generated by a simple context-free grammar was given in Example 6.3 of [19].

The next three examples and Example 10 are products of languages L'_i and $\tilde{L}_i$ constructed according to Eqs. (12) and (15), respectively. Then we can use Theorem 4 to show that $\mathbb{L}_\alpha((L'_i \cdot \tilde{L}_i)^\omega) = 0$. Since $\textbf{\textit{ls}}\,\tilde{L}_i = \{\tilde{d}\}^\omega$, we have $\mathbb{L}_{\alpha'}(\textbf{\textit{ls}}\,(L'_i \cdot \tilde{L}_i)) = \mathbb{L}_{\alpha'}(\textbf{\textit{ls}}\,L'_i)$ for $\alpha' > 0$.

Example 5. Define L'_5 using Eq. (12) and the parameters $W'_5 := W^{(2)}$ and $\gamma' := 2$. This yields $\textbf{t}_1(L'_5) = \frac{1}{\sqrt{2}}$ and $\alpha = \dim L'^\omega_5 = \frac{1}{4}$. Now $\tilde{L}_5 := L_1$ has also $\dim \tilde{L}^\omega_5 = \frac{1}{4}$ and, consequently, $\mathbb{L}_{\frac{1}{4}}((L'_5 \cdot \tilde{L}_5)^\omega) = 0$.

Finally, $\mathbb{L}_\alpha(\textbf{\textit{ls}}\,(L'_5 \cdot \tilde{L}_5)) = \mathbb{L}_\alpha(\textbf{\textit{ls}}\,L'_5) = 1$ and $\mathfrak{s}_{L'_5}(4^{-\alpha}) = \mathfrak{s}_{\tilde{L}_5}(4^{-\alpha}) = 1$ yield $\mathbb{L}_\alpha(\mathcal{C}((L'_5 \cdot \tilde{L}_5)^\omega)) = \infty$. $\qquad\square$

Example 6. Here we use $L'_6 := L_3$ and $\tilde{L}'_6 := L_1$ and argue in the same way as in the preceding example.

Example 7. This example uses the language $L'_7 := L_4$ and concatenates it with $\tilde{L}_7 := L_1$. $\qquad\square$

Because of $\mathbb{L}_\alpha(L^\omega) = 0$ and $\infty > \mathbb{L}_\alpha(\textbf{\textit{ls}}\,L) > 0$ Item 3 requires $\mathfrak{s}_L(\text{rad}\,L) < 1$. This is not possible with languages having a rational structure generating function.

Example 8. Set $W := \{a, b, \tilde{d}\}^* \setminus \{e\}$ and $f(n) := n + 2\lceil\sqrt{n}\rceil$. We obtain $\mathfrak{s}_L(\text{rad}\,L) = \frac{5}{32} < 1$, and consequently $0 = \mathbb{L}_\alpha(L^\omega) < \mathbb{L}_\alpha(\mathcal{C}(L^\omega)) = \frac{32}{27} < \infty$ for $\alpha := \dim L^\omega = \dim \textbf{\textit{ls}}\,L = \log_4 3$. $\qquad\square$

In view of $\alpha < \hat{\alpha}$ and $\mathbb{L}_{\hat{\alpha}}(\mathcal{C}(L^\omega_i)) = 0$ the final two examples require $\mathbb{L}_{\hat{\alpha}}(\textbf{\textit{ls}}\,L) = 0$. Following Lemma 4.3 of [19] $\textbf{\textit{ls}}\,L$ cannot be a regular ω-language and, as the considerations in [12, 16] show, therefore L cannot be a linear language.

Example 9. Let $F := \{a, b\} \cdot \prod_{i=0}^{\infty}(\{a, b\}^{2^i - 1} \cdot a)$ and set $W_9 := \textbf{A}(F) \setminus \{e\}$. Then $\mathfrak{s}_W(n) = 2^{n - \lfloor \log_2 n \rfloor}$ for $n > 0$.

Since F is closed in (X^ω, ρ) and $\mathfrak{s}_{\textbf{A}(F/w)}(n) = \mathfrak{s}_{\textbf{A}(F/v)}(n)$ whenever $w, v \in \textbf{A}(F)$ and $|w| = |v|$, Theorem 4 of [18] shows $\dim F = \liminf\limits_{n \to \infty} \frac{\log_4 \mathfrak{s}_{\textbf{A}(F)}(n)}{n} = \frac{1}{2}$.

Moreover, it is easy to calculate that $\mathbb{L}_{1/2}(F) = 0$.

Choose $\gamma_9 = 3$ and use the construction of Eq. (12). Then $\textbf{\textit{ls}}\,L_9 = F$,

$$-\ln(1 - 2t^3) = \sum_{i=1}^{\infty} \frac{(2t^3)^n}{n} < \mathfrak{s}_{L_9}(t) < 2 \cdot \sum_{i=1}^{\infty} \frac{(2t^3)^n}{n} - 2t^3 = -2(\ln(1 - 2t^3) + t^3)$$

for $0 < t \leq \frac{1}{\sqrt[3]{2}}$, and we obtain $\mathfrak{s}_{L_9}(\frac{1}{\sqrt[3]{4}}) < 1 < \mathfrak{s}_{L_9}(\frac{1}{\sqrt[3]{3}}) < \infty$. Therefore, $\mathfrak{s}_{L_9}(\textbf{t}_1(L_9)) = 1$ and $\alpha = \dim L^\omega_9 = -\log \textbf{t}_1(L_9) < \frac{1}{2}$. This allows us to show $\mathbb{L}_\alpha(L^\omega) > 0$ using Theorem 3 in the following way: We have

$$\mathfrak{s}_{L_9/w}(t) = t^{2|w|} \cdot \sum_{i=1}^{\infty} \frac{\mathfrak{s}_{W_9}(n+|w|)}{\mathfrak{s}_{W_9}(|w|)} t^{3n}$$

$$= t^{2|w|} \cdot \sum_{i=1}^{\infty} 2^{n-\log_2(n+|w|)+\log_2|w|} \cdot t^{3n} \leq t^{2|w|} \cdot |w| \cdot \mathfrak{s}_L(t) \, ,$$

for $w \in \mathbf{A}(W_9)$, $w \neq e$. Hence $\mathfrak{s}_{L_9/w}(t) \leq 1$ for $0 \leq t \leq \frac{1}{\sqrt[3]{3}}$.   □

Example 10. Let $L'_{10} := L_9$ and let $\tilde{L}_{10}$ be constructed according to Eq. (15) with $\tilde{W}_{10} := W_9$ and $\tilde{\gamma}_{10} := \gamma_9 = 3$.

Arguing in the same way as in Examples 1 and 5 we calculate that Corollary 4 is applicable and obtain $\alpha = \dim(L'_{10} \cdot \tilde{L}_{10})^\omega < \hat{\alpha} = \frac{1}{2}$, and $\mathbb{L}_\alpha((L'_{10} \cdot \tilde{L}_{10})^\omega) = 0$ as well as $\mathbb{L}_{\hat{\alpha}}(\mathcal{C}((L'_{10} \cdot \tilde{L}_{10})^\omega)) = \mathbb{L}_{\hat{\alpha}}(\mathcal{C}(L'^\omega_{10})) = 0$.   □

References

1. J.-M. Autebert, J. Berstel and L. Boasson, Context-Free Languages and Pushdown Automata, in: [17], Vol. 1, pp. 111–174.
2. C. Bandt. Self-similar sets 3. Constructions with sofic systems. *Monatsh. Math.*, 108:89–102, 1989.
3. M. F. Barnsley. *Fractals Everywhere.* Acad. Press, 1988 (2nd edition 1993).
4. J. Berstel and D. Perrin. *Theory of Codes.* Academic Press, 1985.
5. K. Čulik II and S. Dube, Affine automata and related techniques for generation of complex images, *Theor. Comp. Sci.*, 116:373–398, 1993.
6. G. A. Edgar. *Measure, Topology, and Fractal Geometry.* Springer, 1990.
7. K.J. Falconer, *Fractal Geometry.* Wiley, 1990.
8. H. Fernau. *Iterierte Funktionen, Sprachen und Fraktale.* BI-Verlag, 1994.
9. H. Fernau. Infinite IFS. *Mathem. Nachr.*, 169:79–91, 1994.
10. H. Fernau and L. Staiger. Iterated Function Systems and Control Languages *Inf. & Comp.*, 168:125–143, 2001.
11. W. Kuich. On the entropy of context-free languages. *Inf. & Contr.*, 16:173–200, 1970.
12. R. Lindner and L. Staiger. *Algebraische Codierungstheorie; Theorie der sequentiellen Codierungen.* Akademie-Verlag, 1977.
13. R. D. Mauldin. Infinite iterated function systems: Theory and applications, in: *Fractal Geometry and Stochastics*, vol. 37 of *Progress in Probability*, Birkhäuser 1995.
14. R. D. Mauldin and M. Urbański. Dimensions and measures in infinite iterated function systems. *Proc. Lond. Math. Soc.*, III. Ser. 73, No.1, 105–154, 1996.
15. R. D. Mauldin and S. C. Williams. Hausdorff dimension in graph directed constructions. *Trans. AMS*, 309(2):811–829, Oct. 1988.
16. W. Merzenich and L. Staiger. Fractals, dimension, and formal languages. *RAIRO Inf. théor. Appl.*, 28(3–4):361–386, 1994.
17. G. Rozenberg and A. Salomaa (eds.) *Handbook of Formal Languages*, Springer, 1997.
18. L. Staiger. Combinatorial properties of the Hausdorff dimension. *J. Statist. Plann. Inference*, 23:95–100, 1989.

19. L. Staiger. Kolmogorov complexity and Hausdorff dimension. *Inf. & Comp.*, 103:159–194, 1993.
20. L. Staiger. Codes, simplifying words, and open set condition. *Inf. Proc. Lett.*, 58:297–301, 1996.
21. L. Staiger. ω-languages, in: [17], Vol. 3, pp. 339–387.
22. L. Staiger. On ω-power languages, in: *New Trends in Formal Languages, Control, Cooperation, and Combinatorics*, vol. 1218 of *Lecture Notes in Comput. Sci.* pp. 377–393. Springer, 1997.
23. L. Staiger. The Hausdorff measure of regular ω-languages is computable, Bull. EATCS 66:178–182, 1998.

A Method for Deciding the Finiteness of Deterministic Tabled Picture Languages

Bianca Truthe

Fakultät für Informatik, Otto-von-Guericke-Universität Magdeburg,
Magdeburg, Germany
truthe@isg.cs.uni-magdeburg.de

Abstract. Chain code picture systems based on LINDENMAYER systems can be used to generate pictures. In this paper, synchronous, deterministic tabled chain code picture systems ($sDT0L$ systems) and their picture languages are considered. A method is given for deciding whether an $sDT0L$ system generates a finite picture language or not. This proves that the finiteness of picture languages of $sDT0L$ systems is decidable.

1 Introduction

Chain code picture systems provide a possibility for describing pictures. They are based on generating words over a special alphabet and interpreting these words as pictures. H. FREEMAN introduced chain code picture languages ([Fr61]). A picture is formed by a sequence of drawing commands represented by symbols (letters). A string describes a picture, which is built by the drawing commands of its letters. FREEMAN uses the alphabet $\{0, \ldots, 7\}$, whose elements are interpreted according to the following sketch.

The picture to the right, for example, is generated by the word 2012 331 577 00 250 67 32 0 670 26 1223 62 456 73 1:
(The drawing 'starts' at the circle.)

For language theoretical investigations, only the four directions 0, 2, 4, 6 are considered [DH89]. The directions *right, up, left, down* are written as r, u, l, d. The connection of strings and pictures suggests to search for relations between formal languages and picture sets. The first paper in that topic is [Fe68]. Chain code picture languages based on the CHOMSKY-hierarchy have been investigated since the 1980s ([MRW82], [SW85]). However, hardly any theoretical investigations on LINDENMAYER-based chain code picture languages have been carried out so far ([DHr92]). Following the paper [DHr92], it occurred that there are so called length constant systems with a finite picture language as well as such with an infinite one. This is contrary to a statement in [DHr92].

C.S. Calude, E. Calude and M.J. Dinneen (Eds.): DLT 2004, LNCS 3340, pp. 406–417, 2004.

According to [RS80], context-free LINDENMAYER systems are divided up in *D0L* systems (deterministic rewriting of letters), *0L* systems (rewriting non-deterministically), *DT0L* systems (selecting a replacement table non-deterministically, rewriting deterministically), and *T0L* systems (selecting a replacement table non-deterministically, rewriting non-deterministically).

In [T02] and [T03], the finiteness of the picture languages of special systems – synchronous *D0L* systems and synchronous *0L* systems (*sD0L* systems, *s0L* systems) – is proved to be decidable. The present paper follows [T02] and [T03]; the decidability of the finiteness of chain code picture languages of synchronous, deterministic tabled, context-free LINDENMAYER systems (*sDT0L* systems) is investigated and proved.

2   Fundamentals

The finiteness investigations on picture languages of *sDT0L* systems in this paper are based on the hierarchy of abstractions developed in [T02]. In this section, the fundamental notations are gathered.

2.1   Structures Over an Alphabet

Let $\mathcal{A} = \{\, r, l, u, d \,\}$ be an alphabet and $(\mathcal{A}^*, \cdot)$ be the free structure over $\mathcal{A}$ with the operation of concatenation. The empty word is denoted by λ, the set of all words without λ is denoted by $\mathcal{A}^+$, the set of all finite, non-empty subsets of $\mathcal{A}^*$ is denoted by $\mathbb{A}$. The set $\mathbb{A}$ with the operations $\cup$ (union) and $\cdot$ (concatenation) forms a semiring $(\mathbb{A}, \cup, \cdot)$ because $(\mathbb{A}, \cup)$ and $(\mathbb{A}, \cdot)$ are semigroups and the distributive laws are valid.

The set of the natural numbers containing 0 is denoted by $\mathbb{N}_0$. All words w of the length $|w| = n$ $(n \in \mathbb{N}_0)$ form the set $\mathcal{A}^n$. A word $w \in \mathcal{A}^n$ is composed of letters $w_1, \ldots, w_n$. In this context, $\overrightarrow{w_i}$ is the word $\overrightarrow{w_i} = w_1 \cdots w_i$ $(0 \leq i \leq n,$ $\overrightarrow{w_0} = \lambda)$. The number $|w|_x$ is the number of occurrences of the letter x in the word w. The set of all letters occurring in w is shortly written as $[w]$:

$$[w] = \{\, x \mid |w|_x \geq 1 \,\}.$$

The elements of $\mathcal{A}^*$ can be interpreted as mappings on $\mathbb{Z}^2$. The empty word corresponds to the identity mapping. The atomic mappings r, l, u, d assign, to a point $q \in \mathbb{Z}^2$, its neighbours:

$$r(q) = q + (1, 0), \qquad l(q) = q - (1, 0),$$
$$u(q) = q + (0, 1), \qquad d(q) = q - (0, 1).$$

The function names r, l, u, d are taken from the directions *right, left, up, down*. A compound word $vw \in \mathcal{A}^*$ stands for the composed mapping $v \circ w$:

$$v \circ w : \mathbb{Z}^2 \longrightarrow \mathbb{Z}^2 \quad \text{with } q \mapsto w(v(q)).$$

The zero point of $\mathbb{Z}^2$ is denoted by $o = (0, 0)$. The translation of any point $q \in \mathbb{Z}^2$ to its neighbour $x(q)$ $(x \in \mathcal{A})$ is denoted by $\mathfrak{v}_x \in \mathbb{Z}^2$: $\mathfrak{v}_x = x(q) - q$. Consequently, $\mathfrak{v}_r = (1, 0)$, $\mathfrak{v}_l = -(1, 0)$, $\mathfrak{v}_u = (0, 1)$, $\mathfrak{v}_d = -(0, 1)$.

The interpretation of words as mappings on $\mathbb{Z}^2$ is a homomorphism from the free structure $(\mathcal{A}^*, \cdot)$ into the free structure $(\mathcal{A}^*, \circ)$. The operator $\circ$ does not need to be written if the context shows which operation is meant.

The mappings r and l as well as u and d are inverse to each other. The mappings ru and ur as well as ld and dl assign, to a point $\mathfrak{p}$, its diagonal neighbours:

$$ru(\mathfrak{q}) = ur(\mathfrak{q}) = \mathfrak{q} + (1,1), \quad ld(\mathfrak{q}) = dl(\mathfrak{q}) = \mathfrak{q} - (1,1).$$

The mapping which leads, together with a mapping x, to a diagonal neighbour is denoted by $x^{\perp}$. The inverse mappings of two mappings x and $x^{\perp}$ are denoted by $\bar{x}$ and $\bar{x}^{\perp}$ respectively. The table to the right shows the corresponding mappings.

x	$\bar{x}$	$x^{\perp}$	$\bar{x}^{\perp}$
r	l	u	d
l	r	d	u
u	d	r	l
d	u	l	r

2.2   Graphical Embedding

A lattice graph is a graph with the following properties: the vertex set is a subset of $\mathbb{Z}^2$, and each edge is incident to two neighbours $\mathfrak{q}$ and $x(\mathfrak{q})$ with $\mathfrak{q} \in \mathbb{Z}^2$ and $x \in \{\, r, l, u, d\,\}$.

In [T02], we define functions that assign to each word $\mathsf{w} \in \mathcal{A}^n$ and a start point $\mathfrak{a} \in \mathbb{Z}^2$

- the vertex set $\odot^{\mathfrak{a}}(\mathsf{w}) = \{\, \overrightarrow{w_i}(\mathfrak{a}) \mid i = 0, \ldots n \,\}$,
- the directed lattice graph (possibly with multiple edges)

$$g^{\mathfrak{a}}(\mathsf{w}) = \left(\odot^{\mathfrak{a}}(\mathsf{w}), \{\, (\overrightarrow{w_{i-1}}(\mathfrak{a}), \overrightarrow{w_i}(\mathfrak{a})) \,\}_{i=1,\ldots,n} \right),$$

- the simple, directed lattice graph $s^{\mathfrak{a}}(\mathsf{w})$ of $g^{\mathfrak{a}}(\mathsf{w})$ (without multiple edges)

$$s^{\mathfrak{a}}(\mathsf{w}) = \left(\odot^{\mathfrak{a}}(\mathsf{w}), \{\, (\overrightarrow{w_{i-1}}(\mathfrak{a}), \overrightarrow{w_i}(\mathfrak{a})) \mid i = 1, \ldots, n \,\} \right),$$

- the edge set $\|^{\mathfrak{a}}\mathsf{w}$ of $s^{\mathfrak{a}}(\mathsf{w})$

$$\|^{\mathfrak{a}}\mathsf{w} = \{\, (\overrightarrow{w_{i-1}}(\mathfrak{a}), w_i) \mid i = 1, \ldots, n \,\},$$

    where an edge is described by a pair of a start point and a direction rather than a pair of start and end points (the set of all x-edges is denoted by $\|_x^{\mathfrak{a}}\mathsf{w}$),
- the picture (the shadow of $s^{\mathfrak{a}}(\mathsf{w})$)

$$p^{\mathfrak{a}}(\mathsf{w}) = \left(\odot^{\mathfrak{a}}(\mathsf{w}), \{\, (\overrightarrow{w_{i-1}}(\mathfrak{a}), \overrightarrow{w_i}(\mathfrak{a})), (\overrightarrow{w_i}(\mathfrak{a}), \overrightarrow{w_{i-1}}(\mathfrak{a})) \mid i = 1, \ldots, n \,\} \right)$$

- and the picture area

$$\boxdot^{\mathfrak{a}}(\mathsf{w}) = \left\{\, (x, y) \; \middle| \; \begin{array}{l} \underline{x}^{\mathfrak{a}}(\mathsf{w}) \leq x \leq \overline{x}^{\mathfrak{a}}(\mathsf{w}) \text{ and} \\ \underline{y}^{\mathfrak{a}}(\mathsf{w}) \leq y \leq \overline{y}^{\mathfrak{a}}(\mathsf{w}) \end{array} \right\},$$

$\underline{x}^{\mathfrak{a}}(\mathsf{w})$, $\underline{y}^{\mathfrak{a}}(\mathsf{w})$, $\overline{x}^{\mathfrak{a}}(\mathsf{w})$, and $\overline{y}^{\mathfrak{a}}(\mathsf{w})$ being the border coordinates of the vertices of $\odot^{\mathfrak{a}}(\mathsf{w})$:

$$\underline{x}^{\mathfrak{a}}(\mathsf{w}) = \min\{\, x \mid (x,y) \in \odot^{\mathfrak{a}}(\mathsf{w}) \,\}, \quad \underline{y}^{\mathfrak{a}}(\mathsf{w}) = \min\{\, y \mid (x,y) \in \odot^{\mathfrak{a}}(\mathsf{w}) \,\},$$

$$\overline{x}^{\mathfrak{a}}(\mathsf{w}) = \max\{\, x \mid (x,y) \in \odot^{\mathfrak{a}}(\mathsf{w}) \,\}, \quad \overline{y}^{\mathfrak{a}}(\mathsf{w}) = \max\{\, y \mid (x,y) \in \odot^{\mathfrak{a}}(\mathsf{w}) \,\}.$$

The upper index will be omitted if the mappings relate to the zero point ($\mathfrak{a} = \mathfrak{o}$).

The picture areas are rectangle sets. A rectangle set $\mathfrak{P}$ is determined by two points, the 'lower left corner' $\mathfrak{a}_\mathfrak{P}$ and the 'upper right corner' $\mathfrak{c}_\mathfrak{P}$ or the 'upper left corner' $\mathfrak{d}_\mathfrak{P}$ and the 'lower right corner' $\mathfrak{b}_\mathfrak{P}$. The notation for a rectangle set is $[\mathfrak{a}_\mathfrak{P}, \mathfrak{c}_\mathfrak{P}]$. The following descriptions are equivalent: $[\mathfrak{a}_\mathfrak{P}, \mathfrak{c}_\mathfrak{P}], [\mathfrak{c}_\mathfrak{P}, \mathfrak{a}_\mathfrak{P}], [\mathfrak{d}_\mathfrak{P}, \mathfrak{b}_\mathfrak{P}], [\mathfrak{b}_\mathfrak{P}, \mathfrak{d}_\mathfrak{P}]$.

Scaling of a picture area $\mathfrak{P} = [\mathfrak{p}, \mathfrak{q}]$ by a factor $s \in \mathbb{N}_0$ produces the picture area

$$s\mathfrak{P} = \{\, s\mathfrak{x} \mid \mathfrak{x} \in \mathfrak{P} \,\} = [s\mathfrak{p}, s\mathfrak{q}].$$

The union of two picture areas is not a rectangle set in general. An extended union of two picture areas is the picture area of the union:

$$\mathfrak{P}_X \uplus \mathfrak{P}_Y = \mathfrak{P}_{X \cup Y}.$$

2.3 Special Endomorphisms

Let κ, μ be two natural numbers, $\kappa, \mu \in \mathbb{N}_0$. An endomorphism h on the semiring $(\mathbb{A}, \cup, \cdot)$ is called a (κ, μ)-endomorphism on the semiring $(\mathbb{A}, \cup, \cdot)$ if, for all $x \in \mathcal{A}$, the following conditions (called synchronization conditions) are satisfied:

If $x' \in h(\{\, x \,\})$, then

1. $x'(\mathfrak{o}) = \kappa \mathfrak{v}_x$ and
2. $\Box(x') \subseteq \kappa[\mathfrak{o}, \mathfrak{v}_x] \uplus \mu[\mathfrak{v}_{x\perp}, \mathfrak{v}_{\bar{x}\perp}]$.

Applying h to a set of words W is called deriving; the set $h(W)$ is obtained in one derivation step. Every element of $h(\{\, \mathsf{w} \,\})$ is called a derivative of the word w. The parameter κ is a factor of the length changing in one derivation step; the parameter μ is an upper bound of the width changing in one derivation step. For example, if r' is a derivative of r, then the picture corresponding to r' starting at the origin has its end point at $(\kappa, 0)$ and fits into the rectangle with corners $(0, -\mu)$ and (κ, μ). If u' is a derivative of u, then the picture corresponding to u' starting at the origin has its end point at $(0, \kappa)$ and fits into the rectangle with corners $(-\mu, 0)$ and (μ, κ). In the case of $\kappa = 0$, the endomorphism is called length contracting, in the case of $\kappa = 1$, length constant and in the case $\kappa > 1$, length expanding. A derivative of a picture (the picture of the derivation of the underlying word) has the same proportions as the picture itself (the 'shape' remains the same). This can be seen immediatly if κ is much greater than μ. Since the picture sizes increase in all directions equally, we speak of synchronization.

The n-ary composition of a (κ, μ)-endomorphism h is shortly written as h^n. Every element of $h^n(\{\, \mathsf{w} \,\})$, $\mathsf{w} \in \mathcal{A}^*$, is called an n-th derivative of w.

Let w be a word of $\mathcal{A}^n$. Then, $h(\{\, \mathsf{w} \,\})$ is composed of $h(\{\, w_1 \,\}), \ldots, h(\{\, w_n \,\})$:

$$h(\{\, \mathsf{w} \,\}) = h(\{\, w_1 \cdots w_n \,\}) = h(\{\, w_1 \,\}) \cdots \{\, w_n \,\}) = h(\{\, w_1 \,\}) \cdots h(\{\, w_n \,\}).$$

For a finite word set $W \in \mathbb{A}$, the invariance of h with respect to union implies

$$h(W) = \bigcup_{\mathsf{w} \in W} h(\{\, \mathsf{w} \,\}).$$

Let $\mu_0 \in \mathbb{N}_0$ be a natural number. For any natural number $\mu \geq \mu_0$, the picture area $\mu_0[\mathfrak{v}_{x\perp}, \mathfrak{v}_{\bar{x}\perp}]$ is a subset of $\mu[\mathfrak{v}_{x\perp}, \mathfrak{v}_{\bar{x}\perp}]$.

Proposition 1. *Each (κ, μ)-endomorphism is also a $(\kappa, \mu+1)$-endomorphism.*

If all atomic values of a (κ, μ)-endomorphism h have only one element, then each word has exactly one derivative; the set signs are omitted in that case: $h(\mathsf{w}) = \mathsf{w}'$.

2.4   Chain Code Picture Systems

An *sDT0L* system is a triple

$$G = (\mathcal{A}, h, \omega)$$

with the alphabet $\mathcal{A} = \{\, r, l, u, d \,\}$, a non-empty word ω over $\mathcal{A}$ (referred to as the axiom), and a finite, non-empty set $h = \{\, h_1, \ldots, h_m \,\}$ where each h_i, $i = 1, \ldots, m$, is a (κ_i, μ_i)-endomorphism and $h_i(\{\, x \,\})$ is a singleton set for any $x \in \mathcal{A}$.

The set of all n-ary compositions of elements of h is denoted by h^n

$$h^n = \{\, h_{i_1} \circ \cdots \circ h_{i_n} \mid i_j \in \{\, 1, \ldots, m \,\};\ j = 1, \ldots, n \,\};$$

applying all those compositions to a set of words W yields the set

$$h^n(W) = \left\{\, h^{(n)}(\mathsf{w}) \mid h^{(n)} \in h^n, \mathsf{w} \in W \,\right\}.$$

The picture language P_G generated by an *sDT0L* system G is defined to be the set of all pictures of derivatives of the axiom ω:

$$P_G = \{\, p(\mathsf{w}) \mid \mathsf{w} \in h^n(\{\, \omega \,\}), n \in \mathbb{N}_0 \,\}.$$

An *sDT0L* system is called length expanding if at least one (κ_i, μ_i)-endomorphism $h_i \in h$ has this property. An *sDT0L* system is called length contracting if all endomorphisms of h are length contracting. In the other cases, at least one endomorphisms of h is length constant while the others are length contracting. Hence, the maximum length of the picture of an atomic derivative stays constant. Those *sDT0L* systems are called length constant.

Let G be an *sDT0L* system. If there are two natural numbers κ, μ such that all endomorphisms of G are (κ, μ)-endomorphisms, G is said to be pure otherwise G is said to be mixed.

A *sD0L* system is an *sDT0L* system with only one endomorphism ([T02]). In order to use results about *sD0L* systems, deterministic subsystems are defined below. An *sD0L* system $U = (\mathcal{A}, h_o, \omega)$ is called a deterministic subsystem of an *sDT0L* system $G = (\mathcal{A}, h, \omega)$ (written $U \sqsubseteq G$) if each derivative of any $\mathsf{w} \in \mathcal{A}^*$ with respect to U is also a derivative with respect to G:

$$U \sqsubseteq G \Longleftrightarrow \forall \mathsf{w} \in \mathcal{A}^* : h_o(\mathsf{w}) \in h(\{\, \mathsf{w} \,\}).$$

Proposition 2. *An sDT0L system $G = (\mathcal{A}, h, \omega)$ with $h = \{\, h_1, \ldots, h_m \,\}$ has exactly the deterministic subsystems $U_i = (\mathcal{A}, h_i, \omega)$ for $i = 1, \ldots, m$.*

A deterministic subsystem $U_i = (\mathcal{A}, h_i, \omega)$ of an $sDT0L$ system generates the picture language

$$P_{U_i} = \{\, p(h_i^n(\omega)) \mid n \in \mathbb{N}_0 \,\}.$$

The following proposition follows immediately from the definitions.

Proposition 3. *The picture language generated by a deterministic subsystem of an $sDT0L$ system G is a subset of the picture language generated by G.*

An $s0L$ system is similar to an $sD0L$ system, but with an arbitrary (κ, μ)-endomorphism ([T03]). In order to use results about $s0L$ systems, simple non-deterministic supersystems are defined below. An $s0L$ system $S = (\mathcal{A}, h_\circ, \omega)$ is called a simple non-deterministic supersystem (shortly $s0L$ supersystem) of an $sDT0L$ system $G = (\mathcal{A}, \{\, h_1, \ldots, h_m \,\}, \omega)$ (written $S \sqsupseteq G$) if each derivative of a word $w \in \mathcal{A}^*$ with respect to G is also a derivative with respect to S. Hence, $h_\circ$ must satisfy $h_\circ(\{\, x \,\}) \sqsupseteq \{\, h_1(x), \ldots, h_m(x) \,\}$ for any $x \in \mathcal{A}$. An $s0L$ supersystem $S \sqsupseteq G$ is said to be minimal in the case of equality.

Lemma 1. *Let $G = (\mathcal{A}, \{\, h_1, \ldots, h_m \,\}, \omega)$ be an $sDT0L$ system. It has an $s0L$ supersystem if and only if G is a pure $sDT0L$ system.*

The following proposition follows from the definitions.

Proposition 4. *In the case that an $sDT0L$ system G has an $s0L$ supersystem, the picture language P_G generated by G is a subset of the picture language P_S generated by the minimal $s0L$ supersystem S of G.*

3   Finiteness Investigations

In this section, conditions are derived for deciding whether an $sDT0L$ system generates a finite picture language or not.

3.1   Length Contracting Systems

Let $G = (\mathcal{A}, h, \omega)$ be a length contracting $sDT0L$ system with a finite, non-empty set $h = \{\, h_1, \ldots, h_m \,\}$ of $(0, \mu_i)$-endomorphisms h_i. Let μ be the maximum of all μ_i: $\mu = \max\{\, \mu_i \mid i = 1, \ldots, m \,\}$. According to Proposition 1, every h_i is also a $(0, \mu)$-endomorphism.

The minimal $s0L$ supersystem $S = (\mathcal{A}, h_\circ, \omega)$ of G is a length contracting $s0L$ system and its endomorphism $h_\circ$ is also a $(0, \mu)$-endomorphism. According to Theorem 3.1 of [T03], the picture language P_S contains $(\mu + 1)^4 + 1$ elements at most. Following Lemma 1, also P_G has $(\mu + 1)^4 + 1$ elements at most.

Theorem 1. *Any length contracting $sDT0L$ system $G = (\mathcal{A}, \{\, h_1, \ldots, h_m \,\}, \omega)$ with $(0, \mu)$-endomorphisms h_i $(i = 1, \ldots, m)$ generates a finite picture language P_G with $(\mu + 1)^4 + 1$ elements at most:*

$$|P_G| \leq (\mu + 1)^4 + 1 < \infty.$$

3.2   Length Expanding Systems

Let $G = (\mathcal{A}, h, \omega)$ be a length expanding $sDT0L$ system with $h = \{\, h_1, \ldots, h_m \,\}$. Since G is length expanding, one of the endomorphisms h_i is length expanding. Hence, a deterministic subsystem $U = (\mathcal{A}, h_i, \omega)$ is length expanding. In [T02] is shown that the picture language of any length expanding $sD0L$ system is infinite. Since the picture language generated by U is a subset of the picture language generated by G (Proposition 3), also the picture language of G is infinite.

Theorem 2. *Every length expanding $sDT0L$ system generates an infinite picture language.*

3.3   Length Constant Systems

Length constant $sD0L$ systems can genrate finite or infinite picture languages ([T02]). They are special length constant $sDT0L$ systems. Hence, also length constant $sDT0L$ systems can genrate finite or infinite picture languages.

As some examples show, the criterion for decision derived in [T02] for $sD0L$ systems and in [T03] for $s0L$ systems is not applicable to $sDT0L$ systems. In order to find a criterion for $sDT0L$ systems, pure and mixed systems are considered seperately.

Pure Length Constant Systems. Let $G = (\mathcal{A}, h, \omega)$ be an $sDT0L$ system with $h = \{\, h_1, \ldots, h_m \,\}$ being a set of $(1, \mu)$-endomorphisms h_i. According to Lemma 1, the $sDT0L$ system G has a minimal $s0L$ supersystem $S = (\mathcal{A}, h_S, \omega)$ where $h_S(\{\, x \,\}) = \{\, h_1(x), \ldots, h_m(x) \,\}$ for all letters $x \in \mathcal{A}$.

According to Proposition 4, the picture language P_G generated by G is a subset of the picture language P_S generated by S.

Proposition 5. *The picture language of a pure, length constant $sDT0L$ system is finite if the picture language of its minimal $s0L$ supersystem is finite.*

In the sequel, let G be such an $sDT0L$ system that its minimal $s0L$ supersystem S generates an infinite picture language. Following [T03], S has a deterministic subsystem $U = (\mathcal{A}, h_U, \omega)$ also generating an infinite picture language. For any $x \in \mathcal{A}$, one has $h_U(x) \in h_S(\{\, x \,\})$. Hence, each $x \in \mathcal{A}$ has an index $i_x \in \{\, 1, \ldots, m \,\}$ such that $h_U(x) = h_{i_x}(x)$. Because the picture language P_U generated by U is infinite, there is a letter $x \in [h_U^2(\omega)]$ such that one of the edge sets $\|h_U(x)$, $\|h_U^2(x)$, $\|h_U^3(x)$ contains an x-edge different from (o, x) (proved in [T02]):

$$\exists x \in [h_U^2(\omega)] \exists l \in \{\, 1, 2, 3 \,\} : \|_x x \neq \|_x h_U^l(x).$$

Under this condition, one can show that also P_G is infinite ($sDT0L$ systems are given that generate infinite picture languages but subsets of P_G).

Proposition 6. *The picture language of a pure, length constant $sDT0L$ system G is infinite if the picture language of its minimal $s0L$ supersystem $S \sqsupseteq G$ is infinite.*

Propositions 5 and 6 lead to the following result.

Theorem 3. *Let* $G = (\mathcal{A}, h, \omega)$ *be a pure, length constant sDT0L system. The picture language generated is finite if and only if the minimal s0L supersystem of* G *generates a finite picture language.*

Mixed Length Constant Systems. In the sequel, mixed length constant sDT0L systems are considered. Those systems contain at least one $(0, \mu)$-endomorphism and at least one $(1, \mu)$-endomorphism. Let $G = (\mathcal{A}, h, \omega)$ be an sDT0L system with a set $h = g \cup f$ of length contracting and length constant endomorphisms. The length contracting endomorphisms are gathered in the set $g = \{g_1, \ldots, g_{m_0}\}$, the length constant ones are gathered in $f = \{f_1, \ldots, f_{m_1}\}$. The system $(\mathcal{A}, f, \omega)$ is the maximal pure length constant subsystem of G and is referred to as the 1-system of G.

Proposition 7. *Let* T *be the 1-system of* G. *Then,* P_T *is a subset of* P_G.

From this inclusion, one concludes immediately the next one.

Proposition 8. *Let* G *be an sDT0L system and* T *be its 1-system. If* T *generates an infinite picture language, also the picture language of* G *is infinite.*

If the picture language of T is finite, however, the picture language of G is not necessarily finite. But then, a letter $x \in \mathcal{A}$ exists which is not produced by the system T but leads to the infinity of the picture language of G. By examining the possibilities of a letter to be produced, one can prove the following lemma.

Lemma 2. *Any letter produced by* G *also occurs in a second derivative.*

If $T = (\mathcal{A}, f, \omega)$ is the 1-system of G, any sDT0L system $T_x = (\mathcal{A}, f, x)$ is called an x-1-system if $x \in [h^2(\{\omega\})]$ is a letter occurring in a second derivative. It will be proved that the sDT0L system G generates an infinite picture language if and only if an x-1-system of G has an infinite picture language.

Let $\omega'' = \widehat{x} x \widetilde{x} \in h^2(\{\omega\})$ be a second derivative of the axiom. The x-1-system generates the picture language

$$P_x = \{p(\mathsf{w}) \mid \mathsf{w} \in f^n(\{x\}), n \in \mathbb{N}_0\} = \left\{p(f^{(n)}(x)) \mid n \in \mathbb{N}_0, f^{(n)} \in f^n\right\}.$$

Let $P_{\omega''}$ be the picture language of the system $(\mathcal{A}, f, \omega'')$:

$$P_{\omega''} = \{p(\mathsf{w}) \mid \mathsf{w} \in f^n(\{\omega''\}), n \in \mathbb{N}_0\} = \left\{p(f^{(n)}(\omega'')) \mid n \in \mathbb{N}_0, f^{(n)} \in f^n\right\}.$$

Because $\omega'' \in h^2(\{\omega\})$ is a second derivative of the axiom and $f^n \subseteq h^n$ for any derivation step $n \in \mathbb{N}_0$, every n-th derivative $\mathsf{w} \in f^n(\{\omega''\})$ of ω'' by length constant endomorphisms is a derivative of ω with respect to G:

$$\mathsf{w} \in f^n(\{\omega''\}) \subseteq f^n(h^2(\{\omega\})) \subseteq h^{n+2}(\{\omega\}).$$

Hence, $P_{\omega''}$ is a subset of P_G:

$$P_{\omega''} \subseteq P_G.$$

The picture of an arbitrary derivative of ω'' by the length constant endomorphisms is, according to Proposition 2.4 of [T02],

$$p(f^{(n)}(\omega'')) = p(f^{(n)}(\tilde{x})) \cup p^{(f^{(n)}(\tilde{x}))(o)}(f^{(n)}(x)) \cup p^{(f^{(n)}(\tilde{x}\,x))(o)}(f^{(n)}(\tilde{x})).$$

Every picture $p(f^{(n)}(x))$ is a subpicture of $p(f^{(n)}(\omega''))$ (a subpicture is a subgraph, possibly shifted). If the picture language P_x is infinite, the picture language $P_{\omega''}$ is infinite, because any picture has only finitely many subpictures. Hence, also P_G is infinite. This result is stated in the next lemma.

Lemma 3. *If G has an x-1-system T_x generating an infinite picture language, also the picture language P_G is infinite.*

Now, let all x-1-systems of G generate finite picture languages.

If $\mathcal{K}$ is a set of edge sets, $\mathcal{K}^{\mathfrak{a}}$ denotes the set of edge sets obtained by translating the edge sets belonging to $\mathcal{K}$ by $\mathfrak{a}$: $\mathcal{K}^{\mathfrak{a}} = \{\,\|^{\mathfrak{a}}\mathsf{w} \mid \|\mathsf{w} \in \mathcal{K}\,\}$. A set of edge sets is called an edge system. On edge systems, a binary operation is defined as follows and called the direct union $\uplus$:

$$\mathcal{K} \uplus \mathcal{L} = \{\,V \cup W \mid V \in \mathcal{K} \text{ und } W \in \mathcal{L}\,\}.$$

This construction implies that the direct union of two finite edge systems is finite. For any word $\mathsf{w} \in \mathcal{A}^l$ such that each of its letters occurs in a second derivative of ω, let T_w be the pure $sDT0L$ system $(\mathcal{A}, f, \mathsf{w})$ and $\mathcal{K}_\mathsf{w}$ be its edge system. Let $\mathsf{w} \in \mathcal{A}^l$ be a word consisting of only those letters occurring in a second derivative of the axiom ω. The edge system $\mathcal{K}_\mathsf{w}$ is

$$\mathcal{K}_\mathsf{w} = \left\{\, \|(f^{(n)}(\mathsf{w})) \mid n \in \mathbb{N}_0, f^{(n)} \in f^n \,\right\}$$
$$= \left\{\, \|(f^{(n)}(w_1) \cdots f^{(n)}(w_l)) \,\right\}$$
$$= \left\{\, \|(f^{(n)}(w_1)) \cup \|^{(f^{(n)}(w_1))(o)}(f^{(n)}(w_2)) \cup \cdots \cup \|^{(f^{(n)}(\overrightarrow{w_{l-1}}))(o)}(f^{(n)}(w_l)) \,\right\}$$
$$= \left\{\, \|(f^{(n)}(w_1)) \cup \|^{w_1(o)}(f^{(n)}(w_2)) \cup \cdots \cup \|^{\overrightarrow{w_{l-1}}(o)}(f^{(n)}(w_l)) \,\right\}$$
$$\subseteq \mathcal{K}_{w_1} \uplus \mathcal{K}_{w_2}^{w_1(o)} \uplus \cdots \uplus \mathcal{K}_{w_l}^{\overrightarrow{w_{l-1}}(o)}.$$

For any picture, there are only finitely many edge sets that yield that picture. Since the x-1-systems of G generate finite picture languages, the edge systems $\mathcal{K}_{w_i}^{\overrightarrow{w_{i-1}}(o)}$ are finite. Hence, also the edge system $\mathcal{K}_\mathsf{w}$ is finite.

Let the edge system $\mathcal{K}_f$ consist of the edge sets of those derivatives of ω resulting from length constant endomorphisms only:

$$\mathcal{K}_f = \{\,\|\mathsf{w} \mid \mathsf{w} \in f^n(\{\,\omega\,\}), n \in \mathbb{N}_0\,\}.$$

Let the edge system $\mathcal{K}_\gamma$ consist of the edge sets of those derivatives of ω that are obtained by applying arbitrary endomorphisms before the last one, which is a length contracting endomorphism:

$$\mathcal{K}_\gamma = \{\,\|\mathsf{w} \mid \mathsf{w} \in (h^m \circ g)(\{\,\omega\,\}), m \in \mathbb{N}_0\,\}.$$

Furthermore, let $\mathcal{K}_g$ consist of the edge sets of those derivatives of ω where at least one applied endomorphism is length contracting:

$$\mathcal{K}_g = \{\, \|w\| \mid w \in (h^m \circ g \circ f^n)(\{\,\omega\,\}), m, n \in \mathbb{N}_0 \,\}.$$

The edge system $\mathcal{K}_f$ is the same as $\mathcal{K}_\omega$. Since every letter of ω also occurs in a second derivative, the edge system $\mathcal{K}_\omega$ is finite. Hence, $\mathcal{K}_f$ is finite.

For each edge set K of $\mathcal{K}_\gamma$, there are two words v, w with v being an m-th derivative of the axiom, w being a derivative of v by a length contracting endomorphism and $\|w = K$:

$$v \in h^m(\{\,\omega\,\}), \quad w = g_i(v).$$

For any word v and $(0, \mu)$-endomorphism g_i, one has $\|g_i(v) = \bigcup\limits_{x \in [v]} \|g_i(x)$.
Hence, the edges of each edge set belonging to $\mathcal{K}_\gamma$ lie on the following cross

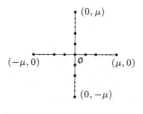

Thus, the edge system $\mathcal{K}_\gamma$ is finite.

If an edge set K belongs to $\mathcal{K}_g$, two words v, w exist such that $\|v = K$, v is an n-th derivative of w by h and w is an $(m + 1)$-th derivative of the axiom where the last applied endomorphism is length contracting. Then, the edge set of w belongs to $\mathcal{K}_\gamma$:

$$
\begin{aligned}
K \in \mathcal{K}_g \Longrightarrow\ &\exists v : \|v = K \\
&\text{and } v \in (h^m \circ g \circ f^n)(\{\,\omega\,\}) \\
&\text{and } v \in f^n((h^m \circ g)(\{\,\omega\,\})) \\
&\text{and } \exists w : v \in f^n(\{\,w\,\}) \wedge w \in (h^m \circ g)(\{\,\omega\,\}) \wedge \|w \in \mathcal{K}_\gamma.
\end{aligned}
$$

If K is an edge set of $\mathcal{K}_\gamma$, it is the edge set of some word $w \in (h^m \circ g)(\{\,\omega\,\})$. For any derivative v of w by f, one obtains $v \in f^n(\{\,w\,\}) \subseteq (h^m \circ g \circ f^n)(\{\,\omega\,\})$. Thus, the edge set of v belongs to $\mathcal{K}_g$:

$$
\begin{aligned}
K \in \mathcal{K}_\gamma \Longrightarrow\ &\exists w : \|w = K \\
&\text{and } w \in (h^m \circ g)(\{\,\omega\,\}) \\
&\text{and } \forall v : v \in f^n(\{\,w\,\}) \Longrightarrow \|v \in \mathcal{K}_g.
\end{aligned}
$$

Hence, the edge system $\mathcal{K}_g$ can be written as

$$\mathcal{K}_g = \{\, \|v \mid v \in f^n(\{\,w\,\}), w \in (h^m \circ g)(\{\,\omega\,\}), n, m \in \mathbb{N}_0 \,\}.$$

The edge set of a derivative is obtained by replacing the initial edges by the edge sets of the corresponding derivatives:

$$\|f^{(n)}(\mathsf{w}) = \bigcup_{(\mathsf{q},x)\in\|\mathsf{w}} \|^{\mathsf{q}}f^{(n)}(x).$$

This yields the following proposition.

Proposition 9. *If the edge sets of two words* v *and* w *coincide, the edge sets of the derivatives* $f^{(n)}(\mathsf{v})$ *and* $f^{(n)}(\mathsf{w})$ *are equal for any* $f^{(n)} \in f^n$, $n \in \mathbb{N}_0$.

According to this proposition, $\mathcal{K}_g$ can be written as

$$\mathcal{K}_g = \left\{ \|f^{(n)}(\mathsf{w}) \;\middle|\; n \in \mathbb{N}_0, f^{(n)} \in f^n, \|\mathsf{w} \in \mathcal{K}_\gamma \right\}$$

$$= \bigcup_{\|\mathsf{w}\in\mathcal{K}_\gamma} \left\{ \|f^{(n)}(\mathsf{w}) \;\middle|\; n \in \mathbb{N}_0, f^{(n)} \in f^n \right\}$$

$$= \bigcup_{\|\mathsf{w}\in\mathcal{K}_\gamma} \mathcal{K}_\mathsf{w}.$$

If the edge sets of two words u, v coincide, the edge systems $\mathcal{K}_\mathsf{u}$, $\mathcal{K}_\mathsf{v}$ are equal. Since $\mathcal{K}_\gamma$ is finite and every edge system $\mathcal{K}_\mathsf{w}$ is finite, also $\mathcal{K}_g$ is finite.

The edge system $\mathcal{K}_G$ generated by G consists of the edge sets belonging to derivatives of the axiom. The applied endomorphisms are all length constant or at least one of them is length contracting. Hence, $\mathcal{K}_G$ is the union of $\mathcal{K}_f$ and $\mathcal{K}_g$:

$$\mathcal{K}_G = \mathcal{K}_f \cup \mathcal{K}_g.$$

Since $\mathcal{K}_f$ and $\mathcal{K}_g$ are finite, also $\mathcal{K}_G$ is finite. This implies that also the picture language P_G is finite. This proves the following lemma.

Lemma 4. *If every* x-1-*system* T_x *of* G *generates a finite picture language, the picture language* P_G *is finite as well.*

This lemma and Lemma 3 yield together the following criterion for the finiteness.

Theorem 4. *Let* $G = (\mathcal{A}, h, \omega)$ *be a mixed, length constant sDT0L system and* f *be the set of the length constant endomorphisms involved. The picture language* P_G *generated by* G *is finite if and only if every* x-1-*system* $T_x = (\mathcal{A}, f, x)$ *of* G *generates a finite picture language.*

By all the stated theorems, one obtains the following.

Theorem 5. *It is decidable whether an sDT0L system generates a finite picture language or not.*

4   Conclusion

In the present paper, synchronous, deterministic tabled, context-free chain code picture systems based on LINDENMAYER systems (*sDT0L* systems) are studied with respect to the finiteness of the picture languages generated. It is shown that it is decidable whether an *sDT0L* system generates a finite picture language or not. The systems considered are divided up in length contracting, constant and expanding systems. Furthermore, pure and mixed systems are distinguished. Length contracting systems generate finite, length expanding systems generate infinite picture languages. Among the length constant systems are those with finite picture languages and those with infinite ones. A finiteness criterion is developed and proved.

References

[DH89]   DASSOW, J.; HINZ, F.: *Kettencode-Bildsprachen. Theorie und Anwendungen.* Wiss. Zeitschrift der Techn. Univ. Magdeburg, 1989.

[DHr92]   DASSOW, J.; HROMKOVIČ, J.: *On Synchronized Lindenmayer Picture Languages.* Lindenmayer Systems, 253–261. Springer-Verlag, Berlin 1992.

[Fe68]   FEDER, J.: *Languages of encoded line patterns.* Inform. Control, 13:230–244.

[Fr61]   FREEMAN, H.: *On the encoding of arbitrary geometric configurations.* IRE Trans. EC, 10:260–168, 1961.

[MRW82]   MAURER, H.; ROZENBERG, G.; WELZL, E.: *Using string languages to describe picture languages.* Inform. Control, 54:155–185, 1982.

[RS80]   ROZENBERG, G.; SALOMAA, A.: *The Mathematical Theory of L Systems.* Academic Press, 1980.

[SW85]   SUDBOROUGH, I. H.; WELZL, E.: *Complexity and decidability for chain code picture languages.* Theoretical Computer Science, 36:173–202, 1985.

[T02]   TRUTHE, B.: *On the Finiteness of Picture Languages of Synchronous Deterministic Chain Code Picture Systems.* To appear. (see also *Zur Endlichkeit von Bildsprachen synchroner deterministischer Ketten-Code-Bild-Systeme.* Otto-von-Guericke-Universität Magdeburg, Preprint Nr. 8/2002.)

[T03]   TRUTHE, B.: *On the Finiteness of Picture Languages of synchronous, simple non-deterministic Chain Code Picture Systems.* Fundam. Inform., 56:389–409, 2003.

Tissue P Systems with Minimal Symport/Antiport

Sergey Verlan

LITA, University of Metz, France,
verlan@sciences.univ-metz.fr

Abstract. We show that tissue P systems with symport/antiport having 3 cells and symport/antiport rules of minimal weight generate all recursively enumerable sets of numbers. Constructed systems simulate register machines and have purely deterministic behaviour. Moreover, only 2 symport rules are used and all symbols of any system are present in finite number of copies (except for symbols corresponding to registers of the machine). At the end of the article some open problems are formulated.

1   Introduction

P systems were introduced by Gh. Păun in [11] as distributed parallel computing devices of biochemical inspiration. These systems are inspired from the structure and the functioning of a living cell. The cell is considered as a set of compartments (membranes) nested one in another and which contain objects and evolution rules. The base model does not specify neither the nature of these objects, nor the nature of rules. Numerous variants specify these two parameters by obtaining a lot of different models of computing, see [16] for a comprehensive bibliography. One of these variants, P systems with *symport/antiport*, was introduced in [10]. This variant uses one of the most important properties of P systems: the communication. This property is so powerful, that it suffices by itself for a big computational power. These systems have two types of rules: symport rules, when several objects go together from one membrane to another, and antiport rules, when several objects from two membranes are exchanged. In spite of a simple definition, they may compute all Turing computable sets of numbers [10]. This result was improved with respect to the number of used membranes and/or the weight of symport/antiport rules ([4], [6], [8], [12], [3]).

Rather unexpectedly, minimal symport/antiport membrane systems, *i.e.* when one uses only one object in symport or antiport rules, are universal. The proof of this result may be found in [2] and the corresponding system has 9 membranes. This result was improved first by reducing the number of membranes to six [7], after that to five [3] and four [5]. In [15] G. Vaszil showed that three membranes are sufficient to generate all recursively enumerable sets of numbers. Another proof of the same result that was obtained independently may be found in [1].

C.S. Calude, E. Calude and M.J. Dinneen (Eds.): DLT 2004, LNCS 3340, pp. 418–429, 2004.

The inspiration for *tissue P systems* comes from two sides. From one hand, P systems previously introduced may be viewed as transformations of labels associated to nodes of a tree. Therefore, it is natural to consider same transformations on a graph. From the other hand, they may be obtained by following the same reflections as for P systems, but starting from a tissue of cells and no more from a single cell.

Tissue P systems were first considered by Gh. Păun and T. Yokomori in [13] and [14]. They have reacher possibilities and the advantages of new topology have to be investigated. Tissue P systems with symport/antiport were first considered in [12] where several results having different values of parameters (graph size, maximal size of connected component, weight of symport and antiport rules) are presented.

In this paper we consider minimal symport and antiport rules, *i.e.* we fix two of four parameters, the weight of symport and antiport rules, to 1. We show that in this case we can construct a system defined on a graph with 4 nodes, *i.e.* 3 cells, that simulates any (non-)deterministic register machine (or counter automata). Moreover, in the deterministic case, the obtained system is also deterministic and only one evolution is possible at any time. Therefore, if the computation stops, then we are sure that the corresponding register machine stops on the provided input. Another difference from previous proofs is that we use a very small amount of symbols present in an infinite number of copies in the environment.

We also introduce *weak tissue P systems with symport/antiport* that are obtained by combining ideas of P systems with symport/antiport and tissue P systems. These systems are a particular variant of tissue P systems with symport/antiport and we present the proof of the main theorem with respect to these systems.

We also remark that we obtain low complexity values for several parameters. More exactly, the systems that we construct use a small amount of symbols and have only 2 symport rules. We remark that because we deal with antiport rules of size one, we need at least one symport rule, otherwise we cannot change the number of objects in the system.

2   Definitions

We denote by $\mathbb{N}$ the set of all non-negative natural numbers: $\{0, 1, \ldots\}$ and by $\mathbb{N}'$ the set $\{1, 2, \ldots\}$.

A *multiset* S over O is a mapping $f_S : O \longrightarrow \mathbb{N}$. The mapping f_S specifies the number of occurrences of each element of S. The size of the multiset S is $|S| = \sum_{x \in O} f_S(x)$. The multiset defined by the mapping $a \to 3, b \to 1, c \to 0$ will be specified as $\{a^3, b\}$ or $a^3 b$.

Multisets as defined above are called *finite* multisets. If we consider that mapping f_S is of form $f_S : O \longrightarrow \mathbb{N} \cup \{\infty\}$, *i.e.* elements of S may have an infinite multiplicity, then we obtain *infinite* multisets.

A deterministic *register machine* is the following construction:

$$M = (Q, R, q_0, q_f, P),$$

where Q is a set of states, $R = \{r_1, \ldots, r_k\}$ is the set of registers (called also counters), $q_0 \in Q$ is the initial state, $q_f \in Q$ is the final state and P is a set of instructions (called also rules) of the following form:

1. $(p, A+, q) \in P, p, q \in Q, p \neq q, A \in R$ (being in state p, increase register A and go to state q).
2. $(p, A-, q, s) \in P, p, q, s \in Q, A \in R$ (being in state p, decrease register A and go to q if successful or to s if A is zero).
3. STOP (may be associated only to the final state q_f).

We note that for each state p there is only one instruction of the type above.

A configuration of a register machine is given by the $k+1$-tuple $(q, n_1, \ldots, n_k)$ describing the current state of the machine as well as contents of all registers. A transition of the register machine consists in updating/checking the value of a register according to an instruction of one of types above and by changing the current state to another one.

We say that M computes a value $y \in \mathbb{N}$ on the *input* $x \in \mathbb{N}$ if starting from the initial configuration $(q_0, x, 0, \ldots, 0)$ it reaches the final configuration $(q_f, y, 0, \ldots, 0)$.

We say that M recognises the set $S \subseteq \mathbb{N}$ if for any input $x \in S$ the machine stops and for any $y \notin S$ the machine performs an infinite computation. It is known that register machines recognise all recursively enumerable sets of numbers [9].

We may also consider non-deterministic register machines where the first type of instruction is of the form $(p, A+, q, s)$ and with the following meaning: if the machine is in state p, then the counter A is increased and the current state is changed to q or s non-deterministically. In this case the result of the computation is the set of all values of the first counter when the computation is halted. We assume that the machine empties all counters except the first counter before stopping. It is known that non-deterministic register machines generate all recursively enumerable sets of non-negative natural numbers starting from empty counters.

A *tissue P system with simport/antiport* (tPsa system) of degree $m \geq 1$ is a construct

$$\Pi = (O, G, w_1, \ldots, w_m, E, R, i_0),$$

where O is the alphabet of objects and G is the underlying directed labeled graph of the system. The graph G has $m+1$ nodes and the nodes are numbered from 0 to m. We shall also call nodes from 1 to m cells and node 0 the environment. There is an edge between each cell i, $1 \leq i \leq m$ and the environment. Each node has a label that contains a multiset of objects. The environment is a special node whose label may contain an infinite multiset. The symbols of the multiset labeling the environment which are present with an infinite multiplicity are given by the set E. The symbols $w_1, \ldots, w_m$ are multisets over O that give initial labels of nodes of G. The symbol i_0 is the output node, and R is a finite set of rules (associated to edges) of the following forms:

1. (i, x, j), $1 \leq i \leq m, 0 \leq j \leq m, i \neq j$, $x \in O^+$ (symport rules for the communication).
2. $(i, x/y, j)$, $0 \leq i, j \leq m, i \neq j$, $x, y \in O^+$ (antiport rules for the communication).

The first rule sends a multiset of objects x from node i to node j. The second rule exchanges multisets x and y situated in nodes i and j respectively. The weight of symport rule (i, x, j) is equal to $|x|$, while the weight of an antiport rule is equal to $\max\{|x|, |y|\}$.

A computational step is made by applying all rules in a non-deterministic maximal parallel way. A configuration of the system is the following $m + 1$-tuple $(z_0, z_1, \ldots, z_m)$ where each $z_i, 1 \leq i \leq m$ represents the label of vertex i and z_0 represents objects that appear with a finite multiplicity in the environment (initially z_0 is an empty multiset). The computation stops when no rule may be applied. The result of a computation is given by the number of objects situated in cell i_0, $i.e.$ by the size of the multiset labeling vertex i_0.

We denote by $NOtP_{n,m}(sym_p, anti_q)$ the family of all sets of numbers computed by tissue P systems with symport/antiport of degree at most m with the maximal size of the connected component equal to n and which have symport rules of weight at most p and antiport rules of weight at most q.

Now we introduce a weaker variant of the previous systems where rules are associated to nodes.

A *weak tissue P system with simport/antiport* (wtPsa system) of degree $m \geq 1$ is a construct

$$\Pi = (O, G, w_1, \ldots, w_m, E, R_0, R_1, \ldots, R_m, i_0),$$

where O is the alphabet of objects, G is the underlying graph of the system, $w_1, \ldots, w_m$ are strings over O representing the multisets of objects initially labeling the cells of the system, E is the set of objects present in arbitrarily many copies in the environment, i_0 is the output cell, and $R_1, \ldots, R_m$ are sets of rules associated to cells (R_0 to the environment) of the following forms:

1. $(x, in; y, out)$ (or $(y, out; x, in)$)
2. (x, in) (or (y, out))

The first rule (antiport rule) exchanges multisets of objects x and y between two cells having a bidirectional connection. The second rule (symport rule) moves the multiset of objects x to the current cell from a cell which have a connection with it. The other variant sends the multiset y to a cell connected to the current one. If there are several connections, one of them is chosen non-deterministically.

At each step all rules that can be applied are applied in parallel in a non-deterministical manner. The result of the computation is the number of objects in the output cell.

More precisely, in weak tissue P systems with symport/antiport rules are associated to graph nodes and not to edges. We remark that each rule of type $(x, in; y, out) \in R_i$ of such system, corresponds to the following set of rules of an ordinary tissue P system with symport/antiport: $\{(i, x/y, j)\}$ for all nodes j that have a bidirectional connection with i. Similarly, rules $(x, in) \in R_i$ and $(y, out) \in R_i$ correspond to the following set of rules $\{(k, x, i)\}$ and $\{i, y, j\}$ for all nodes k that have an edge to node i and for all nodes j that have an incoming edge from node i.

Therefore, a weak tissue P system with symport/antiport is a particular case of tissue P systems with symport/antiport.

We denote by $NOwtP_m(sym_p, anti_q)$ the family of all sets of numbers computed by weak tissue P systems with symport/antiport of degree at most m which have symport rules of weight at most p and antiport rules of weight at most q.

We also remark that there is a similarity between wtPsa systems and P systems with symport/antiport, see [12], which are a special case of tPsa systems defined on trees. However, they differ by the fact that in the second model communications are allowed only with upper cells.

3   The Power of Tissue P Systems with Symport/Antiport

Lemma 1. *For any deterministic register machine M and for any input I_n there is a weak P system with symport/antiport of degree 3 having symport and antiport rules of weight 1, which simulates M on this input and produces the same result.*

Proof. We consider an arbitrary deterministic register machine $M = (Q, R, q_0, q_f, P)$ and we construct a weak tissue P system with symport/antiport that will simulate this machine on the input I_n. We consider a more general problem: we shall simulate M on any initial configuration $(q_0, N_1, \ldots, N_k)$. By commodity, we renumber the nodes of the graph and we consider nodes from 1 to 4 instead of 0 to 3. We also assume, by commodity, that we may have objects initially present in a finite number of copies in the environment and we denote this multiset by w_1. We shall show later that this assumption is not necessary.

We define the system as follows.

$\Pi = (O, G, w_1, w_2, w_3, w_4, E, R_1, R_2, R_3, R_4, 2)$.

$O = \{q : q \in Q\} \cup \{A_{pq}^+ : \exists(p, A+, q) \in P\} \cup \{X, Y, Z_1, Z_2\}$
$\cup \{A_{pqs}^-, Q_{pqs}^-, Q_{pqs}^0, q' : \exists(p, A-, q, s) \in P\} \cup \{A : A \in R\}$.

$E = \{A : A \in R\}$.

We consider the following underlying graph G (cell 1 is the environment):

Below we give in tables rules and objects of our system. In fact, each w_i and R_i, $1 \leq i \leq 4$ is the union of corresponding cells in all tables.

Rule numbers follow the following convention: the first number is the number of cell (environment) where the rule is located, the second number indicates which instruction is simulated using this rule (2 for incrementing, 3 for decrementing, 4 for stop and 1 for common) and the third is the number in group.

Encoding of initial configuration of M ($1 \leq j \leq k$):

Node	Object(s)
1.	
2.	$r_j^{N_j}, q_0$
3.	
4.	

Common rules and objects ($q \in Q \setminus \{q_f\}$, $A \in R$, $p \in Q \setminus \{q_0\}$):

Node	Object(s)	Rules	
1.	Y, A^∞	$1.1.1 : (q, out; X, in)$	$1.1.2 : (Y, in)$
2.	X	$2.1.1 : (X, in)$	
3.			
4.	p		

For any rule $(p, A+, q) \in P$ we have following rules and objects:

Node	Object(s)	Rules	
1.		$1.2.1 : (q, in; A_{p,q}^+, out)$	
2.		$2.2.1 : (p, out; A_{pq}^+, in)$	$2.2.2 : (A, in; A_{pq}^+, out)$
3.	$A_{p,q}^+$		
4.		$4.2.1 : (p, in; A_{pq}^+, out)$	

For any rule $(p, A-, q, s) \in P$ we have following rules and objects:

Node	Object(s)	Rules	
1.		$1.3.1 : (A, in; Q_{pqs}^0, out)$ $1.3.2 : (q', out; q, in)$	$1.3.3 : (Q_{pqs}^-, out; s, in)$
2.		$2.3.1 : (p, out; A_{pqs}^-, in)$ $2.3.2 : (Y, in; A_{pqs}^-, out)$ $2.3.3 : (Q_{pqs}^-, in; A, out)$	$2.3.4 : (A, in; Q_{pqs}^0, out)$ $2.3.5 : (Q_{pqs}^-, out; q', in)$ $2.3.6 : (q', out; Y, in)$
3.	A_{pqs}^-	$3.3.1 : (p, out; Q_{pqs}^-, in)$	$3.3.2 : (Q_{pqs}^0, out; q', in)$
4.	Q_{pqs}^0, Q_{pqs}^-, q'	$4.3.1 : (Q_{pqs}^0, out; A_{pqs}^-, in)$ $4.3.2 : (A_{pqs}^-, out; Q_{pqs}^-, in)$	$4.3.3 : (Q_{pqs}^-, out; Q_{pqs}^0, in)$

Rules and objects associated to the STOP instruction:

Node	Object(s)	Rules	
1.		$1.4.1 : (q_f, out; Z_1, in)$	
2.		$2.4.1 : (q_f, out; Y, in)$	
3.	Z_1	$3.4.1 : (q_f, out; X, in)$	$3.4.2 : (X, out; Z_2, in)$ $3.4.3 : (Z_1, in)$
4.	Z_2		

We organise system Π as follows. Node 2 contains the current configuration of machine M. Node 4 contains one copy of objects that correspond to each state. In the same node, there are additional symbols used for the simulation of the decrementing operation and which are present in one copy. Node 3 contains symbols used for simulation of the incrementing and decrementing. Similarly, these symbols are present only in one copy. Node 1 contains symbols r_j, $1 \leq j \leq k$ which are used to increment registers. These symbols are present in an infinite number of copies. Finally, node 1, 2, 3 and 4 contain symbols Y, X, Z_1 and Z_2 respectively, which are used for technical purposes.

Each configuration $(p, r_1^{n_1}, \ldots, r_k^{n_k})$ of machine M is encoded as follows. Cell 2 contains objects r_j of the multiplicity n_j, $1 \leq j \leq k$ as well as the object p. It is easy to observe that the initial configuration of Π corresponds to an encoding of the initial configuration of M.

Now we shall discuss the simulation of instructions of M.

Incrementing

Suppose that M is in configuration $(p, r_1^{n_1}, \ldots, r_k^{n_k})$ and that there is a rule $(\mathbf{p}, \mathbf{A}+, \mathbf{q})$ in P $(A = r_j)$. Suppose that the value of A is n $(n_j = n)$. This corresponds to the following configuration of Π (see at the right) where we indicate only symbols that we effectively use.

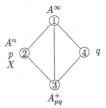

Now we present the evolution of the system in this case.

The symbol p in the second node that encodes the current state of M triggers the application of rule 2.2.1 and it exchanges with A_{pq}^+ which comes to the second node.

After that, the last symbol brings an object A to node 2, which corresponds to an incrementing of register A. Further, the symbol A_{pq}^+ goes to node 4 and brings from there to node 1 the new state q.

Finally, the symbol A_{pq}^+ returns to its original location (node 3) and it brings the symbol p to node 4 where all symbols that correspond to states of the machine are situated. At the same time, the symbol q moves to node 2 (if $q \neq q_f$). This configuration is shown at the right.

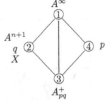

The last configuration differs from the first one by the following. In node 2, there is one more copy of object A and the object p was replaced by the object q. All other symbols remained on their places and the symbol p was moved to node 4 which contains as before one copy of each state of the register machine that is not current. This corresponds to the following configuration of M: $(q, r_1^{n_1}, \ldots, r_j^{n_j+1}, \ldots, r_k^{n_k})$, i.e. we simulated the corresponding instruction of M.

Decrementing

Suppose that M is in configuration $(p, r_1^{n_1}, \ldots, r_k^{n_k})$ and that there is a rule $(\mathbf{p}, \mathbf{A}-, \mathbf{q}, \mathbf{s})$ in P $(A = r_j)$. Suppose that the value of A is n $(n_j = n)$. This corresponds to the following configuration of Π (see at the right) where we indicate only symbols that we effectively use.

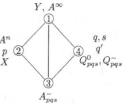

Below we present the evolution of the system in this case.

The symbol p in the second node that encodes the current state of M triggers the application of rule 2.3.1 and it exchanges with A_{pqs}^- which comes to the second node. After that, it goes to node 4 bringing at the same time the symbol Q_{pqs}^- in node 3. In the meanwhile, A_{pqs}^- is moved to node 1. The obtained configuration is shown at the right.

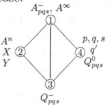

Now there are two cases, $n > 0$ and $n = 0$, and the system behaves differently in each case.

CASE A: $n > 0$

First, suppose that $n > 0$. In this case, Q_{pqs}^- goes to node 2 and brings a symbol A to node 3, which is further exchanged with the symbol Q_{pqs}^0 brought previously by A_{pqs}^- to node 1. Because rule 1.3.1 may exchange the symbol Q_{pqs}^0 with any A situated in any node connected to node 1, the symbol Q_{pqs}^0 may arrive in node 2 if there are enough symbols A in that node. After that, this symbol may either go to node 3, in this case we obtain the desired configuration, or it may return to node 1, in this case we obtain the same configuration as the one from two steps ago. So, an infinite computation may happen at this place, but in this case the system Π does not produce any result. Therefore, we have to do at some moment the exchange of Q_{pqs}^0 with symbol A from node 3.

After that the symbol Q_{pqs}^0 brings the symbol q' to node 3 which exchanges with Q_{pqs}^-. Finally, the symbol q' moves to node 1 and brings after that the symbol q which is moved to node 2 (if $q \neq q_f$) afterwards. At the same time, symbols A_{pqs}^- and Q_{pqs}^- are exchanged and return to their original places (see configuration at the right).

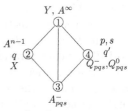

We can see that the obtained configuration differs from the first one by the following. In node 2, there is one copy of object A less and the object p was replaced by the object q. All other symbols remained on their places and the symbol p was moved to node 4 which contains as before one copy of each state of the register machine that is not current. This corresponds to the following configuration of M: $(q, r_1^{n_1}, \ldots, r_j^{n_j-1}, \ldots, r_k^{n_k})$, *i.e.* we simulated the corresponding instruction of M.

CASE B: $n = 0$

Now suppose that $n = 0$. In this case, Q^-_{pqs} remains in node 3 for one more step. After that, it exchanges with A^-_{pqs}.

Now symbols Q^-_{pqs} and Q^0_{pqs} are exchanged and after that Q^-_{pqs} brings in node 1 the symbol s which moves after that to node 2 (if $s \neq q_f$) (see configuration at the right). We remark that the first exchange takes place only if the value of counter A is equal to zero, otherwise rules 3.3.1 and 1.3.1 are applied and the system never reaches the configuration above. Similarly, rule 1.3.1 cannot be applied because there are no objects A in node 2 or 3.

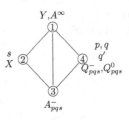

We can see that the obtained configuration differs from the first one by the following. In node 2, the object p was replaced by the object s. All other symbols remained on their places and the symbol p was moved to node 4 which contains as before one copy of each state of the register machine that is not current. This corresponds to the following configuration of M: $(s, r_1^{n_1}, \ldots, r_j^0, \ldots, r_k^{n_k})$, $i.e.$ we simulated the corresponding instruction of M.

Stop

If we simulate an instruction that leads to a final state q_f, then symbol q_f remains in node 1 and starts the halting procedure. We present below the evolution of the system in this case.

It is easy to see that firstly q_f goes to node 3 and after that it exchanges with X from node 2. After that X either return to node 2, in this case we obtain the first configuration and we can continue one more time, or it moves to node 4 using the rule 3.4.2 and no rule can be applied any more.

Final Remarks

It is clear that we simulate the behaviour of M. Indeed, we simulate an instruction of M and all additional symbols return to their places what permits to simulate the next instruction of M. Moreover, this permits to reconstruct easily a computation in M from a successful computation in Π. For this it is enough to look for configurations which have a state symbol p in node 2. We stop the computation when rule 3.4.2 is used and symbol X goes to node 4. In this case, node 2 contains the result of the computation. We remark that symbol X may arrive in node 3 only if the simulation reached the final state q_f.

We remark that the assumption that $w_1 = \{Y\}$ is not necessary. Indeed, we may initially place Y in node 2.

We remark that we used 3 symport rules. If we permit an encoding of the result we may eliminate symport rule 2.4.2. In this case, the system Π will compute $n + 1$ if M computes n.

We also remark that if we are sure that M always computes a positive number, then we may compute the same result using only 2 symport rules. For this, starting from M we construct a new machine M' which has a new halting state q'_f and an additional instruction $(q_f, A-, q'_f, q'_f)$ where A is the resulting register r_1 of M. After that we construct the system Π as above and we eliminate rules from the group associated to the STOP instruction. It is easy to see that Π will compute n.

Theorem 1. $NOwtP_3(sym_1, anti_1) = NRE$.

Proof. It is easy to observe that the system of previous lemma may simulate a non-deterministic register machine. Indeed, in order to simulate a rule $(p, A+, q, s)$ of such machine, we use the same rules and objects as for rule $(p, A+, q)$. We only need to add a rule $1.2.1' : (s, in; A^+_{pq}, out)$ to node 1.

An easy corollary of these results is the following theorem:

Theorem 2. $NOtP_{3,3}(sym_1, anti_1) = NRE$.

However, we remark that we may improve the previous result by using new possibilities that are present in this case. Indeed, we always used the same edge for symport or antiport during the transition between configurations. Therefore, we may associate the corresponding symport or antiport rule to that edge. Consequently, we will not have any more a possible infinite computation during the simulation of decrementing because we place the rule 1.3.1 on the edge going from node 1 to 3. Similarly, we can place the rule 2.1.1 on the edge going from node 1 to 2 which permits us to avoid the possible infinite computation at the end. In this case, we can also eliminate the rule 2.4.2 what reduces the number of used symport rules down to 2. Therefore, the obtained system will have a deterministic evolution in case when we simulate a deterministic machine.

Descriptional Complexity

Let M be a register machine having n states, k registers and n_1 incrementing instructions, *i.e.* M has $n - n_1 - 1$ decrementing instructions. In this case the system constructed as in Lemma 1 need at most $n + n_1 + 4 * (n - n_1 - 1) + 4 = 5n - 3n_1$ symbols present in one copy, k symbols present in infinite number of copies, $n - 2 + 4 * n_1 + 14 * (n - n_1 - 1) + 4 = 15n - 10n_1 - 12$ antiport rules and 3 symport rules. If we consider tPsa systems, then the number of symport rules is decreased to 2.

4   Conclusions

In this article we studied tissue P systems with symport/antiport. We introduced a weaker variant of these systems and we showed that it is able to generate any recursively enumerable set of numbers. We remark that contrarily to other proofs concerning P systems with symport/antiport, the system that we constructed is deterministic. More exactly, in Lemma 1 there are only two places where a non-determinism may occur and if we consider ordinary tPsa systems, then these computations may be avoided. Consequently, if we have a deterministic register machine M and an initial configuration $(q_0, r_1^{n_1}, \ldots, r_k^{n_k})$ of this machine, then the corresponding tPsa system constructed as described in Lemma 1 and using remarks given after Theorem 2 will have a deterministic behaviour, *i.e.* there will be only one possible evolution at each step and it will halt if and only if M halts on the above configuration. We highlight this deterministic evolution because it makes the behaviour of the system more predictable and makes such systems good candidates for possible implementations.

Another advantage of our proof technique with respect to other proofs is that we need only a finite number of symbols present in one copy, except symbols encoding counters.

Weak tissue P systems with symport/antiport are almost deterministic and if we impose evolutions where a rule cannot be applied 2 and 4 steps after its previous application, then these systems also become deterministic.

An open problem naturally raised by the results of this article is to know if 3 or less nodes are sufficient in order to get the universality in the case of tPsa systems. We conjecture that 3 nodes permit to do such computations.

Another open problem concerns the number of symport rules used. In our tPsa system we used 2 symport rules. The question if we may obtain the same result using a smaller number of symport rules remains open. We remark that because we deal with antiport rules of size one, we need at least one symport rule, otherwise we cannot change the number of objects in the system.

We may consider a generalisation of tissue P systems with symport/antiport by taking off restrictions on the graph that they impose. In this case we may label the communication graph of the system by infinite multisets, hence erasing the difference between environment and cells.

Another possibility is to change the definition of tPsa systems and to consider as result a vector of numbers. In order to do this we count the number of elements corresponding to each register of the machine that are present in resulting node at the end of the computation. The resulting vector will contain these numbers as values of its components.

Acknowledgments. The author wants to thank M. Margenstern and Yu. Rogozhin for their very helpful comments and for their attention for the present work. The author acknowledges also the "MolCoNet" IST-2001-32008 project and the *Laboratoire d'Informatique Théorique et Apliquée de Metz* which provided him the best conditions for producing the present result. And finally the author acknowledges the Ministry of Education of France for the financial support of his PhD.

References

1. A. Alhazov, M. Margenstern, V. Rogozhin, Y. Rogozhin, and S. Verlan. Communicative P systems with minimal cooperation. In G. Mauri, G. Paun, M. Perez-Jimenez, G. Rozenberg, and A. Salomaa, editors, *Membrane Computing, International Workshop, WMC5, Milano, Italy, June, 14-16, 2004, Revised Papers,* Lecture Notes in Computer Science, 2005. In publication.

2. F. Bernardini and M. Gheorghe. On the power of minimal symport/antiport. In A. Alhazov, C. Martín-Vide, and G. Păun, editors, *Preproceedings of the Workshop on Membrane Computing*, pages 72–83, Tarragona, July 17-22 2003.

3. F. Bernardini and A. Păun. Universality of minimal symport/antiport: Five membranes suffice. In C. Martín-Vide, G. Mauri, G. Păun, G. Rozenberg, and A. Salomaa, editors, *Membrane Computing, International Workshop, WMC 2003, Tarragona, Spain, July, 17-22, 2003, Revised Papers*, volume 2933 of *Lecture Notes in Computer Science*, pages 43–54. Springer-Verlag, 2003.

4. R. Freund and A. Păun. Membrane systems with symport/antiport: Universality results. In G. Păun, G. Rozenberg, A. Salomaa, and C. Zandron, editors, *Membrane Computing: International Workshop, WMC-CdeA 2002, Curtea de Arges, Romania, August 19-23, 2002. Revised Papers.*, volume 2597 of *Lecture Notes in Computer Science*, pages 270–287, Curtea de Arges, Romania, 2003. Springer-Verlag.

5. P. Frisco. About P systems with symport/antiport. In G. Pıaun, A. Riscos-Núñez, A. Romero-Jiménez, and F. Sancho-Caparrini, editors, *Second Brainstorming Week on Membrane Computing*, number TR 01/2004 in Technical report, pages 224–236. University of Seville, 2004.

6. P. Frisco and H. Hoogeboom. Simulating counter automata by P systems with symport/antiport. In G. Păun, G. Rozenberg, A. Salomaa, and C. Zandron, editors, *Membrane Computing: International Workshop, WMC-CdeA 2002, Curtea de Arges, Romania, August 19-23, 2002. Revised Papers.*, volume 2597 of *Lecture Notes in Computer Science*, pages 288–301, Curtea de Arges, Romania, 2003. Springer-Verlag.

7. L. Kari, C. Martín-Vide, and A. Păun. On the universality of P systems with minimal symport/antiport rules. In N. Jonoska, G. Păun, and G. Rozenberg, editors, *Aspects of Molecular Computing: Essays Dedicated to Tom Head, on the Occasion of His 70th Birthday*, volume 2950 of *Lecture Notes in Computer Science*, pages 254–265. Springer-Verlag, 2004.

8. C. Martín-Vide, A. Păun, and G. Păun. On the power of P systems with symport rules. *Journal of Universal Computer Science*, 8(2):317–331, 2002.

9. M. Minsky. *Computations: Finite and Infinite Machines*. Prentice Hall, Englewood Cliffts, NJ, 1967.

10. A. Păun and G. Păun. The power of communication: P systems with symport/antiport. *New Generation Computing*, 20(3):295–305, 2002.

11. G. Păun. Computing with membranes. *Journal of Computer and System Sciences*, 1(61):108–143, 2000. Also TUCS Report No. 208, 1998.

12. G. Păun. *Membrane Computing. An Introduction*. Springer-Verlag, 2002.

13. G. Păun, Y. Sakakibara, and T. Yokomori. P systems on graphs of restricted forms. *Publicationes Mathematicae*, 60, 2002.

14. G. Păun and T. Yokomori. Membrane computing based on splicing. volume 54 of *DIMACS Series in Discrete Mathematics and Theoretical Computer Science*, pages 217–232. American Mathematical Society, 1999.

15. G. Vaszil. On the size of p systems with minimal symport/antiport. In *Preliminary proceedings of WMC 2004, Workshop on Membrane Computing, Milano, Italy, June, 14-16, 2004*, pages 422–431, 2004.

16. C. Zandron. The P systems web page. http://psystems.disco.unimib.it/.

Author Index

Vol. 3285: S. Manandhar, J. Austin, U.B. Desai, Y. Oyanagi, A. Talukder (Eds.), Applied Computing. XII, 334 pages. 2004.

Vol. 3284: A. Karmouch, L. Korba, E.R.M. Madeira (Eds.), Mobility Aware Technologies and Applications. XII, 382 pages. 2004.

Vol. 3283: F.A. Aagesen, C. Anutariya, V. Wuwongse (Eds.), Intelligence in Communication Systems. XIII, 327 pages. 2004.

Vol. 3282: V. Guruswami, List Decoding of Error-Correcting Codes. XIX, 350 pages. 2004.

Vol. 3281: T. Dingsøyr (Ed.), Software Process Improvement. X, 207 pages. 2004.

Vol. 3280: C. Aykanat, T. Dayar, İ. Körpeoğlu (Eds.), Computer and Information Sciences - ISCIS 2004. XVIII, 1009 pages. 2004.

Vol. 3278: A. Sahai, F. Wu (Eds.), Utility Computing. XI, 272 pages. 2004.

Vol. 3275: P. Perner (Ed.), Advances in Data Mining. VIII, 173 pages. 2004. (Subseries LNAI).

Vol. 3274: R. Guerraoui (Ed.), Distributed Computing. XIII, 465 pages. 2004.

Vol. 3273: T. Baar, A. Strohmeier, A. Moreira, S.J. Mellor (Eds.), <<UML>> 2004 - The Unified Modelling Language. XIII, 454 pages. 2004.

Vol. 3271: J. Vicente, D. Hutchison (Eds.), Management of Multimedia Networks and Services. XIII, 335 pages. 2004.

Vol. 3270: M. Jeckle, R. Kowalczyk, P. Braun (Eds.), Grid Services Engineering and Management. X, 165 pages. 2004.

Vol. 3269: J. Lopez, S. Qing, E. Okamoto (Eds.), Information and Communications Security. XI, 564 pages. 2004.

Vol. 3268: W. Lindner, M. Mesiti, C. Türker, Y. Tzitzikas, A. Vakali (Eds.), Current Trends in Database Technology - EDBT 2004 Workshops. XVIII, 608 pages. 2004.

Vol. 3266: J. Solé-Pareta, M. Smirnov, P.V. Mieghem, J. Domingo-Pascual, E. Monteiro, P. Reichl, B. Stiller, R.J. Gibbens (Eds.), Quality of Service in the Emerging Networking Panorama. XVI, 390 pages. 2004.

Vol. 3265: R.E. Frederking, K.B. Taylor (Eds.), Machine Translation: From Real Users to Research. XI, 392 pages. 2004. (Subseries LNAI).

Vol. 3264: G. Paliouras, Y. Sakakibara (Eds.), Grammatical Inference: Algorithms and Applications. XI, 291 pages. 2004. (Subseries LNAI).

Vol. 3263: M. Weske, P. Liggesmeyer (Eds.), Object-Oriented and Internet-Based Technologies. XII, 239 pages. 2004.

Vol. 3262: M.M. Freire, P. Chemouil, P. Lorenz, A. Gravey (Eds.), Universal Multiservice Networks. XIII, 556 pages. 2004.

Vol. 3261: T. Yakhno (Ed.), Advances in Information Systems. XIV, 617 pages. 2004.

Vol. 3260: I.G.M.M. Niemegeers, S.H. de Groot (Eds.), Personal Wireless Communications. XIV, 478 pages. 2004.

Vol. 3259: J. Dix, J. Leite (Eds.), Computational Logic in Multi-Agent Systems. XII, 251 pages. 2004. (Subseries LNAI).

Vol. 3258: M. Wallace (Ed.), Principles and Practice of Constraint Programming - CP 2004. XVII, 822 pages. 2004.

Vol. 3257: E. Motta, N.R. Shadbolt, A. Stutt, N. Gibbins (Eds.), Engineering Knowledge in the Age of the Semantic Web. XVII, 517 pages. 2004. (Subseries LNAI).

Vol. 3256: H. Ehrig, G. Engels, F. Parisi-Presicce, G. Rozenberg (Eds.), Graph Transformations. XII, 451 pages. 2004.

Vol. 3255: A. Benczúr, J. Demetrovics, G. Gottlob (Eds.), Advances in Databases and Information Systems. XI, 423 pages. 2004.

Vol. 3254: E. Macii, V. Paliouras, O. Koufopavlou (Eds.), Integrated Circuit and System Design. XVI, 910 pages. 2004.

Vol. 3253: Y. Lakhnech, S. Yovine (Eds.), Formal Techniques, Modelling and Analysis of Timed and Fault-Tolerant Systems. X, 397 pages. 2004.

Vol. 3252: H. Jin, Y. Pan, N. Xiao, J. Sun (Eds.), Grid and Cooperative Computing - GCC 2004 Workshops. XVIII, 785 pages. 2004.

Vol. 3251: H. Jin, Y. Pan, N. Xiao, J. Sun (Eds.), Grid and Cooperative Computing - GCC 2004. XXII, 1025 pages. 2004.

Vol. 3250: L.-J. (LJ) Zhang, M. Jeckle (Eds.), Web Services. X, 301 pages. 2004.

Vol. 3249: B. Buchberger, J.A. Campbell (Eds.), Artificial Intelligence and Symbolic Computation. X, 285 pages. 2004. (Subseries LNAI).

Vol. 3246: A. Apostolico, M. Melucci (Eds.), String Processing and Information Retrieval. XIV, 332 pages. 2004.

Vol. 3245: E. Suzuki, S. Arikawa (Eds.), Discovery Science. XIV, 430 pages. 2004. (Subseries LNAI).

Vol. 3244: S. Ben-David, J. Case, A. Maruoka (Eds.), Algorithmic Learning Theory. XIV, 505 pages. 2004. (Subseries LNAI).

Vol. 3243: S. Leonardi (Ed.), Algorithms and Models for the Web-Graph. VIII, 189 pages. 2004.

Vol. 3242: X. Yao, E. Burke, J.A. Lozano, J. Smith, J.J. Merelo-Guervós, J.A. Bullinaria, J. Rowe, P. Tiño, A. Kabán, H.-P. Schwefel (Eds.), Parallel Problem Solving from Nature - PPSN VIII. XX, 1185 pages. 2004.

Vol. 3241: D. Kranzlmüller, P. Kacsuk, J.J. Dongarra (Eds.), Recent Advances in Parallel Virtual Machine and Message Passing Interface. XIII, 452 pages. 2004.

Vol. 3240: I. Jonassen, J. Kim (Eds.), Algorithms in Bioinformatics. IX, 476 pages. 2004. (Subseries LNBI).

Vol. 3239: G. Nicosia, V. Cutello, P.J. Bentley, J. Timmis (Eds.), Artificial Immune Systems. XII, 444 pages. 2004.

Vol. 3238: S. Biundo, T. Frühwirth, G. Palm (Eds.), KI 2004: Advances in Artificial Intelligence. XI, 467 pages. 2004. (Subseries LNAI).

Vol. 3237: C. Peters, J. Gonzalo, M. Braschler, M. Kluck (Eds.), Comparative Evaluation of Multilingual Information Access Systems. XIV, 702 pages. 2004.

Lecture Notes in Computer Science

For information about Vols. 1–3236

please contact your bookseller or Springer